REMOTE SENSING
Theory and Applications

LICENSE, DISCLAIMER OF LIABILITY, AND LIMITED WARRANTY

Remote Sensing

Theory and Applications

P. K. Garg

Mercury Learning and Information
Boston, Massachusetts

ISBN : 978-939315-989-2.

Publisher: David Pallai
MERCURY LEARNING AND INFORMATION
121 High Street, 3rd Floor
Boston, MA 02110
info@merclearning.com
www.merclearning.com
800-232-0223

P. K. Garg. *Remote Sensing: Theory and Applications*
ISBN: 978-1-68392-748-8

Library of Congress Control Number: 2024932123

242526321 This book is printed on acid-free paper in the United States of America.

Dedicated to the loving memory of my parents

CONTENTS

PREFACE

Remote sensing is the acquisition of information about an object or phenomenon without making physical contact with the object. It deals with the data collection of natural and man-made objects on the Earth to show their geographic locations as well as characteristics on the maps. Remote sensing systems, which provide a global perspective and a wealth of data (the panchromatic, multispectral, thermal, hyperspectral, stereo, LiDAR, etc.) about the Earth systems at various resolutions, allow users to capture, visualize, and analyze objects and features on the Earth's surface for faster decision making, based on the current and future state of land resources. Remotely sensed data today can be applied for a large number of applications, and is especially important for studies of the Earth that require frequent monitoring, such as inventories and surveys in agriculture, urban, disaster, floods, hydrography, geology, mineralogy, land use, etc. In addition, it is highly valuable as it can provide information about regions that are difficult or dangerous to access directly. With the introduction of computer and automation, algorithms are available which can analyze the digital remote sensing data quickly, economically, and accurately, and provide useful information, as per the need of applications.

Addressing the need for updated information in remote sensing, this book consists of two major divisions: theory covering principal topics and different applications. The book discusses the basic concept of remote sensing, remote sensing systems, processing, classification, and applications of data. In addition, it provides the latest techniques of satellite image classification, such as Artificial Neural Networks, Convolution Neural Networks, k-NN, Object-based Image Analysis, Fuzzy C-Means, Artificial Intelligence, Machine Learning, Deep Learning, Support Vector Machine, Random Forest, Decision Tree, etc., which are not only useful in accurate classification of an extraction of specific objects from images, but also developing innovative solutions. The book also provides a framework for utilizing the vast amount of remote sensing data and software freely available on an open platform. It introduces widely used forms of remote sensing imagery and their applications in land use, agriculture, urban areas, hydrology, disaster mapping, monitoring, and management, including case studies.

The book covers the course contents required for civil engineering and geoinformatics as well as computer science, information technology, data science, geology, geography, environmental science, and earth science. Today, remote sensing has found a substantial number of applications in civil engineering, natural resources management, disaster management, environment, geology, soil, forestry, agriculture, urban, infrastructure development, 3D mapping, public health, visualization, etc. It provides various classification techniques along with case studies so readers can easily understand the analysis and utility of various satellite images. The book also discusses the limitation, opportunities, and applications of remote sensing in the future. The readers who have an interest in remote sensing will find the book useful, as it has been written in simple language and the topics are discussed with numerous tabular data and figures.

Color figures from the text are available for downloading by writing to the publisher at info@merclearning.com.

P.K. Garg
April 2024

ABBREVIATIONS

AI	Artificial Intelligence
ALI	Advanced Land Imager (on EO-1 satellite)
ALS	Airborne Laser Scanner
ANN	Artificial Neural Networks
AR	Augmented Reality
ASTER	Advanced Spaceborne Thermal Emission and Reflection Radiometer
AVHRR	Advanced Very High-Resolution Radiometer
AVIRIS	Airborne Visible/infrared Imaging Spectrometer
AWiFS	Advanced Wide Field Sensor
BIL	Band Interleaved by Line
BIM	Building Information Modelling
BIP	Band Interleaved By Pixel
BSQ	Band Sequential
CAGR	Compound Annual Growth Rate
CASI	Compact Airborne Spectrographic Imager
CNN	Convolutional Neural Network
DEM	Digital Elevation Model
DL	Deep Learning
DN	Digital Number
DSM	Digital Surface Model
DTM	Digital Terrain Model
ERS-1	Earth Remote Sensing Satellite
ERTS	Earth Resources Technology Satellite
EMR	Electro-magnetic Radiations
EMS	Electromagnetic Spectrum
ENVISAT	Environmental Satellite

EOS	Earth Observing System
ETM	Enhanced Thematic Mapper
FCC	False Color Composite
GA	Genetic Algorithm
GCP	Ground Control Point
GIS	Geographical Information System
GNSS	Global Navigation Satellite System
GOES	Geostationary Operational Environmental Satellite
GPR	Ground Penetrating Radar
GPS	Global Positioning System
HIS	Intensity, Hue (color), Saturation
HRG	High Resolution Geometrical
HRV	High Resolution Visible
IDW	Inverse Distance Weighted
IMU	Inertia Measurement Unit
INS	Inertial Navigation System
IoT	Internet of Things
IR	Infrared
IRS	Indian Remote Sensing Satellites
LAI	Leaf Area Index
LANDSAT	Land Satellites
LiDAR	Light Detection and Ranging
LISS	Linear Imaging Self Scanning Sensor
MERIS	Medium Resolution Imaging Spectrometer
MIR	Middle Infrared
ML	Machine Learning
MLS	Mobile Laser Scanner
MODIS	Moderate Resolution Imaging Spectroradiometer
MSS	Multispectral Scanner
NASA	National Aeronautics and Space Administration
NDVI	Normalized Difference Vegetation Index
NIR	Near Infrared
NOAA	National Oceanic and Atmospheric Administration
OCM	Ocean Color Monitor
OLI	Operational Land Imager
PAN	Panchromatic
PCA	Principal Component Analysis
RADAR	RAdio Detection And Ranging
RBV	Return Beam Videcon
RGB	Red, Green, Blue

rms	Root Mean Square
SAR	Synthetic Aperture Radar
SPOT	Système Pour l'Observation de la Terre
SRTM	Shuttle Radar Topography Mission
SST	Sea Surface Temperature
SVM	Support Vector Machine
SWIR	Short Wave Infrared
TIR	Thermal Infrared
TLS	Terrestrial Laser Scanner
TM	Thematic Mapper
UAV	Unmanned Aerial Vehicle
VR	Virtual Reality
WiFS	Wide Field Sensor

CHAPTER 1

Basics of Remote Sensing

1.1 INTRODUCTION

While the origins of remote sensing can be traced to World War II with the use of radar, sonar, and thermal detection technologies, the use of the term "remote sensing" was coined by Evelyn Pruitt, a geographer with the U.S. Office of Naval Research, in the late 1950s (Fussell et al., 1986). Pruitt promoted the use of emerging imaging capabilities of multispectral cameras, infrared (IR) films, and nonphotographic scanners. *Remote sensing* can be defined as "the science and art of obtaining information about an object, area, or phenomenon through the analysis of data acquired by a device that is not in contact with the object, area, or phenomenon under investigation." The word *remote* means "from a distance" and *sensing*, in this case, means "to record." So a basic definition of remote sensing is to record the data or collect the information about the objects from a distance (Garg, 2019).

Aerial photography is the earliest form of remote sensing which began with the invention of the camera in the 1800s. The first successful photograph was produced in the early 1800s by French inventor Nicéphore Niépce (Moore, 1979). The first known aerial photograph was taken in 1858 by French photographer and balloonist, Gaspar Felix Tournachon, known as "Nadar." In 1855, he had patented the idea of using aerial photographs in mapmaking and surveying. Soon after the development of photography, people became interested in taking aerial photographs. The earliest aerial photographs were taken from balloons. In 1860, James Wallace Black took the image of Boston from a hot-air balloon. In the early twentieth century, remote sensing images were captured using kites, and even with cameras mounted on pigeons. Small lightweight cameras were attached to the birds, and photos were automatically taken using a timing mechanism. The pigeon photography was successful but didn't become widely used due to the rapid development of aviation technology.

During World War I (1914–1918), aerial photography soon replaced sketching and drawings, and the battle maps were produced from aerial photographs. During World War II (1939–1945), there was a paucity of basic geographical data about the European, Asian, and African land masses. In addition to taking the photographs of the battlefields, aerial surveys were made of vast and worldwide logistical networks, extending from Latin America to the Arctic and from the Himalayas to the Pacific Islands. Aerial photogrammetry was used primarily for military purposes until the end of World War II, thereafter its use expanded rapidly. Nowadays, with the availability of digital aerial cameras and outstanding image quality with high-resolution views, these are used for diverse applications, such as commercial, industrial, agricultural, governmental, and private uses.

The development of satellite-based remote sensing began in the 1950s and 1960s. In 1957, the Soviet Union launched Sputnik-1, the world's first artificial satellite. The United States followed in 1958 with the successful launch of Explorer-1. The first successful meteorological satellite (TIROS-1) was launched in 1960. In 1972, Landsat-1, the first Earth resource satellite, was launched by the United States to collect data of the Earth's surface (Moore, 1979). Since then, there are a myriad of satellites in operation today; many of them are used for varied remote sensing applications. A brief summary is presented in Table 1.1. Developments in the second half of the twentieth century included satellite images, very large scale photographs, automatic visual scanning, high-quality color photographs, use of films sensitive to radiations beyond the visible spectrum, and computational photogrammetry. A majority of these satellites are launched by governments to monitor the Earth's resources, but private commercial companies also came forward in launching Earth observation satellites.

In remote sensing, characteristics of the Earth's surface are acquired by a device (sensor) that is not in contact with the objects being measured. The result is usually, though not necessarily, stored as image data. Human beings can see only through the visible light which forms only a very narrow band of the electromagnetic spectrum (EMS). It is because human eyes

Table 1.1. Evolution of Remote Sensing Satellites

Evolution	Time	Salient Features
Airborne remote sensing	During World War I and II	The use of photographs for surveying, mapping, reconnaissance, and military surveillance.
Rudimentary spaceborne satellite remote sensing	In the late 1950s	The launch of Sputnik-1 by Russia in 1957, and Explorer-1 by the U.S. in 1958.
Spy satellite remote sensing	During the Cold War (1947–1991)	Remote sensing for military use spilled over into mapping and environment applications.
Meteorological satellite remote sensing	1960s	The launch of the first meteorological satellite (TIROS-1) by the U.S. in 1960. Since then, data available in digital formats and the use of computer hardware and software expanded.
Landsat by U.S. Agency	1972	Landsat-1 to 8 carried various sensors providing high-resolution images of global coverage. Initial applications were local, and then have become global.
ERS-1 by European Space Agency (ESA)	1991	The ESA launched the first European Remote Sensing (ERS-1) satellite in 1991, which carried a variety of Earth observation instruments. A successor, ERS-2, was launched in 1995.
Earth observing system (EOS)	Since the launch of the Terra satellite in 1999	Terra/Aqua satellites carrying sensors, such as MODIS-provided global coverage, frequent repeat coverage, a high level of processing, easy and mostly free access to data.
New millennium	Around the same time as EOS	Next generation of satellites and sensors, such as Earth Observing-1, acquired first spaceborne hyperspectral data.
Private industry/commercial satellites	2000	A revolutionary means of very high-resolution data acquisition from, such as IKONOS and QuickBird satellites. The launch of GeoEye-1 in 2008 provided very high-resolution imagery (0.41 m). Google streaming technology allowed rapid data access to very high-resolution images.

(Contd.)

Evolution	Time	Salient Features
Microsatellite era and satellite constellations	2008-present	Small satellites and satellite constellations, such as RapidEye and Terra Bella, are launched. RapidEye satellites launched in August 2008 are the first commercial satellites to include the Red-Edge band, which is sensitive to changes in chlorophyll content. On March 8, 2016, Skybox imaging was renamed to Terra Bella, which provided the first-ever commercial high-resolution video of Earth and high-resolution color and near-infrared imagery. For the first time, Russia carried out a single mission to launch thirty-seven satellites in June 2014. The ESA launched the first satellite of the Sentinel constellation in April 2014. Space X reusable rocket capacity since December 2015. Current satellites have a high revisiting period, large coverage, and high spatial resolution up to 31 cm.

are only sensitive to the visible part of the wavelengths of EMS (Gibson, 2000). The characteristics of the objects are measured by the sensor in the form of reflected electromagnetic energy (EME) or emitted by the Earth's surface.

Exploration of the Earth's natural resources is vital for the socioeconomic development of any country. The satellite images differ with respect to their modes of sensing and types of resolution. Remote sensing systems provide a very important source of spatiotemporal information of the Earth's surface processes at scales ranging from regional to global. A wide range of parameters can be measured from remote sensing images, including land use, vegetation types, surface temperatures, soils, water, geology, forestry, surface elevation, and snow. Sequential remote sensing images are useful in monitoring floods, water pollution, deforestation, extent of forest fires, snow cover, urban sprawl, crop acreage and yield estimation, drought monitoring and assessment, wasteland mapping, mineral prospects, forest resource surveys, and so on. The state-of-the-art, technology of space-based Earth observation systems offers timely, accurate, and high-resolution information on various natural resources, such as land, water, forests, mineral resources, and so forth required for various applications.

Remote sensing is a powerful tool for mapping, inventorying, monitoring, and managing of natural resources due to the inherent advantages of their synoptic view, repetitive imaging, capability to provide images of inaccessible areas, relatively low cost, and near real-time availability of data. Before the image data can yield the required information about the objects or phenomena of interest, they need to be processed for various errors. The advancement in image processing software in remote sensing would help not only removing these errors but also enhancing the application of image data in various domains. Today, remote sensing images have become an integral part of the resource management systems in government and private sectors. The analysis and extraction or production of information (mapping) is an important part of the overall remote sensing process. Remote sensing applications are growing very rapidly with the availability of high-resolution and very high-resolution data from the state-of-the-art satellites like Landsat, IRS-1C/1D/P4, IKONOS, WorldView, QuickBird, GeoEye, and so forth (Garg, 2019). Integration of remote sensing data with other thematic layers generated from various sources provides an added advantage for complex modeling and applications in geographic information systems (GIS; Lillesand et al., 2004).

1.2 PRINCIPLES OF REMOTE SENSING

Remote sensing is mainly concerned with the measurement or acquisition of information about an object without being in physical contact with the object under study. It uses EME as the means for collection of information about the objects. Depending on their properties, the Earth's surface features reflect or radiate or emit different kinds and amounts of EME in various wavelengths (Figure 1.1). The measurement of reflected or radiated or emitted electro-magnetic radiation (EMR) forms the basis for understanding the characteristics of the Earth's surface features (Colwell, 1983).

All features on the Earth's surface reflect and absorb the Sun's radiant energy. They also emit and reflect back parts of that energy. Various objects emit and reflect different amounts of energy in different parts of the EMS, depending on the properties of the material of objects (structural, chemical, and physical), surface roughness, angle of incidence, intensity, and wavelength of radiant energy (Gibson, 2000). These signals are recorded by the sensors, and are transmitted to the ground receiving stations where these are processed before storing them digitally as satellite images. These data/images are distributed to the users as per their needs.

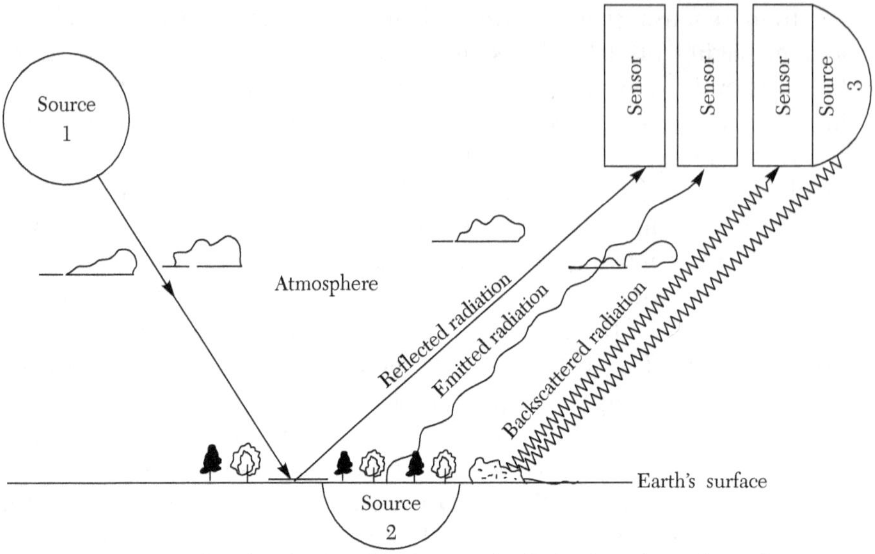

Figure 1.1. Reflected, emitted, and backscattered radiation from the Earth's surface.

1.3 REMOTE SENSING DATA ACQUISITION

The data acquisition involves seven main processes as illustrated in Figure 1.2, and described as follows.

1.3.1 Energy Source

The energy source is the first and foremost requirement for remote sensing data acquisition which is used to illuminate the Earth's surface features/targets. There are two modes of acquiring remote sensing data; passive and active means of remote sensing. The *passive remote sensing systems* rely on the Sun as the source of energy; however *active remote sensing systems* emit their own source of energy by the sensors (such as radar or a camera with a flash gun). Passive systems take images during daylight only, while the active systems are always ready to take images, anytime, anywhere (Colwell, 1983). Most of the remote sensing systems for natural resource applications take daylight images with the Sun as the source of illumination of ground objects, and the corresponding reflected radiation/backscattered energy is recorded.

1.3.2 Energy Propagation Through the Atmosphere

The EME travels through the atmosphere from its source (the Sun in the case of passive remote sensing) to the objects/targets. The atmosphere

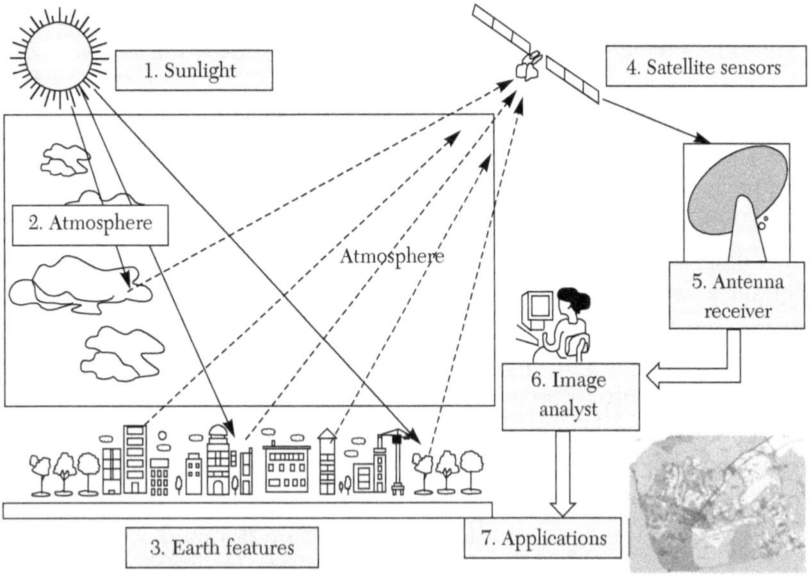

Figure 1.2. Remote sensing data acquisition processes.

absorbs, scatters, or transmits the EME to the objects on Earth, and then energy travels back from the objects to the sensor, following Snell's Law. During both trips (source to object and object to sensor), scattering, absorption, and transmission processes take place, affecting the intensity and quality of EMR. For details of EMR, refer to Chapter 2.

1.3.3 Energy Interaction with Objects/Targets

The EME interacts with the Earth's surface features and gets reflected, absorbed, or transmitted, depending on the physical and chemical characteristics of the objects/targets as well as the wavelength region of observations.

1.3.4 Energy Received by the Sensors

The EME reflected/emitted from the object/target again travels through the atmosphere, and is finally recorded by the sensor(s). The most important component of a remote sensing system is the sensor onboard satellite which records the variation of EME reflected or emitted by objects on the Earth's surface. Different types of sensors are sensitive to different parts of the EMS (Garg, 2019). The data or image is formed of the object in accordance to the response of the detector to the incident energy. These sensors fall in two broad categories: image forming and nonimage forming sensors. The function of a sensor is to convert analog signals into digital signals so that these are machine readable. These signals are transmitted back to the ground receiving station. For details about sensors, refer to Chapter 3.

1.3.5 Data Transmission, Reception, and Processing

The EME recorded by the sensor is transmitted to the ground receiving station in electronic form through telemetry. Ground communication stations provide the radio-telemetry links between satellites and the Earth. The images are raw and unprocessed, and therefore are required to be digitally enhanced to highlight a particular Earth feature (Jain, 1989). After various levels of processing and corrections, these signals are processed into images. The recording of data is done by dividing the incoming energy through filters into different wavelength bands, and then converting energy in each wavelength band into an electrical signal. The electrical signal is processed to give radiometric data for each band, which is then recorded on digital media. From the ground receiving station, data are sent to the archive center for storage and given to users on demand.

1.3.6 Data Interpretation and Analysis

The processed images can be interpreted visually/digitally by the experts to extract the information about the Earth's surface features. Various image processing software is available to process and analyze the remote sensing data and covert them into useful thematic maps needed for various applications.

1.3.7 Applications

The information extracted from imagery can be utilized for several applications, such as land use/land cover change, natural resource mapping and management, crop mapping and monitoring, urban planning, damage assessment due to disaster, and so on.

1.4 TYPES OF REMOTE SENSING SYSTEMS

There are various remote sensing systems providing images in optical, infrared, and microwave regions of EMS. The Landsat Multispectral Scanner System (MSS), launched in 1972, began the modern era of land remote sensing from space (Castleman, 1996). Today, operational remote sensing systems sample in several parts of the EMS with dozens of spectral bands and spatial resolutions better than 1 m. Figure 1.3 presents various remote sensing systems, launched by various countries over a period of time. The essential differences in the several satellite imagery missions are characterized by the number and coverage of the spectral bands, the spatial resolution, the coverage area, and the temporal resolution. A brief explanation of various remote sensing systems is given as follows, while the details can be found in Chapter 3 and Chapter 4.

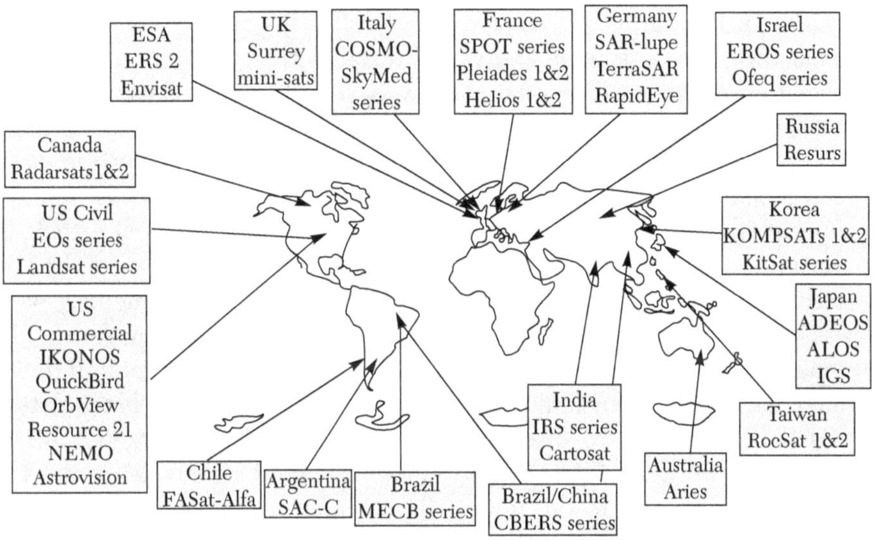

Figure 1.3. Various remote sensing systems launched worldwide (Stoney, 1997).

1.4.1 Optical Remote Sensing System

In optical remote sensing, optical sensor detect the solar radiations reflected or scattered from the Earth, forming images resembling photographs taken by a camera in space. The wavelength region (0.40–0.70 μm) usually extends from the visible and near infrared (VNIR) to the short-wave infrared (SWIR). Various classes, such as water, soil, vegetation, forests, snow, buildings, and roads reflect visible and infrared light in different ways.

1.4.2 Thermal Infrared Remote Sensing System

This system makes use of infrared sensors to detect infrared radiation emitted from the Earth's surface. The wavelength of infrared radiation is between 0.70 μm and 1 mm. The middle-wave infrared (MWIR) and long-wave infrared (LWIR) are within the thermal infrared (TIR) region. These radiations are emitted from warm objects on the Earth's surface, and are used for measurements of the Earth's land and sea surface temperature as well as the detection of forest fires, volcanoes, and so forth.

1.4.3 Microwave Remote Sensing System

The microwave portion of the spectrum covers the range from approximately 1 cm to 1 m in wavelength. Some remote sensing satellites carry passive or active or both microwave sensors. The active sensors emit pulses of microwave radiation to illuminate the areas to be imaged, as shown in Figure 1.4. Images of the Earth's surface are formed by measuring the microwave energy scattered by the ground or sea, reaching back to the sensors. The microwave wavelength region is used in remote sensing to provide useful information about the Earth's atmosphere, land, and ocean. These have several advantages as they can penetrate clouds and moisture, and can be acquired day and night. A microwave imaging system which can produce a high-resolution image of the Earth is the synthetic aperture radar (SAR). The most common form of imaging active microwave sensors is radar (radio detection and ranging), which essentially characterizes the function and operation of a radar sensor. The sensor transmits a microwave (radio) signal toward the target and detects the backscattered portion of the signal. The strength of the backscattered signal is measured to discriminate between different targets, and the time delay between the transmitted and reflected signals determines the distance (or range) to the target. Radar images are particularly used to map landforms and geologic structure, soil types, vegetation and crops, and ice and oil slicks on the ocean's surface.

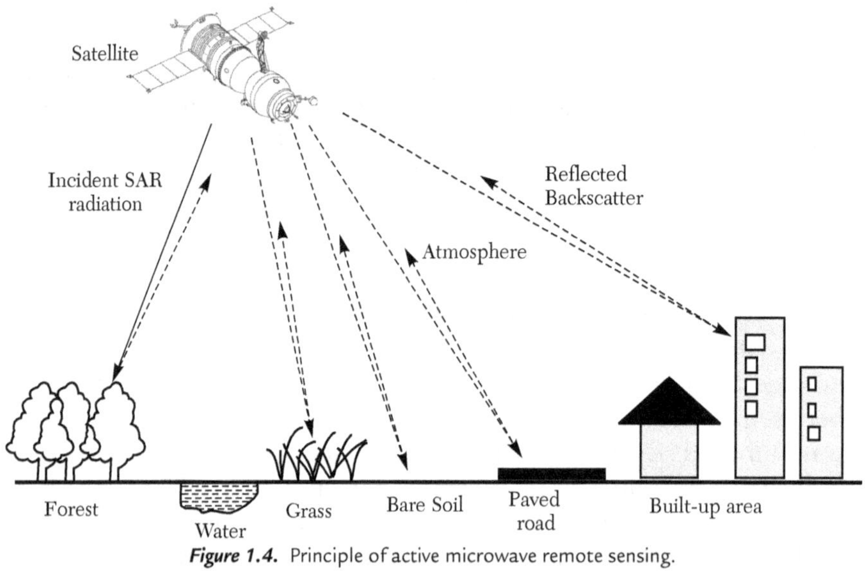

Figure 1.4. Principle of active microwave remote sensing.

1.5 SOME TECHNICAL TERMS

It is essential to understand the popular technical terms frequently used in remote sensing. Some of them are given as follows:

Electromagnetic spectrum (EMS)

The range of energy which contains parts such as the gamma ray, x-ray, ultraviolet, visible, infrared, microwave (radar), and radio waves is called the EMS. Different parts of the EMS have different wavelengths and frequencies which travel with the same speed as the velocity of light (v = 2.98×10^8 m/s).

Reflected energy

Electromagnetic energy that strikes the objects on the Earth's surface can be reflected, absorbed, or transmitted. The part of the incident energy that is returned back from the objects and measured by the sensor is called *reflected energy*.

Absorption

It is the process by which radiant energy is absorbed and converted into other forms of energy.

Transmission

It is the amount of radiation of different wavelengths that a medium (e.g., atmosphere) will transmit or allow to pass through without any change.

Platform

A remote sensing platform is usually a satellite or an airplane carrying sensors.

Sensor

A sensor is an electronic device that detects EME, and converts it into a signal that can be recorded as numbers and displayed as an image.

Spectral band

The sensors are designed to operate in several wavelength ranges to gather the EM radiations reflected from or emitted by the ground features/objects. The wavelength range is called a *channel* or a *spectral band* or simply a *band*. The sensors can collect the data in a number of spectral bands, and may be grouped as: panchromatic (single band), multispectral (more than one band), or hyperspectral (usually over 100 bands) sensors.

Image

The picture resulting from the sensing process is called an *image*. A remote sensing image can either be in paper format or digital format displayed on a computer monitor. A digital satellite image is also called a *raster image*.

Pixel

The word *pixel* is derived from "picture element." Since a digital image is made up of grids arranged in rows and columns, it is the smallest element (grid) in an image. Normally, a pixel is square in shape with its dimensions representing the spatial resolution of the image. The number of pixels in an image may be computed by multiplying the number of rows with the number of columns present in the image. Similarly, the ground area covered by the entire image is computed by multiplying the total number of pixels with a one-pixel ground area.

Gray scale

It is a medium to calibrate the variations in the brightness of an image that range from black to white with intermediate gray values.

Digital number (DN)

The DN in remote sensing systems is a value assigned to a pixel, usually in the form of a binary integer, say in the range of 0–255 (i.e., a byte) in an 8-bit image. It means the range of EME captured by a remote sensing system is divided into 256 gray levels (2^8), where digit 0 indicates a perfect black pixel and digit 255 indicates a white pixel in a B&W image. It is the variation of DN values that produces various shades in an image and helps identification of various objects/features. The DN value is very important as digital analysis of remote sensing images is based on the distribution and pattern of these DN values.

Histogram

A graphical representation of DN values in a set of data is called a *histogram*. In a histogram, individual DN values are displayed along the x-axis, and the frequency of their occurrence is displayed along the y-axis.

Brightness of an image

It can be defined as the amount of energy output by a source of light relative to the source we are comparing it to. Brightness is a relative term, and it depends on our visual perception. For example, in some cases we can easily say that the image is bright or the image is dull.

Contrast of an image

Contrast can be simply explained as the difference between maximum and minimum pixel intensity in an image. Contrast is an important factor in any subjective evaluation of image quality. It is the difference in visual properties that makes an object distinguishable from other objects and the background.

Classification

It is the computational process of assigning pixels or objects into a set of categories, or classes, having common spectral characteristics.

Thematic map

A map that displays the spatial distribution of an attribute related to a single topic, theme, or subject of discourse. Usually, a thematic map displays a single attribute (a "univariate map") such as soil type, vegetation, rivers, geology, land use, or habitation.

1.6 VARIOUS FORMS OF REMOTE SENSING DATA

There are various types and forms of remotely sensed data; these may be either photographic data products, also called hard copy images, or digital data, also called soft copy data (Garg, 2019). The photographic data may be obtained by camera systems installed in aircraft, balloons, drones/unmanned aerial vehicles (UAVs), and so on, while the digital data can be obtained from sensors, scanners, radiometers, scatterometers, and LiDAR (laser-based devices) on board satellites, aircraft, drones/UAVs, and so forth. The digital data collected by sensors/scanners, which is the primary product, can be converted into hard copy product (secondary product) in the laboratory or data processing center. The digital data may be in the form of single bands, multi-bands, or hyperspectral bands. The optical ranges in which these data are collected by various sensors is shown in Figure 1.5.

Generally, manual interpretation is carried out with hard copy (photographic) images to extract information or create thematic maps, whereas digital interpretation methods (computer and image processing software) require digital data to analyze. Hard copy images may be black and white or colored, while the digital images from satellites are in raster format or grid format. The hard copy images may also be converted to digital form (raster) by the process of scanning it through a scanner. Both types of images are in raster format but their nature and characteristics are

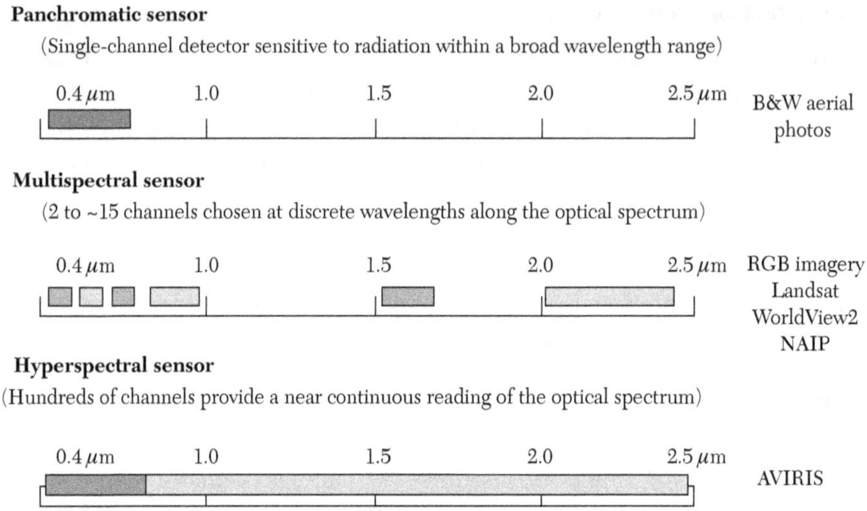

Figure 1.5. Wavelength ranges used by various sensors.

different from an analysis point of view. The latter images can be used for digitization or vectorization purposes.

Various remote sensing data types are briefly presented as follows:

1.6.1 Black and White Images

The individual images from a multispectral (MS) camera/sensor/scanner are black and white (B&W) taken in several spectral wavelength regions of the EMS. A large number of natural resource satellites, such as Landsat, SPOT, IRS, Sentinel, and so forth, provide data in MS bands. These individual band B&W images can be processed to create color images. Many satellites, for example, SPOT-6/7, Cartosat, IKONOS, and so forth, also carry a panchromatic (PAN) camera/sensor with them, providing a single B&W image of the area taken in a larger visible portion of the EMS, as compared to the MS sensors. The PAN data is usually a representative of a range of wavelengths and bands, for example, the thermal infrared or visible; it combines different colors, hence the name "pan" chromatic. This image is more of a combination of blue, green, and red wavelengths to measure the reflectance. In other words, the band is formed by using the total light energy in the visible spectrum (instead of partitioning it into different spectra). It therefore renders a single intensity (DN) value per pixel that is commonly visualized in a gray scale image. Information contained in each pixel of a PAN image is, therefore, directly related to the total intensity of solar radiation that is reflected by the objects in the pixel, and is detected by the PAN sensor.

The B&W images from PAN cameras provide higher spatial resolution as compared to spatial resolution of individual B&W images from MS sensors, and are therefore preferred in many applications. Due to the higher amount of solar radiation collected per pixel, PAN sensors are able to detect brightness changes at smaller spatial extents (i.e., pixel size) than the MS sensors. As an example, a PAN image from a Cartosat image is shown in Figure 1.6. The bandwidth enables it to hold a high signal noise, making the PAN data available at a high spatial resolution. This capability allows the smaller portion to be seen, while still acquiring strong signals.

Figure 1.6. Cartosat-3 panchromatic image.

1.6.2 Multispectral Images

The multispectral sensors that capture data in multiple bands are called multispectral images. Normally, the more spectral bands, the more information is gathered by these sensors, such as Landsat, SPOT, RapidEye, and Worldview-2 and -3. The multispectral sensors typically provide less than fifteen bands of data. In such sensors, due to relatively small amount of energy available for each band, the detectors need to sample a larger area (pixel size) in order to collect the minimum amount of light energy required for detecting the brightness differences. Thus, the multispectral images tend to have larger pixel sizes (i.e., representing the sampled area) than the PAN images, which due to the high amount of energy sample a smaller area and therefore have a smaller pixel size. For example, the PAN band of Landsat has a spatial resolution (pixel size) of 15 m, which is smaller than the 30 m pixel size of its multispectral bands. have a smaller Thematic Mapper (TM) sensor collects data in seven spectral bands, and their spectral resolution is higher than the early sensors onboard Landsat, such as MSS. Figure 1.7 shows six bands data of TM sensors.

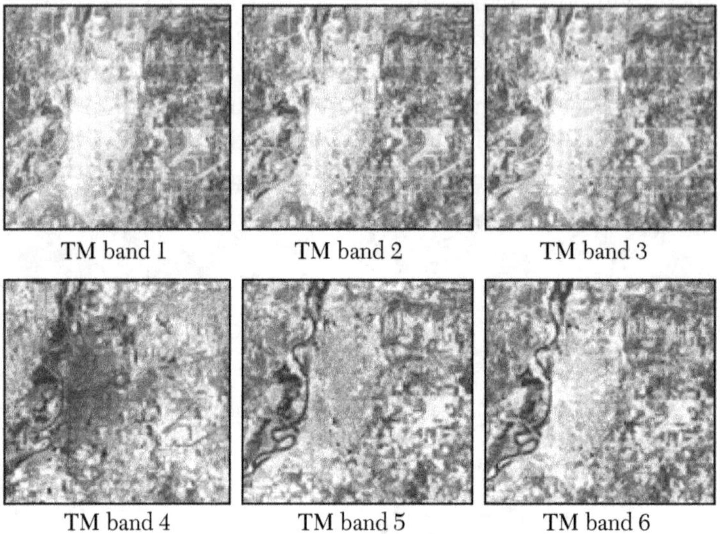

Figure 1.7. Six multispectral bands of Landsat TM data.

1.6.3 Hyperspectral Images

Hyperspectral images consist of more than hundreds of narrower spectral bands (10–20 ηm) where images are recorded by an imaging spectrometer, such as AVIRIS (airborne visible-infrared imaging spectrometer). Hyperspectral imaging is a form of spectral imaging that uses multiple bands across the EMS. Utilizing the technique of spectroscopy, which is used to identify materials based on how light behaves when it hits an object, hyperspectral imaging obtains multiple spectra of data for each pixel in the image of a scene. The wavelength ranges measured depend on the type of camera (or sensor) used.

Further information about these three broad categories of sensor systems is summarized in Table 1.2.

Table 1.2. Details of Panchromatic, Multispectral, and Hyperspectral Sensor Systems.

	Panchromatic systems	**Multispectral systems**	**Hyperspectral systems**
Spectral range (μm)	0.43–0.72	0.43–0.72 and 0.75–0.95	0.47–2.50

	Panchromatic systems	Multispectral systems	Hyperspectral systems
Satellites	QuickBird, SPOT, IKONOS	Landsat, IRS, SPOT, QuickBird, IKONOS	ASTER, Hyperion, MODIS, EO-1
Spectral band	Mono-spectral, B&W, gray-scale image	Several spectral bands	10 to 100 spectral bands
Spatial resolution	Submeter	Up to 0.50 m	Up to 2 m
Applications	Earth observation and reconnaissance applications	Analysis of three bands—Green-red-infrared for vegetation and camouflage detection; Blue-NIR-MIR for visualizing water depth, vegetation coverage, soil moisture content, and the presence of fires, etc.	Agriculture, Mineralogy, Geology, Chemical imaging, Environment, etc.
Advantages	High applicability in (*i*) imaging multiple targets; (*ii*) mosaic strips to large area; (*iii*) stereo acquisition; (*iv*) linear feature acquisition, such as coastlines, pipelines, roads, buildings, etc.		
Disadvantages	Affected by sun illumination and cloud coverage: polar areas with seasonal changes in sun illumination and the equatorial belt with persistent cloud coverage		

1.6.4 Color Composite Images

The human visual system can differentiate several colors but only a few gray levels (B&W shades). Therefore, color images can be generated with a minimum of three multispectral bands of data, as each spectral band is associated with a color resulting in a color composite image. Each band of a multispectral image can be displayed on a computer one band at a time as a gray-scale image, or as a combination of three bands at a time as a color composite image. In addition, a pseudo color image can be produced with a single band B&W image, where the range of DN values is segmented into various homogenous classes and a different color may be assigned to each class. Three primary colors (i.e., red, green, and blue) are used to generate a color composite image, as shown in Figure 1.8. The three primary colors, superimposed in different proportions, can produce a number of colors. There are two types of color composites, natural color, and false color, described as follows.

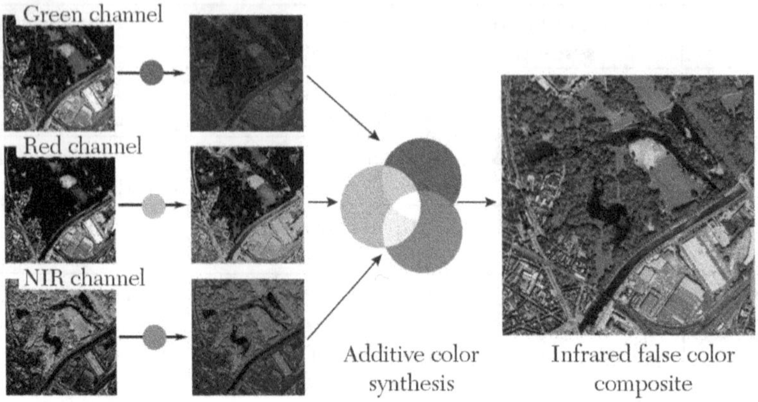

Figure 1.8. Generation process of color image (please see color figures in companion files).

1.6.4.1 Natural color composite

A natural color composite image displays a combination of visible red, green, and blue bands with the corresponding red, green, and blue channels. For example, in the case of Landsat TM, if band 3 (red band) is displayed in red, band 2 (green band) in red, and band 1 (blue band) in a blue color, it results in a natural color composite (Figure 1.9[*a*]) that corresponds to how we usually see the ground (e.g., vegetation appears green, and impervious surfaces appear light gray and brown). Many analysts prefer natural color composites, because colors that seem natural to our eyes are often seen on natural color images. In this combination, channels of the visible range are used; hence it is useful to analyze the state of water bodies, water penetration, and sedimentation processes. In addition, it is also used for studying anthropogenic objects and urban regions. However, sparse vegetation and types of vegetation are poorly detected; in addition, clouds and snow both appear as a similar white, and are therefore difficult to distinguish. It is also difficult to distinguish shallow water from soil. Natural color images can be low in contrast and somewhat hazy, mainly due to the scattering of blue light in the atmosphere.

1.6.4.2 False color composite (FCC)

It is an artificially generated color image in which blue, green, and red colors are assigned to the wavelength regions to which they do not belong in nature. The FCCs allow us to visualize the wavelengths the human eye can't see (e.g., NIR range and beyond). The use of bands, such as NIR, increases the spectral separation, and can enhance the interpretability of data. For example, in a standard FCC the green wavelength (0.5 to 0.6 μm) is assigned a blue color, the red wavelength (0.6 to 0.7 μm is assigned to green, and the near infrared wavelength (0.7 to 0.8 μm) is assigned to red. The color of an object in the FCC image does not have any resemblance to its actual color; therefore, the resulting product is known as a false color composite (Figures 1.8 and 1.9[b]). In FCC images, vegetation appears in different shades of red depending on the types, health, leaf structure, conditions of the vegetation, and moisture content of the plants, since it has a high reflectance in the NIR band.

The FCC using an IR band combination is very common in remote sensing when identifying vegetation, crops, soils, and wetlands. Vegetation appears red, with healthier vegetation being more vibrant in this band combination. In general, a dark red color shows broadleaf and/or healthier vegetation, whereas the lighter red color indicates sparsely vegetated or grasslands areas. Soils vary from dark to light brown and urban areas are cyan blue or at some times can appear yellow or gray, depending on their composition. Clouds, snow, and ice are light cyan or white. Hardwood trees appear as light red than the coniferous. Clear water appears dark-bluish, while turbid water appears cyan. Densely populated urban areas are indicated in light blue. There are many possible combinations of producing FCC images, however, a common scheme for displaying in the case of a SPOT multispectral image is band 3 (NIR) in red, band 2 (red) in green, and band 1 (green) in blue color.

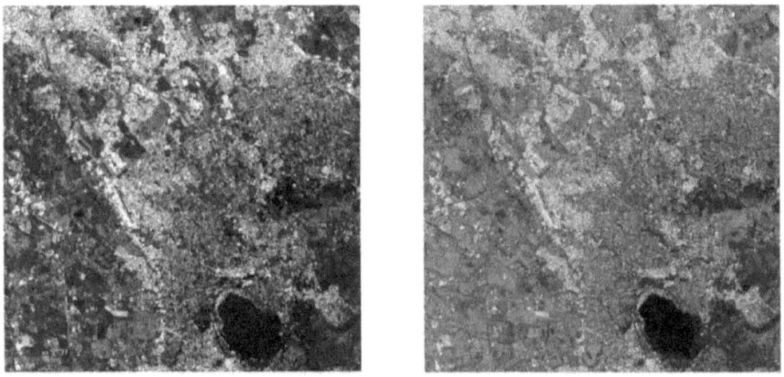

RGB = 432 RGV = 543

(a) (b)

Figure 1.9. (a) Natural color and (b) FCC images (please see color figures in companion files).

1.7 THE MULTI-CONCEPT IN REMOTE SENSING

The multi-concept is an important approach and one of the key concepts in remote sensing data acquisition and analysis (Colwell, 1983). It includes multistage (height), multi-resolution (pixel size), multispectral (bands), multi-detector (sensors), multi-temporal (date), multiuse (applications), multi-direction, multidisciplinary, and multi-thematic, described as follows.

1.7.1 Multistage

The remote sensing data can be collected from different platforms at different altitudes, such as satellite data, high altitude data, low altitude data, drone data, balloon data, and ground observations. Each of the platforms will provide data/images at different scales with different ground coverage (Figure 1.10). The lower platform will cover a smaller area with greater details, as compared to the higher platform. The selection of data/image will depend on the details required to be mapped and studied. Many times, a combination of data from various platforms helps in a more reliable and accurate extraction of information, particularly where ground verification is carried out for accuracy estimation.

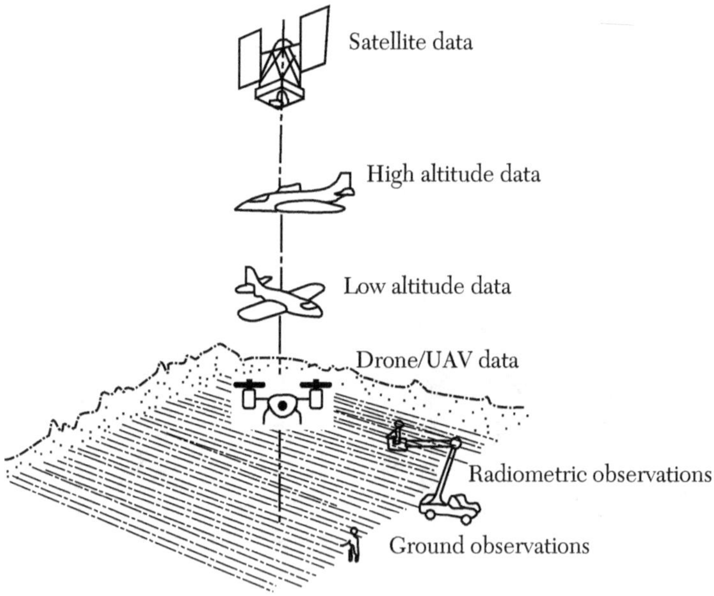

Figure 1.10. Multistage remote sensing platforms.

1.7.2 Multi-Resolution

The images collected at different pixel sizes define the spatial resolutions. The higher the resolution, the more detailed the information about the ground features, but less of the area is covered. Lower resolution images would cover a large area showing only broad/generalized features. The cost of the images is also normally associated with the spatial resolution. The higher the spatial resolution, the higher the cost of the image, and therefore the selection of such an image for the study may be done carefully in order to ensure economy for the project work. In addition, higher resolution images would be required in large number as they cover a small area as compared to lower resolution images, so these would require more storage space and computational requirements. For example, if the forested region is to be demarcated, a medium resolution image would be sufficient, but if the tree species are to be identified, then very high-resolution images are needed. More details about spatial resolution are given in Chapter 3.

1.7.3 Multi-Band

Multi-band indicates individual spectral bands within a given region of the EMS, for example, visible, infrared, thermal, microwave, and so forth. Such images are needed to correctly identify various objects on the ground, as they might look different in each wavelength region, depending on the reflected and emitted energy in that wavelength region. Moreover, multi-band images are also required to generate color composites for better interpretation, as human eyes are more sensitive to colors for identification of features than individual B&W images. Please see Chapter 2 for further details of various wavelength regions of the EMS and their utilization.

1.7.4 Multi-Sensors

The satellite images taken from various sensors are helpful in accurate identification and mapping of features. These sensors provide data at different resolutions and different wavelength regions of the same area, and are thus helpful in recognition of objects. Moreover, data fusion techniques are very popular for merging data from two sensors and getting the resultant image, which uses the best information from both the images. More details about data fusion are given in Chapter 4.

1.7.5 Multi-Temporal

Images taken from the same satellite of the same area can be acquired repetitively at regular intervals, and used to study the change scenario in the area, but images taken from different satellites can also be used to study the change scenario by having a reduced time interval. These images are useful particularly for mapping and monitoring dynamic events and features such as floods, crop growth, forest fires, and so on. Please refer to Chapter 3 for further details on temporal resolution of the images.

1.7.6 Multi-Direction

There are times when more information can be obtained about an area using viewing angles other than the vertical. Such images are useful to create 3D models of the terrain (e.g., see the characteristics of the SPOT satellite in Chapter 4) as well as monitor the changes in the area.

1.7.7 Multi-Disciplinary

By using teams of interpreters and experts with expertise in different disciplines, more information can be extracted for a given application from remote sensing images. Remote sensing is an interdisciplinary subject where experts from various specializations are required to get optimum results from these images.

1.7.8 Multi-Thematic Maps

Digital remote sensing images can be thought of as one-time written, and many times read. Many different themes (e.g., water, vegetation, land use, agricultures, road network, urban, etc.) can be extracted from the same set of satellite images by various users.

1.7.9 Multi-Uses

The same set of images can be used to extract the information related to various applications, such as global change studies, watershed analysis, and vegetation monitoring. For example, a highway engineer may be interested to use images for extracting road networks and assessing the road condition, whereas an irrigation engineer may require information on the spatial distribution of rivers, canals, ponds, reservoirs, and so forth in an area from the same satellite scene.

1.8 ADVANTAGES AND DISADVANTAGES OF REMOTE SENSING

Remote sensing gives us the ability to observe and collect data across large regions. It makes it possible to collect data in dangerous or inaccessible areas or snow-covered areas. For example, it would be extremely difficult to collect field data from the arctic year round, but satellite images can be used to measure the sea ice throughout the year. Remote sensing also provides a unique opportunity to map and monitor global change. The repetitive measurements from satellites and other data provide ways to monitor and quantify changes in the environment. Remote sensing can also be a cost-effective way to collect a large amount of data to model the environment and broaden our understanding of the Earth's systems. Another important aspect of remote sensing is that sensors can measure energy at wavelengths which are beyond the range of human vision, allowing us to measure and visualize phenomena that are beyond the human senses. The advantages and disadvantages are discussed as follows (Garg, 2019):

1.8.1 Advantages of Remote Sensing

1. Provides data of large areas.
2. Provides data of remote and inaccessible regions, without physically visiting the regions.
3. Easy and faster collection of data.
4. Availability of images in different wavelength regions for better identification of features.
5. Temporal images allow natural changes in the landscape and environment to be monitored and analyzed.
6. Relatively economical preparation of maps, particularly for large areas.

1.8.2 Disadvantages of Remote Sensing

1. The optical images may have cloud cover, obscuring the ground information.
2. The interpretation of imagery, especially digital methods, requires specialized skill.
3. Needs field verification of analysis from a satellite image to ascertain its accuracy.
4. Data from multiple sources may create confusion.
5. Objects can be misclassified if sufficient training/ground reference data are not available.
6. Sometimes, due to fixed temporal resolution, the images are not available for a specific date to study the happening of the event that day.
7. Erroneous results may be obtained without applying atmospheric and geometric corrections.

1.9 APPROACHES OF REMOTE SENSING DATA ACQUISITION

Remote sensing images and data are acquired to derive information for mapping, monitoring, and management of water, land, resources, infrastructure, and so on. To collect the required information, a wide variety of methods are to be used: land surveying, analysis of information from various maps, laboratory measurements of samples, interpretation of satellite images, measurements by in situ sensors, using aerial photographs, applying numerical models, and so forth. Although remote sensing data can be interpreted and processed without other information, often the best results are obtained by linking remote sensing measurements to ground (or surface) measurements and observations. *Data* means the representations that can be manipulated using a computer/software. For example, an agricultural specialist may require data on the areal extent with different crops and data on soil and biomass production to estimate the yield. An urban planner needs to identify the types of houses and their configuration as well as illegal constructions. An engineer needs to determine the optimal location for siting the structure, based on the information derived on the terrain features and topography. A geologist may explore an area on a satellite image and provide a map of the surface mineralogy and geology. Many of these applications deal with spatiotemporal phenomena since time is an important dimension to assess the changes.

There are various sources to acquire remote sensing data/information, described as follows:

1.9.1 Topographic/Thematic Maps

Topographic maps throughout the world are the best source to provide basic ground information, including true representation of natural and cultural features on the Earth's surface with their latitudes and longitudes as well as contours (elevations with mean sea level), quickly at a reasonably low cost. In India, topographic maps are produced by the Survey of India (SOI), Dehradun at three different scales; 1: 250,000, and 1: 25,000. These maps are frequently used to georeference the satellite images, which is a prerequisite to carry out any quantitative analysis from satellite images. Topographic maps are very popular in India as they are used to compute the coordinates of control points for georeferencing of satellite images and other digital data, and to create a DEM from the digitized elevation contours of the area (Zhang & Gruen, 2006). For this purpose, these paper maps are normally scanned with a high resolution scanner and converted into a raster image.

Many government organizations worldwide produce their own thematic maps, such as roads, agriculture, soil, forests, geology, water, rails, urban, and so on. In India, Survey of India (SOI) Dehradun and National Atlas Thematic Mapping Organisation (NATMO) Kolkata, and large number of government organizations create various thematic maps for the entire country at different scales. These maps can be digitized and used along with satellite images to obtain a better classification as well as be used as a good source of information in GIS-based analysis.

1.9.2 Advanced Surveying Instruments

The advanced surveying instruments like total stations, and laser-based devices (LiDAR) are very useful to capture spatial digital data of the Earth and its environment. These instruments provide three-dimensional coordinates of ground objects (x, y, and z) which can be uploaded into many image-processing and vector-based GIS systems. These data are helpful in georeferencing of satellite images as well as overlaying on satellite images for added values. The z information (elevation) can be utilized to segment the satellite image into various elevation zones (landscapes) for segmentation or improvement of classification accuracy, particularly in hilly regions.

1.9.3 Global Positioning System (GPS)

The GPS is a satellite-based navigation system made up of a network of twenty-four satellites placed into orbit by the U.S. Department of Defense. The GPS works in any weather conditions, anywhere in the world, twenty-four hours a day, 365 days a year. The GPS was originally invented for military applications, but in the 1980s, it was made available for civilian use. It receives signals from GPS satellites, and computes accurate locations (latitude, longitude, and elevation) of objects on the Earth's surface, along with the time parameter. The elevations of objects are computed with respect to the WGS-84 (World Geodetic System 1984) datum, which is a different datum than the mean sea level. However, many GPS or image processing software are capable to convert WGS-84 datum to mean sea level of the country.

The GPS receivers are of two types, handheld or navigational receivers and geodetic receivers. With handheld GPS receivers, the 3D locations of objects can be determined within medium to good accuracy. However, the geodetic GPS can provide the coordinates with greater accuracy than the navigational GPS. Figure 1.11 shows a geodetic GPS receiver. Several land, sea, and air vehicles are using GPS to determine the point positions with an accuracy better than 10 m. The accuracy can further be improved using two geodetic-type GPS receivers simultaneously (called Differential GPS or DGPS) or using a Wide Area Augmentation System (WAAS)-enabled GPS receiver, which uses satellites and ground stations that provide GPS signal corrections (Carter, 1997). Positioning with DGPS, which uses radio links between mobile and fixed GPS receivers, provides relative position in real time with an accuracy of 10 cm to 1 m for many applications.

Figure 1.11. Geodetic GPS (please see color figures in companion files).

The GPS can be used for multiple applications where the exact location of any object or phenomena on the Earth is required, such as locating a park, restaurant, school, and so on. The GPS data are superimposed on satellite images for various applications, such as finding the shortest route, precision farming, forest and resource management, flood warning systems, and natural resource exploitation. The GPS data along with high-resolution satellite images are also used extensively by government agencies, emergency services, rescue and relief personals, surveyors, transporters, hikers, and trekkers (Kaplan, 1996). It provides a major data input tool for image processing and GIS-related work for location-based applications. The GPS offers the advantages of one person survey, accuracy, speed, versatility, and economy for location-based survey. For further details of the working of GPS, readers can refer to Garg (2019).

1.9.4 Ground Penetrating Radar (GPR)

The GPR, invented in the 1970s for military purposes, can also be used for mapping buried objects, pipelines, and utility lines, and locating them accurately. The GPR consists of a transmitter to transmit the radio waves and a receiver to receive the reflected radio-waves from the underground objects. This unit is kept at the base of a moving cart. The reflected signals can be viewed at the display unit, which is fitted at the handle of the cart (Figure 1.12).

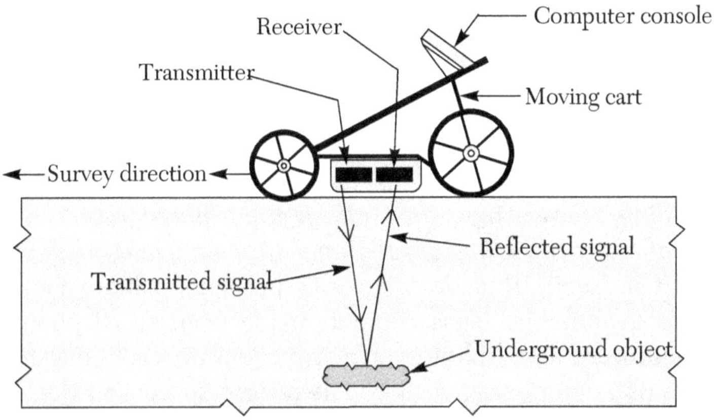

Figure 1.12. Working of a GPR.

Underground utility mapping has become an important concept in modern or smart cities. All modern (smart) cities today have underground utility services, such as water supply, power lines, telephone lines, Internet, gas, and so forth. Detailed maps of underground infrastructure are needed for infrastructure maintenance and its development. The main advantage of having such utility services underground is their low maintenance cost and almost no supporting infrastructure required over the ground. The utilities overground generally require regular maintenance due to wear and tear owing to high wind, storms, heavy rainfall, felling of trees, and so on.

The GPR employs radio waves, typically in the range of 1 to 1000 MHz frequency, to map the man-made structures and features buried in the ground (Bigman, 2018). The device emits radio waves and detects the reflected signals from these structures. The GPR can locate both metallic and non-metallic objects, and other buried structures, such as pipes or cables. The depth and position of metallic and non-metallic pipes, and plastic conduits, such as gas and water lines, can be determined fairly accurately. The GPR is also useful for soil stratification, delineation of rock profile, locating reinforcing and post-tensioning in concrete, delineating the voids under the surface of concrete and asphalt roads, data collection for structural safety, groundwater studies, detecting unexploded land mines, surveying and mapping, and producing 2D/3D data of buried utilities (GSSI, 2016). The accuracy of results from GPR depends on local soil conditions, as some soils (clays, saline) absorb the radio waves, greatly limiting the exploration depth. However, improvements in GPR technology have led to higher resolution scanning and greater penetration depths. For utility mapping, a combination of GPS with the GPR provides more effective results. Further details of GPR are given in Garg (2021).

1.9.5 Photogrammetry

Photogrammetry is the "*art, science and technology of obtaining reliable information about physical objects and the environment through processes of recording, measuring, and interpreting photographic images and patterns of recorded radiant electromagnetic energy and other phenomena*" (Fussell et al., 1986). It incorporates all aspects of qualitative interpretation as well as quantitative measurements made from aerial photographs. In photogrammetry, aerial photographs are taken normally using an aircraft, but these photographs can also be taken from space shuttles, unmanned aviation vehicles, balloons, or even kites. These photographs are found to be very useful to carry out large-scale mapping.

Aerial photographs are examples of analog images, whereas satellite images typically belong to digital images. These photographs are normally recorded over the wavelength range from 0.3–0.9 μm, that is from ultraviolet to visible and near-infrared spectra. In aerial photogrammetry, the camera mounted in an aircraft usually points vertically toward the ground, as shown in Figure 1.13. Multiple overlapping photos of the ground are taken as the aircraft moves along a flight path. These photos are either processed in a stereoplotter or used in a computer for mapping and creating digital elevation model (DEM) of the ground.

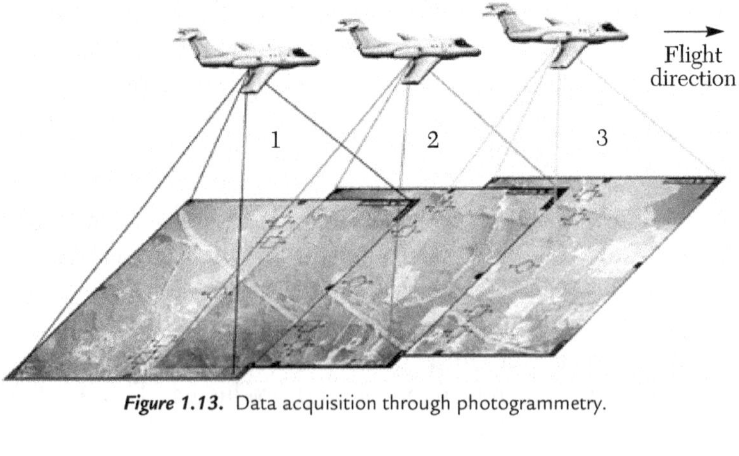

Figure 1.13. Data acquisition through photogrammetry.

The other form of photogrammetry is close-range photogrammetry, where the camera is installed closer to the object, and is typically handheld or mounted on a tripod looking horizontal or near-horizontal. Usually this type of photogrammetry work is non-topographic, that is the output is not topographic products, like terrain models or topographic maps, but instead these are used for drawings of buildings and structures and creating 3D models (Egels & Kasser, 2001).

Photogrammetric techniques can provide ground distances and directions, heights of features, and terrain elevations that can be used in image processing and GIS software. The digital photogrammetric technique combines computer vision and photogrammetry algorithms to extract 3D geometries from a series of 2D digital images. It detects and matches features (arrays of pixels) between overlapping images taken from different locations and is used for creating 3D models or ortho-photomosaics. In addition, stereo-photogrammetric techniques are helpful to get 3D coordinates of ground objects to create a DEM of a area which can be immediately be superimposed on satellite images to create a DTM for better visualization of terrain features. The DEMs and DTMs along with satellite images are widely uused in hydrologic and geologic analyzes, hazard monitoring, natural resources exploration, agricultural management, groundwater modeling, estimation of the volume of proposed reservoirs, determining probability of landslide, flood prone area mapping, terrain visualization, virtual sites, and so forth. Additional details of photogrammetric techniques are given in Garg (2019).

1.9.6 Unmanned Aerial Vehicle (UAV)/Drone

The UAV/drone is a lightweight flying device similar to an aircraft; however, it is different from an aircraft, since it operates remotely without a pilot. A UAV is the aerial platform component of a UAS (Unmanned Aerial System) which has a UAV, sensors, such as NIR/SWIR, MIR, TIR, multispectral/hyperspectral, LiDAR (light detection and ranging), and so forth, camera, GPS, radio-controlled device, a ground control station, and communication components. Historically, the UAVs were mainly used in military applications deployed for remote surveillance and armed attack to reduce pilot losses. In recent years, the use of UAVs in civilian and commercial applications has increased manifold due to technological development, reduced cost, and making them more easily accessible to the public. Nowadays, UAVs have numerous applications in a large number of fields, such as aerial inspection, photography, traffic management, disaster management, precision agriculture, traffic control, search and rescue, package delivery, telecommunications, and so on.

There are many types of UAVs due to their numerous and diversified applications. While there is no single standard for UAV classification, the UAVs can be practically classified into different categories according to their functionality, weight/payload, size, endurance, wing configuration, control methods, cruising range, flying altitude, maximum speed, energy supplying methods, and so forth. Based on how its movement is controlled, it can be an autonomous or a remotely piloted type. Broadly classified, small UAVs use a platform less than 10 kg in weight (Vergouw et al., 2016). Based on the types of wing, UAVs might broadly be classified as fixed-wing or rotary-wing types. The fixed-wing UAVs (Figure 1.14[a]) move horizontally and need a larger open space for takeoff and landing. On the other hand, the rotary-wing UAVs (Figure 1.14[b]) move vertically; thus, they require a small open space for landing and can remain static at a hovering location. Typically, fixed-wing UAVs have a higher maximum flying speed and can carry greater payloads for traveling longer distances as compared to rotary-wing UAVs. In general, selecting a suitable type of UAV is crucial for accomplishing the mission efficiently, which needs to take into account their specifications as well as the requirements of practical applications. A detailed classification for different types of UAVs and their applications has been provided in Garg (2021).

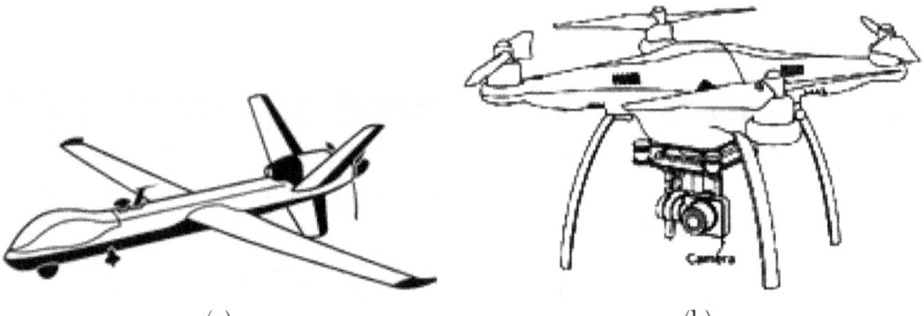

(a) (b)

Figure 1.14. (a) Fixed-wing, and (b) Rotary-wing UAV/Drone.

The UAV datasets are taken with a wide range of sensors, such as RGB cameras, multispectral, hyperspectral, thermal cameras, and lightweight LiDAR. The characteristics of these sensors and their specifications for various remote sensing tasks are given in Table 1.3. Due to their low flying altitude, UAVs can easily acquire very detailed information of objects with a spatial resolution of 1 cm, which allows for accurate geometrical and semantic analysis for a reasonably broader area. The autonomous UAV aided

Table 1.3. Characteristics of Some UAV Sensors (Yao et al., 2019)

Common cameras/ sensors	Example camera	Spectral range	Resolution (MP-megapixel)	Payload (g)	Applications pixel)
RGB cameras	Sony A9	0.40–0.70 μm	24.2 MP	588	Visual analysis, mapping, land cover/ land use, classification, pedestrians and vehicles detection, and tracking, etc.
	Canon EOS 5D mark IV	0.40–0.70 μm	30.4 MP	800	
	Nikon D850	0.40–0.70 μm	45.7 MP	915	
Lightweight multispectral cameras	Sentera Quad multispectral sensor	0.40–0.70 μm	1.2 MP	170	Visual analysis, vegetation detection and analysis, crop monitoring, mining, soil moisture estimation, fires detection, water level measurement, land cover/land use mapping, etc.
	Quest Condor5-UAV	0.40–1.0 μm	2048 × 1088 (2.2 MP)	1450–1950	
	Phaseone iXU/iXU-RS 1000 aerial cameras	0.40–0.70 μm	100 MP	1430–1700	
Hyperspectral sensors	Rikola hyperspectral camera	0.50–0.90 μm	1.05 MP	< 600	Land cover/land use mapping, vegetation indices estimation, biophysical, physiological, or biochemical parameters estimation, agriculture and vegetation disease detection, disaster damage assessment, etc.
	MP Resonon Pika NIR-640	0.90–1.70 μm	640 pixels	2700	
	High-efficiency hyperspectral SWIR	1.00–2.50 μm	384 pixels	4400	
Lightweight thermal infrared sensors	FLIR Vue Pro	7.5–13.5 μm	640 × 512 pixels	72	Tracking creatures, volcanos detection, forest fire detection, hydrothermal studies, urban heat island measurement, etc.
	Workswell WIRIS 640	7.5–13.5 μm	640 × 512 pixels	< 390	
	YUNEEC CGOET thermal imaging camera and low-light camera	8–14 μm	2.1 MP	278	
UAV LiDAR	RIEGL VUX-240	Near-infrared	Up to 1,500,000 per second	3800	Vegetation canopy analysis, estimation of forest carbon absorption, mapping cultural heritage, building information modeling, etc.
	Velodyne Puck LITE	0.903 μm	Up to 600,000 per second	590	
	Livox Mid-40	0.905 μm	100,000 per second	760	

by GPS/IMU systems are able to easily capture either full motion videos or high-resolution photogrammetric images. Avanced photogrammetric processing, orthophotomosaics, 3D models, and DEMs are becoming standard products for UAV-based remote sensing missions. In addition, multi-sensor data from UAVs has brought new opportunities for data fusion with much higher spatial resolution.

Furthermore, the development of UAV platforms and various sensors have given rise to remote sensing applications such as object detection and real-time monitoring at finer scales. The advanced data analysis techniques using computer vision and machine learning enhance the capability of automated UAV data analysis. UAV technology and autonomous flights are expected to have an exponential increase in using UAV images for various mapping applications. Vector and raster layers from UAVs/drones can be used in GIS along with other data (Jessica & Warren, 2018). Nonetheless, the UAV/drone based technology is revolutionizing both spatial data collection and geographic analysis at a large scale. Compared with conventional aerial photography, UAVs have the advantages of rapid takeoff and landing, repetitive operations, low cost for image collection, and high spatial resolution of collected images.

1.9.7 Light Detection And Ranging (LiDAR)

The LiDAR is a remote sensing method ued to collect high-resolution data from ground-based or mobile or airborne systems to examine the surface of the Earth (Buczkowski, 2018). Airplanes, drones, and helicopters are the most commonly used platforms for acquiring airborne LiDAR data, while ground-based systems use tripods or vehicles. The LiDAR sensors have been known as one of the most accurate ways for geometric data acquisition. A LiDAR instrument principally consists of a laser, a scanner, and a specialized GPS receiver. During a LiDAR survey, an active optical sensor transmits laser beams toward a target/object while moving along or rotating across defined survey routes or fixed objects. The laser energy is reflected by the target/object and is detected and analyzed by receivers in the LiDAR sensor. The receiver records the precise time from when the laser pulse left the system to when it is returned to the sensor. Using precise pulse time, the range distance between the sensor and the target may be calculated. These light pulses, combined with the other data recorded by the LiDAR system, generate precise 3D information about the Earth and its surface features. Figure 1.15 shows the concept of point data collection using a terrestrial laser scanner.

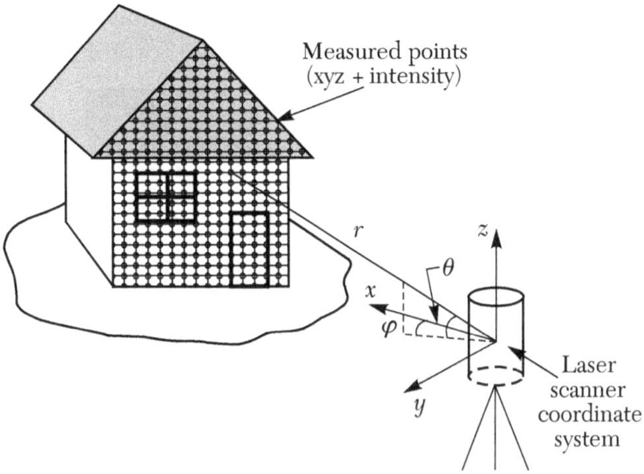

Figure 1.15. Data collection by terrestrial laser scanner.

There are two types of LiDAR: topographic and bathymetric. The topographic LiDAR typically uses a near-infrared laser to map the land, while bathymetric LiDAR uses water-penetrating green light to measure seafloor and riverbed elevations. The LiDARs are widely used in forestry, cultural heritage, and building information modeling (BIM). The LiDAR data can also be used to map structures, including vegetation height, density, and buildings (Cheng et al., 2015). Engineers and Earth scientists use LiDAR to accurately and precisely map and measure natural and constructed features on the Earth's surface, within buildings, underground, and in shallow water. These systems have lot of potential to provide data for proper planning of urban areas or transforming cities into smart cities.

The LiDAR sensors operate on the same principle as that of laser equipment. It is similar to other types of radar, but uses light instead of radio waves. The LiDAR uses a sharp beam with high energy, and hence high resolution data can be achieved (Ren et al., 2016). When laser ranges are combined with position and orientation data generated from integrated GPS and IMU (inertial measurement unit) systems, scan angles, and calibration data, the result is a dense, detail-rich group of elevation points, called a *point cloud*. Each point in the point cloud has 3D spatial coordinates (latitude, longitude, and height) that correspond to that particular point on the Earth's surface from where a laser pulse was reflected. The point cloud data produced by the LiDAR systems can result in thematic layers depicting ground features better than traditional remote sensing and radar methods. The point clouds can also be used to generate other products, such as DEMs, canopy models, building models, and contours. Figure 1.16 shows point cloud data collected through an airborne LiDAR survey.

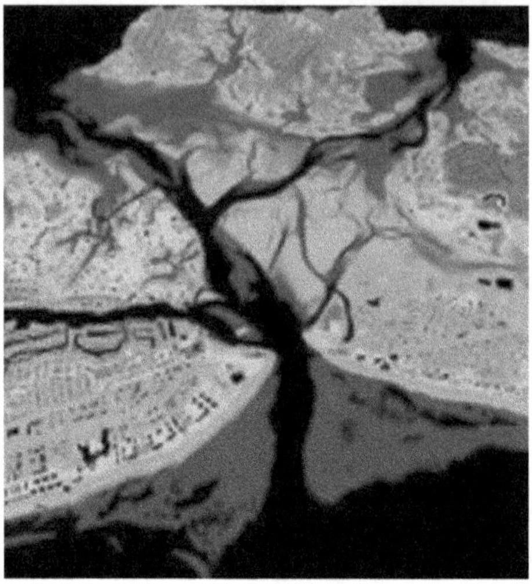

Figure 1.16. LiDAR point cloud data (please see color figures in companion files).

LiDAR-based systems are gaining popularity to provide useful input to GIS, as data collection is much faster and accurate with their all-weather operation capability. Both vector and raster data models in GIS use LiDAR data. Many GIS programs have functions to convert points data (x, y, and z) into vector layers, which may also be converted later into raster layers. In vector systems, LiDAR points frequently generate vector layers, called triangular irregular networks (TIN), to represent terrain. The DEM of a large area can be generated quickly from these data.

The advantages of LiDAR systems over photogrammetry are high reliability and the ability to penetrate thin forests through multiple returns. The LiDAR sensors, even those with relatively low cost, are still orders-of-magnitude higher than the RGB cameras, and require higher payloads (up to a few kg). Therefore, in terms of the cost and needed sensors for integration, the UAV LiDAR system is not yet as popular as UAV-based photogrammetric mapping systems. The disadvantage of LiDAR systems is that the technology is expensive for a country, like India, and the volume of data collected is huge even for a small area. Therefore, it requires a very high-speed computer to do several editing works before processing the data to transform it into thematic layers. Despite the needed consideration of payload and cost, potentials for the use of both RGB and LiDAR sensors are still very promising. Further information about the LiDAR system can be obtained from Garg (2021).

1.9.8 Remote Sensing Images

Satellite imagery may be viewed as an extension of aerial photography because satellite remote sensing relies on the same physical principles to interpret and extract the information content. Satellite images, acquired by sensors and linear/area arrays, are taken at a higher altitude, covering a larger area of the Earth's surface at one time. Satellite sensors can usually detect and record a much wider range of the EMS, from ultraviolet radiation to microwaves. For example, infrared imagery usually consists of images that include the visible channels as well as some range of the infrared spectrum. Multispectral data include up to 7–12 channels of data, and hyperspectral can be more than 100 bands of data collected over discrete bandwidths of the EMS (Birk & McCord, 1994).

Satellite images (photographic or digital) are an important and preferred source of data in Earth-related study. There is a range of remote sensing products available to users, and it may require deeper understanding to determine the suitability of data for a particular application. These images include optical images, panchromatic images, multispectral images, thermal images, hyperspectral images, and microwave images. The quality of satellite images is influenced by the type of sensors used and weather conditions leading to poor image quality. In addition, selection of appropriate spectral bands is an important aspect.

Manual analysis of remotely sensed images provides vector data, while digital image processing produces raster data which is obtained through specialized software packages, like ERDAS Imagine, ENVI, ILWIS, and so on. After processing the raster images, various thematic layers can be generated and used directly in GIS with other raster data (Richards & Jia, 2013). Remote sensing data provides essential information that helps in mapping and monitoring various activities, such as change detection and land use and land cover classification, urban mapping, and so forth. The most popular thematic layer in the analysis of satellite images which is generally used in a large number of applications is land use and land cover maps and associated changes with time (Wulder et al., 2008). Satellite imagery with 5–15 m resolution (e.g., IRS-1C & 1D, SPOT-5/HRS, SPOT-2-4/HRV, ASTER) is found to be suitable for mapping from 1: 25,000–1: 50,000 scales.

In addition to thematic layers, another important source of information from satellite images is the DEM. On a global scale, DEMs from SRTM (Shuttle Radar Topography Mission; radar), ASTER (Advanced Spaceborne Thermal Emission and Reflection Radiometer; optical) and

ALOS (Advanced Land Observing Satellite; optical) are freely available with 1 arc second resolution. Recently, the global DEM from Tandem-X mission based on radar interferometry has also become available at 0.4 arc second resolution. The SRTM data, produced by NASA originally, is a major breakthrough in digital mapping of the world, and is available free of cost by the USGS. The ASTER provides DEM data with a vertical accuracy of approximately ±15m (Cuartero et al., 2004). The DEMs from various stereo satellite sensors, including Cartosat, WorldView-2, IKONOS, and SPOT-5 stereo-images (Deilami & Hashim, 2011), have been successfully created and used for various applications.

Elevation data is used in topographic maps for generation of contours. The DEM is also critical for performing geometric and radiometric corrections for terrain on remotely sensed imagery, and allows the generation of contour lines and terrain models, thus providing another source of information for analysis (Jenson, 2007). The use of DEM is increasing in GIS with improvement in information extracted using elevation data (e.g., discriminating wetlands, flood mapping, forest classification, snow-melt studies, landslides, etc.). The incorporation of elevation and terrain data is crucial to many applications, particularly if radar data is used, to compensate for layover effects and slope-induced radiometric effects. Multi-temporal DEMs can provide valuable information in studying dynamic phenomena such as glacier, floods, avalanches, and landslides. The DEM can be integrated into the GIS environment for landscape representation, and textured images can be used for photorealistic visualization. The satellite image draped on a DEM is shown in Figure 1.17, which gives a realistic view of the terrain in 3D. Animation models and fly-through models of an area can be created from such images by superimposing several multi-temporal images of the area.

Figure 1.17. The satellite image draped on a DEM (please see color figures in companion files).

1.10 SOURCES OF REMOTE SENSING DATA

The Internet is a great place to start looking for data. users can find satellite images and other associated geospatial datasets on the Internet that may save a significant amount of time and funds. A large amount of base map data (countries, states, districts, forests, lakes, major roads, rails, tourist places, rivers, townships, etc.) as well as spreadsheets already exist on the Internet. A large quantity of raw data (e.g., satellite images, maps, digital files, etc.) may be available for free download. Some of these freely available data from various sources is given in Table 1.4. However, due to security constraints, the images of highly sensitive regions are not freely available to public. In addition, the latest or multiple images of the same area or same seasons may not be available as per our requirements in the study. Therefore, much of the available data can be purchased from a number of commercial sources/government sites. For example, satellite data in India can be purchased from the Data Center of NRSC (National Remote Sensing Centre), Hyderabad (*www.nrsc.gov.in*).

Table 1.4. Freely Available Remote Sensing Data from Various Sources (Compiled from GIS Geography, 2020)

Source	Data types	Website
USGS Earth Explorer	It is the largest free source of images, including Landsat, Sentinel-2, and land cover data, digital elevation models (DEMs) such as NASA ASTER and SRTM, as well as LiDAR data.	*https://earthexplorer.usgs.gov/*
Sentinel Satellite Data	It houses 10m resolution free download Sentinel-2 satellite data in 11 spectral bands. It also stores Sentinel-1 Synthetic Aperture Radar (SAR) data.	*https://scihub.copernicus. eu/dhus/#/home*
OpenStreet-Map (OSM)	GIS users are harnessing the power of OSM as a means to crowdsource data with the biggest inventory of buildings in the world. It contains high spatial resolution cultural vector data, such as buildings, roads, and waterways.	*https://www. openstreetmap. org/#map=5/21.843/82.795*

(Contd.)

Source	Data types	Website
NASA World-view	It showcases scientific satellite-derived products across disciplines, e.g., hazards and disasters, including free download of satellite imagery.	*https://worldview.earth-data.nasa.gov/*
BHUVAN Geo-Platform of ISRO, India	It has IRS images since 1998. Thesa includes IMS-1 (Hyper-spectral), Cartosat, OceanSat, and ResourceSat. The products available include normalized difference vegetation index (NDVI), Global Coverage, and CARTODEM, a 30m elevation model generated using overlapping Cartosat-1 stereo images.	*https://bhuvan-app3.nrsc.gov.in/data/download/index.php*
NOAA Data Access Viewer	It holds satellite, aerial, and LiDAR imagery. Overall, this is a great option for coastal areas, like California, Florida, and New York.	*https://coast.noaa.gov/dataviewer/#/*
NOAA Digital Coast	It has coastal data, such as LiDAR data, elevation, land cover, and socioeconomic data. Also there is a range of satellite imagery, such as radar, infrared, and color composites.	*https://coast.noaa.gov/digitalcoast/*
DigitalGlobe Open Data Program	For any natural disaster, such as fires, floods, hurricanes, typhoons, and earthquakes, DigitalGlobe's Open Data Program supplies satellite imagery for relief in the public domain. It also houses sample high-resolution imagery.	*https://www.digitalglobe. com/ecosystem/open-data https://www.digitalglobe. com/samples*
Geo-Airbus Defense	It has a collection of more than 140 sample images for experimentation. For optical imagery, it gets as fine as 1.5m with SPOT. Radar imagery from TerraSAR-X is as sharp as 3m and a 12m WorldDEM, which is much finer and accurate than ASTER and SRTM global DEMs.	*https://www.intelligence-airbusds.com/en/8262-sample-imagery*

Source	Data types	Website
VITO Vision	It offers a coarser resolution PROBA-V, SPOT-Vegetation, and METOP satellite imagery to study vegetation patterns on the Earth's surface.	*http://www.vito-eodata.be/PDF/portal/Application. html#Home*
Satellite Land Cover	It has the best global land cover data sources from Landsat, MODIS, and AVHRR, quantifying land cover change such as vegetation, geologic, agriculture, hydrologic, and urban areas around the world.	*http://www.globallandcover.com/GLC30Download/index.aspx*

One of the biggest developments in geospatial data history was the launch of Google Maps in 2005, which made mapping technology available to people worldwide. Google Earth Engine, Sentinel Hub, Open Data Cube (ODC), OpenEO, and so on, provide free satellite images and other geospatial data. The Google Earth Engine platform provides petabytes of satellite imagery for large-scale applicability and analysis. Sentinel Hub is a cloud-based engine for browsing, visualizing, and processing multi-petabytes of satellite data. The ODC, an open-source project of geospatial data management and analysis, provides facilities to work with raster data using Python libraries and PostgreSQL. The geographic databases consisting of vector and raster data are also provided to the public for numerous applications by OpenStreetMap data, which is the largest crowd-sourced GIS database in the world. Landsat satellite images may be freely downloaded with the USGS website. The BHUVAN site of the Department of Space, Government of India, provides a large amount of free satellite images and other thematic data. These platforms provide facilities to store and process satellite images both in commercial and open-source solutions. In fact, however, there are some policies of organizations or governments that do not allow to storing EO data in the cloud, for example, high-resolution satellite imagery from military areas or restricted areas.

In recent years, open remote sensing resources have made great progress. Beginning on April 1, 2016, all Earth imagery from a widely used Japanese remote sensing instrument operating aboard NASA's Terra spacecraft since late 1999 has been made available to users free of cost. On April 8, 2016, ESA announced that WorldView-2 data at 40 cm resolution would be available for European cities for free download through the Lite Dissemination Server. This dataset, acquired between February 2011 and October 2013, was collected by ESA in collaboration with European Space

Imaging, over the most populated areas in Europe. The dataset is available to ESA member states (including Canada) and European Union Member states. In open resources, NASA is a pioneer in sharing its imagery data, and many NASA projects are open-source. In addition, some commercial companies like DigiGlobal (USA) have also partly opened their data to the public. In the future, more and more open resources will become available. Near real-time monitoring provided by open-source satellite images can provide useful information necessary to evaluate the impact on ecosystem processes. It is expected that the openness and sharing remote sensing resources will further promote the utilization of remote sensing in various applications.

CHAPTER 2

Electromagnetic Radiations and Interaction with Atmosphere

2.1 INTRODUCTION

Remote sensing techniques allow taking images of the Earth's surface in various wavelength regions of the electromagnetic spectrum (EMS). One of the major characteristics of a remotely sensed image is the wavelength region which it represents in the EMS. Some of the images are captured using reflected solar radiation in the visible and the near infrared (NIR) regions of the EMS, while other images are the measurements of energy emitted by the Earth's surface in the thermal infrared (TIR) wavelength region (Garg, 2019).

Electromagnetic energy (EME) can be modeled in two ways. by waves or by energy-bearing particles, called photons. In the wave model, the EME propagates through space in the form of sinusoidal waves (Jenson, 2007). The term EME is used as these sinusoidal waves are characterized by electrical and magnetic fields, which are perpendicular to each other, as shown in Figure 2.1(*a*). Remote sensing systems depend on the measurement of EME, which can take several forms. The most important source of EME at the Earth's surface is the Sun, which provides (visible) light and heat. The EMS forms a very broad spectrum varying from very low frequency to very high frequency or from very long wavelength to very short wavelength. It does not require a material medium to propagate, but travels in the atmosphere with the velocity of light ($v = 2.98 \times 10^8$ m/s).

As all EME travels in sinusoidal form with different wavelengths and frequencies, the distance between the two successive wave crests in the same phase is called the *wavelength* (Figure 2.1[*a*]). Wavelength is usually represented by the Greek letter lambda (λ). Wavelength is measured in meters (m) or some factor of meters such as nanometers (μm = 10^{-9} m), micrometers (μm = 10^{-6} m), or millimeters (mm). The *frequency* (*f*) is defined as the number of peaks passing through a given point at any given instant of time, and is measured in Hz or GHz. The frequency (*f*), wavelength (λ), and velocity of light (v) of the EMR are related to each other as

$$l = v/f \tag{2.1}$$

It indicates that the wavelength of radiation is inversely proportional to its frequency; the lower the wavelength, the higher the frequency, as shown in Figure 2.1(*b*). Many sensors used in remote sensing measure reflected sunlight. Some sensors, however, detect energy emitted by the Earth's objects, whereas some sensors provide their own energy to measure reflected energy. A basic understanding of EME, its characteristics and interactions are therefore required to understand the principle of remote sensing (Richards & Jia, 2013). This knowledge is also needed to interpret remote sensing data correctly.

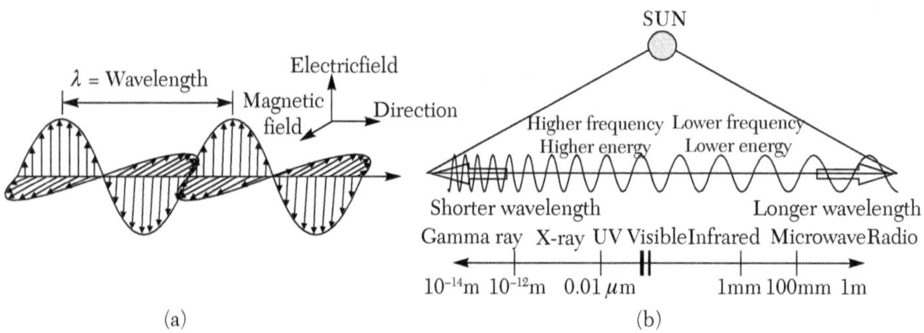

Figure 2.1. (a) Components of EMS (b) Relationship between wavelength and frequency of EMS.

2.2 COMPONENTS OF EMS

The EMS ranges from the shorter wavelengths to the longer wavelengths, and consists of g-ray and *x*-ray, ultraviolet (UV) region, visible region, infrared region, microwaves, and radio waves (Jenson, 2007). Table 2.1 presents the broad parts of the EMS along with their wavelength regions. This wavelength interval in the EMS is commonly referred to as a band, channel, or region. It is to be noted here that the EMS has much beyond the parts which are shown in Figure 2.2.

Table 2.1. Various Regions of EMS

Region	Wavelength	Remarks
Gamma ray	< 0.03 mm	Incoming radiation is completely absorbed by the upper atmosphere, and is not available for remote sensing.
x-ray	0.03–3.0 mm	Completely absorbed by atmosphere. Not employed in remote sensing.
Ultraviolet (UV)	0.3–0.4 μm	Incoming wavelengths less than 0.30 μm are completely absorbed by ozone in the upper atmosphere.
Photographic UV	0.3–0.4 μm	Transmitted through atmosphere. Detectable with film and photodetectors, but atmospheric scattering is severe.
Visible	0.4–0.7 μm	Imaged with film and photodetectors. Includes reflected energy peak of the Earth at 0.5 μm.
Infrared (IR)	0.7–100 μm	Interaction with matter varies with wavelength. Atmospheric transmission windows are separated.
Reflected IR	0.7–1.3 μm	Reflected solar radiation that contains information about thermal properties of materials. The wavelength from 0.7 to 0.91 μm is detectable with film and is called the photographic IR band.
Middle IR	1.3– 3.0 μm	Middle IR is well suited for materials identification, and yields a higher intensity, less convoluted spectra than the overtone and combination bands measured with near-IR and visible radiation.
Far or Thermal IR	3–14 μm	Images acquired by optical mechanical scanners and special systems but not by film. Principal atmospheric windows occur in 3 to 5 μm and 8 to 14 μm region.
Microwave	0.1–100 cm	Longer wavelengths can penetrate clouds, fog, and rain. Images may be acquired in active or passive mode.
Radar	0.1–100 cm	Radar images are an active form of microwave remote sensing, acquired at various wavelengths.
Radio	> 100 cm	Longest wavelength portion of EMS. Some classified radars with very long wavelengths operate in this region.

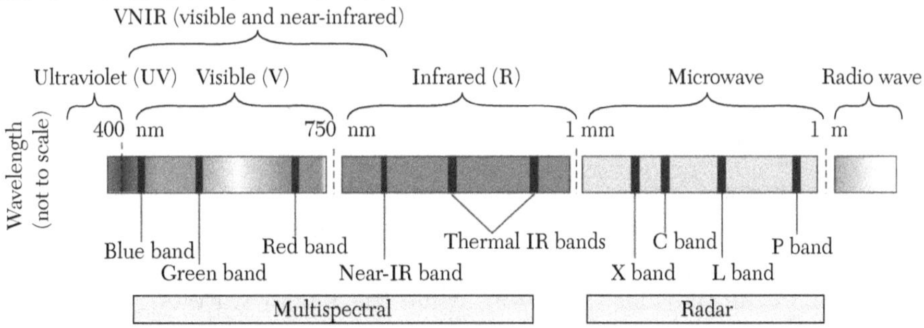

Figure 2.2. Various parts of the electromagnetic spectrum (please see color figures in companion files).

For remote sensing purposes, the UV portion of the spectrum (0.3–0.4 μm) has the shortest wavelengths. Some of the Earth's surface materials, primarily rocks and minerals, fluoresce or emit visible light when illuminated by UV radiation. UV radiation is absorbed by ozone at an altitude of between 20 and 40 km.

Our eyes can detect light which is called the *visible spectrum* that ranges from approximately 0.4–0.7 μm; the longest visible wavelength is red and the shortest is violet. This is a very small part relative to the remainder of the EMS. The common wavelengths to perceive a particular color from the visible portion of the spectrum are: violet (0.40–0.446 μm), blue (0.446–0.500 μm), green (0.50–0.578 μm), yellow (0.578–0.592 μm), orange (0.592–0.620 μm), and red (0.620–0.70 μm) (Lillesand & Kiefer, 1994). This is the only portion of the spectrum which can be associated with the concept of colors. Its component colors can be visualized when sunlight is passed through a prism. Blue, green, and red are the three *primary colors* or wavelengths of the visible spectrum, and all other colors can be formed by combining these three in various proportions.

The Sun produces 41% of its energy in the visible region, while the other 59% of the energy is in wavelengths shorter than blue light (< 0.4 μm) and longer than red (> 0.7 μm). There is a lot of radiation around us which is "invisible" to our eyes, but can be detected by other remote sensing instruments. Remote sensing sensors can be made sensitive to energy in the non-visible regions of the spectrum. The next portion of the spectrum is the *infrared* (IR) region which covers the wavelength range from approximately 0.7–14 μm. The infrared can be divided into three categories based on their radiation properties: *reflected near-IR; middle-IR* and *thermal IR*. The reflected near-IR covers wavelengths from approximately 0.7–1.3 μm and is commonly used to expose black and white and color-infrared sensitive film. The middle-IR region includes energy with a wavelength of 1.3–3.0 μm. The thermal IR region is quite different than the visible and reflected IR portions, as this energy is essentially the radiation that is emitted from the Earth's surface in the form of heat. The thermal IR covers wavelengths from approximately 3.0–14 μm. Principal *atmospheric windows* occur in the thermal region (Garg, 2019).

The portion of the spectrum of more recent interest to remote sensing is the *microwave* region ranging from about 1 mm to 100 cm, covering the longest wavelengths. The shorter wavelengths of this region have properties similar to the thermal IR region, while the longer wavelengths are used for radio broadcasts. This portion of the EMS is used for active remote sensing. Radar images are acquired at various wavelength bands. Longer wavelengths can penetrate clouds, fog, and rain (Vane, 1993). The energy measured in the microwave region is the measure of the relative return from the Earth's surface.

2.3 INTERACTION OF EMR WITH ATMOSPHERE

The word *atmosphere* refers to the gas layers surrounding the Earth. Nitrogen, oxygen, carbon dioxide, ozone, water vapor, and other gases are present in the atmosphere. All EMR travels through the Earth's atmosphere and during travel several phenomena can happen that alter the radiations in some way or the other, either by scattering or change in energy level. The EMR travels through the Earth's atmosphere twice; once on its onward travel from the Sun to the Earth, and a second time on its backward travel after being reflected/emitted by the Earth's surface to the sensor. Particles and gases present in the atmosphere interact with the EMR twice. When EMR strikes a material/object, it is called *incident radiation*. For Earth, the strongest source of incident radiation is the Sun (Sabins, 1996). The full Moon is the next strongest source, but its radiant energy is only about a millionth of that of the Sun.

The interaction of EMR with the atmosphere is thus an important phenomenon in remote sensing as information carried by reflected/emitted radiations from the Earth's surface is modified while traversing through the atmosphere. These changes are caused by the mechanisms of refraction, reflection, scattering, absorption, and transmission, as shown in Figure 2.3. Theoretically, it may not be possible to compute the emission, reflection, and absorption coefficients, and the values have to be determined experimentally. For Earth observations, it is easier to measure reflectance

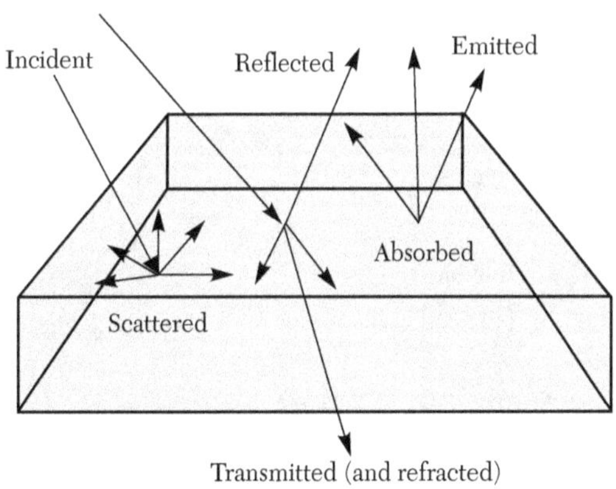

Figure 2.3. Interaction of EMR and processes.

in terms of the proportion of incident EMR that is reflected for a particular wavelength, which is termed as *spectral reflectance*. The interactions of EMR however depend upon the compositional and physical properties of the medium, the wavelength or frequency of the incident radiation, and the angle at which the incident radiation strikes a surface (Castleman, 1996). In remote sensing, the amount of energy is measured that is reflected back from a surface to the recording sensor. In terms of an energy balance relationship, the reflectance can be expressed as:

$$Er(\lambda) = Ei(\lambda) - [Es(\lambda) + Et(\lambda) + Ea(\lambda)] \quad (2.2)$$

Where $Er(\lambda)$ is reflected energy, $Ei(\lambda)$ is incident energy, $Es(\lambda)$ is scattered energy, $Et(\lambda)$ is transmitted energy, and $Ea(\lambda)$ is absorbed energy (all for a specific wavelength λ).

2.3.1 Reflection

Reflection is caused by surfaces that are smooth relative to the wavelengths of incident radiation. In remote sensing, reflectance is also described with respect to the direction of the returned energy which is recorded by the sensor system. These are four types (Gibson, 2000), given as follows:

1. *Specular:* This is where where the angle of reflection is equal and opposite to the angle of incidence. Specular reflection is the process whereby incident radiation bounces off the surface of an object in a single, predictable direction. The amount of reflected energy will depend upon the nature of the material of the object, wavelength region, and the atmospheric condition. Smooth surfaces act as *specular reflectors* (i.e., mirror-like objects) where the direction of scattering is predominantly away from the incident direction as shown in Figure 2.4(*a*). Consequently, such objects appear dark to black in image data.

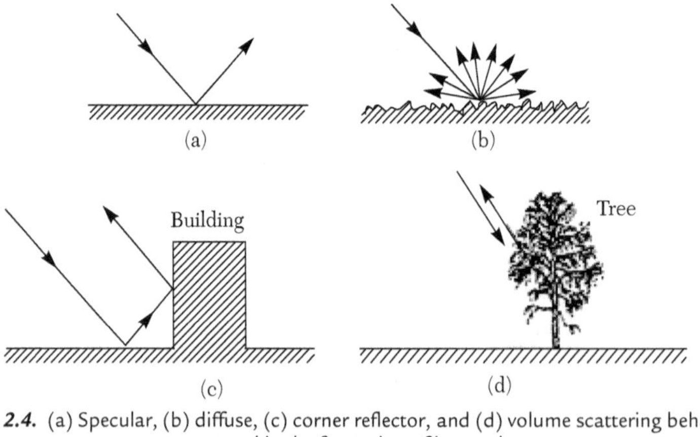

Figure 2.4. (a) Specular, (b) diffuse, (c) corner reflector, and (d) volume scattering behavior, encountered in the formation of image data.

2. *Diffuse (or Lambertian)*: This is where where reflectance is equal in all directions. Objects with rough surfaces act as diffuse reflectors, and they scatter the incident energy in all directions as shown in Figure 2.4(*b*), including toward the sensor. As a result, such surfaces would appear with a light tone in image data. Indeed, most natural surface features have a surface reflectance somewhere between true specular and diffuse reflector. The range of reflectance (specular to diffuse) that occurs naturally has implications on how we record certain surfaces and how we use (geometrically) certain sensors.
3. *Corner reflector:* The surface reflection mechanism of the corner reflector effect is often common in microwave image data, particularly associated with features such as buildings, as shown in Figure 2.4(*c*). The scattering is a result of the right angle formed between a vertical structure, such as a building, pillar, or ship, and a horizontal plane, such as the surface of the Earth or sea, giving a very bright response.
4. *Volume scattering:* Another type of reflection is called volume scattering, and objects, such as vegetation canopies and sea ice, exhibit volume scattering behavior. Backscattered energy emerges from many hard to define sites, as depicted in Figure 2.4(*d*). This leads to a light tonal appearance in radar imagery.

In interpreting image data acquired in the microwave region, it is important to recognize that all four reflection mechanisms shown in Figure 2.4 are present, which can substantially modify the tonal differences resulting from the complex surface variations. By comparing the images taken in the visible/infrared range where the Sun is the energy source, the interpreter can concentrate on tonal variations.

2.3.2 Transmission

Transmission is the process by which incident radiation passes through the atmosphere and reaches the surface; the substance is thus transparent to the radiation (Jenson, 1986). Refraction or deflection from a path with change in its velocity may take place during transmission of EMR passing through materials of different densities (e.g., air to water). Visible wavelengths and radio wavelengths are able to penetrate the Earth's surface with little atmospheric interaction.

2.3.3 Absorption

Absorption occurs when atmospheric particles and gases do not allow EMR to be fully transmitted through the atmosphere. Absorption of EMR also takes place on the surface of objects; the amount would depend on the physical and chemical properties of the object as well as the wavelength region. For example, iron objects would absorb more energy than wooden objects, and therefore would appear hotter if exposed for some time in the sunlight. Absorption is the process by which radiant energy is absorbed and converted into other forms of energy. Certain wavelengths are affected far more by absorption than scattering. This is particularly true of infrared and wavelengths shorter than visible light, like the UV, where nearly all incoming energy is absorbed (Sabins, 1996). In the far infrared region, most of the radiation is absorbed by the atmosphere.

The portions of the EMS that are absorbed by atmospheric particles and gases are known as *absorption bands*. The amount of energy absorbed by the atmosphere depends upon the wavelength of incoming radiation and characteristics of particles and gases present in the atmosphere, such as ozone, water vapor, oxygen, carbon dioxide, oxides of nitrogen, and so on. Three gases present in the atmosphere cause most of the absorption in the Earth's atmosphere. CO_2 is concentrated in the lower atmosphere, absorbing high because of the burning of fossil fuels by human beings. Ozone is concentrated in the upper stratosphere, and is responsible for much of the absorption of the high energy of short wavelengths such as UV solar radiation from entering the Earth's lower atmosphere. Water vapor varies dramatically throughout the Earth's atmosphere, and is an absorber of EMR which varies with time and location. The water and CO_2 molecules have absorption bands centered at the wavelengths from near to far infrared (0.7–15 μm; Garg, 2019).

Atmospheric absorption may alter the apparent spectral signature of the target/object being observed. The atmosphere is practically transparent to microwave radiation. The wavelength regions where absorption is at a minimum and transmittance is at a maximum are considered to be best for the acquisition of images. The concept has been explained as follows through *atmospheric windows*.

2.3.3.1 Atmospheric Windows

Some EMRs easily pass through the atmosphere, while others do not. The ability of the atmosphere to allow radiations to pass through it is referred to as its *transmissivity*, and it and varies with the wavelengths of the radiations. In optical remote sensing, many wavelengths within the visible and IR portions of the EMS (0.40–2.50 μm) transmit majorly through the Earth's atmosphere where they can interact with the Earth's surface and reflect back to the sensor. Those regions in the atmosphere where transmission of

EMR from the source to the object and back to the sensor is maximum, and absorption and scattering processes are minimum, are called *atmospheric windows* (Colwell, 1983). The 8–14 μm region has been of greatest interest for thermal remote sensing due to transmission of a maximum of radiations. Poorer atmospheric windows lie in 3–5 μm and 17–25 μm. Interpretation of the data in 3–5 μm is complicated due to overlap with solar reflection in day imagery, and the 17–25 μm region is still not well investigated. It is therefore important to design a sensor in atmospheric window regions that will detect reflected light in those specific regions where transmission is maximum (Lillesand et al., 2004). The remote sensing systems are usually designed to fall within these windows to minimize atmospheric absorption effects. These windows are found in the visible, near-infrared, certain bands in TIR, and the microwave regions, as given in Table 2.2.

Table 2.2. Major Atmospheric Window Regions Used in Remote Sensing

Atmospheric window	Wavelength	Characteristics
Upper ultraviolet, visible, and photographic	0.3–1.0 μm approx.	95% transmission
Reflected infrared	1.3, 1.6, and 2.5 μm	Three narrow bands
Thermal infrared	3.0–5.0 μm and 8.0–14.0 μm	Two broad bands
Microwave	1 mm–1 m	Atmosphere is mostly transparent

Figure 2.5 presents the graph of wavelength verus transmission. Some wavelengths cannot be used in remote sensing because the atmosphere absorbs majorly these wavelengths produced by the Sun. In certain parts of the spectrum such as the visible region (0.4–0.7 μm), the atmosphere does not absorb all of the incident energy but transmits it effectively. In particular, the molecules of water, carbon dioxide, oxygen, and ozone in the atmosphere block solar radiation. O_2 and O_3 absorb almost all wavelengths shorter than 0.30 μm. Water (H_2O) absorbs many wavelengths above 0.70 μm, but this depends on the amount of water vapor present in the atmosphere. It is

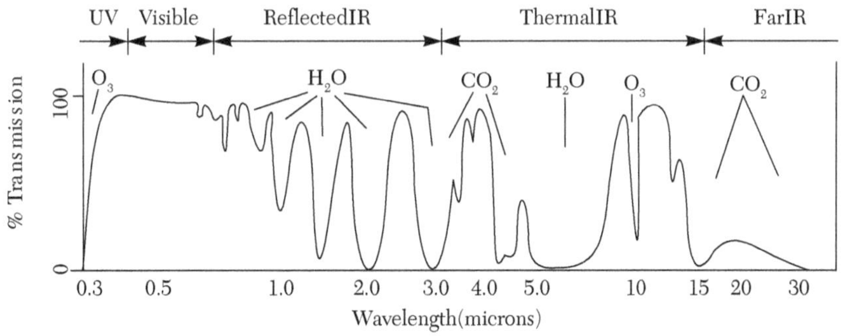

Figure 2.5. Atmospheric windows regions (Lillesand & Kiefer, 1994).

observed that there is a maximum of atmospheric transmission of radiations at 0.5 μm, 2.5 μm, and 3.5 μm, while it is minimum at about 2.0, 3.0, and 7.0 μm (Sabins, 1996). Both passive and active remote sensing technologies can provide the best images operating within the atmospheric windows.

2.3.4 Scattering

Scattering is the redirection of EMR by suspended particles and gases in the atmosphere. In the real world, scattering is much more common than reflection. It is dependent upon the number of particles and size of the particles present in the atmosphere, the wavelength of incoming radiations, and the depth of the atmosphere through which radiations will travel. The velocity and wavelength of electromagnetic waves are not affected by scattering. Scattering will not transform any energy, but will change the spatial distribution of energy. Scattering of radiation by the constituent gases and aerosols in the atmosphere causes degradation of remotely sensed images. It therefore can produce blurred objects in images. Most noticeably, the solar radiation scattered by the atmosphere toward the sensor produces a hazy appearance of the image.

Essentially there are three types of scattering taking place in the atmosphere; *Rayleigh scattering*, *Mie scattering*, and *nonselective scattering* (Richards & Jia, 2014). These depend on the wavelength of the incident radiant energy, and the size of the gas molecule, dust particle, and/or water vapor droplet encountered. The summary of these three types of scattering in the order of their importance in remote sensing is given in Table 2.3, and a brief explanation is given as follows.

Table 2.3. Summary of Three Types of Atmospheric Scattering Processes

Types	Wavelength dependence	Size (μm) of effective particles	Types of effective atmospheric particles	Scatter of particles	Effects of scatter
Rayleigh	λ^{-4}	Smaller than the wavelength of radiations (<0.1)	Gas molecules	Molecules absorb high energy radiation and reemits. Scatter is inversely proportional to the fourth power of the wavelength	Affects short visible wavelengths, resulting in blue sky and haze in images/photos

Types	Wavelength dependence	Size (μm) of effective particles	Types of effective atmospheric particles	Scatter of particles	Effects of scatter
Mie	λ^0 to λ^{-4}	Same size as the wavelength of radiations (0.1 to 10)	Spherical particles, smoke, fumes, haze, and dust	Physical scattering under overcast skies	Affects all visible wavelengths
Non-selective	λ^0	Larger than the wavelength of radiations (>10)	Water droplets, dust, fog, and cloud	Physical scattering by fog and clouds	Affects all visible wavelengths equally, resulting fog and clouds appearing as white

2.3.4.1 *Rayleigh scattering*

Rayleigh scattering mainly consists of scattering from atmospheric gases. It is primarily caused by air particles such as O_2 and N_2 molecules. It takes place when the dimension of the particles is much smaller than the size of the wavelength λ.

$$\text{Scattering}\, \alpha 1/\lambda^4 \tag{2.3}$$

Since scattering is inversely proportional to the fourth power of the wavelength, it indicates that the radiations in the shorter blue wavelengths scatter much more strongly than the radiations in the red wavelengths. It is due to this reason that Rayleigh scattering causes the sky to appear blue, as shown in Figure 2.6. In addition, the red color of the sky during the sunset is caused by this scattering. As the sun approaches the horizon and its rays follow a longer path through the denser atmosphere, the shorter wavelength radiations are comparatively scattered out strongly, leaving only the radiations in longer wavelengths (i.e., red and orange) to reach our eyes. It is primarily because blue light is scattered around four times as much as red light, and UV light is scattered about sixteen times as much as red light (Jenson, 1986). Because of Rayleigh scattering, multispectral remote

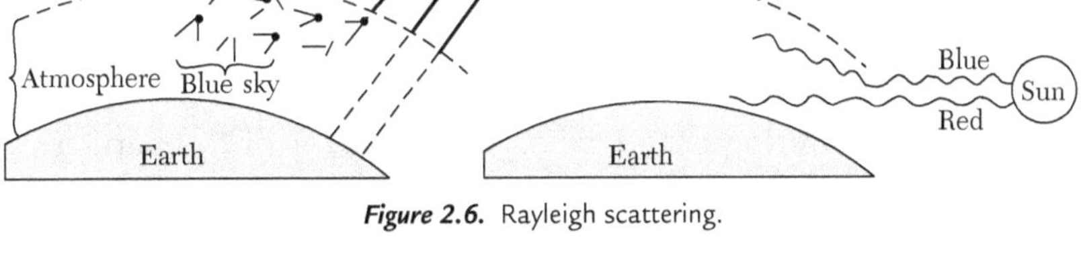

Figure 2.6. Rayleigh scattering.

sensing data taken in the blue portion of the spectrum is of relatively limited use. In the case of aerial photography, special filters are used to cut out the scattered blue radiations due to haze present in the atmosphere.

2.3.4.2 *Mie scattering*

Mie scattering occurs when the diameter of the particle is roughly equal to the size of wavelength (λ). It is caused by many factors, like pollen, dus.t, smoke, water droplets, and other particles in the lower atmosphere (4–5 km). Mie scattering may be dependent on the wavelength characteristics of the scattering surfaces. In remote sensing, clear atmosphere has both Rayleigh and Mie scattering (Figure 2.7). Mie scattering usually causes a general deterioration of multispectral images in the optical wavelength region under heavy atmospheric haze conditions. The greater the amount of smoke and dust particles in the atmosphere, the more violet light and blue light will be scattered and only the longer orange and red wavelengths reach our eyes (Lillesand et al., 2004).

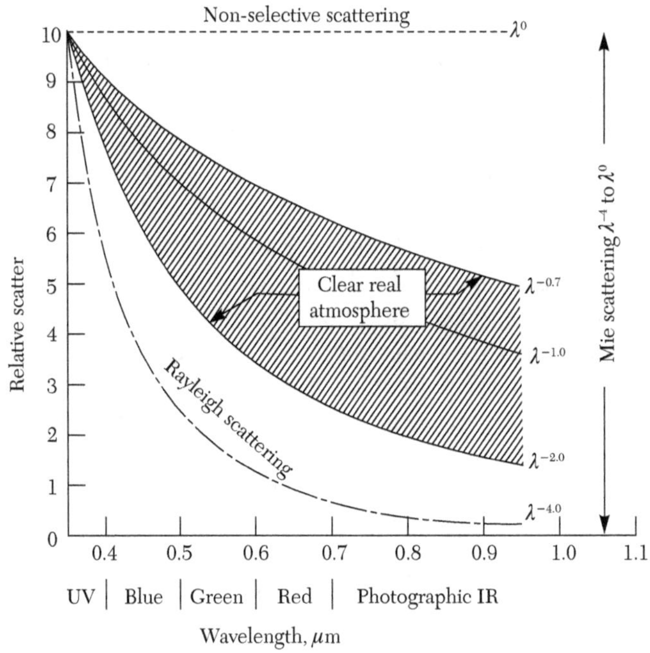

Figure 2.7. Various types of scattering with wavelength (Cowell, 1983).

2.3.4.3 *Nonselective scattering*

Nonselective scattering occurs when the diameter of the particles in the lower atmosphere are much larger than the wavelength of the radiations being transmitted. It is primarily caused by water droplets in the atmosphere.

It scatters all radiations evenly throughout the visible and IR portions of the spectrum (as shown in Figure 2.7), hence it is called nonselective scattering. Nonselective scattering is not wavelength dependent and scatters all wavelengths equally, hence fog and clouds appear as white. Since clouds scatter all wavelengths of light, these would block most energy from reaching the Earth's surface. Optical remote sensing, therefore, cannot penetrate clouds. Clouds also have a secondary effect, that is shadowed regions on the Earth's surface (Figure 2.8). This makes interpreting and analyzing the remotely sensed imagery difficult in areas having cloud and fog cover. Scattering can severely reduce the information content of remotely sensed data that may lead to loss of contrast, and at times it becomes difficult to differentiate one object from another.

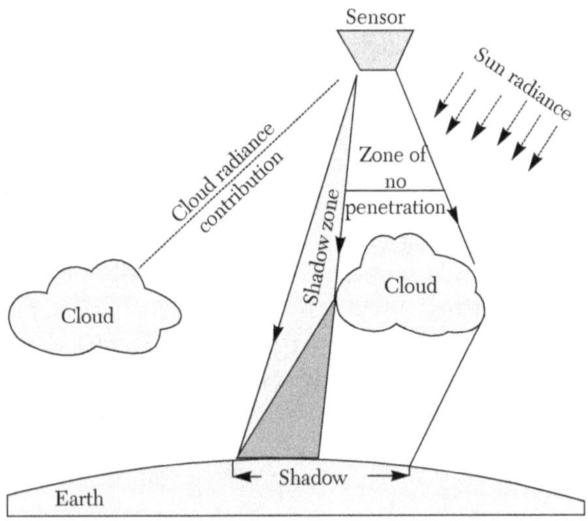

Figure 2.8. Effects of clouds in optical remote sensing.

2.4 BLACK BODY RADIATION

Black body radiation refers to the spectrum of light emitted by any heated object. The spectral intensity of black body radiation peaks at a frequency that increases with the temperature of the emitting body. General thermodynamic considerations allowed the derivation of a series of important laws controlling the emission of heated bodies, given as follows.

2.4.1 Black Body

The term *black body* was introduced by Gustav Kirchhoff in 1860 (Colwell, 1983). A black body is an idealized object that absorbs all EMR that falls on it; no EMR passes through it and none is reflected. Because no light is reflected or transmitted, the object appears black when it is cold. The black body is an ideal source that transforms heat energy into radiant energy with the maximum rate permitted by thermodynamic laws at a given temperature (T) and for a given λ. A black body emits a temperature-dependent spectrum of light. This thermal radiation from a black body is termed as *black-body radiation* (Richards and Jia, 2013). At room temperature, the black bodies emit mostly infrared light, but as the temperature increases past a few hundred °C, these start to emit visible wavelengths, from red, through orange, yellow, and white, before ending up at blue, beyond which the emission includes increasing amounts of UV. All objects with a temperature above absolute zero (K, or –273°C) emit energy in the form of EMR. An ideal black body is a perfect emitter and a perfect absorber of radiation; the body does not have to necessarily appear "black." The Sun and Earth behave approximately as black bodies.

On the contrary, a white body is a non-absorber and non-emitter, and is a perfect reflector (Jenson, 1986). Natural bodies show behavior intermediate between a perfect black body and a perfect white body; the gray body. The *emissivity* (ε) is a measure of a material's ability to absorb and radiate energy, and has a value from 0–1; zero for a perfect white body and one for a perfect black body. Emissivity describes the actual absorption and emission properties of white bodies, or gray bodies, and is wavelength dependent. It is the ratio of the energy radiated (emitted) by the material to the energy radiated by a black body at the same temperature. The emissivity is dependent on wavelength, temperature, and geometry and composition of the surface. See Kirchhoff's law in the next section to compute the emissivity.

The Sun acts as a black body at 6,000 K, which radiates energy at the maximum possible rate per unit area at each wavelength for any given temperature. In wavelength ranges between about 3 and 14 μm, the level of solar energy actually reaching the Earth's surface is small due to both the small amount of energy leaving the Sun in this range as compared to the visible and NIR range, and the presence of strong atmospheric absorption bands between 2.6 and 3.0 μm, 4.2 and 4.4 μm, and 5 and 8 μm (Castleman, 1996). Consequently, the emitted energy from the objects on the Earth's surface is measured rather than that of reflected solar radiation. Figure 2.9 shows the relative amount of energy radiated from perfect black bodies at different temperatures. The Sun at 6000 K radiates the maximum in the visible and near infrared region but generates little radiation in the range around 10 μm. The Earth, at a temperature of about 300 K, has its maximum emission around 10–12 μm, all thus a sensor operating in this range will measure the amount of heat being radiated from the Earth. Hot bodies on the Earth's surface, such as wildfires, and volcanic activities at around 800 K, have a maximum emission in the range of about 3–5 μm, and therefore sensors operating in this range would be useful to map the fire/temperature (Colwell, 1983).

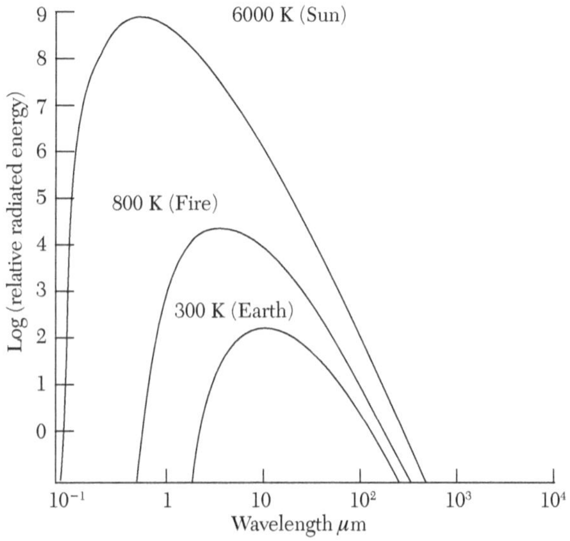

Figure 2.9: Energy from black bodies as a function of wavelength (Castleman, 1996).

2.4.2 Radiation Laws

The EMR consists of many discrete units called *photons* or *quanta*. The energy carried by a waveform in Joules (J) is computed for a single photon as (Richards & Jia, 2013):

$$E = h * f \tag{2.4}$$

where h is Planck's constant (6.626×10^{-34} Joule seconds) and f is the frequency of EMR.

This can be simply defined as the following:

The capacity of electromagnetic radiation to do work is directly proportional to its frequency. Taking a clue from Equation 2.1, the wavelength and the energy of a photon is related by:

$$E = h * \nu/\lambda \tag{2.5}$$

As this is an inverse relationship indicating that the short wavelengths carry more energy. The implication for Earth observations is that at longer wavelengths it is more difficult to detect a discernible energy signal.

(1) Planck's law

It determines the spectral energy density of the emission at each wavelength (E_λ) at a particular absolute temperature (T). The spectral radiance can also be measured per unit wavelength instead of per unit frequency. In this case, spectral energy is given by (Colwell, 1983):

$$E_\lambda = \frac{8\pi h v^2}{\gamma^5 (e^{(h_v/\lambda KT)} - 1)} \tag{2.6}$$

where

h = Planck's constant = 6.626×10^{-34} Js (joule-second)
v = Velocity of light = 2.9979×10^8 m/s = 2.9979×10^{14} m/s
λ = Wavelength of radiation
k = Boltzmann's constant = 1.38×10^{-23} J /K
T = Absolute temperature of the radiating body (K)

(2) Kirchhoff's law

It states that for all black bodies at the same temperature, the ratio of emitted radiation to absorbed radiation is the same. Emissivity (ε) can be computed as the ratio of the emittance of an object and the emittance of a black body at the same temperature, given as follows (Sabins, 1996).

$$\varepsilon = M/M_b \tag{2.7}$$

where M = emittance of a given object, and M_b = emittance of a black body.

(3) Stefan-Boltzmann law

It states that the energy that a black body emits per unit area increases as the temperature (T) of the black body increases. In principle, this law states that the hotter the body the higher the radiant energy from the body. Total emitted radiation is calculated by (Jenson, 1986):

$$E = \sigma T^4 \tag{2.8}$$

where, σ is the Stefan-Boltzmann constant = 5.67×10^{-8} W/(m^2 K^4)

(4) Wien's displacement law

It describes the relationship between the wavelength of emitted radiation and the temperature (T) of the object, given as follows. This law states that as the temperature of an object increases the wavelength of the maximum emittance increases (Colwell, 1983).

$$\lambda_{max} = \frac{b}{T} \tag{2.9}$$

where b is a constant of proportionality, known as Wien's displacement constant = 2.898×10^{-3} mK.

2.5 SPECTRAL SIGNATURE OF OBJECTS

Different features on the Earth's surface reflect and absorb the Sun's EMR in different ways. The reflectance properties of an object depend on its physical and chemical properties, the surface roughness, and the angle and wavelength of the EMR. The amount of reflectance from a surface can be measured as a function of wavelength, which is known as *spectral reflectance* or *spectral signature*. It is a measure of energy (in terms of percentage) a surface reflects at a specific wavelength. These signatures from objects over different wavelengths help in separating distinctly various distinct based on their reflected response in a given wavelength (Garg, 2019).

Two objects may produce different reflectances in a certain wavelength region, and due to this, these two objects are easily identifiable from an image due to their appearance in different tone (color) and texture. In multispectral images, the objects tend to show more separation among their reflectances. The ability of multispectral sensors to detect small changes in reflectance pattern provides a basis for identification of various targets/objects on remote sensing data.

Spectral reflectance curves of objects can be created by plotting the wavelengths on the x-axis and reflectance of objects on the y-axis. Figure 2.10 shows the typical spectral response curves for vegetation, water, and dry soil. By comparing the reflectance patterns of different objects, useful information may easily be derived. For example, water and vegetation

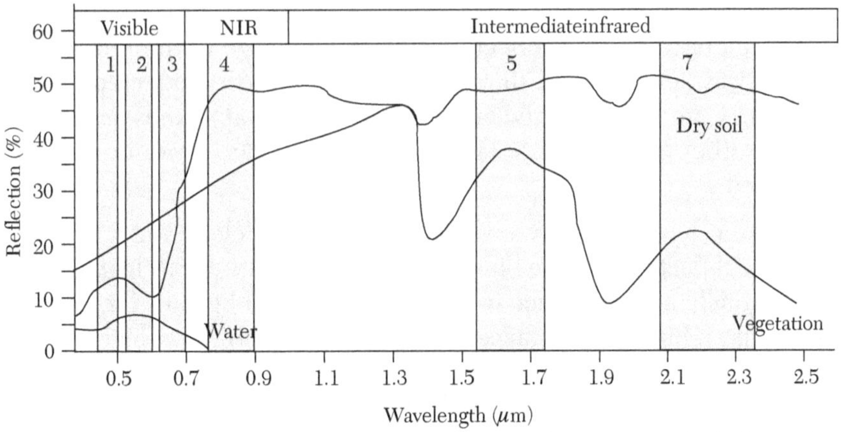

Figure 2.10. Spectral reflectance curve of vegetation, soil, and water features (Colwell, 1983).

may reflect somewhat similarly in the visible wavelengths but are more easily separable in the IR region. Spectral response can vary with the wavelength, even for the same object, and also can vary with time (e.g., greenness of leaves; Colwell, 1983).

The strong absorption by leaf pigments (particularly chlorophyll for purposes of photosynthesis) in the blue and red regions of the EMR leads to the characteristic green appearance of healthy vegetation. However, while this *signature* is distinctively different from most non-vegetated surfaces, it is not able to distinguish between species of vegetation, as most of the vegetation will have a similar green color at full maturation. In NIR, however, a much higher reflectance is observed from vegetation because of scattering phenomenon within the fleshy mesophyllic layer of the leaves (Sabins, 1996). Plant pigments can't absorb energy in this region, and thus the scattering, combined with the multiplying effect of a full canopy of leaves, leads to high reflectance in this region. However, the extent of this reflectance will highly depend on the internal structure of leaves (e.g., broadleaf versus needleleaf). As a result, significant differences between species can be detected in this region. Similarly, in the MIR region, a significant drop in the spectral response pattern is observed that is associated with leaf moisture. In this region, significant differences can arise between mature species. Optimal differentiation between species, therefore, will typically require both the near and middle infrared images to study the development cycle of vegetation.

The reflectance of soils increases approximately with the wavelength, although it is dependent on many factors, such as the color, constituents, and especially the moisture content. As soil moisture content increases, the overall reflectance of soil tends to decrease (Lillesand et al., 2004). Soils rich in iron oxide reflect proportionally more in the red than other visible wavelengths, and therefore appear reddish (rust color) on color images. A sandy soil, on the other hand, tends to appear as a bright white color in the image because visible wavelengths are more or less equally reflected.

Water has relatively low reflectance, with clear water having the greatest reflectance in the blue portion of the EMS. It has high absorption and virtually no reflectance in the NIR range and beyond 1.2 μm. Turbid water has a higher reflectance in the visible region than clear water. This is also true for water containing high chlorophyll concentrations. In longer wavelengths, visible and NIR radiations are absorbed more by water than shorter visible wavelengths. Thus, water typically looks blue or blue-green in color due to stronger reflectance at these shorter wavelengths, and darker in red or NIR wavelengths. Chlorophyll in algae absorbs more of the blue wavelengths and reflects the green, making water appear greener in color when algae are present (Gibson, 2000). Vegetation may reflect up to 50 percent, soils up to 30 to 40 percent, while water reflects at most 10 percent of the incoming radiation.

2.6 MEASUREMENT OF SPECTRAL REFLECTANCE

There are many ways that the spectral reflectance of an object or surface can be measured. Reflectance measurements of objects can be made either in the laboratory, or in the field using a field spectroradiometer or captured by other remote sensors, including those mounted on aircraft and satellites (Colwell, 1983). Data measurements in the laboratory and field through spectroradiometers is a time-consuming task. Instruments/sensors mounted on satellites detect and record the energy that has been reflected from various ground objects. The detectors in these sensors measure the intensity of the reflected energy and record it. The number of data points for which spectral reflectance is collected depends on the spectral resolution of the sensor. For example, Landsat TM provides three bands in blue, green, and red wavelengths; three bands in the near and mid infrared part, and one band in the TIR part of the EMS.

A *spectroradiometer* or *radiometer* is a field-based, portable, easy-to-operate, and environmentally friendly instrument that measures the intensity of radiation in several wavelengths of the EMS, such as IR or microwave (Figure 2.11). It also helps in measuring the differences in the reflectance pattern (or spectral signatures) of various objects as a function of wavelength. The reflected radiations of various objects in different wavelength regions can be measured in the field and plotted on a graph showing the spectral signature of objects.

A radiometer produces consistent and reliable datasets after the calibration. To do a calibration, measurement of the Sun's energy reaching the Earth can be made in each of the bands using a Barium Sulphate ($BaSO_4$)-coated plate, and this value is taken as the reference value for computation of percentage reflectance. The reference plate, being a Lambertian surface, reflects radiations in all directions evenly, and the values are considered as 100% reflected radiation. The reflectance of an object for various wavelengths is collected. This process is repeated for all the sample sites to take several readings at each sample location, and mean value is adopted. The percentage reflectance for each sample in a wavelength band is computed with respect to the values from the reflectance plate. This process can be repeated at several locations of the same sample for better accuracy.

Figure 2.11. Field spectroradiometer for collecting spectral reflectances.

The readings from a spectroradiometer are normally taken between 8:00 a.m. and 10:30 a.m. and 3:00 p.m. and 5:00 p.m. to avoid very low or high Sun elevation. Thus, extreme values as a result of high intensity due to high Sun elevation or low intensity due to low Sun elevation may be avoided. The presence of cloud cover can also make a significant difference in the reflectance values, and thus cloudy days are avoided as far as possible. The spectral response of the object would also depend on the orientation of the Sun (solar azimuth), look angle of the spectroradiometer, and the distance between the instrument and the object (Garg, 2019). The reflected radiance value can be measured by holding the radiometer vertically over the object at around 1.5 m high from the ground.

The measurement of reflectance values has several applications in remote sensing. It provides useful inputs to the design and calibration of remote sensing sensors. The databases created in the field would be helpful in deciding the sensor parameters, such as optimal number of spectral bands, optimal ground resolution, actual required dynamic range of input flux reflected from the terrain surface, and so forth. Such measurements can also provide input data for radiometric and atmospheric corrections of satellite images, which requires field reflectance measurements of the targets/objects with illumination and viewing geometry equivalent to that of the imaging sensor of interest. The most important applications include collection of ground truth for remote sensing image analysis, and subsequently accuracy assessment of classification for developing vegetation growth models, and estimating crop biomass and crop yield.

2.7 ATMOSPHERIC CORRECTIONS TO REMOTE SENSING IMAGES

The EME passes twice through the atmosphere when traveling from the Sun to the Earth and back to the sensor. The atmosphere affects the spatial and spectral distribution of the EMR originating from the Sun and finally recorded by a satellite sensor. Gas absorption, as well as molecule and aerosol scattering, influence the incident and reflected radiations. The procedure for removing the atmospheric effects from the satellite-measured signal top of atmosphere (TOA) reflectance to retrieve the true surface reflectance values is called *atmospheric correction*. Atmospheric correction removes the scattering and absorption effects from the atmosphere to obtain the surface reflectance characterizing (surface properties). The main atmospheric effects (scattering by molecules and aerosols, absorption by water vapor, carbon dioxide, oxygen, and ozone) on remote sensing data are downwelling diffuse irradiance, upwelling atmospheric radiance (or path radiance), and atmospheric transmittances (Cui et al., 2004).

The light reaching the satellite sensor is not equal to the light reflected by an object on Earth. The pure radiation is called *radiance*, and the radiation arriving at the sensor is called *reflectance* as it is influenced by the atmosphere. Radiance is the amount of radiation, so it is an absolute value. Reflectance is the proportion of the amount of radiation hitting an object, and the amount reflected off of it, so this is a ratio of two values. To draw information about the nature of the Earth's surface, the influence of the atmosphere has to be subtracted from the reflectance. Atmospheric correction enables the data to remove these effects and to reduce the surface spectral reflectance. It involves a complex step used in the extraction of information and to produce surface reflectance. Remote sensing images are always a coupling of atmosphere and land surface information, which makes it very difficult to decompose the accurate atmospheric information and land surface information respectively from remote sensing. Figure 2.12 shows an example of a Landsat TM image after applying atmospheric correction.

(a) (b)

Figure 2.12. (a) Landsat TM band 1 (raw image), and (b) After atmospheric correction.

Atmospheric correction is essentially required in atmospheric studies for exploration of terrestrial surfaces (land and ocean) using airborne and spaceborne observations. As the atmospheric effects become more quantitative, the retrieval of accurate surface reflectances becomes increasingly important. Atmospheric correction may not be important for all applications (e.g., land cover classification for a single year), but it is absolutely crucial when performing a time-series analysis of crop growth. For example, in comparing spectral characteristics of a pixel or group of pixels acquired on different dates, removal of the effect of atmospheric conditions is necessary.

Several methods have been developed to remove or reduce atmospheric effects and to retrieve surface reflectance. These methods include direct digital number (DN) to reflectance transformation, the image-based dark object subtraction (DOS) method, and radiative transfer models (Cui et al., 2004). For DN to reflectance transformation, the image pixel values (i.e., DNs) are converted to a physically interpretable measure, often referred to, and interpreted as, *surface reflectance* (Baisantry et al., 2012). This transformation generally involves two steps. The first step is radiometric calibration, which involves (*a*) the conversion of DNs to radiance, and then (*b*) to TOA radiance. The obtained values can be interpreted as radiance observable just outside of the Earth's atmosphere; their derivation from the DN values can normally be done with just the metadata that is delivered along with the image. In the next step, the TOA reflectance is converted to surface reflectance (also known as *bottom-of-atmosphere reflectance* or *top-of-canopy reflectance* in vegetation studies). Top-of-canopy reflectance can be understood as reflectance as would be measured from just above the vegetation. This phase requires knowledge of atmospheric conditions present during the image acquisition time.

Atmospheric correction methods can be further categorized into physically based models, relative calibration methods, and image-based DOS based on model characteristics and complexity. Physically—based models require atmospheric information acquired at the time of remote sensing data acquisition. They are often very complex and time-consuming but more accurate than the image-based models. This kind of model incorporates 6S (Second Simulation of the Satellite Signal in the Solar Spectrum), MODTRAN (MODerate resolution atmospheric radiance and TRANsmission model), and LOWTRAN. Relative calibration methods, including histogram adjustment, dark-pixel subtraction, and the multi-date normalization with regression model approach, cannot be used for quantitative remote sensing analysis because their results are frequently inaccurate. The DOS technique is the simplest and most used for image atmospheric correction. This method assumes the existence of zero or small surface reflectance. The minimum DN value in the histogram from an entire scene is subtracted from all the pixels. The image-based DOS models have the main advantage that they purely depend on image information and take little account of the variability of atmospheric parameters, without the requirement of field-based atmospheric parameters. It is a widely used method, and is a better option for analysis of time-series image data.

Some agencies (e.g.., USGS) now provide atmospherically corrected images free of charge (e.g., Earth Explorer website, *https://earthexplorer.usgs.gov/*). Atmospheric correction modules are scarce in open-source remote sensing software, and even in proprietary software (like ATCOR and FLAASH for ERDAS), atmospheric correction is a separate license module. Chrysoulakis et al. (2010) compared several atmospheric correction methods to study the effects of atmospheric correction on land cover classification and change detection. Mather (1992) mentioned that the there is no universal atmospheric correction method available that works for all the regions. Accordingly, a suitable atmospheric correction method may be selected based on data quality, data availability, and objectives of the study.

CHAPTER 3

Various Remote Sensing Sensors and Data Characteristics

3.1 INTRODUCTION

A major advance in aerial remote sensing has been the development of digital aerial cameras for extensive coverage, high geometric and radiometric resolution, high accuracy, multispectral imagery, and stereo-capability (Al-Tahir et al., 2006). Until the 1960s, camera systems dominated photographic image collection using visible and NIR radiations reflected from the Earth. Beginning in the 1960s, electronic sensor systems were increasingly used for collection and storage of the Earth's reflected radiation, and satellites were developed as an alternative to aircraft platforms. These images provided greater capabilities for image analysis and interpretation. Aerial photographs and remote sensing images have been used successfully in agriculture, forestry, land use planning, fire detection, mapping wetlands, erosion, oceanography, and many other applications (Garg, 2019).

There are several broad categories of sensor system types aboard satellites, such as passive and active sensors, which are further subdivided as per their imaging characteristics and wavelengths used. A variety of different sensors fit in these categories, which are not mutually exclusive. Passive sensors measure solar light reflected or emitted naturally from surfaces and objects. Active systems supply their own illumination energy, and do not normally require perfect weather conditions to collect the data. In addition, they can also be deployed at night or during haze, clouds, or light rain, depending on the wavelength of the system.

Various sensor devices record wavelengths of energy, which are mounted/fixed on a "platform" that can be ground-based, airborne, or spaceborne. Advances in electronic sensors and platforms were accompanied by an increase in applications and use of EMR not only from the visible and NIR wavelength regions, but also from the TIR and microwave regions. For instance, TIR sensors operating in this wavelength region primarily detect thermal radiative properties of ground material. Thermal images help in identifying surface materials and features, such as rock types, soil moisture, geothermal anomalies, and so on. Radar images from microwave sensors represent landscape and ocean surface features that differ significantly as observed by aerial photography or multispectral sensors. Hyperspectral images provide detailed information about any object because of narrowband information acquisition. They have more advantages for the identification and discrimination of targets/objects than the multispectral data (Zhu et al., 2017). As the hyperspectral sensors provide images with high spectral resolution, a single pixel may contain spectra from more than one material/object. There are many significant challenges that need to be addressed when performing image classification from thermal, microwave, and hyperspectral sensors.

3.2 IMAGE CHARACTERISTICS

3.2.1 Image Acquisition

In a broad sense, an image is a picture or photograph. They are the most common and convenient means of storing, conveying, and transmitting information. Satellite images are full of useful and interesting information. Remote sensing images are representations of the Earth's surface as seen from space. These images may be analog or digital. Aerial photographs are examples of analog images, while satellite images acquired using electronic sensors are examples of digital images. Remote sensing images can be divided into two broad types, depending on the media: hard copy or photographic products (B&W, color) and soft copy (digital product; Garg, 2019). In the digital world, the digital image is the original product from the sensor, and hard copy products are the by-products of digital data.

For the collection of image data, the satellite moves north to south in the orbit, but sensors collect the reflected/emitted EMR radiations from the ground objects in an east-west direction covering a certain width on the ground. The sensor would cover a certain strip of uniform width on the ground as the satellite moves in the orbit. The width of the area covered on the ground when a satellite sensor scans the Earth's surface while moving in an orbit is called a *swath*, as shown in Figure 3.1 (*a*). The sensor of the satellite will thus collect the reflected signals from west to east, called the *scan line*. The swath width may generally vary from sensor to sensor; for example, the swath width 185 km for an earlier Landsat MSS. The swath width generally is taken as a standard dimension to define the size of the image (e.g., one Landsat MSS scene would cover 185 km × 185 km on the ground). The swath width of various satellites is shown in Figure 3.1 (*b*).

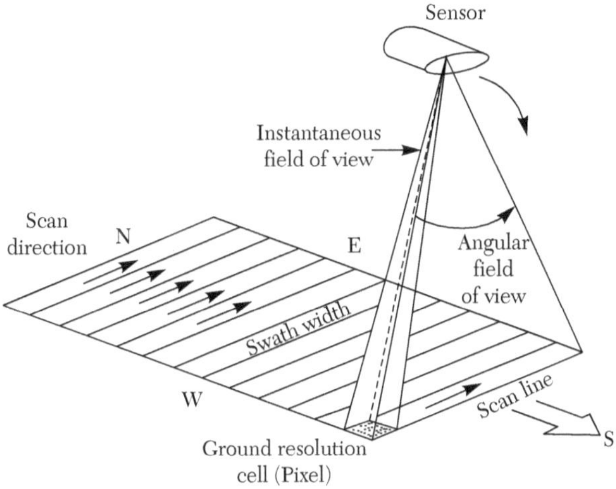

(a)

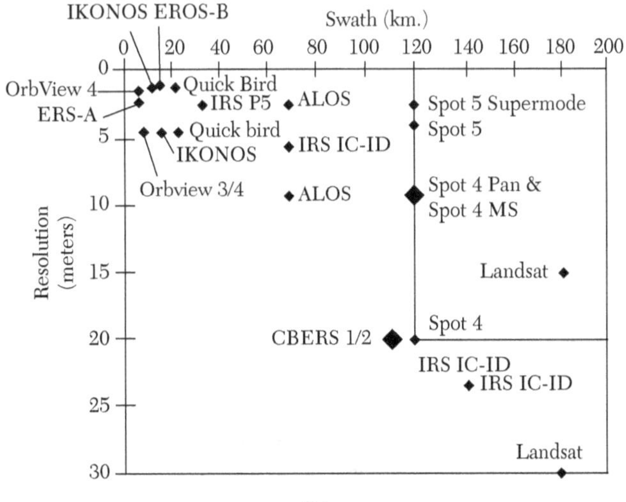

(b)

Figure 3.1. (a) Imaging of a scene/image by sensor and (b) swath width of various satellites.

3.2.2 Bits, Bytes, and Digital Number

In digital data, photographic information is stored as an array of discrete numbers. Each number corresponds to a discrete dot, that is one image element in an image, generally known as a *picture element* or *pixel*. The number of pixels in an image depends upon the image size (swath width of the image). At each pixel location, the image brightness is sampled and quantized by the sensor, and is called the *brightness value* or *gray value* or *digital number* (DN).

Any bitmap image can be represented by giving the locations (line number, pixel number) and brightness (DN) value of all its pixels. A digital photo is an example of a bitmap image as it is represented as a two-dimensional array of pixel values (Figure 3.2). The depth of a pixel in a digital image is determined by the number of bits of information it contains. A *bit* is an irreducible discrete unit of information used by computers. The appearance and quality of a digital photo seen on the computer screen is determined by the depth of the pixels.

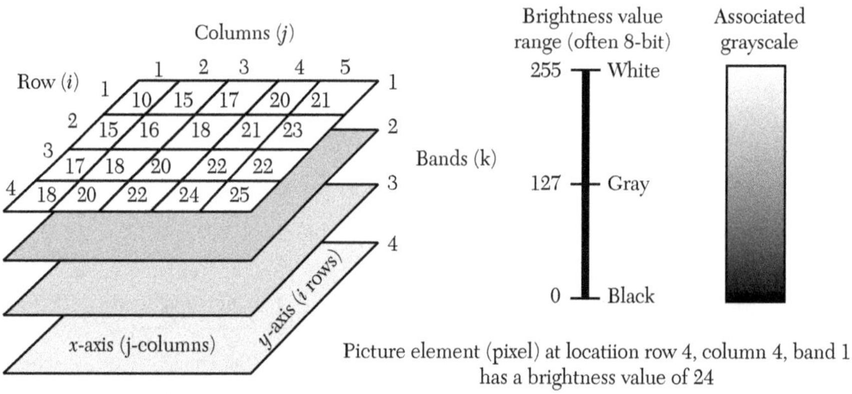

Figure 3.2: Image as a 2D array of pixels with brightness values.

A 1-bit pixel can take on $2^1 = 2$ different values, either 0 or 1 (i.e., bit is off or bit is on). If the bit is off, the color of the pixel is black and nothing is recorded, while the color is white when the bit is on, so this means the color of a 1-bit pixel is black or white. A 2-bit pixel can take on $2^2 = 4$ different values corresponding to a two-digit binary number which are 00, 01, 10, and 11, producing shades of black, two shades of gray, and white colors, respectively. A 3-bit pixel can take on $2^3 = 8$ different values and can generate black, 6 different shades of gray, and white colors. Similarly, a 8-bit pixel can display images with $2^8 = 256$ colors, covering the entire range between 0–255 (Christiansen, 2021).

A byte is an ordered set of 8 bits with given values. A byte is also called a pixel, which can have any one of 256 values, depending on the combinations of 8-bit values (0 and 1). Examples of different values of a byte in 8-bit data are:

00000000 (binary number for 0; darkest pixel, black)

00000001 (binary number for 1)

00000010 (binary number for 2) and so on

11001000 (binary for 208) and so on

11111111 (binary number for 255; brightest pixel, white)

This conversion step generates an integer at each pixel representing the brightness or darkness of the image at that point and is represented by a two-dimensional integer array, and the digitized brightness value is called the gray level. These numbers vary from place to place within the image depending upon the tonal variation. In any image, bright areas are represented by higher values whereas dark areas are represented by lower values. The values are known as digital numbers.

The reason that there are 256 values is that there are 256 different 8-digit binary numbers made out of combinations of 0's and 1's. A byte can have one of 256 values, and it is known as a *discrete unit* of information. In a color 24-bit image, each pixel is represented by three bytes, usually representing RGB (red, green, and blue) colors. A 24-bit or RGB color makes the most colors available to pixels. Thus the red, green, and blue phosphor brightness can each separately take any of 256 different brightness levels. Therefore, the total number of different colors each pixel can exhibit in 24-bit color = 256 × 256 × 256 ≈ 16.8 million colors. The visualization of such a large number of colors is not possible, but their mathematical values can be recognized by the computer dealing with the analysis of such a color image. Each of the different colors is defined by a different set of 24 bits. For example, bright yellow is 11111111 11111111 00000000, meaning full brightness red (11111111), full brightness green (11111111), and no blue (00000000).

3.2.3 Image Representation

A digital satellite image can be represented in three ways, as shown in Figure 3.3: (*a*) pictorially in the form of an image, (*b*) numerically in the form of DN values, and (*c*) graphically through a histogram. An ideal satellite image should have a good contrast to identify various features present on

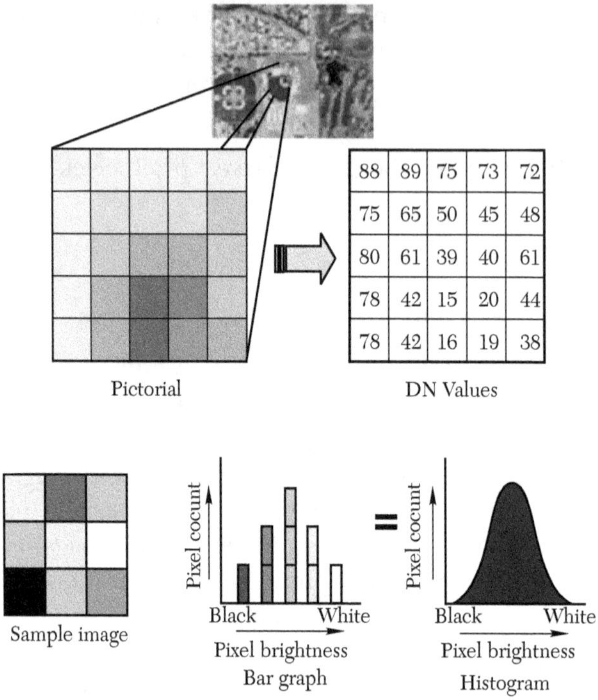

Figure 3.3. Representation of an image.

it, and should be free from cloud cover. It should cover the entire range of gray levels (DN values) in order to allow identification of various objects/features. In addition, the histogram should have a single peak bell-shaped figure which conveys well distributed gray levels in the image. However, in reality it is difficult to get an ideal image of the area with a bell-shaped histogram (Garg, 2019).

A histogram is a graphical representation showing the distribution of data; DN values are shown along the *x*-axis, while their frequencies (the number of occurrences of these DN values) are shown along the *y*-axis. When applied to imagery, histograms display the distribution of brightness values from pure black (zero DN value) to pure white (maximum DN value), and are often defined by a curved line. The image histogram can provide some inherent characteristics of an image. In raw imagery, the useful data often occupies only a small portion of the available range of DN values.

3.2.4 Image Formats

Multispectral satellite images may be stored in one of three formats, BSQ (band sequential), BIL (band interleaved by line), and BIP (band interleaved by pixel; Jenson, 1986). The BSQ, BIL, and BIP files are binary files, and have an associated ASCII header file containing metadata of the image. In the BSQ format, all the information related to one particular band including the header information is stored at one place in one file. So, the contents of multiple bands are written sequentially, as depicted in Figure 3.4 (lower left image); the header information is followed by the image content of the first band, and the sequence is repeated for all the multispectral bands. This method of image storage and retrieval is the most practical, but it is complicated to deal with extracting the sub-image from the entire data. The BSQ format has the advantage in certain image analyses where individual band data is dealt with separately, but it may not be suitable for some image classification in which multiple pixels at the same location need to be taken up simultaneously. In such cases, the other two formats (i.e., BIL and BIP) of image storage are more efficient.

In the BIL format, multiple bands are stored line by line, that is the first line of the first band is recorded, followed by the first line of the second band, followed by the first line of the third band, and so on. Thereafter, the

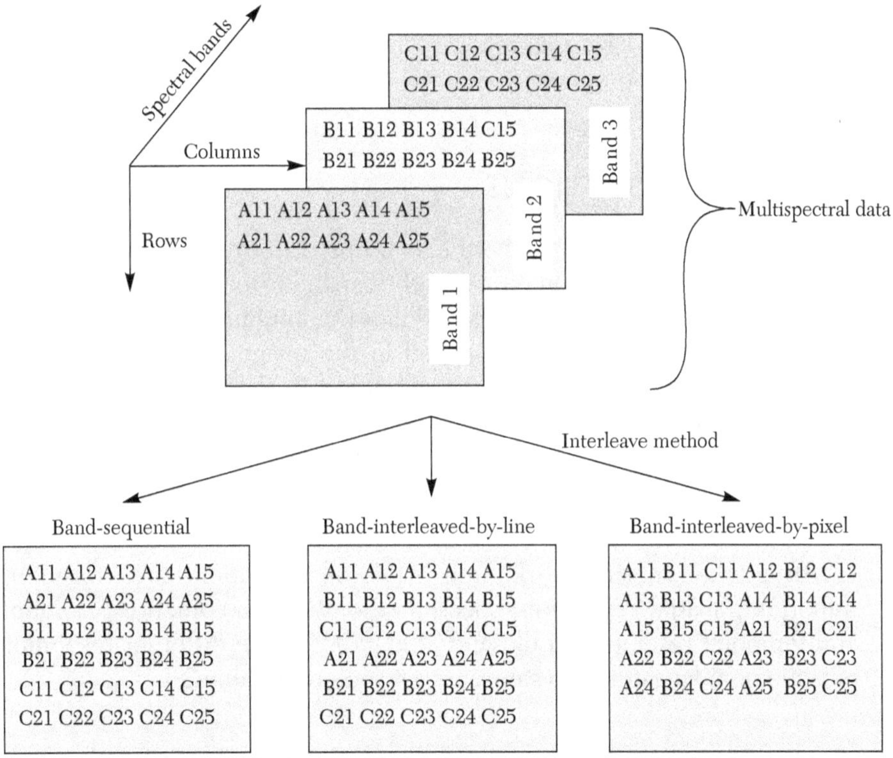

Figure 3.4. Examples of BSQ, BIL, and BIP formats of image storage with three band of data.

second line of the first band is stored, followed by the second line in the second band, and so on, as shown in Figure 3.4 (lower middle image). The BIP form is very similar to that of BIL, except the value associated with a pixel is stored sequentially in all bands. This means, the first pixel of the first line of the first band is recorded, followed by the first pixel of first line of the second band, followed by the first pixel of the first line of the third band, and so on. Thereafter, the second pixel of the first line of the first band is stored, followed by the second pixel of the first line in the second band, and so on, as shown in Figure 3.4 (lower right image). In this format, all the pixels of a certain location are stored close to one another. This kind of data storage is preferred when all pixels of a certain location are to be examined at the same time during a classification. However, it is a difficult format if only one band is to be examined, such as during image contrast enhancement.

Satellite images are taken by various sensors to gather information in different wavelength ranges. The images taken in various channels or bands allow production of color images for human viewing, interpretation, and analysis. Because these images are collected in digital format, they can be effectively analyzed by a computer. The most commonly used images in land use and land cover mapping are visible, near-infrared, and middle-infrared images.

3.3 IMAGE RESOLUTIONS

An image can be described better in terms of its resolution. In remote sensing, the term *resolution* is used as the capability to identify the presence of two objects. In general, the resolution is the minimum distance between two objects that can be distinguished in the image. Objects closer than the resolution would appear as a single object in the image. In qualitative terms, resolution is the amount of detail that can be observed in an image. An image showing finer details will have finer resolution as compared to an image that shows coarser details. This information from an object is gathered by using the multispectral sensors/scanners.

In remote sensing work, four types of resolutions are considered important: spatial resolution, spectral resolution, radiometric resolution, and temporal resolution (Lillesand et al., 2004). The different resolutions of some popular satellite systems are presented in Table 3.1.

Table 3.1. Resolutions of Some Satellite Sensor Systems

Satellite	Sensors	Resolution Spectral bands (μm)	Spatial (m)	Temporal (days)
LANDSAT-1, 2 and 3*	RBV	Blue-green (0.475–0.575) Orange-red (0.580–0.68) Red to Near-Infrared (0.69–83)	80 40*	18
	MSS	Green (0.5–0.6) Red (0.6–0.7) NIR (0.7–0.8) NIR (0.8–1.1) TIR (10.4–12.6)*	80	18
LANDSAT-4, and 5	MSS	Same as LANDSAT-1 and 2	80	16
	TM	Blue (0.45–0.52) Green (0.52–0.60) Red (0.63–0.69) NIR (0.76–0.90) NIR (1.55–1.75) Thermal (10.40–12.50) Mid-IR (2.08–2.35)	30 120	16
LANDSAT 7	ETM+	Blue (0.45–0.52) Green (0.52–0.60) Red (0.63–0.69) NIR (0.77–0.90) NIR (1.55–1.75) Thermal (10.40–12.50) Mid-IR (2.08–2.35) PAN (0.52–0.90)	30 60 15	16
LANDSAT 8	Operational Land Imager (OLI)	(0.43–0.45) Blue (0.450–0.51) Green (0.53–0.59) Red (0.64–0.67) NIR (0.85–0.88) SWIR 1(1.57–1.65) SWIR 2 (2.11–2.29) PAN (0.50–0.68) Cirrus (1.36–1.38)	30 15	16
	Thermal Infrared Sensor (TIRS)	TIRS 1 (10.6–11.19) TIRS 2 (11.5–12.51)	100	16

(Cont.)

Satellite	Sensors	Resolution Spectral bands (μm)	Spatial (m)	Temporal (days)
SPOT-1, 2 and 3	HRV	Green (0.50–0.59) Red (0.61–0.68) NIR (0.78–0.89)	20	26 (4–5 days by using steering capabilities)
	PAN	Pan (0.50–0.73)	10	
SPOT-4	HR-VIR	In addition to HRV and PAN MIR (1.58–1.75)	20	26 (4–5 days by using steering capabilities)
SPOT-5	2 HRG sensors	Green (0.50–0.59) Red (0.61–0.68) NIR (0.78–0.89) SWIR (1.58–1.75) Pan (0.51–0.73)	10 5	26 (4–5 days by using steering capabilities)
SPOT-6 and 7	2 HRG sensors	Pan (0.45–0.75) Blue (0.45–0.53) Green (0.53–0.59) Red (0.62–0.69) NIR (0.76–0.89)	1.5 8.0	26 (4–5 days by using steering capabilities)
IRS-1A	LISS-I, and LISS-II A/B (3 sensors)	Blue (0.45–0.52) Green (0.52–0.59) Red (0.62–0.68) NIR (0.77–0.86)	72.5 for LISS-I; 36 for LISS-II	22
IRS-1B	LISS-I and LISS-II	same as for IRS-1A		22
IRS-1C & 1D	LISS-III	Green (0.52–0.59) Red (0.62–0.68) NIR (0.77–0.86) NIR (1.55–1.70)	23.5 70	24
	PAN	(0.50–0.75)	5.8	24 (5)
	WiFS	Red (0.62–0.68) NIR (0.77–0.86)	188	5
IRS-P3	WiFS	Red (0.62–0.68) NIR (0.77–0.86) NIR (1.55–1.70)	188	5
	MOS-A MOS-B MOS-C	(0.75–0.77) (0.41–1.01) (1.595–1.605)	1500 520 550	Ocean surface

Satellite	Sensors	Resolution Spectral bands (μm)	Spatial (m)	Temporal (days)
IRS-P4 (Oceansat-1)	OCM MSMR	(0.40–0.90) 6.6, 10.65, 18, 21 GHz (frequencies)	360 × 236 105 × 68, 66 × 43, 40 × 26, 34 × 22 (km for frequency sequence)	2 2
IRS-P6 Resourcesat-1	LISS-IV	Green (0.52–0.59 Red (0.62–0.68 NIR (0.77–0.86	5.8	24 (5)
	LISS-III	Green (0.52–0.59) Red (0.62–0.68) NIR (0.77–0.86) NIR (1.55–1.70)	23.5	24
	AWiFS	Red (0.62–0.68) NIR (0.77–0.86) NIR (1.55–1.70)	70	5
IRS-P5 Cartosat-1	PAN-F PAN-A	(0.50–0.75) (0.50–0.75)	2.5	2-line stereo camera
Cartosat-2	PAN camera	(0.50–0.85)	< 1	
Oceansat-2	OCM SCAT ROSA	(0.40–0.90) 8 bands 13.515 GHz GPS occultation	360 × 236 25 km × 25 km	2
ADEOS-1	POLDER	B1–B9 (0.443. 0.443. 0.49, 0.565, 0.67, 0.763, 0.765, 0.865 and 0.91)	6000–7000	4
TERRA and AQUA	MODIS	B1 (0.62–0.67) B2 (0.841–0.876)	250	1–2
		B3 (0.459–0.479) B4 (0.545–0.565) B5 (1.23–1.25) B6 (1.628–1.652) B7 (2.105–2.155)	500	
		B8–B36 (0.405–14.385)	1000	
NOAA	AVHRR	B1–B5 (red, NIR, MIR, 2 × TIR)	1100 at nadir	1

(Cont.)

Satellite	Sensors	Resolution Spectral bands (μm)	Spatial (m)	Temporal (days)
TERRA	ASTER	B1 B2–B9 B11–B14	15 30 90	16
IKONOS	OSA	PAN (0.45–0.9)	0.82 at nadir	14
		B1–B4 (blue, green, red, VNIR)	3.2 at nadir	
QuickBird	BGIS 2000	PAN (0.45–0.9)	0.61	1–3.5
		Blue: 0.45–0.52 Green: 0.52–0.6 Red: 0.63–0.69 Near IR: 0.76–0.9	2.4–2.8	
WorldView	WV60	PAN (0.45–0.9)	0.5 at nadir	1.7–5.9
Geoeye-1	GIS	PAN (0.45–0.9)	0.45 at nadir	2.1–8.3 at 40 degree latitude
		Blue: 0.45–0.52 Green: 0.52–0.6 Red: 0.63–0.69 Near IR: 0.76–0.9	1.65	

∗Landsat 3 MSS has 5 bands of data and RBV data has 40 m spatial resolution.

3.3.1 Spatial Resolution

Spatial resolution measures the smallest angular separation between the two objects that a sensor can record distinctly. The geometric properties of the imaging system are usually described by the *instantaneous field of view* (IFOV). The IFOV of the ground seen from the detector of a sensor is also called a *pixel*, which is the smallest unit that makes up an image [Figure 3.1(*a*)]. Spatial resolution refers to the size of one pixel on the ground. The IFOV is dependent on the altitude of the satellite; the higher the altitude, the larger is the IFOV (Colwell, 1983).

A digital image consists of several pixels arranged in rows and columns, and each pixel contains information about the feature covered on the land surface. Spatial resolution is a measure of the area or size of the smallest dimension on the Earth's surface over which an independent measurement has been made by the sensor. Spatial resolution can be coarse when it is greater than 1000 m, medium when between 100 and 1000 m, high when between 5 and 100 m, and very high when it is less than 5 m. This subdivision into low, medium, high, and very high resolution is provisional; as the sensor technology advances all the time, it is subject to change. What was considered high resolution back in the 1980s, has become low to medium in today's mapping world. The finest resolution as of now is 30 cm provided by very high resolution commercial satellites (e.g., WorldView-4).

The spatial resolution basically determines how detailed the information will be. For example, Landsat TM data has a 30 m resolution, covering a 30 m × 30 m area on the ground by one pixel. It is considered to be a medium resolution image, as one image can cover an entire medium-sized city area, but the level of detail won't be fine enough to distinguish individual objects like houses or streets. However, an object of sufficient contrast with respect to its background can change the overall appearance on an image, and due to this reason, the features smaller than the spatial resolution, such as roads, canals, railway lines, and drainages are sometimes identifiable on such images. Figure 3.5 shows the effect of spatial resolution on details represented on an image; 1m resolution provides clarity of features, and features become blurred or unidentified as the resolution increases. Some examples of spatial resolutions include: MODIS and AVHRR sensors of more than 1 km, IRS WiFS (188 m), TIR band 6 of the Landsat TM (120 m), Landsat ETM+ (30 m), bands 1–7 of MODIS having a resolution of 250–500 m, IRS LISS-III (23.5 m MSS and 5.8 m PAN), AWiFS (56–70 m), SPOT 5 (2.5–5 m PAN), GeoEye (0.45m for PAN and 1.65m for MSS), IKONOS (0.82–1 m PAN), and QuickBird (2.4–2.8 m). For additional details, refer to Table 3.1.

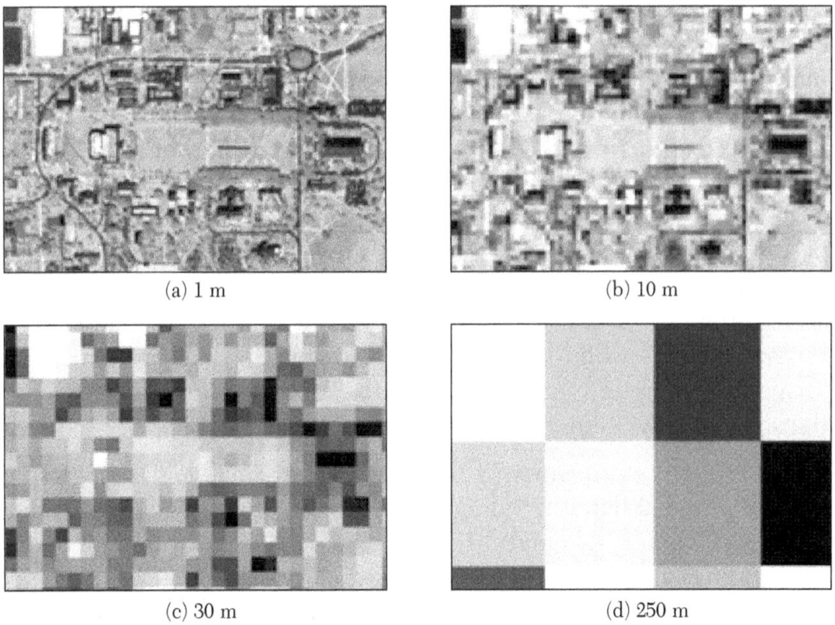

(a) 1 m (b) 10 m

(c) 30 m (d) 250 m

Figure 3.5. Information available at different spectral resolutions.

Table 3.2. Suitability of Spatial Resolutions for Mapping at Various Scales

Scale	Approx. pixel size (m)	Sensor
1 : 10,000	1	IKONOS panchromatic
1 : 50,000	5	Aircraft MSS, IKONOS Multispectral
1 : 100,000	10	SPOT HRG
1 : 250,000	25	SPOT HRVIR, Landsat TM
1 : 500,000	50	Landsat TM, MSS, IRS-LISS
1 : 5,000,000	500	OCTS, OCM
1 : 10,000,000	1000	NOAA AVHRR, MODIS GMS TIR band

The digital images provided by various remote sensing sensors will find application at different scales due to their different spectral resolutions. As a guide, Table 3.2 relates the scale of mapping to spatial resolution; this has been derived somewhat by considering an image pixel to be too coarse if it approaches 0.1 mm in size on a photographic product at a given scale. Thus, the Landsat MSS data is suggested as being suitable for scales smaller than about 1:500,000, whereas NOAA AVHRR data is suitable for scales below 1:10,000,000.

3.3.2 Spectral Resolution

Spectral resolution is the number and size of bands in the EMS that a sensor can capture. While the human eye only recognizes the visible part of the EMS, a sensor, depending on the type, can capture radiations in many spectral regions. The spectral resolution is thus the ability of a sensor to resolve the energy received in a spectral bandwidth to characterize different features of the Earth's surface. The narrower the wavelength range for a particular band, the finer the spectral resolution (Garg, 2019). Many remote sensing systems are multispectral in nature and record energy over several wavelength ranges, that is at various spectral resolutions. For example, the earlier Landsat satellites used a multispectral scanner (MSS) and captured images using four spectral bands (green, red, and two near-infrared bands), while hyperspectral platforms (e.g., Hyperion) captured images in hundreds of bands of the EMS. Table 3.1 shows further details of the spectral resolution of various sensors.

In remote sensing analysis, images may be available with high spectral resolution: >200 bands, medium spectral resolution: 4–15 bands, and low spectral resolution: <4 bands. High spectral resolution is achieved by narrow bandwidths which are collectively likely to provide a more accurate spectral signature for discrete objects than the broad bandwidths. Generally, surface features can be better distinguished from multiple narrow bands than from a single wide band. With a higher spectral resolution, single objects can be perceived better and spectrally distinguished (Jain, 1989). Broad classes, such as water and vegetation, can easily be separated using broad wavelength ranges, like visible and NIR. However, for more specific class such as vegetation type, rock classification and so forth much finer spectral resolution may be required.

The difference in reflectance characteristics of features in different wavelengths is taken as the main basis for their identification and classification from satellite images. The greater the difference, the better the identification. The concept is made clear in Figure 3.6, which presents a typical example of ice and snow, generally having high reflectance across all visible wavelengths; hence, their appearance is bright white. Reflectance decreases in the near infrared portion, and there is very low reflectance in the SWIR. The low reflection of ice and snow in the SWIR is related to their microscopic liquid water content. Reflectance for snow and ice would differ depending on the actual composition of the material, including impurities and grain size, and can easily be identified in specific wavelength regions.

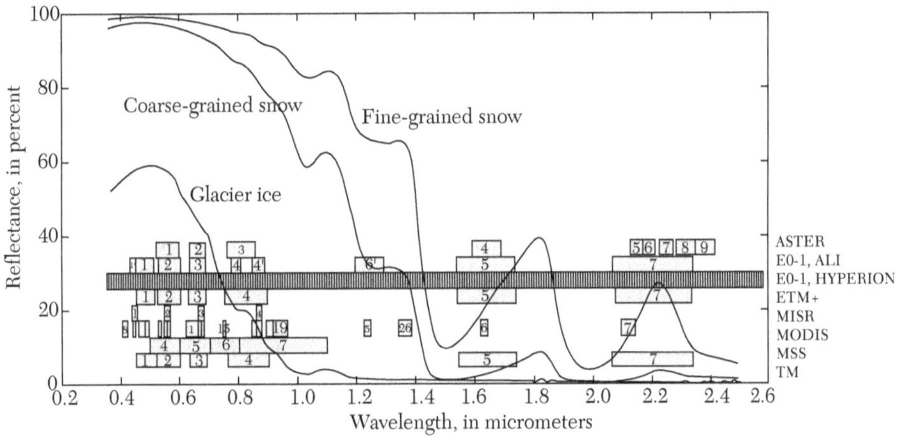

Figure 3.6. Spectral resolution of various sensors.

3.3.3 Radiometric Resolution

Radiometric resolution is the sensitivity of a remote sensing platform to detect slight differences in electromagnetic energy. It is measured in bits (a number to the exponential power of 2). Radiometric resolution refers to the dynamic range, or the number of different output numbers in each band of data, and is determined by the number of bits into which the recorded radiation is divided (Lillesand et al., 2004). For example, in 8-bit data, the DNs can range from 0 to 255 (28 = 256 total possible numbers). The maximum number of brightness levels (DN values) available depends on the number of bits used in recording the energy. The larger this number, the higher the radiometric resolution, and the sharper the image. Figure 3.7 presents an image captured at various radiometric resolutions and sizes, which indicate the importance of higher radiometric resolution.

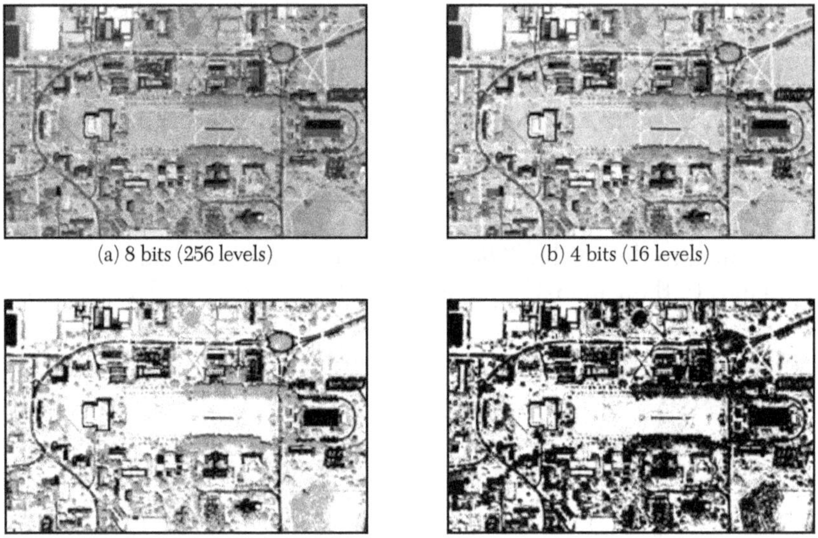

(a) 8 bits (256 levels) (b) 4 bits (16 levels)

(c) 2 bits (4 levels) (d) 1 bits (2 levels)

Figure 3.7. Image represented at various radiometric resolutions.

Advances in remote sensing technology have significantly improved the radiometric resolution or how sensitive an instrument is to small differences in electromagnetic energy. Sensors with high radiometric resolution can distinguish greater detail and variation in light. If the radiometric resolution is finer or higher, the small differences in reflected or emitted radiation can be measured accurately, but the volume of data will be larger. Table 3.3 shows the radiometric resolutions of some sensors. Radiometric resolution depends on the wavelengths and the type of the sensor. Some examples include: 12 bit sensor, 2^{12} or 4,096 levels (MODIS, MISR, Landsat-8); 11-bit sensor, 2^{11} or 2048 (IKONOS and QuickBird); 10-bit sensor, 2^{10} or 1,024 levels (AVHRR); 8-bit sensor, 2^{8} or 256 levels (Landsat-7 TM); 7-bit sensor, 2^{7} or 128 levels (IRS-LISS I-III); and 6-bit sensor, 2^{6} or 64 levels (Landsat 1-3 MSS).

Table 3.3. Remote Sensing Sensors and Their Various Radiometric Resolutions

Satellites/sensors	Radiometric resolution (bit)
POLDER	12
MODIS	12
AVHRR	10
ASTER/Terra	8 and 12
ETM+/Landsat-7	8
HRV/SPOT-5	8
IKONOS	11
QuickBird	11
WorldView	11
Geoeye-1	11

3.3.4 Temporal Resolution

Temporal resolution is a measure of the repeat cycle or frequency with which a sensor revisits the same part of the Earth's surface. Depending on the phenomenon being observed and application in hand, temporal resolution could play an important role. The smaller the revisit time, the better the temporal resolution of the sensor system (Richards & Jia, 2013). The frequency characteristics are determined by the design of the satellite sensor and its orbit pattern. It mostly depends on the swath width of the satellite; the larger the swath, the higher the temporal resolution. High temporal resolution may be < 24 hours–3 days, medium temporal resolution between 4–16 days, and low temporal resolution >16 days. For example, the temporal resolution of IKONOS is 14 days; Landast-7, 16 days; SPOT, 26 days; meteorological satellites, such as Meteosat-7 every half an hour and Meteosat-8 every 15 min; and MODIS satellite every one to two days. Table 3.1 shows further details of temporal resolutions of various sensors.

The ability to collect imagery of the same area at different periods of time is one of the most important elements for applying remote sensing data. There is some degree of overlap in the imaging swaths of adjacent orbits for most satellites, which increases with increasing latitude; therefore, some areas of the Earth tend to be reimaged more frequently. In addition, some satellite systems are able to point/orient their sensors to image the same area between different satellite passes separated by a time period of one to five days. Thus, the actual temporal resolution of a sensor depends on a variety of factors, including the satellite/sensor capabilities, the swath overlap, and latitude. The temporal resolution can also be decreased by studying the features from images acquired by various sensors at different resolutions.

Spectral characteristics of features may change over time, which can be detected by collecting and comparing multi-temporal imagery. By imaging the Earth on a continuing basis at different times, the naturally occurring changes (such as changes in natural vegetation cover or flooding) or changes induced by humans (such as urban development or deforestation) can be monitored. For example, during the growing season, most species of vegetation are in a continual state of change, which can be detected and monitored by frequently collected remote sensing imagery. The temporal images are also important (*a*) to acquire images in tropical areas which are mostly covered by clouds and offer limited clear views of the Earth's surface, (*b*) to capture sudden phenomena (e.g., floods, forest fires, etc.), (*c*) to compare before and after an activity happened (e.g., the spread of a crop disease from one year to the next), and (*d*) to study the changing appearance of a feature over time. Figure 3.8 shows the changes in land use features visible on Landsat-7 ETM + images of May, June, and July 2001.

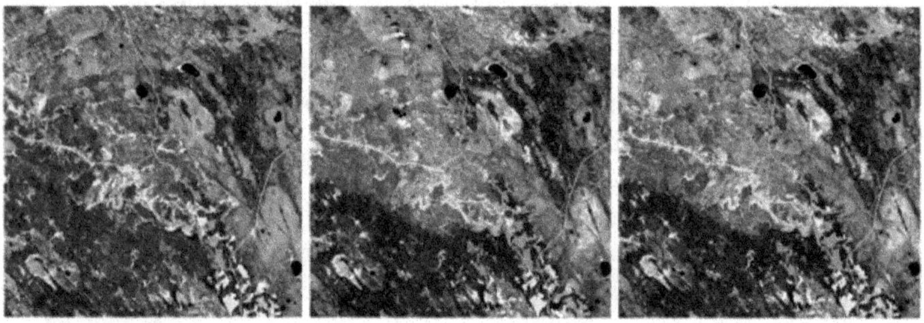

Figure 3.8. Landsat imagery taken on May 24, 2001 (left), July 11, 2001 (middle), and August 12, 2001 (right; Xue et al., 2017) (please see color figures in companion files).

3.4 REMOTE SENSING SENSORS

Remote sensing sensors acquire information about objects situated on the Earth's surface without any physical contact. They detect and measure the changes on the Earth. The main function of these sensors is to convert analog signals into digital signals (Sabins, 1996). Most satellite and airborne imaging sensors produce digital images from several spectral bands; which capture the reflected radiations from various Earth objects. Each band corresponds to a specific wavelength range within the EMS. Our eyes are sensitive to the blue, green, and red bands of light, known as the visible spectrum, but the satellites' sensors can record the reflected radiation not only of the visible spectrum, but also infrared, near infrared, thermal infrared, and microwave bands, which are useful in a variety of remote sensing applications.

There are two broad categories of sensor systems, passive and active systems, as per their source of illumination of the objects. Passive sensors measure the sunlight reflected or emitted naturally from surfaces and objects. Such sensors depend primarily on solar energy as the radiation source illuminating the surfaces and objects. Active sensors (such as radar and LiDAR systems) first emit energy (supplied by their own energy source) and then measure the return of that energy after interaction with the surface (Colwell, 1983). Both types of sensor systems are shown in Figure 3.9.

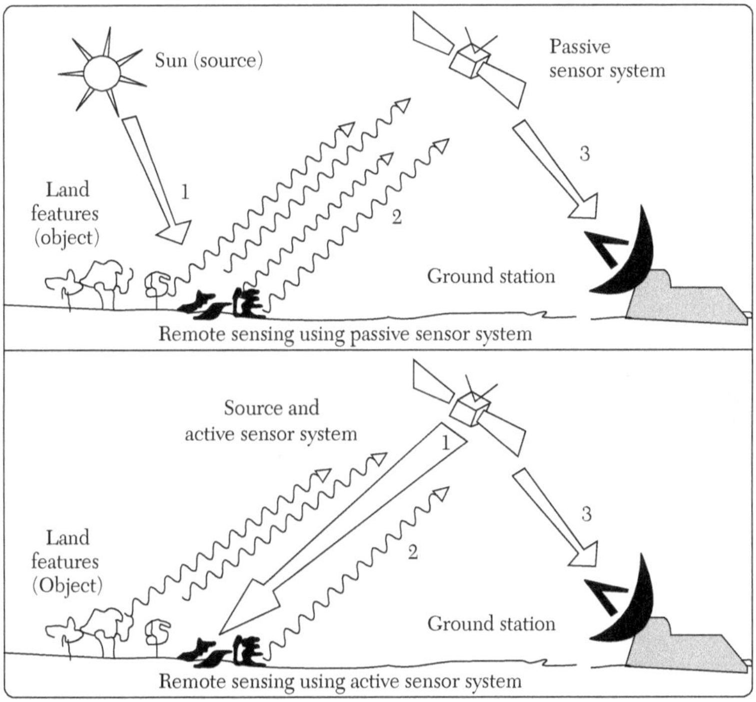

Figure 3.9. Working of passive and active sensors.

For acquiring the information, passive and active sensors are further divided into non-scanning (or non-imaging) and scanning (imaging) systems. The reflected or emitted energy can be measured by either imaging or non-imaging sensors. Each, image and non-image data has specific uses. Image data provides an opportunity to look at spatial relationships, object shapes, and to estimate the physical sizes based on the spatial resolution of data. Data from imaging sensors can be processed to produce an image of an area. Non-imaging sensors record usually a single response value for a specific area, and do not record how the incoming information varies across the field of view. They can be used to characterize the interaction between the received information and the illuminated target. A microwave radiometer sensor, for example, is classified as a combination of a passive and non-imaging sensor, while an aerial camera is classified as a passive and imaging sensor.

Cameras with framing systems have been successfully used as the remote sensors from aircraft, balloons, and manned and unmanned spacecraft. The size of the scene that is framed depends on the apertures and optics of the system that define the field of view. Images produced from remote sensing data can be either analog (such as a photograph) or digital (an array or grids). Photographic cameras are the oldest and most widely used remote sensing sensor, especially in aerial photography. The internal design of a spaceborne multispectral sensor is different from an aerial camera. In the imaging system, the entire frame of an image is acquired instantaneously in the basic image unit; for example, a camera used in photography or a RBV (Return-Beam Vidicon) in Landsat. An imaging sensor produces an image of an area of interest, giving spatial information. Spatial relationships between objects can be identified and used for interpretation.

Figure 3.10 shows six types of remote sensing imaging systems: digital frame area array, scanning mirrors, linear pushbroom arrays, linear whiskbroom areas, and frame area arrays (Lillesand & Kiefer, 2004). The scanning systems use an electrical sensor called a *detector* that records the brightness of the small scene of the terrain within its IFOV to produce an image. In the scanning systems, information is acquired sequentially from the Earth's surface in bits of picture elements or pixels, point by point and line by line, which may be arranged into a frame format after the acquisition. These signals are amplified, converted to digital form to produce an image, and recorded on the magnetic media or stored digitally. Earlier, photographic images were interpreted visually, but nowadays, the scanning technology makes it easier to convert analog photographs to digital images. Digital data can be analyzed through a computer by using the DN values, or processed to produce an image for its visual interpretation. In many cases, image interpretation involves the combination of both visual and digital techniques.

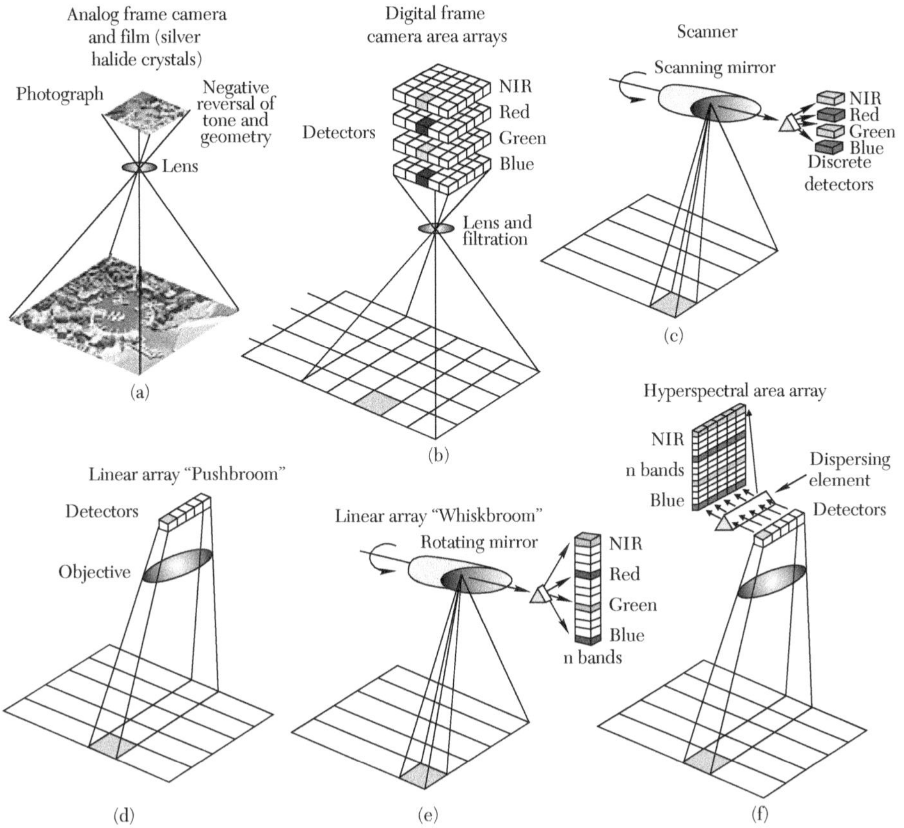

Figure 3.10. Six types of remote sensing imaging systems (Jenson, 2007).

The scanning systems include *whiskbroom scanners* (or across track scanners) and *pushbroom scanners* (along track scanners; Garg, 2021). Whiskbroom scanners use a mirror to reflect light onto a single detector. The mirror moves back and forth in a direction perpendicular to the flight path of spacecraft to collect measurements from one pixel in the image at a time. The whiskbroom system is similar to a LiDAR scanner where each pixel is captured at a unique time, and there is a stack of recording pixels, one for each spectral band, which are required to be precisely co-registered pixel-by-pixel to create an accurate multispectral image. The moving parts make this type of sensor expensive and more likely to wear out. For example, all Landsat sensors prior to Landsat-8 used the whiskbroom scanner.

The pushbroom scanners use a line of detectors arranged perpendicular to the flight direction of the spacecraft. As the satellite moves forward, each line of Earth data is captured at a time, corresponding to an instantaneous position and attitude of the spacecraft. A pushbroom scanner receives a stronger signal than a whiskbroom scanner because it looks at each pixel area for longer time. The OLI (Operational Land Imager) instrument on Landsat-8 uses a pushbroom scanner. The drawback of pushbroom is that the detectors in the scanners can have varying sensitivity, and if they are not perfectly calibrated, these can result in stripes in the data. Examples of satellites that also use the pushbroom design include SPOT, IRS, QuickBird, OrbView, and IKONOS.

The non-imaging sensors do not form images, and as such, are not used in operational remote sensing but give detailed information on spectral characteristics of the targets. These sensors usually are the devices that capture only a single response value, with no finer resolution. The whole area is viewed by these devices, and therefore no image can be made from the data. These single values can be referred to as a type of "point" data, however some small area is typically involved depending on the sensor's spatial resolution/field of view. The non-image data provides information for one specific (usually small) area, and can be used to characterize the reflectance of various materials/surfaces occurring in a larger scene and to learn more about the interactions of EMR and objects.

Non-imaging types of sensors are used to record a spectral quantity or a parameter as a function of time or distance (such as gamma radiation, magnetic field, temperature measurement, etc.). These sensors include (*a*) sounders and altimeters for the measurement of locations and topographic profiles with high accuracy, (*b*) spectrometers and spectroradiometers for the measurement of high spectral resolution along track lines or swath, and (*c*) radiometers, scatterometers, and polarimeters for high accuracy intensity measurements and polarization change measurements along track lines or wide swath (Richards & Jia, 2013). They are mostly used for ground observations and study of atmosphere and meteorology. While, multispectral scanners/thematic mappers are used with limited spectral resolution and high spatial resolution mapping, imaging spectrometers provide high spectral and spatial resolutions. Imaging radiometers and imaging scatterometers (microwave) are used for high accuracy intensity measurement with moderate imaging resolution and wide coverage (Jenson, 1986). Imagers and scanning altimeters/sounders can be used for 3D topographic mapping. The laser sensors are used more frequently for monitoring air pollution by laser spectrometers and for measurement of distance by laser altimeters. Various non-imaging remote sensing sensors are summarized in Table 3.4.

Table 3.4. Summary of non-imaging remote sensing sensors

Sensor	Operational wavelength	Definition	Application
Radiometer	Ultraviolet, IR, microwave	To measure the amount of EMR present within a specific wavelength range	Calculating various surface and atmospheric parameters
Altimeter	IR, microwave/ radiowave, sonic	To measure the altitude of an object above a fixed level	Mapping ocean-surface topography and the hills and valleys of the sea surface
Spectrometer	Visible, IR, microwave	To measure the spectral content of the incident EMR	Multispectral and hyperspectral imaging
Spectro-radiometer	Visible, IR, microwave	To measure the intensity of radiation in multiple spectrums	Monitoring sea surface temperature, cloud characteristics, ocean color, vegetation, and trace chemical species in the atmosphere
LiDAR	Ultraviolet, visible, NIR	To measure distance and intensity	Ocean, land, 3D topographic mapping
		Doppler LIDAR: measures the wave number for speed; polarization effects of LIDAR: shape	Meteorology, cloud measurements, wind profiling, and air quality monitoring
Sonar	Acoustic	Measure the distance to an object; determine the depth of water beneath ships and boats	Navigation, communication, and security (e.g., vessels) and underwater object detection. For example, handheld Sonar for a diver
Sodar	Acoustic	As a wind profiler, Sodar systems measure wind speeds at various heights above the ground and the thermodynamic structure of the lower layer of the atmosphere	Meteorology: atmospheric research, wind monitoring (typically in a range from 50–200 m above ground level)
A radio acoustic sounding system (RASS)	Radio wave and acoustic wave	Measuring the atmospheric lapse rate using backscattering of radio-waves from an acoustic-wave front to measure the speed of sound at various heights above the ground	Added to a radar wind profiler or to a Sodar system

Sensors can be categorized on the basis of:

1. Platforms on which the instrument is deployed: either ground based (e.g., terrestrial laser scanner), airborne (e.g., aircraft, drone), or spaceborne (e.g., satellite).
2. The orbit geometry and altitude (e.g., geo-synchronous, sun-synchronous), as they play an important role, and most often determine the application of the satellite in combination with the deployed sensor (e.g., weather study or Earth observation study).
3. Their observed portion of the EMS (e.g., optical, infrared, thermal, microwave).
4. Instrument used (e.g., imagers, altimeters, spectrometers, radiometers).
5. Instrument precision, for example, in terms of very high spatial resolution *vs*. low resolution sensors; in terms of narrow band spectral resolution (hyperspectral sensors) *vs*. broadband sensors (mono-and multispectral sensors); in terms of very high radiometric resolution *vs*. low resolution sensors. Some applications do not require very high precision instruments, for example, sea surface temperature (SST) measurements, while others, for example, vegetation monitoring, require high spectral and radiometric resolution for good data interpretation and analysis.
6. As per specific applications of each sensor (e.g., weather, environment, urban, land, water, mapping, photogrammetry, structure-from-motion, etc.).

Many satellites normally carry more than one sensor system (Lillesand et al., 2004). Remote sensing sensors are characterized according to several different properties, depending on the interaction between the sensor and the Earth's surface, and between active (e.g., radar) and passive (e.g., optical imagery) sensors. As sensor technology has advanced, the integration of passive and active sensors into one system has also emerged, and some systems use both kinds simultaneously. This trend makes it difficult to categorize sensors in the traditional way into passive sensors and active sensors. The details of some sensors are shown in Figure 3.11 and are briefly explained as follows.

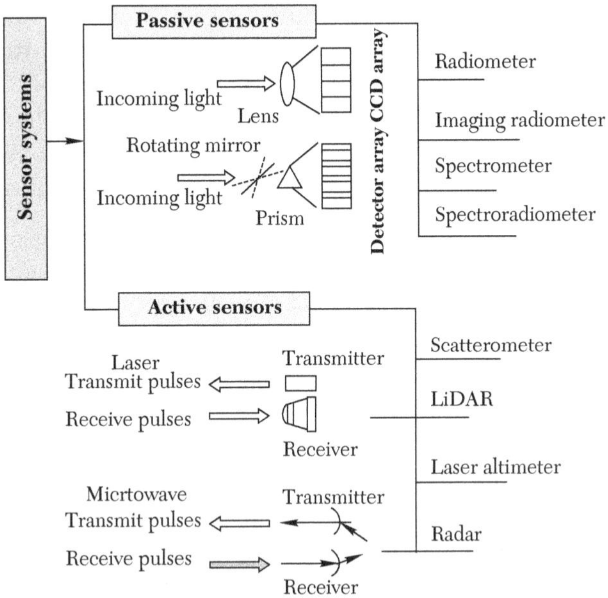

Figure 3.11. Classification of sensor systems.

3.4.1 Passive Sensors

Passive sensors are the most common sensor type for vegetation related study. Passive sensors depend on an external (natural) source (e.g., the Sun) to illuminate the target/object. They do not emit their own radiation. Passive sensors detect the sunlight radiation reflected from the Earth and thermal radiation in the visible and IR of the EMS. In visible light, the Sun illuminates the target, and the reflected light from the target is detected by these sensors (Figure 3.9). Most sun-synchronous satellites employ passive sensors, for example, Landsat MSS, and IRS LISS. These sensors have been applied in geology, mapping, ecology, forestry and agriculture, marine sciences, meteorology, and so forth. (Garg, 2019). In agriculture, remote sensing utilizes the reflectance properties of vegetation, measures them, and assesses crop health with vegetation indices. It is possible because specific values of vegetation indices correlate with certain species at a certain growth stage.

A major limitation of passive systems is that in most cases they require sunlight to acquire the data. Consequently, data acquisition by these sensors is very much dependent on lighting (time of day, time of year, latitude) and weather conditions, since cloud cover can interfere with the path of solar radiation from the Sun to the Earth's surface and then back to the sensor.

The signals detected by passive sensors are affected by the atmosphere, especially in the shorter wavelengths of the EMS that are strongly scattered by the atmosphere (Jain, 1989). These effects can be minimized by collecting the data only under very clear and dry atmospheric conditions, which may not be practically possible. However, various atmospheric correction algorithms are now available to remove atmospheric effects from data acquired by passive sensors (see Chapter 2).

Most passive sensors make use of a scanner for imaging, equipped with spectrometers, and measure signals at several spectral bands simultaneously, resulting in multispectral images needed for digital classification. Passive remote sensing employs multispectral or hyperspectral sensors with multiple band combinations. The most popular examples are various types of radiometers or spectrometers. While the radiometer determines the power of radiation emitted by the object in particular band ranges (visible, IR, microwave), the spectrometer distinguishes and analyzes the spectral bands. Hyperspectral radiometer operates with the most accurate type of passive sensor that is used in remote sensing (Birk & McCord, 1994). Due to extremely high spectral resolution, it differentiates hundreds of narrow spectral bands within visible, NIR, and MIR regions. The imaging radiometer scans the object or a surface to reproduce the image. The sounder senses the atmospheric conditions vertically. The accelerometer detects changes in speed per unit of time (e.g., linear or rotational). Various passive sensors are briefly described as follows.

3.4.1.1 Photographic systems

The historic developments in remote sensing are directly related to the development of photographic systems. The most common photographic sensor system is the photographic camera—a simple passive sensor. Aerial photography is one of the oldest forms of remote sensing, and it is still used extensively today to capture great detail. Aerial photography is also used as a reconnaissance tool to provide overview information for a particular area.

3.4.1.2 Electro-optic radiometers

A radiometer is an instrument designed to measure the intensity of EMR in wavebands ranging from the ultraviolet to microwave (Calla, 2010). Electro-optic radiometers measure electromagnetic energy using optical techniques and electronic detectors. Though they are only capable of recording a single data value for their viewed area, images can also be produced when they are mounted on a scanner device. Radiometers are similar in design to a camera in that they have an opening for the light to enter, lenses and mirrors for the light to pass through, and an electronic detector to record the intensity of EMR. As energy hits the detector, a signal proportional to the incoming irradiance is processed to either a digital or analog output that can be recorded.

Detectors in radiometers are designed to measure wavelengths from 0.4–14 μm (Colwell, 1983). Although some radiometers can detect this entire range of wavelengths, others only measure selected wavelengths in this range. Radiometers that measure more than one wavelength are called *multispectral radiometers*. For this type of radiometer, the incoming radiation is separated into discrete wavebands so that multichannel readings can be taken. This separation can be done using filters, prisms, or other sophisticated techniques. Non-imaging radiometers are commonly used as research tools to better understand how light interacts with objects, for spectral characterization of a variety of surfaces, and for atmospheric measurements, as well as to measure the quantity and quality of solar energy. These measurements can also be used effectively to apply correction to other imaging and non-imaging measurements for atmospheric effects.

3.4.1.3 Passive microwave systems

Passive microwave systems are based on a type of radiometer that detects wavelengths in the microwave region. Because of the advantages of microwave radiation, optical systems can't be used. As with optical systems though, both imaging and non-imaging systems are available. The components of a microwave radiometer are an antenna, receiver, and recording device. Microwave energy emitted from the Earth's surface is collected by an antenna, converted by a receiver into a signal, and recorded. The features of EMR measured by microwave radiometers are polarity, wavelength, and intensity. These properties provide useful information about the structure and composition of an object. Most of the applications of passive microwave radiometers are in the fields of atmospheric and oceanographic research (Cracknell, 2018). It has also proven to be an effective tool for the measurement of soil moisture, an important parameter in studying vegetation.

3.4.1.4 Visible, infrared, and thermal imaging systems

There are three basic designs for imaging sensors: frame, pushbroom, and mechanical scanners. The frame type sensor is a 2D array of detectors that acquires an entire image in one exposure similar to the way a camera captures an image on film. A pushbroom type sensor is a 1D array that obtains an image one line at a time (Lillesand & Kiefer, 1994). In a mechanical scanner system, the sensor acquires only one or several pixels in any given instant, but since the scanner physically sweeps or rotates the sensor (a radiometer) or a mirror back and forth, an image is produced. This category of sensors (passive visible, infrared, and thermal imaging systems) contains numerous instruments that have been deployed on a wide variety of platforms and used for many applications. Most modern imaging systems are multispectral, which acquire data in more than one band.

3.4.2 Active Sensors

Active sensors (e.g., radar and laser scanners) have their own EMR that is transmitted from the sensor toward the target/object, interacts with the targets and reflects back from the target/object energy, and is finally recorded by the sensor (Figure 3.9). The majority of active sensors operate in the microwave portion of the EMS, which are capable to penetrate the atmosphere under any condition. Radar penetrates through vegetation and soil, and can gain information about surface layer depth and moisture content of the soil layer. These sensors have the advantage of obtaining data any time of day or season, and are hardly affected by clouds, dust, fog, wind, and bad weather conditions (Garg, 2019).

Active remote sensing sensors transmit light or waves and determine, for example, distance, height, atmospheric conditions, and so on. The SAR is an example of an active sensor which transmits microwave pulses from its transmitter antenna to illuminate the target and receives the return signals by its receiver antenna. Radar, LiDAR, and LADAR (laser detection and ranging) are examples of active remote sensing where the time delay between emission and return of EMR is measured. In both LiDAR and LADAR, remote sensing data are gathered by illuminating the targets with laser light, and the measured reflection can then be used to determine the distance and height of objects for 3D representation. Other active sensors include: (*a*) A laser altimeter that measures the elevation with LiDAR, (*b*) ranging instruments that estimate the range either with one or two identical devices on different platforms sending signals to each other, (*c*) a sounder that is required to study weather conditions vertically by emitting impulses, and assists in developing weather forecasts with vertical profiles of humidity, precipitation, temperature, and absence/presence of clouds, and (*d*) a scatterometer that is used to measure bounced (backscattered) radiation (Richards, 1990). The SRTM (Shuttle Radar Topography

Mission) collects the Earth's elevation data which is required in many studies. Active remote sensing has been used for a variety of applications: oceanography, hydrology, geology, glaciology, agriculture, forestry, marine and arctic monitoring, and search and rescue missions. Some active sensors are discussed as follows.

3.4.2.1 Radar (active microwave)

Radar systems use microwaves (1 mm to 1 m) which are transmitted at a target or surface, and the timing and intensity of the return signal is recorded (Gibson, 2000). Transmission characteristics of radar depend on the wavelength and polarization of the energy pulse. Common wavelength bands used in pulse transmission are K-band (11–16.7 mm), X-band (24-37.5 mm), and L-band (150–300 mm). Pulses can be transmitted or received in either an H (horizontal) or V (vertical) plane of polarization. The use of letter codes to designate the wavelength range for various radar systems originated when radar was being developed during World War II. These random letter designations were assigned arbitrarily to ensure military security, however their use has continued.

Factors determining the strength of a radar return signal are complex and varied, however the most important are geometric and electrical properties of the surface or object that reflect the signal. Information about the structure and composition of objects and surfaces can be detected with radar. Radar data can be used in a number of fields, including geology, snow and ice studies, oceanography, agriculture, and vegetation studies.

3.4.2.2 LiDAR (active optical)

The LiDAR systems use laser light as an illumination source in wavelengths ranging from about 0.3–1.5 μm, which covers the ultraviolet through NIR. A short pulse of light is emitted from a laser and a detector receives the light energy (photons) after it has been reflected, or absorbed and remitted by an object or surface (Garg, 2021). LiDAR systems emit pulses at specific, narrow wavelengths that depend on the type of laser transmitter used. The simplest LiDAR systems measure the round trip travel time of a laser pulse, which is directly related to the distance between the sensor and the target/object. Basic distance measuring LiDARs are often referred to as rangefinders or as laser altimeters if deployed on an aircraft or spacecraft that can typically measure elevation, slope, and roughness of land, ice, or water surfaces.

The LiDAR systems can also make fluorescence measurements. Fluorescence refers to the process where a material absorbs radiant energy at one wavelength and then emits it at a different wavelength without first converting the absorbed energy into thermal energy. The wavelengths at which absorption and emission occur are specific to particular molecules. Fluorescence data can identify and quantify the amount of plankton and pollutants in the marine environment. For example, leaf fluorescence can also help to identify plant species. More advanced LiDARs measure the received intensity of the backscattered light as a function of travel time (Buczkowski, 2018). The intensity of the signal provides information about the material that reflected the photons. Such backscatter LiDAR systems are often used for atmospheric monitoring applications dealing with the detection and characterization of various gases, aerosols, and particulates. LiDAR methods have recently been adapted to measure the height of objects with great accuracy and precision (Mallet & Bretar, 2009). LiDAR instruments have flown on the space shuttle, Vegetation Canopy Lidar (VCL), and the Ice, Cloud, and land Elevation Satellite (ICESat).

3.4.3 Optical Sensors

Optical imaging sensors operate in the visible and reflective IR of the EMS, such as panchromatic systems, multispectral systems, and hyperspectral systems. Panchromatic (PAN) sensors provide single band data (B&W or gray scale image) within a broad wavelength range. Multispectral (MS) sensors have multichannel detectors with a few spectral bands, where each band is sensitive to radiations within a specific wavelength. MS sensors provide multi-images that contain both the brightness and spectral (color) information of the targets/objects being measured. Hyperspectral sensors collect and process information from more than 100 spectral bands taken in very narrow spectral bands (Chavez et al., 1994). The greater the number of channels, the lower the imaging resolution. Therefore, a PAN image usually presents a higher resolution than a multispectral/hyperspectral image. Table 3.1 gives a more details of these optical imaging systems. Some optical sensors are described as follows.

3.4.3.1 Sensors in LANDSATs

Since the early 1970s, the Landsat program has continuously and consistently captured images of the Earth. The Landsat program offers the longest continuous global record of the Earth's surface. For over forty years, it has collected spectral information from the Earth's surface, creating a historical archive unmatched in quality, detail, coverage, and length. Information on Landsat satellites are given in Chapter 4. Some important sensors/scanners aboard Landsats are explained as follows (*https://Landsat.usgs.gov/*).

Return-beam vidicon camera

The Landsat-1, -2, and -3 each carried a set of three Return-Beam Vidicon (RBV) cameras that operated in the visible and IR bands, providing a spatial resolution of 80 m and covering a 185 km × 185 km area. The RBV of Landsat-1 and -2 provided multispectral images in the green (0.475–0.575 μm), red (0.580–0.680 μm), and IR (0.698–0.830 μm) part of the EMS (Lillesand & Kiefer, 1994). During image processing, these data are taken as blue, green, and red bands to develop the false color image. The Landsat-3 provided an improved resolution of 40 m, but the cameras took images in a single PAN spectral band (0.5–0.75 μm) only. Each of these two RBVs has a covered adjacent ground area of 99 km × 99 km with some overlap. Thus, two pairs of RBV images cover an area of 181 km × 183 km which is equivalent to the area corresponding to the MSS scene (185 km × 185 km).

Multispectral scanner

The Multispectral Scanner (MSS) instrument on board Landsat 1–5 was designed to provide repetitive acquisition of high resolution, multispectral data of the Earth's surface on a global basis. The MSS is an opto-mechanical scanning instrument (with whiskbroom technique, unidirectional operation) consisting of a double reflector-type telescope, scanning mirror, filters, detectors, and associated electronics. The MSS instrument had a spatial resolution of approximately 79 m with four bands ranging from the visible blue to NIR (0.5–0.6 μm, 0.6–0.7 μm, 0.7–0.8 μm, and 0.8–1.1 μm; Richards & Jia, 2013). Since this instrument was developed after the three RBV cameras, these bands were numbered from 4–7.

The MSS of Landsat-3 included an additional spectral band in the TIR band. The MSS consisted of an oscillating mirror scanning the ground in a cross-track direction by using six simultaneous line scans (one line scan per detector per spectral band). The sensor operated by repeatedly scanning a 24-element fiber-optic array (arranged in 6 × 4 elements) from west to east across the Earth's surface (185 km swath), while the satellite motion in the orbit provided the north-south scanning (Figure 3.12). The line array of six detectors was positioned in the along-track direction, thus providing an instantaneous parallel along-track coverage of about 480 m in one cross-track scan with the configuration (Jenson, 1986). This wide along-track coverage permitted sufficient integration time for all cells in each scan sweep. The sensor demonstrated that remote sensing from space can be used to efficiently manage the Earth's resources.

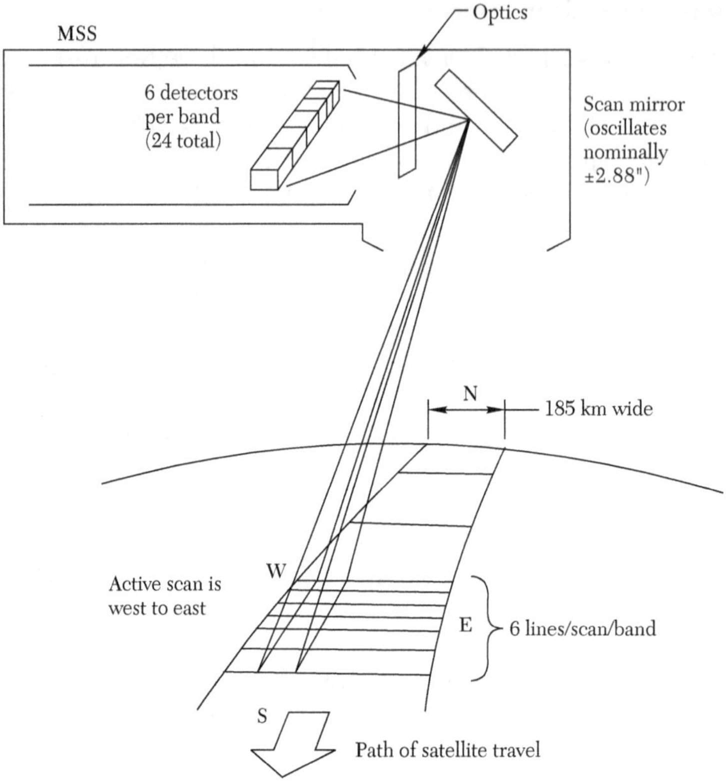

Figure 3.12. Configuration of Landsat MSS (Colwell, 1983).

Thematic mapper

The Thematic Mapper (TM) got its name from the intended "thematic" applications of its data. It is a whiskbroom instrument which is mounted on Landsat-3 and -4, and is regarded as a second generation imaging sensor for monitoring Earth's resources. The TM is an advanced, multispectral scanning, Earth resources sensor designed to capture higher resolution images, sharper spectral separation, improved geometric fidelity, and greater radiometric accuracy than the MSS sensor (Figure 3.13). The TM data are sensed in seven spectral bands simultaneously, and are used to produce maps for different themes, such as agriculture, forestry, water quality, hydrology, urban, geology, and so on (Gibson, 2000).

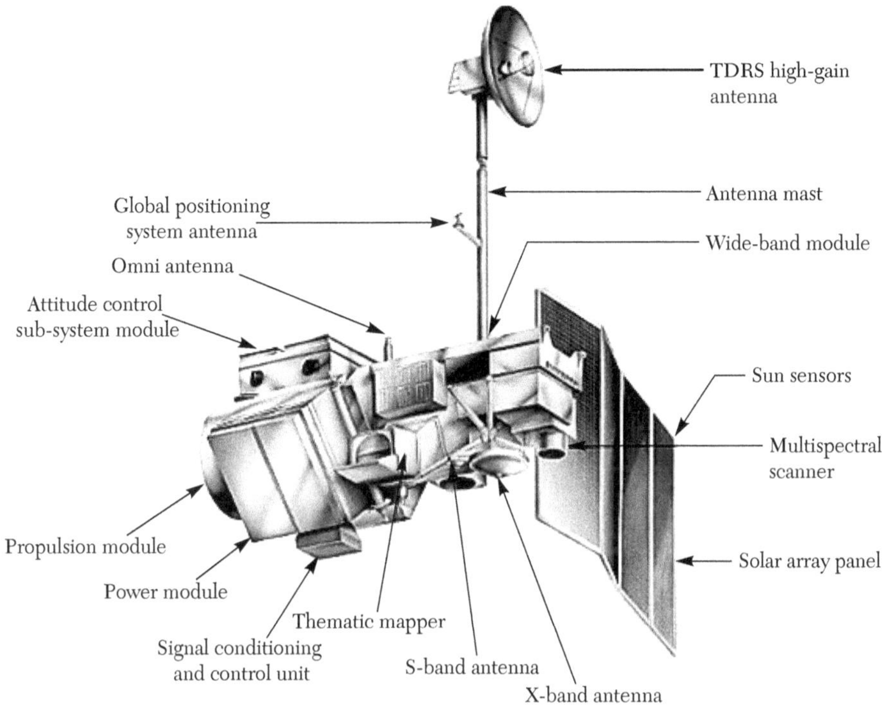

Figure 3.13. Landsat TM sensor (Colwell, 1983).

The TM is a mechanically scanning optical sensor operating in the visible and IR of the EMS. The instrument consists of a SMA (scan mirror assembly), telescope, SLC (scan line corrector), primary focal plane detector array, relay optics, cooled focal plane detector array, and internal calibrator. The scan line corrector compensates for the forward motion of the satellite, allowing the scan mirror to produce usable data (parallel scans) in both scan directions. The bidirectional scan and the use of detector arrays for each spectral band provides scan efficiency. The TM detector array features a total of 96 parallel line arrays (16 each for the seven spectral bands), oriented in the along-track direction. This new arrangement technology provides a parallel coverage of 480 m along-track in one scan sweep (cross-track direction). Band 6 senses the TIR (heat) radiation. A TM scene has a spatial resolution of 30 m in bands 1–5 and 7, while band 6 has a spatial resolution of 120 m on the ground. The image size is 185 km × 172 km with 5760 lines × 6928 pixels. As an example, these images captured in 7 bands are shown in Figure 3.14. Table 3.5 shows the principal applications of TM images.

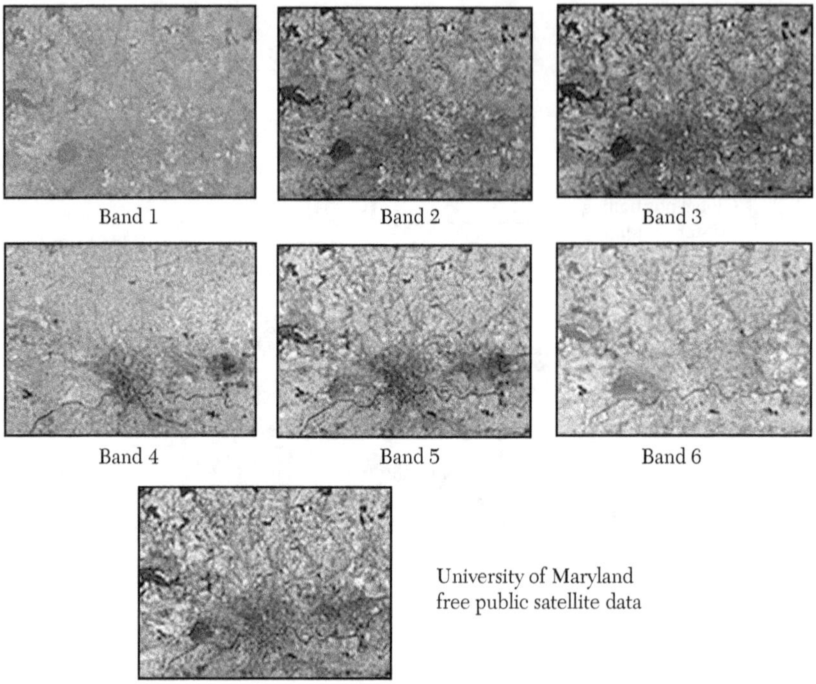

Figure 3.14. Images of Landsat TM bands.

Table 3.5. Principal applications of Landsat TM bands

Band	Spectral band (μm)	Applications
1	0.45–0.52	Coastal water mapping, soil/vegetation differentiation, deciduous/coniferous differentiation, chlorophyll absorption
2	0.52–0.60	Green reflectance, peak of healthy vegetation, plant vigor
3	0.63–0.69	Chlorophyll absorption, plant type discrimination
4	0.76–0.90	Biomass surveys, water body delineation
5	1.55–1.75	Vegetation moisture measurement, snow/cloud differentiation
6	10.41–12.5	Plant heat stress, thermal mapping, soil mapping
7	2.08–2.35	Hydrothermal mapping, geology

Enhanced thematic mapper plus

The Landsat-7 satellite is equipped with Enhanced Thematic Mapper Plus (ETM+), the successor of the TM sensor. The spectral bands are essentially the same seven bands as TM, with a newly added PAN band 8 (0.515–0.896 μm) having high resolution of 15 m. Landsat ETM+ data has been used to support a wide range of applications in areas, such as global change research, agriculture, forestry, mining, land cover, change detection (anthropogenic and natural), desertification, natural disasters, mineral exploration and classification, urbanization, and the development and degradation of water resources (Wulder, et al., 2008).

Operational land imager and the thermal infrared sensor

The Landsat-8 satellite carries two sensors: the Operational Land Imager (OLI) and the Thermal Infrared Sensor (TIRS), as per details given in Table 3.6. The OLI collects images using nine spectral bands in different wavelengths of visible, near-infrared, and shortwave with a 185 km swath in 30 m resolution (Band 8, PAN band, 15 m resolution) providing sufficient resolution to distinguish features, like urban centers, farms, forests, and other land uses (Trinh et al., 2017). The TIRS with two bands (10 & 11) was added to the satellite mission as highly accurate measurements of Earth's thermal energy were important for water resource study. The TIRS is an invaluable tool for managing water consumption.

Table 3.6. Details of OLI and TIRS Sensors

Sensor/ Band	Wavelength (mm)	Resolution (m)	Applications
OLI/1	0.43–0.45	30	Coastal and aerosol studies
OLI/2	0.45–0.51	30	Bathymetric mapping, distinguishing soil from vegetation, and deciduous from coniferous vegetation
OLI/3	0.53–0.59	30	Emphasizes peak vegetation, which is useful for assessing plant vigor
OLI/4	0.63–0.67	30	Discriminates vegetation slopes
OLI/5	0.85–0.88	30	Emphasizes biomass content and shorelines
OLI/6	1.57–1.65	30	Discriminates moisture content of soil and vegetation; penetrates thin clouds
OLI/7	2.11–2.29	30	Improved ability to track moisture content of soil and vegetation and thin cloud penetration
OLI/8	0.50–0.68	15	Sharper image definition
OLI/9	1.36–1.38	30	Improved detection of cirrus cloud contamination
TIRS/10	10.60–11.19	30 (100)	100 meter resolution, thermal mapping, and estimating soil moisture
TIRS/11	11.50–12.51	30 (100)	100 meter resolution, thermal mapping, and estimating soil moisture

3.4.3.2 Sensors in SPOTs

The SPOT (Système Pour l'Observation de la Terre) is a series of high resolution optical imaging Earth satellites, launched by the CNES (Centre National d'études Spatiales—the French Space Agency). The first one was launched on February 21, 1986 in sun-synchronous orbit, and since then seven satellites in the series have been launched. Sensors in SPOT-1, -2, and -3 were identical, including two identical optical HRV (High Resolution Visible) sensors which operated in PAN and MS modes, either simultaneously or individually. The multispectral images cover an area of 3600 km^2 (Garg, 2021).

The SPOT-4 carried two identical optical instruments: HR-VIR (Visible & Infrared High-Resolution) sensors with a new SWIR band (1.58–1.75 μm) at 20 m spatial resolution. It also had on board the first VEGETATION sensor, providing more than 1 km spatial resolution images for monitoring of vegetation at a global scale (Gibson, 2000). The VEGETATION sensor carries a wide-angle radiometric camera operating in four spectral bands (blue, red, NIR, and MIR). The characteristics of these sensors are given in Table 3.7.

Table 3.7. Characteristics of SPOT Sensors

Mode	Band	Spectral band (μm)	Spatial resolution (m)
1. High Resolution Visible (HRV) Sensor			
Multispectral	B1	0.50–0.59 (green)	20
	B2	0.61–0.68 (red)	20
	B3	0.78–0.89 (near IR)	20
Panchromatic	PAN	0.50–0.73	10
2. Visible and Infrared High Resolution (HR-VIR) Sensor			
Multispectral	B1	0.50–0.59	10
	B2	0.61–0.68	10
	B3	0.78–0.89	10
	SWIR	1.58–1.75	20
Monospectral	PAN	0.51–0.73	5 (or 2.5 in supermode)
3. VEGETATION Sensor			
Multispectral	B1	0.43–0.47μm (blue)	1165
	B2	0.61–0.68 μm (red)	1165
	B3	0.79–0.89 μm (near IR)	1165
	MIR	1.58–1.75 μm (middle IR)	1165

The SPOT-5 had two HRG (High Resolution Geometrical) sensors providing a higher resolution of 2.5–5 m in PAN mode and 10 m in MS mode. It had a SWIR band (1.58–1.75 μm) essential for vegetation data at a resolution of 20 m, in addition to the first three multispectral bands of the SPOT-4 HR-VIR sensor. The SPOT-5 also features an imaging instrument HRS (High-Resolution Stereoscopic) operating in PAN mode, pointing forward and backward from the satellite in order to take stereo-images for creating 3-D model (Holland et al., 2003). The aim of the VEGETATION sensor in SPOT-4 and -5 is to provide daily global coverage for accurate measurements of vegetation cover and study the global environmental changes.

3.4.3.3 Sensors in IRSs

Under the Indian Remote Sensing Satellite (IRS) program, the Indian Space Research Organisation (ISRO), Bangalore, has launched a series of land observation satellites. Recognizing the importance of satellite-based remote sensing systems for the management of natural resources, the ISRO has undertaken the design and development of a series of Indian Remote Sensing Satellites (IRS). The IRS system provides remotely sensed data for applications in the areas of agriculture, hydrology, geology, drought and flood monitoring, marine studies, snow studies and land use, and so forth. For details of the IRS series, please refer to Chapter 4. Some of the most popular sensors aboard IRS are explained as follows, and their characteristics are given in Table 3.8.

Table 3.8. Characteristics of IRS Sensors

Sensor	Resolution (m)	Swath width (km)	Spectral bands (μm)
Linear Imaging Self-Scanning System I (LISS-I)	72.5	148	0.45–0.52 (blue) 0.52–0.59 (green) 0.62–0.68 (red) 0.77–0.86 (near IR)
Linear Imaging Self-Scanning System II (LISS-II)	36.25	74	0.45–0.52 (blue) 0.52–0.59 (green) 0.62–0.68 (red) 0.77–0.86 (near IR)
Linear Imaging Self-Scanning System III (LISS-III)	23.5	142	0.52–0.59 (green) 0.62–0.68 (red) 0.77–0.86 (near IR) 1.55–1.70 (SWIR)
Linear Imaging Self-Scanning System IV (LISS-IV)	5.8	23–70	0.52–0.59 (green) 0.62–0.68 (red) 0.77–0.86 (near IR)
Wide Field Sensor (WiFS)	188	774	0.62–0.68 (red) 0.77–0.86 (near IR)

(Cont.)

Sensor	Resolution (m)	Swath width (km)	Spectral bands (μm)
Advanced Wide Field Sensor (AWiFS)	56–70	370–740	0.52–0.59 (green) 0.62–0.68 (red) 0.77–0.86 (near IR) 1.55–1.70 (mid–IR)

Linear imaging self scanner

The IRS-1A and -1B consists of LISS-I and -II (Linear Imaging Self-Scanning System), each providing multispectral images in 4 bands (0.46–0.52 μm blue, 0.52–0.59 μm green, 0.62–0.68 μm red, and 0.77–0.86 μm NIR) at 7-bit radiometric resolution (Garg, 2019). The two LISS sensors provide valuable data that help in medium to large scale mapping applications. Each LISS camera consists of the collecting optics, imaging detectors, inflight calibration system, processing electronics, and data formatting electronics. The LISS-I employs four 2048-element linear CCD detector arrays with spectral filters. It provides 76 m spatial resolution images with a swath of about 150 km, covering a 148 km × 174 km area in a scene. The LISS-II features eight 2048-element linear CCD detector arrays with spectral filters, providing images at 36.25 m resolution. Four LISS-II scenes cover an area equivalent to one LISS-I scene.

For multispectral scanning, the along track scanner accommodates one line of detector array for each spectral band. However, IRS satellites use as many cameras as there are spectral bands, each with a characteristic filter and a linear array of CCD detectors. Under identical setups, it is found that the along track scanner minimizes the detector size corresponding to a smaller IFOV producing higher spatial resolution and narrowing down the spectral bandwidth, thereby increasing the spectral resolution. If an area is to be observed more frequently, the IRS satellites can steer the cameras to look off-nadir at the required scene, some days before to some days after the normal nadir viewing date, thus providing several temporal images of the same area much earlier than its repeat period.

The LISS-III scanner was employed in IRS-1C and -1D satellites. It provided 7-bit multispectral images with 23.5 m ground resolution images in green 0.52–0.59 μm, red 0.62–0.68 μm, NIR 0.77– 0.86 μm, and SWIR 1.55–1.7 μm bands. The LISS-IV scanner, employed in IRS-P6 with 5.8 m resolution and multispectral mode, is useful in large scale mapping of natural resources and urban areas, including infrastructure mapping, such as road/rail networks. This camera can be operated in two modes: mono and multispectral. In the multispectral mode, data is collected in three spectral bands, that is, 0.52–0.59μm (green band 2), 0.62–0.68 μm (red band 3), and 0.76–0.86μm (NIR band 4) corresponding to a 23.5 km swath. In mono mode, the data corresponding to a swath of 70 km is obtained (Muthugadahalli et al., 1996). Nominally, band-3 data will be transmitted in this mode. The LISS-IV camera has the additional feature of off-nadir viewing capability by tilting the camera ± 26°, thus reducing the revisit period to 5 days for any given area.

Wide field sensor/Advanced wide field sensor

The WiFS (Wide Field Sensor), employed in IRS-1C and -1D satellites, provides 7-bit data at 188 m resolution in two spectral bands (i.e., 0.62–0.68 μm and 0.77–6 μm) with a swath width of 774 km. AWiFS (Advanced Wide Field Sensor) provides 56–70 m ground resolution, providing four images in 0.52–0.59 μm (green), 0.62–0.68 μm (red), 0.77–0.86 μm (near IR), and 1.55–1.70μm (mid-IR) with 370–470 km swath width, in addition to LISS-III and LISS-IV sensors (Garg, 2019). The AWiFS camera provides enhanced capabilities as compared to the WiFS camera, in terms of spatial resolution (56 m *vs.* 188 m), radiometric resolution (10-bit vs. 7-bit), and spectral bands (4 *vs.* 2). The AWiFS data with improved spatial and spectral resolution and large swath helped in better classification accuracy of agricultural related applications, monitoring large areas for flood inundation, vegetation stress, and so on.

Figure 3.15 presents a comparison of coverages by LISS-III, LISS-IV, and AWiFS Sensors (*https://earth.esa.int/web/eoportal/satellite-missions/i/irs-p6*).

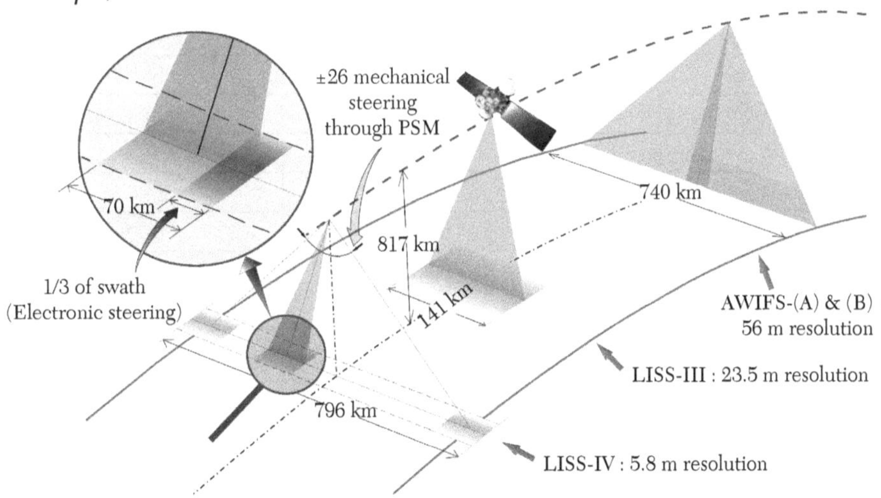

Figure 3.15. Area coverages by LISS-III, LISS-IV, and AWiFS sensors.

3.4.3.4 Sensors in Sentinel Systems

The Sentinel satellite is part of the Copernicus program—the European Space Agency (ESA). The Sentinel-1, launched on April 3 2014, is a polar-orbiting satellite, equipped with a C-band InSAR to observe sea-ice zones, arctic environment, marine environment, land surface motion risks and mapping in crisis situations (ESA, 2019). The Sentinel-2A, launched on June 23 2015, carries a visible and NIR sensor covering a wide swath of 290 km. It provides MS images in 13 spectral bands at different spatial resolutions which include four visible and NIR bands at 10 m resolution, six red-edge and SWIR bands at 20 m resolution, and three atmospheric correction bands at 60 m resolution (Drusch et al., 2012). The details of the Sentinel-2 bands are given in Table 3.9.

The Sentinel-3A, which launched on February 16 2016, and Sentinel-3B, which joined its twin in orbit on April 25 2018, carry several payloads: OCLI (Ocean and Land Color Instrument), SLSTR (Sea and Land Surface Temperature Instrument), and SRAL (SAR Radar Altimeter). The Sentinel-4, to be launched in 2022, on a Meteosat third generation satellite, will carry a UV-Visible-NIR spectrometer that will also use data from a TIR sounder. The Sentinel-5, to be launched in the future, will consist of a UV-VIS-NIR and SWIR spectrometer and TIR sounder and imager.

Table 3.9. Details of Sentinel-2 Bands

Band specification	Wavelength (μm)	Spatial resolution (m)
Band 1-Coastal	0.433–0.453	60
Band 2-Blue	0.458–0.523	10
Band 3-Green	0.543–0.578	10
Band 4-Red	0.650–0.680	10
Band 5-Vegetation red edge	0.698–0.713	20
Band 6-Vegetation red edge	0.734–0.748	20
Band 7-Vegetation red edge	0.765–0.785	20
Band 8-NIR	0.785–0.900	10
Band 8a-Vegetation red edge	0.855–0.875	20
Band 9-Water vapor	0.930–0.950	60
Band 10-SWIR-Cirrus	1.365–1.385	60
Band11-SWIR	1.565–1.655	20
Band 11-SWIR	2.10–2.80	20

3.4.4 Thermal Sensors

Thermal remote sensing deals with the acquisition, processing, and interpretation of data acquired in the TIR region of the EMS, where radiations emitted by ground objects are measured for temperature estimation. It is a well-known fact that all natural targets reflect as well as emit radiations. In thermal remote sensing, radiations emitted from the surface of the target/object are measured, in contrast to optical remote sensing where the radiations reflected by the target/object are measured (Fricker, 2018). All objects, both natural and man-made, with a temperature greater than absolute zero, emit infrared energy as heat. In the TIR region, radiations emitted by the Earth due to its thermal state are far more intense than solar reflected radiations, and therefore, sensors operating in this wavelength region primarily detect thermal radiative properties of the ground material. These measurements give the radiant temperature of a body, which depends on two factors: kinetic temperature and emissivity of the body (Sabins, 1996).

Human eyes cannot detect differences in TIR energy because they are not sensitive to IR (0.7–3 μm) or TIR energy (3–14 μm). The main difference between IR and TIR is that the infrared is reflected energy, whereas the thermal infrared is emitted energy (Rai et al., 2018). As the data acquisitions are made in atmospheric window regions, the TIR provides an excellent atmospheric window between the 8–14 μm wavelength. Various wavelength regions of infrared energy are shown in Figure 3.16.

Thermal images were first developed for military uses, but later these images were adopted by law enforcement, fire and rescue teams, security professionals, maintenance operations, coal fire team, and so forth. Even in complete darkness and challenging weather conditions, thermal imaging has the ability to detect very subtle temperature differences. The high temperature of the burning areas makes the fires detectable from TIR

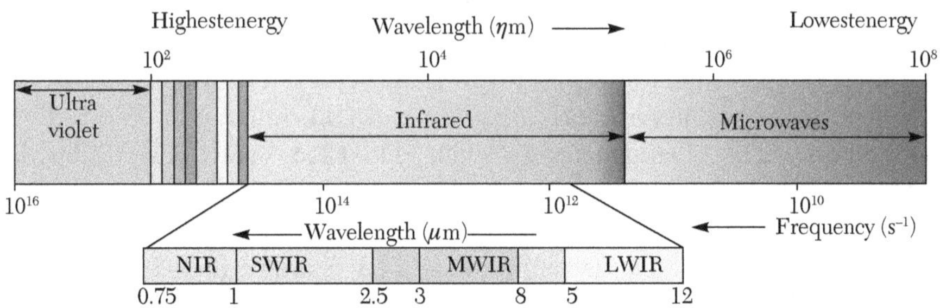

Figure 3.16. Various infrared wavelength regions used in remote sensing.

imaging. Smoke plumes consist of ash particles and other fine combustion products that are readily penetrated by the relatively long wavelengths of TIR radiation.

Thermal remote sensing is very useful to measure the surface temperature and thermal properties of targets for fire detection and thermal pollution studies. Some materials respond to changes in temperature more rapidly or slowly than others (thermal inertia). Water, rocks, soil, vegetation, and the atmosphere; all have the ability to conduct heat directly through them (thermal conductivity) onto another surface and to store the heat (thermal capacity). Absorption by water and other gases in the atmosphere restricts the sensors to record thermal images in two wavelength regions, 3–5 μm and 8 to 14 μm, and that's why the interpretation and processing of thermal IR imagery is different and more difficult than multispectral images (Rai et al., 2018). The details of some of the thermal sensors employed by remote sensing satellites are given in Table 3.10.

Table 3.10. Details of Some Thermal Infrared Sensors

Satellite/Sensor	Wavelength (μm)	Thermal bands (7–14 μm)	Spatial resolution (m)	Temporal resolution (days)
Landsat/TM	10.4–12.5	1	120	16
Landsat/ETM+	10.4–12.5	1	60	16
Thermal Infrared Sensor (TIRS) which has two thermal bands				
Landsat-8/(TIRS)	10.60–11.19 and 11.50–12.51	2	100	16
TERRA/ASTER	8.15–11.65	5	90	16
NOAA/AVHRR	10.3–12.5	2	1090	0.5
TERRA/MODIS	8.4–14.4	8	1000	1
Sentinel-3/SLSTR	10.95–13 2	1	1000	1–2

The TIR data may be collected by: (*a*) across-track thermal scanners, and (*b*) push-broom linear and area array CCD (charge-coupled device) detectors. The Daedalus DS-1260, DS-1268, and AMS (Airborne Multispectral Scanner) provide useful high spatial and spectral resolution thermal infrared data for monitoring the environment. The DS-1260 records data in 10 spectral bands including a TIR channel (8.5–13.5 μm). The DS-1268 incorporates the MIR bands (1.55–1.75 μm and 2.08–2.35 μm).

The AMS contains a TIR detector (3.0–5.5 μm) in addition to standard TIR detector (8.5–2.5 μm).

Thermal sensors detect radiation from the surface of objects, which might not be indicative of the internal temperature of an object. For example, the surface of a water body might be much warmer than the water temperature several feet deep, but a thermal sensor would only record the surface temperature. There are also topographic effects to consider. For example, in the northern hemisphere, north facing slopes will receive less incoming shortwave solar radiation from the Sun, and will therefore be cooler. Clouds and fog will usually mask the thermal radiation from surface features. Clouds and fog are generally cooler and will appear darker on a thermal image. Clouds will also produce shadows, where areas underneath the clouds are cooler than the surrounding areas. Changes in atmospheric moisture and varying emissivities of surface materials can make it difficult to accurately calibrate the thermal data (Tomlinson, et al., 2011).

Quantitative information about forest canopy structure, biomass, age, and physiological condition can be extracted from TIR data. Thermal images help identifying surface materials and features, such as rock types, soil moisture, geothermal anomalies, detection of geology and mineralogy, track potentially deadly patterns of heat in and around the active volcanoes, and so on. Thermal data of hotspots can alert administrators and planners about the forecast of volcanic activity before it becomes hazardous. In high resolution TIR images, active volcanoes stand out as bright spots, and these become brighter in time series images as they begin to erupt (Gillespie et al., 1998).

Images from thermal sensors, due to their higher temporal and spatial resolution, play an important role in various fields of agriculture, such as monitoring irrigation scheduling, soil salinity stress detection, plant disease detection, and so forth. The TIR sensors are used in forestry to map and monitor forest cover in terms of vegetation stress and evapotranspiration, which is important in environmental management, making vegetation a simple and effective way to reduce urban heat islands. Quantitative information about forest canopy structure, biomass, age, and physiological condition can be extracted from TIR data (Schmugge et al., 1998). Information about sea surface temperature (SST) is required by scientific, commercial, and social interest activities, such as weather forecasting, air-sea interaction modeling, climate change studies, fisheries, and coastal zone management, which may be obtained by satellite microwave radiometers and IR radiometers. Table 3.11 presents the application areas of some TIR imaging and non-imaging sensors.

Table 3.11. Application areas of thermal infrared sensors (Zhu et al., 2017)

Sensor	Operational wave band	Definition	Satellite sensors	Applications
IR imaging radiometer	UV, mid-to-far-infrared, or microwave	Measures the intensity of electromagnetic radiation	ASTER	Volcanological, mineralogical, and hydrothermal studies, forest fires, glacier, limnological and climatological, studies and DEM
Imaging spectro-radiometer	Infrared	Measures the intensity of radiation in multiple spectrums	MODIS, ASAS, IRIS	Sea surface temperature, cloud characteristics, ocean color, vegetation, trace chemical species in the atmosphere
Infrared imaging camera	Mid-far infrared	Measures reflected energy from the surface		Volcanology, determining thunderstorm intensity, identifying fog and low clouds

Though still not fully explored, thermal remote sensing has immense potential for various applications, such as identification of geological units and structures, soil moisture studies, heat loss from buildings, hydrology, coastal zones, sea surface temperature, volcanic activities, forest fires, seismology, environmental modeling, meteorology, medical sciences, and intelligence/military. In practice, thermal data prove to be complementary to optical and microwave remote sensing data.

The following are the benefits or advantages of thermal imaging (*https://www.ulis-ir.com/*):

1. It generates very high-contrast images, as thermal sensors are just as effective at night as they are during the day. The amount of daylight has no impact on the quality of the images obtained. Thermal cameras also offer a longer viewing range than visible-light cameras.

2. Thermal imaging does not require any additional lighting. Fog and outdoor lighting do not affect the images. This makes thermal sensors the ideal choice for surveillance and defense applications.

3. Thermal imaging cameras can detect motion over a very long range. This means that motion or the presence of an object can be easily detected.

4. It provides fast and accurate measurements of objects which are difficult to touch/to reach, such as high-altitude power lines.
5. It allows temperature measurements, which is not possible with optical images.
6. It can help in identifying air leakages, documenting irregular heat dispersion and identifying possible irregularities in insulation.

Drawbacks or disadvantages of thermal imaging include:

1. Images are difficult to interpret and process because there is absorption of thermal radiation by moisture in the atmosphere.
2. Accurate temperature measurements are hindered by differing emissivities and reflections from surfaces.
3. Most of the thermal imaging systems have ±2% accuracy for temperature measurement, and are not as accurate as contact methods.
4. Thermal imaging cameras cannot see through water and glass-like surfaces, as thermal energy can reflect off of shiny surfaces.
5. Thermal images are costly.
6. The TIR data can be difficult to calibrate.
7. Most applications of thermal remote sensing are qualitative, and they are not employed to determine the absolute surface temperature but instead to study the relative differences in radiant temperature.

Some of the thermal sensors used in remote sensing are explained as follows:

3.4.4.1 Advanced spaceborne thermal emission and reflection radiometer

The Advanced Spaceborne Thermal Emission and Reflection Radiometer (ASTER) is a sensor onboard the TERRA satellite launched in 1999. It is an optical multispectral sensor providing images at spatial resolutions of 15.0, 30.0, and 90.0 m in 14 spectral bands (Fricker, 2018). It has a revisit time of 16 days with a 60 km swath. The ASTER instrument consists of three separate instruments, each operating in a different spectral region: the visible and near infrared, the SWIR, and the TIR. Visible and NIR images are helpful to study the characteristics of snow cover, water, vegetation, and the degree of oxidation of the surface of objects. Mid-IR range is optimal for the recognition of minerals, especially of hydrated minerals in the clay soils.

Thermal data in five bands ranging from 8.1 to 11.6 μm with 90 m spatial resolution are mainly used to create detailed maps of surface temperature of land, emissivity, reflectance, and elevation. Thermal images are designed to record the temperature of the Earth's surface and decrypt the main types of rocks. The main areas for using ASTER images are: studies of global ecosystem changes; monitoring of natural disasters; geological, soil, climatological, and hydrological studies; and study of changes in vegetation cover. The ASTER data can also be used to get a better understanding of the interactions between the biosphere, hydrosphere, lithosphere, and atmosphere. The ASTER data is available for free download through Earth Explorer.

3.4.4.2 Moderate-resolution imaging spectroradiometer

The Moderate Resolution Imaging Spectroradiometer (MODIS), on board NASA's AQUA and TERRA, collects data in a variety of wavelengths, including thermal data. Image acquisition time of Aqua is ~1330 and 0130 h and Terra is ~1030 and 2230 h, all local time. This is high-temporal resolution data with a one- to two-day return time, having the spatial resolution of 1 km. Similar to ASTER, MODIS collects reflective data and emitted, thermal data. This makes it an excellent resource for detecting and monitoring wildfires. The products generated from MODIS data include land surface temperatures, thermal anomalies, and fire products which detect hotspots and fires.

The MODIS can be processed using a tool in ESRI ArcMap, called the MODIS Reprojection Tool (*https://lpdaac.usgs.gov/lpdaac/tools/modisreprojectiontool*). A strength of the MODIS sensor is the compromise between regular image acquisition and reasonable spatial resolution, in comparison to other sensors that offer higher spatial resolution but lower temporal resolution (e.g., Landsat), or higher temporal resolution but lower spatial resolution (e.g., SEVIRI).

3.4.4.3 Advanced very high resolution radiometer

The Advanced Very High Resolution Radiometer (AVHRR) sensor has been operational on a number of NOAA satellites, and is currently operational on NOAA-15, -16, -17, -18, and 19, offering at least daily coverage, but restricted to daytime images (Tomlinson, et al., 2011). These data have a spatial resolution of ~1.1 km. MetOP, the EUMETSAT satellite platform, also has an AVHRR sensor with a repeat time of 29 days. A major strength of the AVHRR sensor is that a relatively long historical record of data is available, which is required in many different applications. A large number of studies have been carried out using AVHRR sensor data for assessing land surface temperature (LST) and developing the vegetation index (NDVI). A significant weakness of AVHRR includes the lack of availability of nighttime images.

3.4.4.4 *Thermal infrared sensor*

Landsat-4 and -5 included a single TIR band (band 6) on a TM sensor with 120 m spatial resolution. A similar band was included on the Landsat 7 ETM+ sensor. The ETM+ offers some of the highest resolution thermal measurements from space, and data are available freely from the U.S. Geological Survey (USGS; *http://earthexplorer.usgs.gov/* or *http://glovis.usgs.gov*). Landsat 8 includes a separate thermal sensor known as the Thermal Infrared Sensor (TIRS) which has two thermal bands; band 10 (10.60–11.19 μm) and band 11 (11.50–12.51 μm). The TIRS bands are acquired at 100 m spatial resolution, but are resampled to 30 m as the delivered data products. These data are very useful for water resource managers and administrators to assess the irrigation water requirement from individual fields. A disadvantage of Landsat data is that they are not collected at night, and the thermal calibration is limited. Landsat has great strength in terms of spatial resolution, however its 16 days revisit time and lack of nighttime image acquisition is limiting at the temporal scale.

3.4.4.5 *Advanced along track scanning radiometer*

The Advanced Along Track Scanning Radiometer (AATSR) is carried onboard the European Space Agency (ESA) ENVironment SATellite (ENVISAT) which was launched in 2002. This was the third instrument in a series (ATSR-1 and ATSR-2) which started with the Along Track Scanning Radiometer (ATSR-1) in 1991 (Tomlinson, et al., 2011). The primary objective of all missions to date has been for sea surface temperature (SST) collection. The AATSR data can be used for monthly LST mapping, drought prediction, estimating evapotranspiration, and detection of snow covered areas.

3.4.5 Microwave Sensors

The microwave region falls between the IR and radio wavelengths, and has a range extending from approximately 0.1 cm–1 m. Because of their long wavelengths, as compared to the visible and IR, microwave radiations can pass through clouds, tree canopies, haze, dust, and the rainfall because these are not susceptible to atmospheric scattering, which affects the shorter optical wavelengths. These sensors include active and passive types, and therefore have the capability to collect imagery day and night, and under almost all weather and environmental conditions (Richards, 1990).

Passive microwave sensors are typically radiometers or scanners, and they operate in the same manner as other radiometers/scanners, except that an antenna is used to detect and record the microwave energy. Their field of view is large to detect enough energy to record a signal; therefore, most passive microwave sensors have low spatial resolution. These sensors can be used to measure atmospheric profiles and to determine water and ozone content in the atmosphere. This emitted energy is related to the temperature and moisture properties of the emitting objects (Fricker, 2018).

Active microwave sensors radiate their own signal to the target/object and record the reflected radiations. Different backscattering properties of different targets/objects as well as knowing the travel time of wavelengths allow differentiating the objects, and finding out the distance. It also depends on the illumination angle and the surface roughness. The most typical example of such devices is radar (operating with microwaves). The two basic types of remote sensing in this category are imaging (two-dimensional, e.g., radar), and non-imaging (linear, e.g., altimeters or scatterometers; Garg, 2021).

An active, scanning, and imaging sensor, for example, is SAR, which can produce high-resolution, imagery, day or night, independent of cloud coverage. The SAR, whether used airborne or spaceborne, emits microwave radiation in a series of pulses from an antenna. The working of a radar is shown in Figure 3.17. The radar moves along a flight path, and the area illuminated, or footprint, is moved along the surface in a swath. Each pixel in the radar image represents the backscatter for that area on the ground. This backscattered microwave radiation is detected, and the time required for the energy to travel to the target and return back to the sensor determines the distance or range to the target. By recording the range and magnitude of the energy reflected from all targets, a 2D image of the surface can be produced. The intensity in a SAR image depends on the amount of microwaves backscattered by the target and received by the SAR antenna (Castleman, 1996). Since the physical mechanisms responsible for this backscatter is different for microwaves, the interpretation of SAR images requires the knowledge of how microwaves interact with the targets/objects. Because of the speckle and quite different imaging sensor technology, the SAR images cannot be compared directly with the optical images of the same spatial resolution. Due to these differences, radar and optical data can be complementary to each other, as they offer different perspectives of the Earth's surface providing different information content.

The SAR images have been used frequently for the generation of maps of tropical forests. But interferometric SAR (InSAR) images can be used for the generation of height models. So images from ERS-1 and ERS-2, as well as SRTM (Shuttle Radar Topography Mission), launched in 2000, are used to measure the Earth's elevation and to create DEMs of the Earth. The SRTM is a good source of data to provide homogenous DEMs covering the Earth from 56° southern latitude up to 60.25° northern latitude. Frequently used spectrum bands in various microwave remote sensing sensors include the X-band, C-band, S-band, L-band, and P-band. The specific characteristics of each band can be found in Table 3.12.

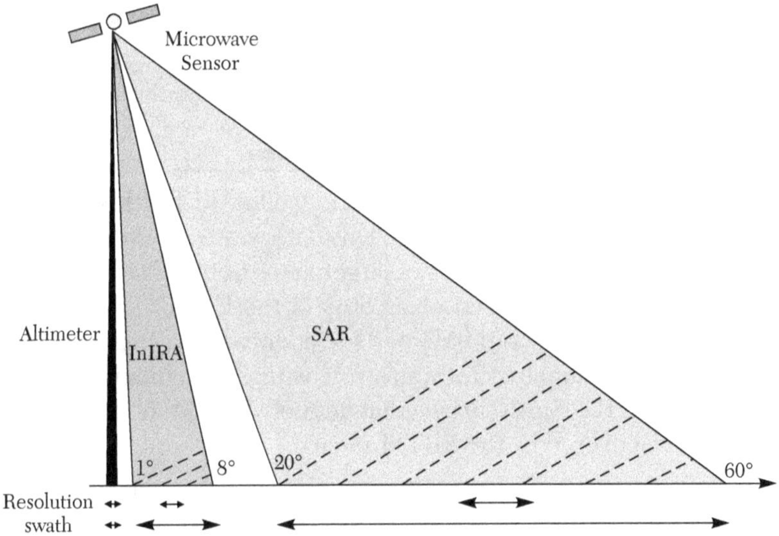

Figure 3.17. Working of a SAR.

Table 3.12. Various Microwave Bands and Their Key Characteristics

Band	Frequency (GHz)	Wavelength (cm)	Key characteristics
Ka	40–27	0.75–1.11	Usually for astronomical observations.
K	27–18	1.11–1.67	Used for radar, satellite communications, astronomical observations, automotive radar.
Ku	18–12	1.67–2.5	Typically used for satellite communications.

(Cont.)

Band	Frequency (GHz)	Wavelength (cm)	Key characteristics
X	12.5–8	2.4–3.75	Widely used for military reconnaissance, mapping, and surveillance.
C	4–8	3.75–7.5	Penetration capability of vegetation or solids is limited and restricted to the top layers. Useful for sea-ice surveillance.
S	4–2	7.5–15	Used for medium-range meteorological applications, e.g., rainfall measurement, airport surveillance.
L	2–1	15–30	Penetrates vegetation to support observation applications over vegetated surfaces and for monitoring ice-sheet and glacier dynamics.
P	1–0.3	30–100	So far, only for research and experimental applications. Significant penetration capabilities regarding vegetation canopy, sea-ice, soil, and glaciers.

The SRTM, mounted on a space shuttle, collected Earth surface data by utilizing a synthetic aperture radar. This data was converted into DEMs to generate more precise 3D maps of larger areas of the Earth. During its 11-day flight, it obtained data covering 80% of the Earth's surface (excluding north and south poles), and 95% of its residential area (Acker et al., 2014). The SRTM was launched into an orbit with an inclination of 57°. This allowed most of the Earth surface that lies between 60° north and 56° south latitude to be covered by the SRTM radars.

In order to generate a 3D map, it is necessary to collect the data from two viewpoints. As shown in Figure 3.18, two antennas, the main antenna mounted in the cargo bay of the shuttle and another one mounted at the end of the 60 m extended mast on the port side of the cargo bay, are used. The observation data of these two viewpoints, and a technique called interferometry is used to generate 3D images (Campbell, 2011). While data is being collected, the antenna mounted at the end of the 60 m mast must be properly positioned within several millimeters, which requires highly advanced techniques. The SRTM used the radar interferometry technique, where two radar images are taken from slightly different locations, and the differences between these images allow for the calculation of surface elevation or change. The SRTM radar mast created a

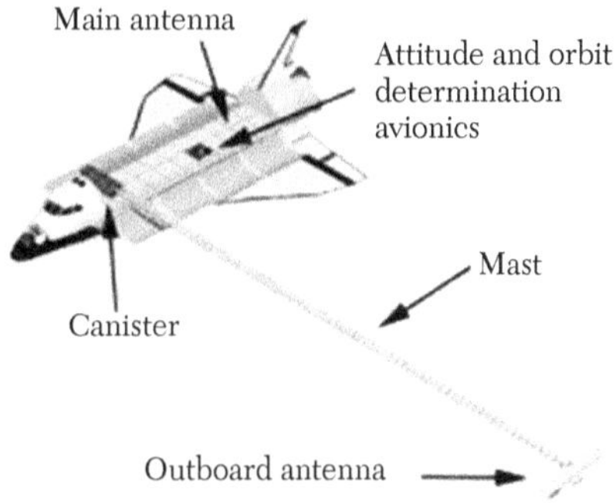

Figure 3.18. Configuration of SRTM.

fixed-length interferometer at both C-band and X-band frequencies. The elevation measurements of the SRTM are expected to have an absolute accuracy of better than 16 m. The SRTM 90 m resolution (3 arc-second) and 30 m resolution (1 arc-second) datasets have been utilized by large number of users worldwide, making it the most popular dataset produced by the space shuttle missions. Potential applications of 3D data include regional weather forecasting that takes account of topography, obtaining an accurate understanding of the distribution of forests in mountains, safe navigation of aircraft, and determining line-of-sight areas for wireless communications.

Non-imaging microwave sensors include altimeters, scatterometers, and LiDARs, which are mostly profiling devices taking measurements in one linear dimension, as opposed to 2D representation of imaging sensors. Radar altimeters transmit short microwave pulses and measure the two-way trip time to targets/objects to determine their distance (range) from the sensor (Kuenzer et al., 2016). Radar altimetry is used on aircrafts and satellites for topographic mapping and sea surface height estimation. The scatterometer is a high-frequency microwave radar, designed specifically to measure the backscattered radiations that can be used to derive maps of surface wind speed and direction over ocean surfaces. LiDAR is an active microwave non-imaging sensor that measures ground height, as shown in Figure 3.19. In addition, it can provide 3D coordinates of the points densely, creating point cloud data. It consists of a sensor that uses a LASER (light amplification by stimulated emission of radiation) to transmit a light pulse, and a receiver with detectors that measure the backscattered or reflected light. The time between transmitted and backscattered pulses is recorded, and the distance to the object is determined by multiplying time with the speed of light.

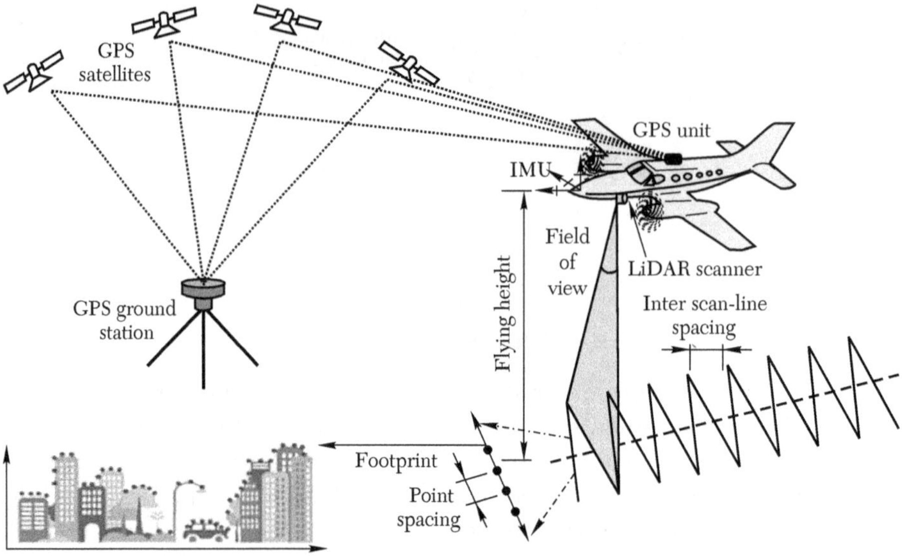

Figure 3.19. Data collection by LiDAR non-imaging active sensor.

Microwave sensors can be used for the study of agriculture, urban land use, and land cover, discrimination of crop types, crop condition monitoring, geology, hydrology, forest cover, snow and ice, soil moisture and soil types, snow studies, hydrocarbons, and so on. (Calla, 2010). Many studies have demonstrated the use of SAR remote sensing to retrieve biophysical characteristics from forests (Richards, 1990) and established useful relationships between the backscattering coefficients and the above-ground biomass (Imhoff, 1995). These relationships may provide a method of monitoring forest ecosystems which play a vital role in carbon storage. For ocean applications, active as well as passive microwave sensors both can be used for determination of salinity, SST, suspended sediment concentration, shoreline detection, sea-ice identification, ocean color and bathymetry studies, wave heights, and the detection of tsunami. For atmospheric applications, microwave remote sensing can be used for profiling the moisture and temperature in the atmosphere, which is essential for delineating mesoscale climatic systems. Microwave sensors are being used for planetary exploration; planets, like Mars and Venus, and satellites like the Moon have been explored to detect the presence of frozen water on Moon (e.g., very successful *Chandrayan* Mission of India) and presence of buried channels under sand dunes on Mars (Calla, 2010).

For details of active and passive microwave sensors, see Section 3.4.

3.4.6 Hyperspectral Sensors

The most significant recent breakthrough in remote sensing has been the development of hyperspectral sensors and software to analyze the resulting image data. The "hyper" in hyperspectral means "over" or "too many," and refers to the large number of measured wavelength bands. Hyperspectral sensors observe the Earth's surface by simultaneously sampling hundreds of fine narrow contiguous spectral bands with a resolution of up to 0.01 μm in the visible and IR spectrum (Birk & McCord, 1994). Hyperspectral sensors, both airborne and spaceborne (also referred to as *imaging spectrometers*) acquire images that allow the characterization of objects of interest (e.g., land cover classes) with high accuracy. Unlike the sensors on aircraft (e.g., AVIRIS), sensors on satellites have the capacity to provide global coverage at regular intervals (e.g., Hyperion EO-1).

The development of hyperspectral sensors has involved the convergence of distinct technologies, such as spectroscopy and the remote imaging of Earth. Instruments called *spectrometers* or *spectroradiometers* are used to make ground-based or laboratory-based measurements of the light reflected from a surface/material. An optical dispersing element in the spectrometer splits light into many narrow, adjacent wavelength bands and the energy in each band is measured by a separate detector. Spectrometers can make spectral measurements of bands as narrow as 0.01 mm over a wide wavelength range by using very large number of detectors (Chavez et al., 1994). Many ground-based and airborne hyperspectral sensors are available, whereas very few spaceborne hyperspectral sensors are in orbit.

Figure 3.20 shows the concept of hyperspectral imaging technology. Each individual pixel is characterized by a complete spectrum of ground

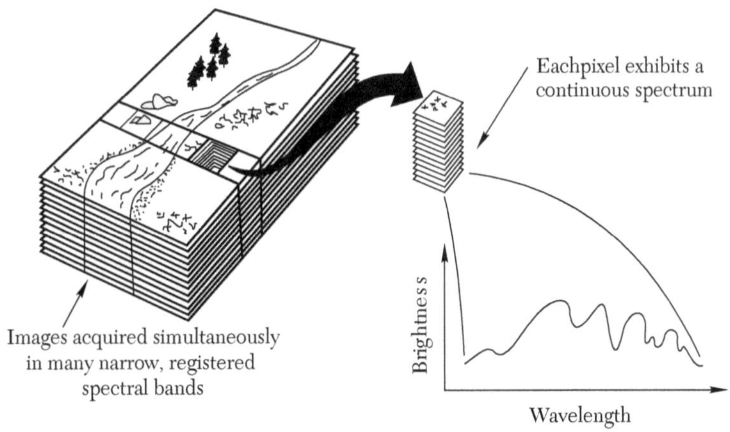

Figure 3.20. The concept of hyperspectral imaging (Chavez et al., 1994).

targets (and their mixtures) that can be quantitatively analyzed within the spatial view. The capability of acquiring quantitative information from many points on the ground at almost the same time provides another innovative aspect of hyperspectral imaging technology. It freezes time for all spatial pixels at almost the same point, subsequently permitting adequate temporal analysis. Hyperspectral imaging technology is thus a promising tool that adds many new aspects to the existing mapping technology and improves our capability to identify materials from remote distances.

The remote sensing hyperspectral appeared in the mid 1980s, and since then it has been widely used by geologists for mapping minerals. Detection of the type of material is dependent on coverage and spectral resolution, relation signal/noise of the spectrometer, the material density, and strength of material absorption in the wavelength in which the measurements are made. The rich spectral information helps to better discriminate surface features and objects than traditional multispectral imaging systems. Land cover classification and target detection are some of the most common hyperspectral remote sensing applications. The analysis of critical spectrum-based signatures is an invaluable tool to the remote sensing community because it can improve our understanding of plant physiology, crop science, geological formations, and the underpinnings of important infrastructure assets, such as pipelines, dams, and railroad beds.

The Hyperion EO-1 sensor was launched in November 2000 by NASA with the purpose of taking hyperspectral images from space in order to create mineralogical mapping (Vorovencii, 2009). Hyperion is a push-broom sensor that takes pictures with a radiometric resolution of 8 bits, and a width of 7.5 km. Table 3.13 presents some of the hyperspectral sensors and their characteristics (Vangi et al., 2021).

Earth-observing satellites are increasing in space, and remote sensing data are undergoing an explosive growth. New and improved sensors flown in a constellation of satellites in a variety of orbits are producing several terabytes of data per day. A variety of advanced and complementary instruments on various satellites, combined with ground-based and aircraft/UAV systems, eventually provide a wealth of data. Such a volume of remote sensing data produced by various sensors gives rise to complexity and diversity in high-dimensional characteristics of data (Sajjad & Kumar, 2019). These data are required to be analyzed to get continuous digital representation of Earth on a global basis. This global database may include a complete and accurate representation of the atmosphere, ocean, and land and ice surfaces, which is essential in order to have much more accurate weather forecasts, to characterize the climate, and to better understand the Earth system (Anthes, 2003).

Currently available data from various Earth observation systems are operationally used in resource mapping, monitoring, and management programs, but still there are many gap areas in observations as well as applications that need to be addressed in the future. In the future, remote sensing data may be acquired at high resolutions (spatial, spectral, radiometric, and temporal), at low cost for the creation of new applications, such as infrastructure development, risk mapping, and providing real-time support for natural and human induced disasters, by integrating spatial/aerial and ground-based sensors (Crowley & Cardille, 2020). The fusion of remote sensing observations from multiple sources into data cubes can increase temporal and spatial resolutions without trading off the spatial extent coverage. The advances in data sources at varying scales and resolutions from very high resolution to large area coverage would be very helpful to carry out analysis of large regions more rapidly and for a longer time.

Table 3.13. Characteristics of some hyperspectral sensors

Platform and sensor	Year of launch	Spectral resolution (μm)	Spectral bands	Spatial resolution (m)	Temporal resolution (days)	Swath width (km)
AVIRIS	1987	0.4–2.5	224	20	–	12
MODIS (USA)	1999	0.4–14.39	36	250–1000	1–2	2330
Hyperion EO-1 (USA)	2000	0.4–2.5	220	20–60	16–30	7.5
CHRIS-PROBA (ESA)	2001	0.415–1.05	19	25		17.5
MERIS (ESA)	2002	0.41–0.90	15	300	3	1150
AIRS (USA)	2002	3.70–15.40	20	13500	0.5	2378
IASI	2006	3.62–15.50	3		29	2200
IMS-1 Mx-Hy (India)	2008	0.4–0.95	84	550	4	128
Shenzhou-8 TianGong-1 (China)	2011	0.4–2.5	128	20 (SWIR) 10 (VNIR)	7–14	10
MSI, Sentinel-2	2015	0.44–2.5	12	10–60	5	275
PRISMA	2019	0.4–2.5	249	30	10	30

(Cont.)

Platform and sensor	Year of launch	Spectral resolution (μm)	Spectral bands	Spatial resolution (m)	Temporal resolution (days)	Swath width (km)
HISUI (Japan)	2019	0.4–2.5	185	30 20 (SWIR)	2–60	30
HyspIRI VS-WIR (USA)	2020	0.39–2.5	210	60	10	145
SHALOM (Italy–Israel)	2021	0.40–2.5	275	10		30
HypXIM (France)	2022	0.40–2.5	210	8		145–600

Note: AIRS: Atmospheric Infrared Sounder; ESA: European Space Agency; HISUI: Hyperspectral Imager Suite; HypXIM (HYPperspectral-X Imagery); IASI: Infrared Atmospheric Sounding Interferometer; IMS-1: Indian Microsatellite-1; MERIS: Medium Resolution Imaging Spectrometer; SHALOM (Spaceborne Hyperspectral Applicative Land and Ocean Mission); PRISMA (Precursore IperSpettrale della Missione Applicativa); PROBA (Project for On Board Autonom); VSWIR (Visible Short Wave Infrared)

As the temporal resolution of satellite image observations gets shorter, landscape changes, whether they are persistent (e.g., fire), ephemeral (e.g., floods), or gradual (e.g., land degradation) can be studied in real time. The finer spatial and temporal resolutions are expected to be made available by future optical satellites that are being built (e.g., Landsat 9, Sentinel constellation). Future studies can use observations from data sources like UAVs and microsatellites (i.e., small satellites from companies like DigitalGlobe and Planet) for very high spatial resolution observations of large scale features, hyperspectral sensors for greater spectral sensitivity, and SAR and LiDAR sensors for reconstructing 3D terrain models. The use of non-visible portions of the spectrum (particularly infrared and microwave) and increase in the number of active sensors, such as LiDAR and radar, will improve the spatial and temporal sampling rate. Particularly important will be new technologies for linking sensors through wireless and traditional means into sensor networks. This will allow the information to be combined so as to support rapid decisions in complex situations (Gail, 2007). In addition, the output of one or several sensors can be used to trigger the observations from others, or even to rapidly reconfigure the other sensors so as to optimize the observations of an event.

Rapidly increasing computing power demands data at higher spatial and temporal resolutions in order to adequately describe the processes in several numerical models related with the Earth's surface. Complex data formats, communication capability, and processing time of large datasets are crucial issues for the wide use of large remote sensing data (Crowley & Cardille, 2020). Emerging new information technologies, including new visualization tools, are to be employed to effectively interpret and use the data/information. Although it is an enormous and challenging task to assimilate the large volume of data expected from future satellite systems, the efficient conversion of these data into usable information still remains the key challenge for the remote sensing community. Preprocessing, data format conversion, open-source software tools, and real-time data processing will help in better utilization of the Earth observation data collected from various remote sensing sensors.

CHAPTER 4

VARIOUS REMOTE SENSING PLATFORMS

4.1 INTRODUCTION

The sensor is the device in remote sensing that records wavelengths of energy. Generally, these sensors are mounted or fixed with a "platform" that can be ground, airborne, or spaceborne based (e.g., UAV, aircrafts, satellites, etc.). The platform is a vehicle where the sensor or camera is mounted and operated to acquire information about a target/object under investigation. Though aircraft and satellites are commonly used platforms, balloons and rockets are also used. For remote sensing applications, sensors should be mounted on suitable stable platforms. The type of platform and its characteristics depend on the type of sensor attached and its application. Since the 1960s, remote sensing has progressed to include aircraft and satellite platforms carrying electro-optical and antenna sensor systems (Campbell, 2011).

There are various remote sensing platforms today which have been launched by many countries around the globe. Essentially these platforms onboard carry various sensors to collect information about the Earth's surface features and transmit the same to the ground receiving station. These platforms can be categorized based on the height of operation and area covered. Ground-based platforms are extensively used, both at laboratories and in the field. Such experimentation greatly helps in the design and development of sensors for the characterization of the Earth surface features and understanding the relationships between signals and objects. Handheld cameras, spectroradiometers, laser-based equipment, and GPS are some of the examples used in laboratory and field experiments to collect information about the Earth's features, which is also known as *ground truth data*.

Airborne platforms are used to collect data as well as test the performance of the sensors before they are actually mounted in the spaceborne platforms. These platforms include balloons, aircraft, and drones/UAVs, which collect data from different heights at different scales and resolutions. Spaceborne platforms operate at greater heights, such as Landsat, SPOT, IRS, IKONOS remote sensing satellites, the NOAA series of meteorological satellites, and the GOES and INSAT series of geostationary satellites. They provide data at various spatial, spectral, radiometric, and temporal resolutions.

Although free data sets and free satellite imagery are available online, many of these resources are at lower resolutions (e.g., MODIS, Landsat), limited by the available dates. High to medium resolution imagery provided via Google Earth or the Indian site BHUVAN is increasingly being used in many scientific researches and helping in the selection of field sampling locations for classification. Image analysis based on Google Earth images, however, has limitations, as multispectral bands are not available for digital analysis. The very high resolution images from IKONOS, QuickBird, OrbView, and EROS-A1 are operated by commercial companies (e.g., GeoEye, DigitalGlobe) that have to be purchased for many applications and research areas.

A side looking airborne radar (SLAR) irradiates a swath along the aircraft flight direction by scanning the terrain with radar pulses at right angles to the flight path. Thus, the radar image is created by the pulse energy reflected from the terrain, and represents primarily the surface topography. The range resolution of SLAR depends on the length of the radar pulse, which can be made quite short with new techniques (Garg, 2019). The SAR employs the Doppler shift technique to narrow down the azimuth resolution even with a small antenna. However, the azimuth resolution is limited by the antenna size and altitude, thus preventing SLAR systems from being used on satellites. The SAR was specifically developed to provide high-resolution images from satellite platforms. Radar can be used for imaging the sea surface, with altimeters to map sea surface height, scatterometers to determine sea surface winds, and so on (Imhoff, 1995). Radar can penetrate fog and clouds, making it particularly valuable for emergency applications and in areas where cloud cover persists (Richards, 1990).

Airborne LiDAR (light detection and ranging) has become quite useful for topographic and bathymetric mapping. Laser profilers are unique in that they confine the coherent light energy in a very narrow beam, providing pulses of very high peak intensity. This enables LiDARs to penetrate moderately turbid coastal waters for bathymetric measurements or gaps in forest canopies to provide topographic data for DEMs (Mallet & Bretar, 2009). The LiDAR is costly but the most accurate method to acquire digital surface models (DSMs) of the Earth surface, and hence it could be used to develop 3D models of buildings, trees, and so forth. (Chang et al., 2015).

The UAV/drone is a miniature remotely piloted aircraft. It is designed to fulfill the requirements for a low-cost platform, with long endurance, moderate payload capacity, and the capability to provide very high resolution data (LiDAR or image), especially to inaccessible sites (Garg, 2021). An onboard computer controls the payload and stores data from different sensors and instruments mounted on the the UAV. The drone was developed in Britain during World War II for military reconnaissance, but now it plays an important role in remote sensing. The unique advantage is that it could be used to locate the area for which data was acquired, and it is capable to provide both night and day data.

Figure 4.1 shows various platforms used in remote sensing for data collection. As the platform height increases, the spatial resolution and observational area increases. Thus, the higher the sensor is mounted, the larger the spatial resolution and view.

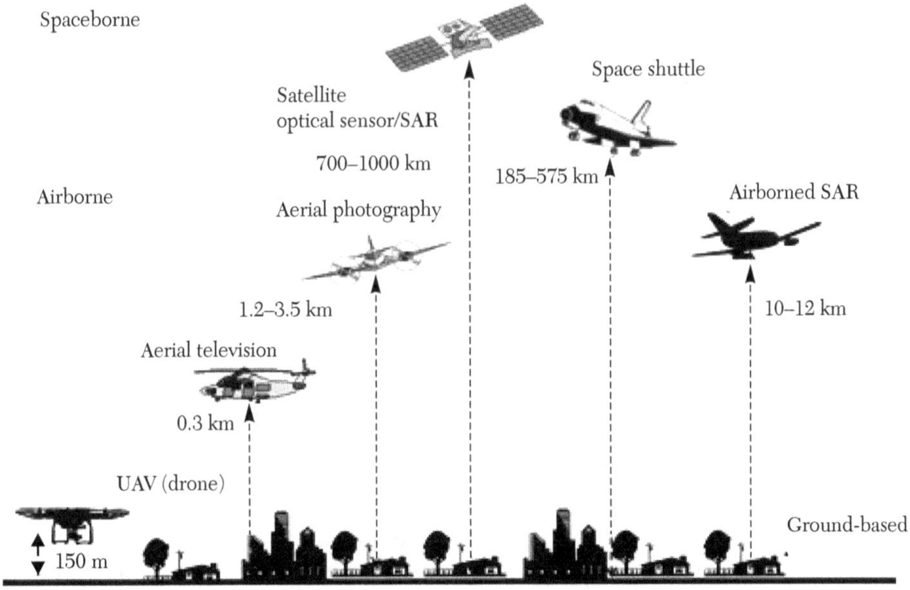

Figure 4.1. Various platforms used in remote sensing (Yamazaki & Liu, 2016).

4.2 TYPES OF SATELLITE ORBITS

An orbit is a curved path of a celestial object around another celestial object due to the force of gravity. The Moon orbits the Earth, and the Earth orbits the Sun, and the Sun orbits around the center of the galaxy. A spaceborne remote sensing platform is placed in an orbit where it moves continuously. From a geometrical characteristic point of view, orbits of the spaceborne platform can be circular, elliptic, parabolic, or hyperbolic. In practice, it is difficult to establish and maintain an exactly circular orbit, therefore, the so-called *nominally circular orbits* are slightly elliptical. The parabolic and hyperbolic orbits are not used for terrestrial remote sensing.

The orbits can be broadly classified as *geosynchronous* and *sun-synchronous*, as shown in Figure 4.2 (Garg, 2019). The images from geosynchronous satellites cover much larger areas at poor resolution (≈1.1 km), and as such these images are more useful in communication and weather-related studies. Remote sensing applications generally use images from near-polar, sun-synchronous satellites, as they provide global coverage of natural resources. Such images provide good spatial resolution, which is required for resource surveys and mapping.

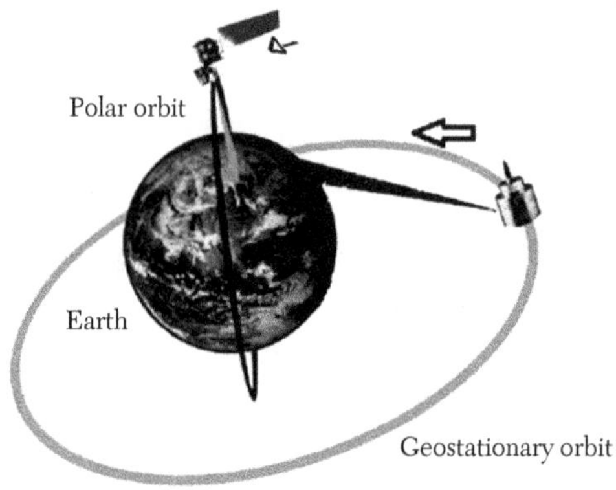

Figure 4.2. Geostationary and polar orbits.

4.2.1 Geosynchronous Orbits

The satellites moving in geosynchronous orbits are also called *geostationary* satellites (Gibson, 2000). These satellites remain fixed over a single spot, and always view the same region for their entire life span, but cannot cover the high latitudes. Geosynchronous satellites are not really stationary, but appear to be stationary when viewed from Earth because they move in the orbit at the same speed as the rotation of the Earth about its polar axis. It is an extremely useful orbit, which may be used for anything where a satellite needs to send or receive signals from the same part of the Earth all the time. These satellites revolve in the same direction as that of the Earth (west to east) at an angular velocity equal to the Earth's rotation rate. In order to achieve this orbital period, geosynchronous satellites move at higher altitude of 36,000 km above the equator, making an inclination of 0° from the equatorial plane (Figure 4.2). The geosynchronous satellites are mainly used for weather forecasting, satellite TV, and communications (Richards & Jia, 2013).

The INSAT (Indian Satellite) satellite series, launched by the Indian Space Research Organization (ISRO), Department of Space, Government of India, is an example of these satellites. Other satellites are Meteosat, GOES, GMS, and so on. Satellites in the geosynchronous orbit are located at any particular longitude to get a constant view of that particular region, thus providing frequent images of that area in a day. For example, the GOES East and GOES West satellites are placed in orbits over North America (normally at 75° W and 135° W, respectively), the INSAT 3-A is located at 93.5o E longitude over India, the Japanese Geo-stationary Meteorological Satellite (GMS) is located over New Guinea, and Meteosat is located over Europe.

4.2.2 Sun-Synchronous Orbits

Satellites in sun-synchronous orbits usually travel past Earth from north to south rather than from west to east, passing roughly over the Earth's poles (Figure 4.2). The satellites in such an orbit are also called *polar* satellites as their orbital plane is passing close to the poles (Colwell, 1983). These satellites orbit at 700–1000 km over the Earth's surface with a rotation period typically varying from 90–103 minutes (Campbell, 2011). The orbit has a small angle of inclination with the north–south line, maintaining nearly a 3° angle. The sun-synchronous satellites often try to maintain the same angle with respect to the Sun to get illumination of the Earth just below it. This means they are synchronized to always be in the same "fixed" position relative to the Sun. The imaging is done during mid-morning, when there is minimum cloud formation and a pollution free environment. They will make about a dozen orbits every day.

Almost all the satellites used for remote sensing work move in polar orbits, and use optical sensors. The NOAA (National Oceanic and Atmospheric Administration) series of satellites, like NOAA-17, NOAA-18,

IRS, Landsat, SPOT, and IKONOS, are all examples of polar orbiting satellites. The sensors in these satellites capture images of the Earth during daylight only, and will always observe a point on the Earth almost at the same time after a few days. The availability of multiple images for a region serves a number of applications, for example, change analysis over time.

The sun-synchronous satellite retraces its path when an orbital cycle is completed. The orbital cycle of a satellite is the number of orbits required to compete the entire coverage of the Earth once. Due to the rotation of the Earth on its own axis and the travel of the satellite from the north to south poles, each time the satellite moves in a new orbit it observes a new area of the Earth down below. The satellite's orbit period, orbit inclination, and the rotation of the Earth together allow complete coverage of the Earth's surface (Garg, 2019). After a few days when the orbital cycle is completed, it again starts collecting repeat images of the Earth, known as the *revisit period*. The revisit period or temporal resolution is the time elapsed between two successive views of the same area by the same satellite.

Satellites with steerable sensors, such as SPOT, can view off-nadir areas before and after the orbit passes over a ground. Due to off-nadir viewing capability of the satellites, the revisit period can be reduced considerably. This is important when frequent images of an area are required to be analyzed, such as during floods, earthquakes, wildfires, and so forth. In near-polar orbits, areas at high latitudes will be imaged more frequently than the equatorial zone due to the increasing overlap in adjacent swaths as the orbit paths come closer together near the poles.

4.3 PATH-ROW REFERENCE

Since the reflected radiations from the ground are continuously recorded by the sensor, each orbit is given a unique number for identification of the ground scene/area, called the *Path number*. The length of the scene across the orbit path defines the *Row number*. Thus, each scene/image can easily be located with its unique *Path-Row number*, as shown in Figure 4.3. Thus, the whole world is divided into unique Path-Row numbers for different satellites as well as different sensors. These continuous signals are transmitted to ground receiving stations where the signals are truncated as per dimensions of the scene/image (defined by the swath width). These signals are processed there before supplying the scenes/image to the users.

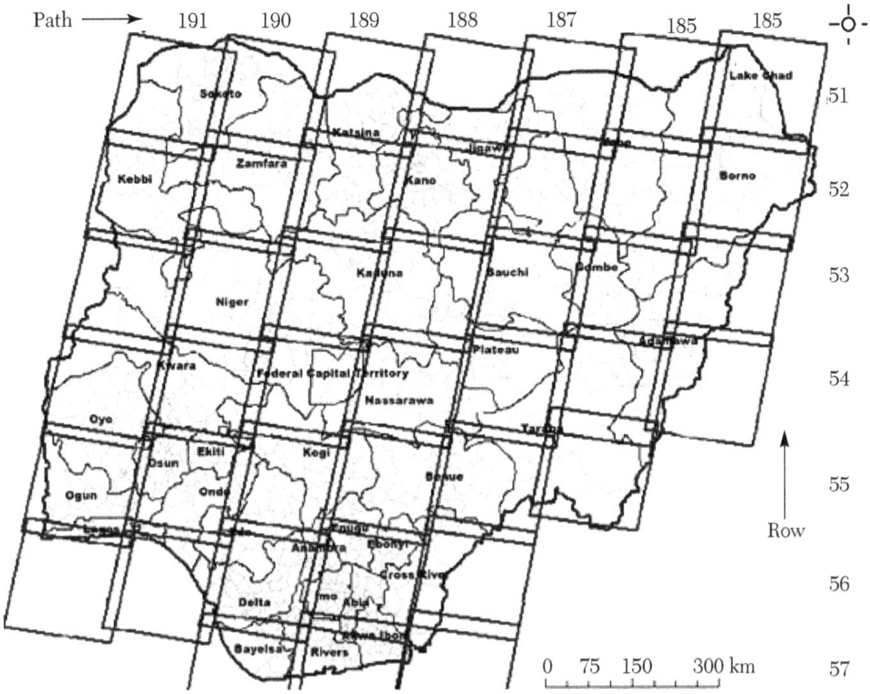

Figure 4.3. Landsat path-row reference map of Nigeria.

These Path-Row numbers are significant and required to be mentioned when placing the order for procurement of satellite scenes from the data receiving agency.

4.4 VARIOUS SATELLITE PLATFORMS

Satellite images from various platforms are obtained at different heights with varying spatial resolution from 1 km to <1 m. These platforms can be classified as per their data resolutions: (*a*) low resolution, (*b*) medium resolution, and (*c*) high/very high resolution. In addition, the platforms are classified as per wavelength employed by their sensors, such as hyperspectral systems, thermal systems, and microwave systems. Figure 4.4 presents various satellite platforms which have been launched by various countries to capture that data for the Earth's surface. Some Earth resource satellites are explained as follows:

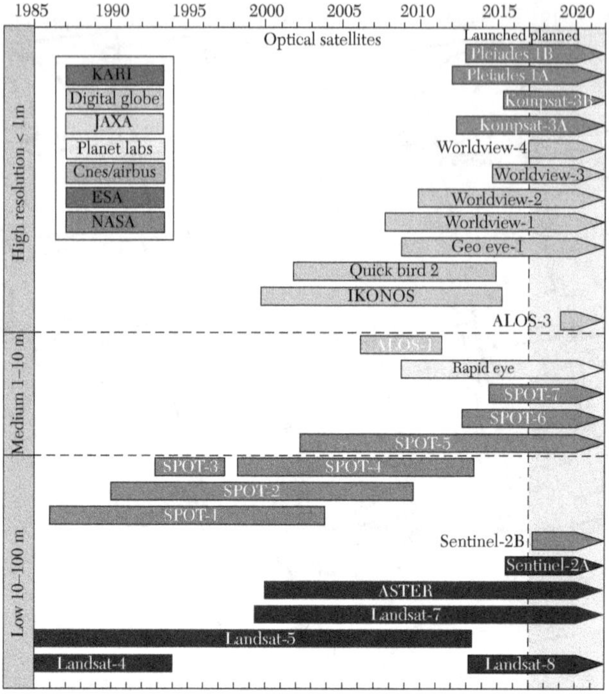

(a)

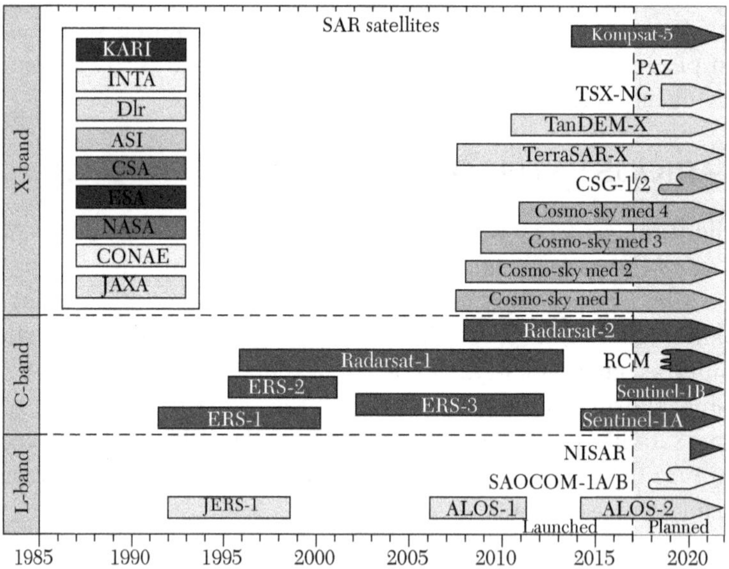

(b)

Figure 4.4. (a) Optical and (b) SAR satellites launched for systematic and global coverage (Elliott et al., 2016).

4.4.1 Low Resolution Satellites

Low-resolutions satellites provide images that cover large areas; the larger the swath width, the poorer the spatial resolution. But, a lower resolution satellite usually gives a higher temporal resolution, that is within a short interval the satellite completes and repeats the same area (Meteosat-8 with 3 km spatial resolution for every 15 min). For example, the NOAA-GOES collects TIR data at a spatial resolution of 8 km for the entire Earth every 30 minutes, both day and night. A lower resolution image may be used to map large or national or global areas for climate-related studies, agricultural mapping, forest fires, weather monitoring, snow cover, observations of land and oceans, ocean's ice cover, and land surface temperatures. Some of the characteristics of low-resolution satellites/sensors are summarized in Table 4.1.

Table 4.1. Low Resolution Images from Various Satellites/Sensors

Sensor/Satellite	Year of launch	Spectral bands	Spatial resolution (km)
GOES (NOAA)	1975	1 TIR	8.0
AVHRR (NOAA)	1978	1 TIR	(LAC) and 4.0 (GAC)
VEGETATION (SPOT-4)	1988	4	1.15
ATSR-1 (ERS-1)	1994	4	1.0
ATSR-2 (ERS-2)	1995	6	1.0
SeaWiFS (OrbView-2)	1997	8	1.1–4.5
MODIS (EOS-1)	2000	30	0.25, 0.50, and 1.0
MERIS (ENVISAT)	2000	15	1.2
GERB (Meteosat MSG)	2002	7	40.0
SEVIRI (Meteosat MSG)	2002	12	1.0, 3.0
EPIC (DSCOVR)	2015	10	25.0
Tropomi (Sentinel-5P)	2017	7	3.5, 7.0
SLSTR (sentinel 3B)	2018	9	0.5, 1.0

Note: LAC-local area coverage, GAC-global area coverage, GOES-Geo-stationary Operational Environmental Satellite.

4.4.1.1 NOAA-AVHRR systems

The U.S. NOAA (National Oceanic and Atmospheric Administration) carried onboard the AVHRR (Advanced Very High Resolution Radiometer) which operates at an average altitude of 833 km. It began with the launch of TIROS-N in 1978. It acquires data along a 1.1 km wide swath each day in five spectral bands: red, near-infrared, and three thermal infrared bands. The AVHRR data are acquired in three formats: (*a*) full resolution image data of 1.1 km, transmitted to a ground station as they are collected, (*b*) LAC (Local Area Coverage) data, which are also full resolution data at 1.1 km that are recorded on an on board tape for subsequent transmission during a station overpass; and (*c*) GAC (Global Area Coverage) data that are derived from a sample averaging of the full resolution AVHRR data to study very large areas (Gibson, 2000). The AVHRR may provide high spatial resolution data for meteorological applications, but the images portray only broad patterns and little detail for terrestrial studies. However, they do have a high temporal resolution, showing wide areas on a daily basis, and are therefore used for monitoring large areas, and developing many early warning activities (Garg, 2019). Table 4.2 shows the development of various NOAA-AVHRR satellites.

Table 4.2. Launch Dates of Various NOAA-AVHRR Satellites

Satellite	Start date	End date
TIROS-N	Oct 19, 1978	Jan 30, 1980
NOAA-6	June 27, 1979	Nov 16, 1986
NOAA-7	Aug 24, 1981	Feb 1, 1985
NOAA-8	May 3, 1983	Oct 31, 1985
NOAA-9	Feb 25, 1985	Nov 7, 1988
NOAA-10	Nov 17, 1986	Sept 16, 1991
NOAA-11	Nov 8, 1988	June 16, 2004
NOAA-12	Sept 17, 1991	Aug 10, 2007
NOAA-14	April 10, 1995	May 23, 2007
NOAA-15	Dec 15, 1998	Present
NOAA-16	March 20, 2001	Present
NOAA-17	Oct 15, 2002	Present
NOAA-18	Aug 30, 2005	Present
MetOp-A	May 20, 2007	Present
NOAA-19	June 2, 2009	Present
MetOp-B	Sept. 17, 2012	Present

4.4.1.2 Terra-MODIS

Terra was launched into a sun-synchronous orbit on December 18, 1999, and started sending data back to Earth in February 2000. It is a series of spacecraft that represent the next landmark step in NASA's role to observe

Earth from the unique vantage point of space. It carries five instruments: ASTER, CERES, MISR, MODIS, and MOPITT (Morisette et al., 2002). The MODIS, (Moderate Resolution Imaging Spectroradiometer) is a key instrument aboard the Terra (EOS AM) and Aqua (EOS PM) satellites. Its orbit around the Earth is synchronized so that it passes from north to south across the equator in the morning, while Aqua passes south to north over the equator in the afternoon. Terra-MODIS and Aqua MODIS view the entire Earth's surface every 1–2 days, acquiring data in 36 multispectral spectral bands, at spatial resolutions of 250, 500, and 1000 m. Terra enables new research into the ways Earth's land, oceans, air, ice, and life function as a total environmental system.

4.4.2 Medium Resolution Satellites

Under this category, there are large number of satellite platforms; such as LANDSAT, SPOT, IRS, Sentinel, and so forth. Image processing started developing with the launch of the American LANDSAT-1 in 1972. The satellite provided global observations of land surfaces, mapping and assessments of tropical rain forests, and land cover changes. The applications were further enhanced with the launch of the French SPOT satellite in 1986, which provided PAN and MS images as well as offered stereoscopic possibilities. It was further improved with the Indian IRS-1A, launched in 1995. Since then, there is a large variety of imaging sensors in space which provide medium resolution images; some of them are described as follows.

4.4.2.1 LANDSAT systems

The LANDSAT (Land Satellite) is one of the longest running space satellite programs of the United States for the acquisition of Earth images. At the time of launch in July 1972, it was called ERTS (Earth Resources Technology Satellite), which was renamed as Landsat-1 in 1975. Since then, a series of Landsats has been launched, which provided a vast amount of medium resolution images for various applications, such as land use/ land cover, vegetation and agricultural crops, water, soils, geology, and other resources (Garg, 2019). It is considered as one of the most successful missions, with demand of images from all over the world (*https://Landsat.usgs.gov/*). The development of Landsat is given in Figure 4.5.

The launches of Landsat-2 and Landsat-3 followed in 1975 and 1978, respectively. The first three satellites were identical and their payloads consisted of an MSS (Multispectral Scanner) and two video cameras called

RBVs (Return-Beam Vidicons). One scene covered an area of 170 km × 185 km, as the satellites operated at an altitude between 907–915 km in sun-synchronous polar orbit with 103 minutes of orbital period and 18 days of revisit time. The Landsat-4 was launched in 1982. The Landsat-5 was launched in 1984, and it continued to deliver high-quality global data of the Earth's surfaces (Wulder et al., 2008). The Landsat-4 and -5 were equipped with two multispectral sensors, that is an MSS and a TM (Thematic Mapper). The altitude of the orbit was 705 km with 99 minutes orbital period, and a revisit time of 16 days. Other parameters were the same as earlier Landsat satellites.

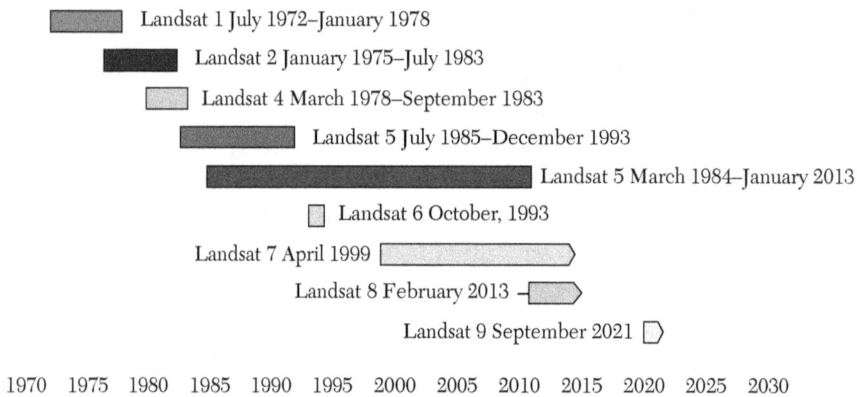

Figure 4.5. History of Landsats.
Source: *https://Landsat.usgs.gov/Landsat-missions-timeline*

The Landsat-6 was launched in October 1993, but was lost during the launch. The Landsat-7 was successfully launched in April 1999 and has 8 separate spectral bands with spatial resolutions ranging from 15–60 m, and a temporal resolution of 16 days. It was equipped with a multispectral sensor known as the ETM+ (Enhanced Thematic Mapper Plus). The ETM+ also provides data in one PAN band. Other specifications of the satellite were the same as the Landsat-4 and -5 satellites.

The Landsat-8 was launched in 2013, which continued to provide global data. It was launched in sun-synchronous orbit at an altitude of 705 km, circling the orbit every 98.9 minutes, covering the entire globe every 16 days (Trinh et al., 2017). It has two sensors: the OLI (Operational Land Imager) operating in nine spectral bands, with 30 m resolution and 15 m in the PAN band, and the TIR sensor operating in two spectral bands (10.6–11.19 μm and 11.5–12.51 μm) with 100 m spatial resolution. The Landsat-9 was launched in September 2021. Further details of Landsats are given in Table 4.3.

Table 4.3. Details of some medium resolution satellites

Satellite	Launch date	Sensors	Spatial resolutioin	Swath width (km)
Landsat-1	07-23-1972	RBV MSS	80 80	185
Landsat-2	01-22-1975	RBV MSS	80 80	185
Landsat-3	03-05-1978	RBV MSS	40 80	185
Landsat-4	07-16-1982	MSS TM	80 30 and 120	185
Landsat-5	03-01-1984	MSS TM	80 30 and 120	185
Landsat-6	10-05-1993	ETM	could not achieve the orbit, and hence, failed	
Landsat-7	04-15-1999	ETM+	30, 60 and 15	185
Landsat-8	02-11-2013	OLI TIRS	30 and 15 100	185
SPOT-1	02-21-1986	HRV PAN	20 10	60
SPOT-2	01-22-1990	HRV PAN	20 10	60
SPOT-3	09-25-1993	HRV PAN	20 10	60
SPOT-4	03-24-1988	HR-VIR	20	60
SPOT-5	05-04-2002	HRG	10 and 5	60
SPOT-6	09-08-2012	HRG	1.5	60
SPOT-7	06-30-2014	HRG	8.0	60
Envisat	03-01-2002	MERIS	300	1150
Sentinel-3 A	02-16-2016	OLCI	300	1270

4.4.2.2 SPOT systems

The SPOT (Système Pour l'Observation de la Terre; i.e., System for Earth Observation) system is a series of high-resolution optical imaging Earth satellites that were initiated by the CNES (Centre National d'études Spatiales–the French Space Agency). The SPOT-1 was launched on February 21, 1986 in the sun-synchronous orbit at an altitude of 832 km with an inclination of 98.7° from the equator. It provided data in the PAN band at 10 m spatial resolution and the MS in three band images at 20 m spatial resolution with a repeat period of 26 days (Holand et al., 2003). The SPOT-2 was launched on January 22, 1990. The SPOT-3, launched on September 25, 1993, could not operate for long due to problems with stabilization. SPOT images have been used to explore the Earth's resources, detecting phenomena, such as climatology and oceanography, monitoring human activities and natural phenomena, updating topographic maps, creating 3D models and DEMs, and so forth.

The SPOT satellites provided improved spatial resolution images over the Landsat images. SPOT has the capability to collect stereo-images for creation of a 3D model of Earth. The overlap images of the area are captured on successive days by the same satellite by viewing the area off-nadir up to an angle of ±27° from the satellite's vertical axis (Welch & Ehlers, 1987). It can be remotely steered from ground control stations, thus reducing the temporal resolution from 26 days to within 10 days. Such stereo-images are very useful to create a 3D model of the area (e.g., DEM). The off-nadir viewing capability was a unique feature of these satellites, as shown in Figure 4.6 where from 4 different days (D-5, D, D+5, D+10) the satellite can capture the images of the same area at an interval of 5 days (Garg, 2019). However, the disadvantage of off-nadir viewing is that the satellite will miss the images of areas just below it. The availability of stereo-imagery is therefore very limited, as their collection requires demand from the users and/or special control on the satellite.

The SPOT-4 satellite, with same geometric imaging characteristics as the earlier SPOT satellites, was launched in March 1988. It operated at an altitude of 822 km with a period of revolution of 101 minutes. The SPOT-5, launched in May 2002, provided further improved images.

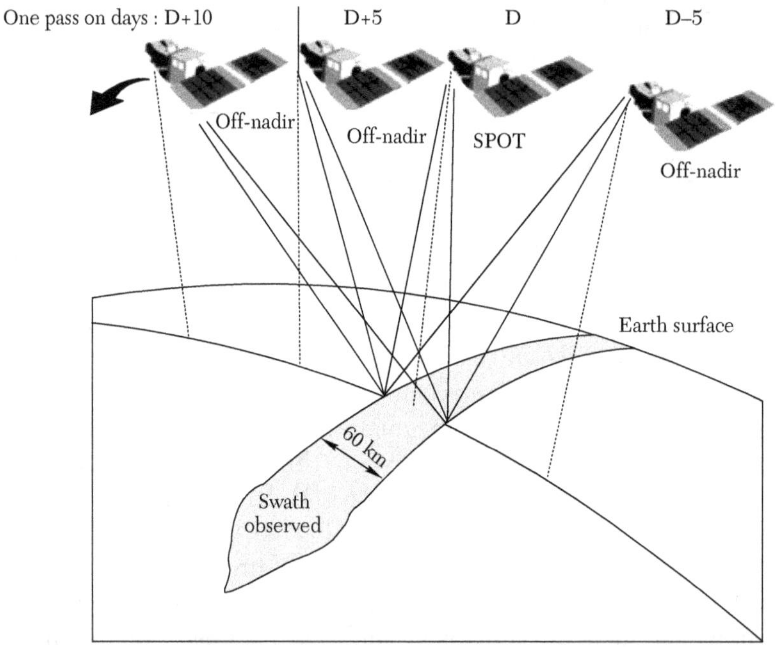

Figure 4.6. Off-nadir viewing capabilities of SPOT.

The SPOT-6 and SPOT-7, launched on September 8, 2012 and June 30, 2014, respectively, provided continuity of high-resolution images. The SPOT-6 satellite provided PAN (0.45–0.75 μm) at 1.5 m resolution, and MS in blue (0.45–0.53 μm), green (0.53–0.59 μm), red (0.62–0.69 μm), and NIR (0.76–0.89 μm) bands at 8 m resolution, covering an area 60 km × 60 km, as well as color merged products at 1.5m, resolution (*https://crisp.nus.edu.sg/~research/tutorial/spot.htm*). The SPOT-6 and SPOT-7 have identical sensors and operating characteristics, and offer the capability to provide a daily revisit everywhere on Earth with a 60 km × 60 km swath. Further details of SPOT satellites are given in Table 4.3.

4.4.2.3 IRS systems

India in the early experimental phase, launched Bhaskara-1 on June 7, 1979 and Bhaskara-2 on November 20, 1981. These satellites carried two types of sensor systems: a television camera with a spatial resolution of 1 km operated in visible and NIR bands, and SAMIR (Satellite Microwave Radiometer) for oceanic and atmospheric applications. Following their success, India initiated an indigenous IRS (Indian Remote Sensing Satellite) program to support the areas of agriculture, soil, water resources (surface and ground), forestry and ecology, geology and mineral resources, cartography, rural and urban development, marine fisheries, watershed and coastal management, and many more (ISRO, 2018). The NRSC (National Remote Sensing Centre), Hyderabad, is the focal point for distribution of various remote sensing satellite data products in India and its neighboring countries. The NRSC has an earth receiving station at Shadnagar, about 55 km from Hyderabad, to receive data from almost all contemporary remote sensing satellites. The NDC (National Data Center) is a one-stop shop for procuring a range of data products with a wide choice of satellite data.

The successful launch of IRS-1A on March 17, 1988 into a sun-synchronous orbit was a major milestone in the IRS program for managing the natural resources of the nation. The IRS-1A was followed by the launch of IRS-1B on August 29, 1991, an identical satellite. The IRSs operated at a nominal altitude of 904 km, crossing the equator at 10:26 a.m. with an orbital period of 103.2 minutes and a repeat cycle of 22 days. These satellites carried two sensors, LISS-I and LISS-II (Linear Imaging Self-Scanning), that operated in visible and NIR bands with spatial resolutions of 72.5 m and 36.25 m, respectively, providing 11 days repetivity (Muthugadahalli, 1996). The LISS-I had a swath width of 148 km on the ground. The LISS-II had two separate imaging sensors, LISS-II A and LISS-II B, with spatial resolution of 36.25 m, that are mounted on the spacecraft in such a way to provide a composite swath of 146.98 km on the ground. These sensors also operated in the visible and NIR bands. These satellites provided very useful data for applications in the fields of land use and land cover mapping, agriculture, forestry, hydrology, oceanography, geology, natural resource management, disaster monitoring, cartography, and many more (Reddy, 2012).

Since then, a series of IRS spacecraft was launched with enhanced capabilities in payloads and platforms. The IRS system has one of the largest constellations of remote sensing satellites in operation in the world today. These satellites with imaging capabilities in visible, IR, thermal, and microwave regions have helped the country in realizing major operational applications. The imaging sensors have been providing spatial resolution ranging from 1 km to better than 1 m and repeat images from 22 days to 2 days and radiometry ranging from 7-bit to 12-bit (Muthugadahalli, 1996). In addition, the Indian space missions based on specific thematic applications, like natural resources monitoring, ocean and atmospheric studies, and cartographic applications, resulted in the development of theme-based satellite series, such as (*a*) land/water resources applications (Resourcesat series and RISAT series); (*b*) ocean/atmospheric studies (Oceansat series, INSAT-VHRR, INSAT-3D, Megha-Tropiques, and SARAL); and (*c*) large-scale mapping applications (Cartosat series; ISRO, 2018). Cloud cover remains a major hurdle in optical remote sensing during monsoon season. Imaging RADAR applications are being served through the RISAT-1 satellite. The C-band SAR sensor onboard the RISAT-1 satellite is found to be very useful for resource monitoring and disaster monitoring. The L-band SAR is suggested on the RISAT-3 mission for applications supporting soil moisture, crops, and forest type discrimination. In addition, RISAT-1A (C band) and RISAT-2A (X band) are planned to meet the needs of monitoring disaster situations.

In the IRS-P2 satellite, the LISS-IIM (Linear Imaging Self-Scanning System-II Modified) sensor was used, which operated in pushbroom scanning mode in four spectral bands in visible and NIR bands. To further improve the quality of data, IRS-1C and -1D, second-generation identical satellites, were launched in 1995 and 1997, respectively (Puckorius, 2018). The IRS-1C and -1D both used three sensors, LISS-III, PAN camera (0.5–0.75 μm) and a WiFS (Wide Field Sensor), with spatial resolution of 23.5 m, 5.8 m, and 188 m, respectively. These sensors operated in pushbroom scanning mode. The LISS-III sensor operates in the VNIR and SWIR bands with a spatial resolution of 23.5 m and 70.5 m, respectively (Garg, 2019). The WiFS camera provides two spectral bands in the VNIR range with spatial resolution of 188 m and temporal resolution of 5 days. The PAN and

WiFS sensors further strengthened application areas like resources survey and management, forest studies and disaster monitoring, environmental studies, cartography, and town planning.

The experimental earth observation satellite IRS-P3 used WiFS and MOS (Multispectral Optoelectronic Scanner) and IXAE (Indian X-ray Astronomy Experiment) experimental sensors. The Oceansat-1 (or IRS-P4), launched on May 26, 1999, carries an OCM (Ocean Color Monitor) and an MSMR (Multifrequency Scanning Microwave Radiometer) to study the physical and biological aspects of oceanography. The OCM operates in 0.402–0.422, 0.433–0.453, 0.480–0.50, 0.50–0.52, 0.545–0.565, 0.66–0.689, 0.745–0.785 and 0.845-0.885 μm bands with 360 m spatial resolution and 1420 km swath, covering the entire country every 2 days (Reddy, 2012). The MSMR is a dual-polarized four-frequency radiometer measuring microwave brightness temperature. The Oceansat-2, launched on September 23, 2009, carries three payloads: OCM, Ku-band Pencil Beam scatterometer (SCAT), and ROSA (Radio Occultation Sounder for Atmosphere). The Oceansat-3 with enhanced imaging capability planned to be launched in 2021. It will carry a TIR sensor, 12 channels OCM, a scatterometer and a passive microwave radiometer for simultaneous measurement of ocean color and sea surface temperature (ISRO, 2018). The IR sensor and OCM would be used for the analysis of operational potential fishing zones.

The Cartosat-1 or IRS-P5 is a state-of-art satellite which carries three or two cameras having different nadir angles in orbit direction for the generation of stereo-models required for cartographic applications. The unique high-resolution along-track stereo-imaging capability, carried out for the first time anywhere in the world, enabled the generation of DEM and other value-added products from Cartosat-1 (ISRO, 2018). It carries two PAN cameras. PAN-F (forward) and PAN-A (after), to obtain two-line stereo-configuration for terrain modeling. Each camera operates in spectral band 0.5 to 0.85 μm, with a spatial resolution of 2.5 m, a swath width of 30 km for stereo-imagery, and a radiometric resolution of 10 bits. These cameras are mounted on the satellite in such a way that near simultaneous imaging of the same area from two different angles is taken, which facilitates the generation of 3D models of the terrain. The Cartosat-1 data products are available as a standard product which is radiometrically corrected, and a georeferenced and precision product which is an ortho-rectified product (Puckorius, 2018). The images are useful for large-scale mapping applications for urban and rural development, land and water resources management, disaster assessment, relief planning and management, environment impact assessment, and various other GIS-based applications.

Besides generation of large-scale topographic maps, the data is also used for updating the topographic maps.

The Cartosat-2 was launched on January 10, 2007 with capability of steering along and across the track up to ±45° to facilitate the imaging of any area more frequently. It carries a single PAN camera capable of providing better than 1 m spatial resolution imagery, with a swath of 9.6 km. It has revisit periods of 4 days and 1 day (Garg, 2019). The Cartosat-2A, launched on April 28, 2008, can also provide scene-specific imagery with the PAN camera, with the same specifications as Cartosat-2. Imageries from this satellite are also used for cartographic applications like mapping, urban and rural infrastructure development, and management in GIS. The Cartosat-2B was launched on July 12, 2010, with a PAN camera of spatial resolution of about 0.8 m. The Cartosat-3 was launched on November 27, 2019, having the capability of imaging with 0.25 m in PAN, and 1 m in MS modes (4 bands) with a swath of 16 km and a resolution of 12 m with 5 km swath in hyperspectral mode (Reddy, 2012). It is a third-generation agile advanced satellite having high resolution imaging capability. It also features a MWIR camera with 5.7 m resolution. Images from this satellite are used for cadastre, infrastructure mapping and analysis, disaster monitoring, and damage assessment. In 2021, there is also a plan to launch two similar devices, Cartosat-3A and 3B.

The Resourcesat-1 or IRS-P6, the tenth satellite in the IRS series, was launched in October 2003. It operated at 820 km height, and carried three pushbroom sensors: LISS-III, LISS-IV, and AWiFS (Advanced Wide Field Sensor). The LISS-III camera is identical to the one used in IRS-1C/1D. The LISS-IV is a MS high-resolution three bands sensor operated in visual and NIR bands with a spatial resolution of 5.8 m. It can be operated in MS mode, that is data collected in 3 bands, and mono-mode, that is data collected in a single selected band corresponding to a swath of 70 km (ISRO, 2018). The AWiFS is a wide-angle medium resolution (56 m nadir and 70 m at swath) camera, with swath width of 804 km, and a temporal resolution of 5 days. With the enhanced multispectral/spatial coverage and stereo-imaging capabilities, studies such as improved crop discrimination, crop yield estimation, forestry and disaster management, and land water resources management can be carried out in advanced areas. The Resourcesat-2 is the eighteenth remote sensing satellite, launched on April 20, 2011, to continue providing remote sensing data to global users provided by Resourcesat-1, with enhanced multispectral and spatial coverage. Important changes in Resourcesat-2 as compared to Resourcesat-1 are: enhancement of LISS-IV, MS swath from 23 km to 70 km, improved radiometric accuracy from 7 bits to 10 bits for LISS-III and LISS-IV and 10 bits to 12 bits for AWiFS (Puckorius, 2018). It also carries an additional payload known as AIS (Automatic Identification System) as an experimental payload for ship surveillance in the VHF band to derive position, speed, and other information about ships. The Resourcesat-3, a medium resolution satellite launched on November 17, 2019 at 795 km altitude, ensures continuity of data required for land and water resources management. It carried an ALISS-III (Advanced Linear Imaging Scanning Sensor-III) payload consisting of visual and NIR and SWIR bands with 20 m spatial resolution, and an ACS (Atmospheric Correction Sensor) for quantitative interpretation and geophysical parameter retrieval. Hyperspectral atmospheric correction bands operate in visual and NIR bands with a spatial resolution of 240 m. The LISS-III sensor would be modified into LISS-III-WS (Wide Swath) having swath around 925 km and revisit capability better than AWiFS in a future Resourcesat-3/3A mission. New configurations in Resourcesat-3/3A will help in overcoming the spatial resolution limitation of AWiFS. Resourcesat-3S, with multi-view panchromatic stereoscopic capability having high spatial resolution of 1.25 m, will facilitate the generation of fused products having great utility.

Additional details of IRS satellites are given in Table 4.4. Presently, India has the world's largest constellation of remote sensing satellites in operation, as shown in Figure 4.7. It provides space-based remote sensing data in a variety of spatial, spectral, and temporal resolutions, meeting the needs of many applications of relevance to national development as well as global applications.

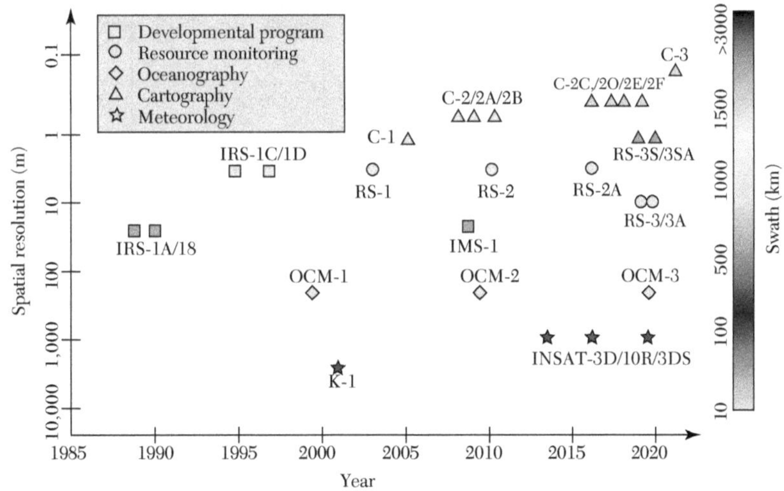

Figure 4.7. Indian remote sensing satellites suitable for various applications.

Table 4.4. Characteristics of IRS satellites

Satellite	Sensors	Resolution			Orbit Characteristics
		Spectral bands (μm)	Spatial (m)	Temporal (days)	
IRS-1A	LISS-I, and LISS-II A/B (3 sensors)	Blue (0.45–0.52) Green (0.52–0.59) Red (0.62–0.68) NIR (0.77–0.86)	72.5 m LISS-I 36 m LISS-II	22	Altitude–904 km; Equatorial crossing–10:26 a.m. Orbit Period–103.2 minutes, Radiometric resolution–7-bit
IRS-1B	LISS-I and LISS-II	Same as IRS-1A	Same as IRS-1A	Same as IRS-1A	Same as IRS-1A
IRS-1C & 1D	LISS-III	Green (0.52–0.59) Red (0.62–0.68) NIR (0.77–0.86) NIR (1.55–1.70)	23.5 23.5 23.5 70	24	Altitude–904 km, Equatorial crossing–10.30 a.m. Orbit Period–101.23 min., Radiometric resolution–7-bit, PAN–6 bit
	PAN	(0.50–0.75)	5.8	24 (5)	
	WiFS	Red (0.62–0.68) NIR (0.77–0.86)	188	5	
IRS-P3	WiFS	Red (0.62–0.68) NIR (0.77–0.86) IR (1.55–1.70)	188	5	Equatorial crossing–10:30 am, Altitude–817 km, Orbit period–101.35 min, Radiometric resolution–7-bit
	MOS-A MOS-B MOS-C	(0.75–0.77) (0.41–1.01) (1.595–1.605)	1500 520 550	Ocean surface	

Satellite	Sensors	Resolution			Orbit Characteristics
		Spectral bands (μm)	Spatial (m)	Temporal (days)	
IRS-P4 (Oceansat-1)	OCM MSMR	(0.40–0.90) 6.6, 10.65, 18, 21 GHz (frequencies)	360 × 236 105 × 68, 66 × 43, 40 × 26, 34 × 22 km for frequency sequence)	2 2	Altitude–720 km, Orbit period–99.31 min, Equator crossing–12:00 noon, Radiometric resolution–12 bit.
IRS-P6 Resourcesat-1	LISS-IV	Green (0.52–0.59) Red (0.62–0.68) NIR (0.77–0.86)	5.8 5.8 5.8	24 (5)	Altitude–817 km, Orbit period–101.35 min, Equator crossing–10:30 a.m. Radiometric resolution–10 bit
	LISS-III	Same as IRS-1C	23.5 23.5 23.5 23.5	24	
	AWiFS	Red (0.62–0.68) NIR (0.77–0.86) NIR (1.55–1.70)	70 70 70	5	
IRS-P5	PAN-F	(0.50–0.75)	2.5	2-line stereo-camera	Altitude–618 km, Orbit period–97 min, Equatorial crossing-10:30 a.m. Radiometric resolution–10 bit
Cartosat-1	PAN-A	(0.50–0.75)	2.5		
Cartosat-2	PAN	(0.50–0.85)	< 1	4–5 days	Altitude–644 km (400. Inclination 98°, period: 97.40 min., Swath-9.6 km

(Contd.)

Satellite	Sensors	Resolution			Orbit Characteristics
		Spectral bands (μm)	Spatial (m)	Temporal (days)	
Cartosat-3	PAN	(0.45–0.9)	0.25		Altitude–509 km at an inclination of 97.5°, Swath = 16 km
	MS	B1: 0.45–0.52 B2: 0.52–0.59 B3: 0.62–0.68 B4: 0.77–0.86	1.0		Swath–16 km
		MWIR	5.7		
	Hyperspectral		12.0		Swath–5 km
Oceansat-2	OCM SCAT ROSA	(0.40–0.90) 8 bands 13.515 GHz GPS occultation	360 × 236 25 km × 25 km	2	1420 1400
Resourcesat -2	LISS III, LISS IV, AWifs	Same as IRS P-6	23.8 m, 5.6 m, and 56 m	2–3 days for AWiFS, 12–13 days for LISS-III and 25–26 days for LISS-IV	Altitude–817 km, inclination–98.78°, period–101.35 minutes, local time of equator crossing at 10:30 AM on descending node.
Resourcesat -3	ALISS-III	NIR and SWIR	20 m		
	ATCOR		240 m		
RISAT-1	ScanSAR	Operating in C-band (5.35 GHz)	FRS-1: 3 m; FRS-2: 6 m; FRS-2: 9 m; MRS: 25 m; CRS: 50 m	25 days	Altitude–536 km, inclination–97.55°, period–95.49 minutes
SARAL	ALTIKA, ARGOS Data Collection System, Solid State C-band Transponder (SCBT)	Ka band Altimeter, C-Band Transponders		35 days	Altitude–781 km, inclination–98.54°, period–100.59 minutes

4.4.2.4 Sentinel systems

The Sentinel satellite is part of the Copernicus program–the European Space Agency (ESA) Earth Observation Programme, which aims to achieve an autonomous and operational Earth observation capacity. The program uses data provided by environmental satellites, and air and ground stations to provide a comprehensive set of land, ocean, emergency response, atmosphere, security, and climate change data in support of environment and security policy needs.

The Sentinel-1, launched on April 3, 2014, is a polar-orbiting, all-weather, day-and-night radar imaging mission for land and ocean applications. It provides global coverage every 12 days with high resolution images of 5 × 20 m for application areas, such as flood mapping, snow cover monitoring, soil moisture, surface topography, marine surveillance, sea-ice classification, vegetation analysis, and agricultural applications (ESA, 2019). For agricultural applications, radar data are an important addition to optical images in crop mapping. Thus, the free data policy of ESA regarding Sentinel-1 data is expected to improve crop mapping activities in cloud-affected areas where the lack of useful images during the cropping season often restricts mapping the spatial distribution of crops. The Sentinel-1 data are delivered within an hour of reception for NRT (near real-time) emergency response, within three hours for NRT priority areas, and within a day for systematically archived data. The data are acquired in different modes and delivered at different processing levels with varying resolutions. All data are delivered in the Standard Archive Format for Europe (SAFE) format. To facilitate processing, the ESA has developed the Sentinel toolbox for processing Sentinel data. Please see Chapter 7 for further details.

The Sentinel-2A, launched on June 23, 2015, is the first optical satellite from the Sentinel series under the Copernicus program. It is a sun-synchronous satellite operating at an altitude of 786 km, carrying a visible and NIR sensor. It provides MS images in 13 spectral bands which include four visible and NIR bands at 10 m resolution, six red-edge and SWIR bands at 20 m resolution, and three atmospheric correction bands at 60 m resolution (Drusch et al., 2012). The Sentinel-2 aims to provide images of global land surfaces, including major islands, with a high revisit frequency over Europe and Africa (ESA, 2019).

Both the satellites (Sentinel-1 and -2A) carry a multispectral optical sensor to provide data in continuity with the Landsat program and SPOT program. Compared to SPOT-4 and Landsat-8, Sentinel-2 has a higher spatial resolution and better spectral sensitivity. With the availability of several spectral bands of Sentinel-2 data, it is being used for a wide range of applications such as agriculture, forestry, land use/land cover analysis, risk and disaster mapping, coastal applications, and retrieval of critical vegetation and biophysical variables, such as LAI (leaf area index), NDVI (normalized difference vegetation index), fPAR (fraction of photosynthetically active radiation), leaf chlorophyll content, and leaf water content (Inglada et al., 2015).

The Sentinel-3A was successfully launched on February 16, 2016, and Sentinel-3B joined its twin in orbit on April 25, 2018. These satellites carry multi-payload packages to provide ocean observation data and land optical observation products. They carry several payloads: OCLI (Ocean and Land Color Instrument), SLSTR (Sea and Land Surface Temperature Instrument), and SRAL (SAR Radar Altimeter), and collect ocean color, surface temperature, sea surface topography, land surface color, temperature, and land ice topography data at a high revisit time of 2 days or less.

Sentinel-4 is to be launched in 2022 on a Meteosat third generation satellite, not an independent satellite. It will carry a UV-Visible-NIR spectrometer that will also use data from a TIR sounder and MTG (Meteosat Third Generation) cloud imager that are part of the Meteosat spacecraft, operated by the European Organization for the Exploitation of Meteorological Satellites (EUMETSAT, 2020). The mission aims to provide continuous monitoring of air quality as well as composition of the Earth's atmosphere at high temporal and spatial resolution to support monitoring and forecasting over Europe.

The Sentinel-5 Precursor, also known as Sentinel-5P, is the forerunner of Sentinel-5 to provide timely data on a multitude of trace gases and aerosols affecting air quality and climate. It was launched on October 13, 2017 to reduce data gaps between the Envisat satellite, in particular the Sciamachy instrument, and the launch of Sentinel-5. Sentinel 5 is to be launched in 2021 on the MetOp second generation spacecraft consisting of a UV-VIS-NIR and SWIR spectrometer and TIR sounder and imager. Sentinel-6, the sixth Sentinel satellite, launched on November 21, 2020, will carry a radar altimeter to measure global sea-surface height, primarily for operational oceanography and climatic studies. It will continue the precise ocean altimetry data provided by the Jason-2 and Jason-3 satellites (EUMETSAT, 2020).

Table 4.5 presents a summary of Sentinel satellites along with their application areas.

Table 4.5. Summary of Sentinel Satellites

Satellite/ Year	Sensor	Spatial resolution (m)	Application areas
Sentinel-1/ 2014	C-Band (SAR)	5–40	Ocean and ice monitoring, glacier flow velocity monitoring, snow classification, maritime monitoring, ship and oil detection, wind and wave information, ground moisture, ground movement (interferometry, et al. in earthquake and volcano regions), flood monitoring
Sentinel-2/ 2015	MSI (Multispectral Instrument), 13 spectral bands	10, 20 and 60	European, wide coverage, high resolution super-spectral mission for land observation data continuity, e.g., agricultural monitoring, areal extent of fires
Sentinel-3/ 2016	OLCI (Ocean and Land Color Instrument); SLSTR (Sea and Land Surface Temperature Radiometer); SRAL (Radar Altimeter); MWR (Microwave Radiometer), 21 bands	300	European global land and ocean mission, global coverage every 2 days. For land and ocean monitoring, e.g., recording of areal extent of fires and sea level fluctuations, water surface temperature mapping
Sentinel-5 Precursor/ 2017	UV-VIS-NIR-SWIR TRO-POMI: 0.27–0.495, 0.71–0.775, and 2.305–2.385 μm	7 × 3.5 km²	Customized global monitoring of the atmosphere and its trace gases at high geometric resolution. Trace gases such as O_3, NO_2, SO_2, HCHO, aerosols. Precursor mission to Sentinel-5
Sentinel-6/ 2020	Poseidon-4 (SAR Radar Altimeter) AMR-C (Climate-quality microwave radiometer)		Continuity of high-precision measurement of ocean surface topography to obtain data on changes to climate and environment
Sentinel-5/ 2021	High resolution spectrometer: 0.27–0.37, 0.37–0.50, 0.685-0.773, 1.590-1.675, and 2.305-2.385 μm	7 × 3.5 km²	Customized global monitoring of the atmosphere and its trace gases at high geometric resolution. Trace gases, such as O_3, NO_2, SO_2, HCHO, aerosols

(Contd.)

Satellite/ Year	Sensor	Spatial resolution (m)	Application areas
Sentinel-4/ 2022 (geo-stationary)	UVN Instrument: high-resolution spectrometer 0.305–0.40, 0.40–0.50, 0.75–0.775 μm	8000	Continuous monitoring of atmospheric trace gases over Europe (geostationary). Trace gases, such as O_3, NO_2, SO_2, HCHO, aerosols with high temporal resolution (~1 hour)

Source: *(https://www.dlr.de/eoc/en/desktopdefault.aspx/tabid-16002/25948_read-66677)*

4.4.3 High/Very High Resolution Satellites

High-resolution data is mainly used for mapping details of smaller areas of the Earth's surface. With the availability of higher resolution satellite images, there is always a competition between aerial images, UAV images, and satellite data, starting at a map scale of 1:5,000. Optical images at a spatial resolution of 1 m can be used to prepare maps at scales between 1:10000–1:20000 (Eugenio & Marcello, 2019). The new generation of Earth imaging satellites, launched in early 1998, marked the beginning of a new era for Earth observations at high/very high resolution (up to 0.82 m or 1 m; Fritz, 1996). A major change was observed with the launch of very high resolution IKONOS in 1999 by a private company. The images at < 5 m resolution, such as IKONOS or QuickBird, are now available for topographic and thematic mapping. High-resolution imagery, as well as long-term, time-series data, are essential for monitoring the activity, studying the growth trend, planning developmental activities, and many other applications (Richards & Jia, 2013). These images are typically used to map regional-to-local areas, but often such images are expensive, prohibiting their use.

DigitalGlobe was founded 1993 under the name WorldView Imaging Corporation, became EarthWatch Incorporated in 1995, and lastly DigitalGlobe in 2002 (Garg, 2019). In October 2001, DigitalGlobe successfully launched the world's highest resolution commercial satellite, QuickBird. This satellite provided high-resolution imagery to the Keyhole Corporation, acquired by Google in 2004, a vital component in the creation and commercial success of Google Earth, and beginning the proliferation of the online mapping portal. DigitalGlobe further launched the next generation, high-resolution imagery satellites WorldView-1 and WorldView-2. In January 2007, the company acquired Globe Xplorer, a leading online imagery provider. By early 2009, DigitalGlobe announced several other strategic agreements, expanding the availability of high-resolution imagery to online portals, navigation, and applications. It has further extended its service agreement with Google to provide wider accessibility of imagery globally. A new partnership with Microsoft provided high-resolution satellite and aerial imagery for its Virtual Earth platform.

Details of very high resolution satellites are given as follows, and their summary is presented in Table 4.6.

Table 4.6. Characteristics of High/Very High Resolution Satellites

Sensor	Resolution/Band*		Stero-capability	Swath (km)
	Panchromatic	Multispectral		
IKONOS-1	0.82 m 0.445–0.9 μm	4 m (blue, green, red, NIR) bands	Yes	11.3
IKONOS-2	1.0 m 0.445–0.9 μm	4 m (blue, green, red, NIR) bands	Yes	11
QuickBird-1	Failed after launch			
QuickBird-2	61 cm 0.45–0.9 μm	2.4 m (blue, green, red, NIR) bands	Along track	16.5
OrbView-1	1 m	4m (blue, green, red, IR)	Yes	8
OrbView-2 (SeaStar)	1 km	SeaWiFS (Sea-viewing Wide Field-of-view Sensor)		
OrbView-3	1 m	4m (blue, green, red, IR)	No	8
GeoEye-1	41 cm 0.45–0.80 μm	1.65 m (blue: blue, green, red, NIR)	Along track	15.2
World-View-1	50 cm 0.40–0.90 μm	Not applicable	Along track	17.6
World-View-2	50 cm 0.45–0.80 μm	2.0 m, Coastal: 0.40–0.45 μm, Blue, green, yellow, Red, Red Edge: 0.705–0.745 μm, NIR 1: 0.77–0.895 μm, NIR 2: 0.86–1.040 μm	Along track	16.4
World-View-3	31cm 0.45–0.80 μm	1.24 m, Same as in WorldView-2 plus 3.7 m resolution, SWIR-1: 1.195–1.225 μm, SWIR-2: 1.55–1.59 μm	Yes	13.1

(Contd.)

Sensor	Resolution/Band*		Stero-capability	Swath (km)
	Panchromatic	Multispectral		
		SWIR-3: 1.64–1.68 μm, SWIR-4: 1.71–1.75 μm, SWIR-5: 2.145–2.185 μm, SWIR 6: 2.185–2.225 μm, SWIR-7: 2.235–2.285 μm, SWIR-8: 2.295–2.365 μm CAVIS bands (30 m resolution) Desert clouds: 0.405–0.42 μm, Aerosol-1: 0.459–0.509 μm, Green: 0.525–0.585 μm, Aerosol-2: 0.635–0.685 μm, Water-1: 0.845–0.885 μm, Water-2: 0.897-0.927 μm, Water-3: 0.930-0.965 μm, NDVI-SWIR: 1.220–1.252 μm, Cirrus: 1.365–1.405 μm, Snow: 1.620–1.680 μm, Aerosol-3: 2.105–2.245 μm, Aerosol-4: 2.105–2.245 μm		
Worldview-4 (formerly GeoEye-2)	31 cm 0.45–0.80 μm	1.24 m (blue: blue, green, red, NIR)	Yes	13.1
GeoEye-1	0.5 m PAN 1	2.0 m MSS 4	Yes	15.2
KOMP-SAT-2	1.0 m PAN 1	4.0 m MSS 4	Yes	15
Pléiades-1A and -1B	0.5 m PAN-HiRI	2.8 m MSS-HiRi 4	Yes	20
Cartosat-1	2.5 m PAN A, PAN F	Both in 0.50–0.85 μm	Yes	29.42 (PAN F), 26.24 (PAN A)
Cartosat-2C, -2D and -2E	0.65 m PAN	2.0 m HRMX 4	Yes	9.6
Cartosat-3	0.28 m PAN MX	1.14 MX 4	Yes	16

∗***Note:*** All data are 11-bit (2048 gray levels) radiometric resolution.

4.4.3.1 IKONOS

GeoEye Inc. (formerly Orbital Imaging Corporation or ORBIMAGE), a commercial company, operates with a growing constellation of high-resolution Earth imaging satellites (e.g., IKONOS, OrbView-2, OrbView-3, GeoEye-1, GeoEye-2) which today are an important source for large-scale mapping. The IKONOS, launched on September 24, 1999 in a sun-synchronous orbit, derives its name from the Greek word *eikōn* for image. It is the first satellite to collect high-resolution multispectral imagery with 3375 pixels and PAN imagery with 13500 pixels, providing a swath width of 11.3 km. It simultaneously collects the PAN image at 1 m (0.82 m) resolution, and MS images (blue, green, red simultaneously and NIR from 0.45–0.90 μm) at 4 m spatial resolution. The IKONOS operates at an altitude of about 680 km, completing the Earth every 98 minutes with a revisit time between 3–5 days off-nadir and 144 days for true-nadir images. The IKONOS-2, developed by the then Space Imaging, then GeoEye, now DigitalGlobe, was launched in September 1999 in a sun-synchronous orbit at an altitude of 681 km. It provides PAN and MS images, the same as IKONOS-1. It covers an 11 km swath having up to a 60° viewing angle, and 2048 (11 bits) radiometric resolution (Eugenio & Marcello, 2019).

4.4.3.2 QuickBird

The QuickBird-1, a very high resolution satellite, developed by then EarthWatch, now DigitalGlobe, was launched in November 2000, but was lost due to launch vehicle failure. It operates at an altitude of 450 km with a 16.5 km swath and 2048 (11-bit) radiometric resolution. It can provide 1–5 days temporal resolution with up to 25° of viewing angle along- and cross-track. The QuickBird-2 was launched on October 18, 2001 by DigitalGlobe, in a sun-synchronous polar orbit, at an altitude of 470 km, with 94 minutes of orbital period. The satellite has 4 MS bands (blue, green, red, and NIR) with 2.4 m resolution and a PAN band with 0.61 m resolution (Alexandrov et al., 2020).

4.4.3.3 OrbView

The OrbView satellite series of ORBIMAGE is launched to acquire affordable high-quality imagery of the Earth for a variety of users that include local governments, telecommunication companies, architects, civil engineers, real estate managers, farmers, and environmental monitoring agencies. ORBIMAGE is employing a uniquely integrated global system of imaging satellites, ground stations, and Internet-based sales channels to collect, process, and distribute imagery products at affordable cost.

OrbView-1 (old name Microlab-1) is an imaging microsatellite, launched on April 1, 1995 (*https://directory.eoportal.org/web/eoportal/satellite-missions/o/*). It carried two sensor systems: OTD (Optical Transient Detector), a lightning imaging system, and GPS/MET (GPS/Meteorology), an atmospheric measurement system by radio occultation techniques. ORBIMAGE renamed its satellites in 1997.

OrbView-2 (old name SeaStar) was launched by GeoEye Inc. on August 1, 1997 in sun-synchronous polar orbit at an altitude of 705 km. It was a low-resolution satellite (1 km), capturing broad-area color imagery of land and ocean surfaces for agricultural and ocean monitoring. It carried a SeaWiFS (Sea-viewing Wide Field-of-view Sensor) to provide quantitative data on global ocean bio-optical properties as well as subtle changes in ocean colors to signify various types and quantities of marine phytoplankton (microscopic marine plants) for both scientific and practical applications. The OrbView-3, launched on June 26, 2003 by GeoEye Inc, now DigitalGlobe, acquired 1 m resolution PAN and 4 m resolution MS imagery (blue, green, red, IR). The satellite moves in the orbit at 470 km height with up to 45° of viewing angle, thus revisiting the entire Earth in less than 3 days. It provides an 8 km wide swath, having 2048 (11-bit) radiometric resolution. OrbView-4 (old name OrbView-3), launched on September 21, 2001, was a launch failure. It had PAN and MS imaging sensors. For details of OrbView-5, see the section on GeoEye (*https://directory.eoportal.org/web/eoportal/satellite-missions/o/*).

4.4.3.4 GeoEye

GeoEye-1 (formerly OrbView-5), developed by then GeoEye, now DigitalGlobe, was launched on September 6, 2008 (GeoEye, 2007). It provides 253 million km^2 of satellite images to Microsoft and Yahoo search engines with Google having exclusive online mapping access from this satellite. It provides 41 cm PAN and 1.65 m MS imagery in 4 bands (blue, green, red, and NIR) at a 15.2 km wide swath with 2048 (11-bit) radiometric resolution. It moves in a sun-synchronous orbit at an altitude of 684 km with 99 minutes orbital period, which can image up to a 60° off-nadir viewing angle (*https://www.satimagingcorp.com/satellite-sensors/geoeye*). This satellite can rotate or swivel forward, backward, or side-to-side with robotic precision, and has a revisit capability of 1–3 days depending on latitude and elevation angle. The PAN and MS datasets can also be used as combined (pan-sharpened) 0.41 m multispectral imagery. The users can map natural and man-made features to within 3 m of their actual locations. Such images are very useful in studies related to telecommunications, infrastructure planning, mapping/surveying, civil engineering, mining and exploration, oil and gas, and DEM generation. For details of GeoEye-2, see the section on WorldView (WorldView-4).

4.4.3.5 WorldView

The WorldView-1 was launched on September 18, 2007 by DigitalGlobe. Operating at an altitude of 496 km, WorldView-1 has an average revisit time of 1.7 days, and is capable of collecting up to 750,000 km^2 data per day. The satellite is also equipped with state-of-art geolocation accuracy capabilities and efficient in-track stereo-data collection. The PAN imaging system captures 0.5 m resolution imagery for creation of precise maps, and is used in telecommunications, infrastructure planning, mapping/surveying, civil engineering, mining and exploration, oil and gas, and DEM generation (Eugenio & Marcello, 2019).

WorldView-2 was successfully launched on October 8, 2009, but was fully operational on January 6, 2010. It provides sub-meter resolution data, and is capable of collecting up to 975,000 km^2 of imagery per day (Garg, 2019). It was the first very high resolution satellite to offer 8-band MS imagery, with increased agility, accuracy and stereo-capability. With the additional four spectral bands, WorldView-2 offers unique opportunities for analysis of vegetation, coastal environments, agriculture, geology, environment, tourism, and many others.

The WorldView-3 was launched on August 13, 2014 in a sun-synchronous orbit, at an altitude of 617 km, with an orbital period of 97 minutes. It provides 31 cm resolution in PAN, 1.24 m resolution in MS, 3.7 m resolution in SWIR, and 30 m resolution in CAVIS (Clouds, Aerosols, Vapors, Ice, and Snow). The CAVIS monitors the atmosphere and provides correction data for haze, soot, dust, or fog that influences the objects on the imagery. It has super-spectral and high resolution, for example, 31 cm resolution in PAN; SWIR bands allow accurate imaging through haze, fog, dust, smoke, and other airborne particles; and eight multispectral bands offer unique opportunities for remote sensing analysis of vegetation, coastal environments, agriculture, geology, environment, tourism, and so on. WorldView-3 has an average revisit time of <1 day, and is capable of collecting up to 680,000 km^2 per day.

WorldView-4 (formerly GeoEye-2) by DigitalGlobe is a third generation spacecraft, that offers the highest resolution imagery of the Earth available for civilian and military applications. It was launched on November 11, 2016 in a sun-synchronous orbit, at an altitude of 617 km, with 97 minutes orbital period. It has an effective revisit time capability of ≤ 3 days. The satellite provides 31 cm resolution in PAN at nadir, 0.34 m at 20° off-nadir, 1 m at 56° off-nadir, and 3.51 m at 65° off-nadir, as well as 1.24 m resolution in multispectral at nadir, 1.38 m at 20° off-nadir, and 4m at 56° off-nadir, 14 m at 65° off-nadir in 4 bands (blue, green, red, and NIR). It offers a 13.1 km swath with 2048 (11-bit) radiometric resolution data. The satellite renders high-resolution images to execute online mapping, disaster evaluation and relief, environmental supervision, and land management missions. It provides accurate geospatial intelligence data to assist the military and decision-makers in maintaining national security.

WorldView Legion is the next evolution of the leading WorldView constellation. It will consist of six satellites planned to launch into a mix of sun-synchronous and mid-latitude orbits. The six satellites will be launched in two phases—in batches of two and four—all in 2021 (Datta, 2020). These satellites will replace imaging capability currently provided by DigitalGlobe's WorldView-1, WorldView-2, and GeoEye-1 Earth observation satellites. The WorldView Legion constellation will initially include six high-performance satellites to collect 30 cm resolution imagery. The constellation will have the ability to capture 5 million km^2 a day, at 30 cm resolution and better than 5 m accuracy. The new constellation will be able to capture 3% of the land mass every day. They will provide dramatically more frequent images (up to 15 times per day) over the most in-demand areas throughout the day for their decision-making processes. This advanced constellation will offer high-agility point collection for monitoring targets and large areas for mapping missions, enabling more persistent monitoring, near-real time change detection, and analysis. The WorldView Legion's frequent revisit rates will enable significantly more accurate, comprehensive, and timely pattern-of-life and human geography analysis (Werner, 2019). Combining the most advanced geospatial analytics with the new datasets will provide meaningful insights to studies such as vehicle navigation, environmental sustainability, global connectivity, disaster response, and national security missions.

4.4.3.6 KOMPSAT

The KOMPSAT (Korea Multi-Purpose Satellite) program was initiated in 1995 as a major space investment in Korea. The KOMPSAT-1 or Arirang-1 is a minisatellite, launched on December 20, 1999. It moves in a sun-synchronous polar orbit, at an altitude of 685 km, with an orbital period of 98.46 min, and a revisit time of 28 days (Kressler et al., 2006). The satellite has three onboard processors: OBC (on-board computer) for command, telemetry and data handling management, RDU (remote drive unit) for attitude and orbit control management including propulsion subsystem, and ECU (electrical power system control unit) for electrical power and thermal control management. It provides high-resolution imagery of the Korean Peninsula using an EOC (electro optical camera), collects wide-swath multispectral imagery of the ocean and coastal zones to support biological oceanography, and provides information on the LEO particle environment and globally on plasma distribution in the ion layer using a SPS (space physics sensor). The objective of KOMPSAT-1 is the development of a national space segment in Earth observation along with an efficient infrastructure and ground segment to provide valuable services to remote sensing users in various fields of applications (*https://www.kari.re.kr/eng/sub03_02_01.do*).

The KOMPSAT 2 (Arirang-2), launched on July 28, 2006, is a sun-synchronous low Earth orbit satellite with a 5.5 days revisit time, which obtains high-resolution imagery of the Korean peninsula. The satellite is equipped with an MSC (multi-spectral camera) working in a strip mode operation to capture defined observation areas one at a time with a swath width of 15 km at nadir. The sensor provides PAN as well as 4 bands of MS imagery with a resolution of 1 m. It is fitted with an optoelectronic linear pushbroom scanner with a single nadir-pointing telescope that provides FOR (field of regard) of ±30° in pitch and up to ±56° in the roll direction, supporting observations even in low-light conditions.

The KOMPSAT-3 (Arirang-3) and KOMPSAT-3A (Arirang-3A) were launched on May 17, 2012 and March 25, 2015, respectively. Both satellites are fitted with a pushbroom scanner, AEISS (Advanced Earth Imaging Sensor System) with a spatial resolution of 0.5 m for KOMPSAT-3 and 0.4 m for KOMPSAT-3A, which was enhanced with IIS (Infrared Imaging System) to operate within the MIR region of 3–5 μm with high resolution. The products obtained are: PAN, four MS, or four Pansharpened bands. It has a pointing capability of ±45° into any direction. The PAN images are useful for application areas such as agriculture, environment, oceanography, and natural disasters. The KOMPSAT-3A temperature-sensitive sensors can assist in the monitoring of wildfires, volcanic and seismic activity, as well as water currents and natural disasters. KOMPSAT-5 (Arirang-5) was launched on August 22, 2013 and is equipped with imaging radar (Earth-i, 2018).

The KOMPSAT-6 (Arirang-6) is an all-weather observation satellite with improved imaging radar performance. It will be launched in 2021 at an orbital height of 505 km and 11 days repeat cycle. The KOMPSAT-7, an ultra-high resolution optical satellite which has higher precision in Earth observation capabilities, is being developed. It will be the first Korean satellite to apply optical transmission technology for the transfer of ultra-high resolution image in real time and use terra-bit or higher storage technology to store large-capacity image data (*https://www.kari.re.kr/eng/sub03_02_01.do*).

4.4.3.7 Pléiades

The Pléiades satellite system comprises two satellites; Pléiades-1A and -1B, launched by the CNES (Centre National d'Études Spatiales), the French space agency, on December 16, 2011 and December 2, 2012, respectively. They are phased 180° apart in the same near-polar sun-synchronous orbit at 694 km height to provide global coverage in 26 days and daily revisit to any place on Earth. Both satellites are equipped with high-resolution optical sensors providing PAN and MS images with a swath of 20 km. It provides along-track stereo and tri-stereo images within an 800 km wide ground strip. The Pleiades-1A satellite features four MS bands (blue, green, red, and IR), as well as image location accuracy of 3 m (CE90) without the GCPS (ground control points). Image location accuracy can be improved to an exceptional 1 m by the use of GCPs. Because the satellite has been designed with urgent tasks in mind, images can be requested from Pleiades-1A less than six hours before they are acquired. This functionality will prove invaluable in situations where the real-time collection of new image data is crucial, such as disasters and crisis monitoring (*https://www.satimagingcorp.com/satellite-sensors/pleiades-1/*).

The Pleiades program will follow the SPOT program satellite series services, with overall objectives to provide an optical high-resolution PAN (0.7 m) and MS (2.8 m) imagery with high-quality product standards (level-2 products pansharpened ortho-rectified on a DTM), make available stereo-imagery (up to 350 km × 20 km or 150 km × 40 km) and mosaic imagery of size up to 120 km × 120 km, and collect >250 images/day from each spacecraft of the constellation. They will support risk management support services in terms of observation coverage (*https://earth.esa.int/web/eoportal/satellite-missions/p/pleiades*).

4.4.4 Thermal Remote Sensing Platforms

The single most important development in infrared technology was the development of the detector element during World War II for military uses, and later these images have been adopted for civilian applications. Besides the measurement of regular surface temperature, infrared sensors can be used for detection of forest fires or other warm/hot objects, or monitoring volcanic activities, hydrology, coastal zones, seismology, environmental modeling, meteorology, medical sciences, vetenary sciences, intelligence/military applications, and heat loss from buildings (Sabins, 1996). Early infrared detectors included lead-salt photodetectors but now include very fast detectors consisting of mercury-doped germanium (Ge:Hg), indium antimonide (InSb), and other substances that are very responsive to infrared radiation.

The first thermal remote sensing data was collected by the TIROS (Television IR Operational Satellite) launched by the United States in 1960. The coarse resolution of TIR data was ideal for monitoring regional cloud patterns and frontal movement. NASA then launched the HCMM (heat capacity mapping mission) on April 26, 1978 that obtained 600 m spatial resolution TIR data (10.5–12.6 μm) both during day (1:30 p.m.) and night (2:30 a.m.). NASA's Nimbus-7 launched on October 23, 1978 carried a CZCS (coastal zone color scanner) that included a TIR sensor for monitoring sea-surface temperature (SST). In 1980, NASA and the Jet Propulsion Laboratory developed the TIMS (Thermal Infrared Multispectral Scanner) that acquired TIR data in six bands at wavelength intervals of <1.0 μm. The success of TIMS resulted in the development of the 15-channel ATLAS (Airborne Terrestrial Applications Sensor). The TM (Thematic Mapper) of Landsat-4 and -5, launched on July 16, 1982 and March 1, 1984, respectively, collected 120 m resolution TIR data (10.4–12.5 μm) along with two bands of MIR data. The NOAA GOES (Geostationary Operational Environmental Satellite) collects TIR data at a spatial resolution of 8 km for weather prediction, and full images of the Earth are obtained every 30 minutes both day and night. The NOAA-AVHRR (Advanced Very High Resolution Radiometer) collects TIR data of LAC (local area coverage) at 1.1 km and GAC (global area coverage) at 4 km resolution which can also be used routinely to forecast night weather (Schmugge et al., 2018). For details of NOAA, see Table 4.2. Sensors such as ERS-ATSR and TERRA-MODIS are equipped at 3.8 μm wavelength that can be used for the detection of fire and hot spots. Table 4.7 presents various satellites/sensors which could be used for land surface temperature measurements.

All objects, both natural and man-made, emit TIR energy as heat that has a temperature greater than absolute zero. Thermal sensors record the

Table 4.7. Sensors and Satellite Used for Land Surface Temperature (Tomlinson et al., 2011)

Sensor	Satellite	Resolution		TIR spectral bands (μm)	Launch year	Website
		Spatial	Temporal			
GOES Imager	GOES network	4 km	30 min	4 (10.2–11.2) 5 (11.5–12.5)	1974	*http://goespoes.gsfc.nasa.gov/goes/index.html*
AVHRR	Multiple NOAA	1.1 km	Twice daily	4 (10.3–11.3) 5 (11.5–12.5)	1979	*http://noaasis.noaa.gov/NOAASIS/ml/avhrr.html*
Landsat ETM+	Landsat-7	60 m	16 days	6 (10.4–12.5)	1999	*http://Landsat.gsfc.nasa.gov/*
ASTER	Terra	90 m	Twice daily	10 (8.1–8.5) 11 (8.5–8.8) 12 (8.9–9.3) 13 (10.2–11.0) 14 (11.0–11.6)	1999	*http://asterweb.jpl.nasa.gov/index.asp*
MODIS	Terra	1 km	Twice daily	31 (10.78–11.28) 32 (11.77–12.27)	2000	*http://modis.gsfc.nasa.gov/*
MODIS	Aqua	1 km	Twice daily	Same as MODIS Terra	2002	*Same as MODIS Terra*
AATSR	Envisat	1 km	35 days	11 and 12	2004	*http://envisat.esa.int/instruments/aatsr/*
SEVIRI	Meteosat-8	3 km	30 min	12 (10.8)	2005	*http://landsaf.meteo.pt/*
AVHRR	MetOP	1.1 km	29 days	4 (10.3–11.3) 5 (11.5–12.5)	2006	*http://www.esa.int/esaLP/ESA7 USVTYWC LPmetop 0.html*

Note: AATSR-Advanced Along Track Scanning Radiometer, and SEVIRI-Spinning Enhanced Visible and Infrared Imager.

energy emitted from the Earth's features in the thermal range of EMS (3–5 μm and 8–14 μm). Most thermal remote sensing of the Earth's features is focused in the 8–14 μm range because peak emission (based on Wien's displacement law) for objects around 300 K (27° C or 80° F) occurs at 9.7 μm. The wavelength range 3-5 μm is related to high temperature phenomena, like forest fires, while the 8–14 μm range is related to general Earth features having lower temperatures. Thermal images have the ability to detect very subtle temperature differences, even in complete dark and challenging weather conditions. The lighters or brighter areas indicate warmer areas, while darker areas are cooler in B&W thermal images (Kahle, 1998).

Thermal data can be acquired during the day and night, which is useful for some applications. However, for many applications nighttime or more specifically predawn images are preferred, as during this time the effect of differential solar heating is minimal. Figure 4.8 shows the diurnal radiant temperature curves for water, vegetation, and soils. The thermal inertia of water is similar to that of soil and rocks, but during the day, water bodies have a cooler surface temperature than the soil and rocks. Water generally appears warmer than its surroundings in nighttime imaging. At night, the surface temperatures are reversed with water becoming warmer than the soil and rocks. The reason is that convention currents maintain a relatively uniform temperature at the surface of a water body. Vegetation has a warm signature during nighttime, and a cooler signature during daytime as compared to adjacent soils during daytime (Sabins, 1996).

The thermal property of a material is representative of several upper centimeters of the surface. The ability to record variations in infrared

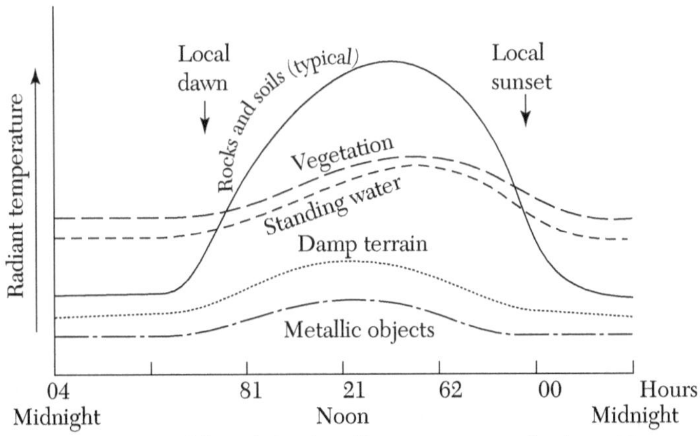

Figure 4.8. Day-night variation in radiant temperature of various objects.

radiation has the advantage in understanding phenomena where minor temperature variations may be significant in the environment. Single band thermal images can also be displayed in pseudo-color to better display the variations in temperature. Paved areas appear relatively warmer during the day and night. Pavement heats up quickly and to higher temperatures than the surrounding areas during the day. Paved areas also lose heat relatively slowly at night so they are relatively warmer than surrounding features (Tomlinson et al., 2011).

4.4.5 Microwave Remote Sensing Platforms

The microwave region falls between the IR and radio wavelengths, and has a range extending from approximately 0.1 cm to 1 m with frequencies between 3 GHz–30 GHz. Microwave ranges are often referred to as the different frequency bands: S, C, X, Ku, K, and Ka. Microwaves have special properties that are important for remote sensing applications because of their long wavelengths, as compared to the visible and IR wavelengths (Colwell, 1983). Potential applications of microwave data are given in Table 4.8.

Table 4.8. Potential Applications of Microwave Data (Calla, 1983)

Agriculture	Crop classification, Crop yield monitoring, Soil moisture measurement, Cultivated and forest area mapping, Crop stress
Geology	Petroleum and mineral prospecting, Large scale mapping of nearly forested geological sites
Water resources	Mapping of water bodies, Snow cover mapping, Drainage pattern study
Flood mapping	Flood prediction
Land use survey	Forest biomass, tree height
Oceanography	Coastal monitoring, Surface wind speed and direction, Wave height, and wave spectrum
Environmental studies	Landslide prediction, Rainfall rate

SAR images do not have the same information content as an optical image with the same resolution, but these have the advantage of penetrating clouds. Table 4.9 presents the characteristics of some microwave satellites/sensors. The European Space Agency's ERS and Envisat, India's RISAT, Canada's Radarsat, and Shuttle Radar Topography Mission (SRTM) and other satellite altimeters measure the distance traveled by microwave pulses transmitted to provide elevation data which can be used for DEM generation. Other examples of active sensors include ERS-1 launched in 1991, J-ERS satellite launched in 1992, ERS-2 in 1995, Radarsat-1 launched in 1995, SRTM in 2000, Radarsat-2 launched in 2007, TerraSAR-X Radar Satellite launched

Table 4.9. Characteristics of microwave systems

System	Launch year	Spatial resolution (m)	Swath (km)
NIMBUS	1978	18 km–148 km	800
SEASAT (USA)	1978	25	100
Seasat	1978	0.8 to 4.2 degree	600
NOAA-5	1978	110 km	2300
NOAA-6	1979	110 km	2300
SIR-A (USA)	1981	40	50
SIR-B (USA)	1984	30	20–50
ALMAZ (RUSSIA)	1987	15–30	20–45
ALMAZ-1 (RUSSIA)	1991	15–30	20–45
ERS-1	1991	25	80
JERS-1	1992	18	75
SIR	1994	30	15–90
ERS-2	1995	10–30	100
RADARSAT-1	1995	9–100	50–100
EOS	1998	30,15	30–20, 700
Space Shuttle (SRTM)	2000	30	225
ENVISAT	2002	30–1000	100–405
ALOS-1	2006	2.5	
RISAT-2	2009	1–8	10–120
Megha Tropiques	2011	6–9 and 40–60	1700
SARAL	2011	35.75 GHz Ka-band altimeter satellite laser ranging	
RISAT-1	2012	1–50	10–225
ALOS-2	2014	1–100	25–490
RADARSAT-2	2007	1–100	

in 2007, IRS-P4 (Oceansat-1) launched in 1999, Oceansat-2 launched in 2009, RISAT-1 launched in 2012, and RISAT-2 launched in 2009. The SRTM uses InSAR which measures the Earth's elevation with two antennas to create DEMs of Earth.

Microwave images are useful in many applications (Garg, 2021), like meteorology, hydrology, agriculture, ecology, oceanography, assessing soil moisture, atmospheric water and ozone concentrations, distinguishing oil spills, and addressing water pollution. Microwaves are sensitive to the moisture content in the soil that can be assessed by measuring the dielectric constant of the soil. The depth of penetration of microwaves is a function of moisture content in the soil. Table 4.10 summarizes the advantages and disadvantages of microwave and thermal infrared images for soil moisture studies.

Table 4.10. The Advantages and Disadvantages of Microwave and Thermal Infrared Images for Soil Moisture Studies (Kuenzer et al., 2013)

Parameters	Microwave data	Thermal infrared data
Spatial resolution	Not so useful	Useful
Temporal resolution	Useful	Useful
Perturbance by clouds	No	Yes
Perturbance by relief	Strong	Weaker
Easy to understand–user acceptance	Not so useful	Useful
Retrieval for wet conditions	Useful	Not so useful
Retrieval for dry conditions	Useful	Useful
Retrieval of root zone soil moisture under dense vegetation	Not so useful	Useful

4.4.6 Hyperspectral Imaging Platforms

Hyperspectral sensors (also known as *imaging spectrometers*) collect images of a scene in hundreds of narrow contiguous spectral bands (nearly 5–15 μm) simultaneously. Hyperspectral images provide large spectral information to identify and distinguish spectrally unique materials. For example, AVIRIS with 224 spectral bands provides 4 m resolution data, while Hyperion and CHRIS Proba have spatial resolutions between 15–30 m (Garg, 2019).

A hyperspectral image looks very much like a spectrum measured in a spectroscopy laboratory and provides much more information about the surface than a multispectral image. It is a 3D data cube which contains 2D spatial information (image feature) and 1D spectral information (spectral bands). The spectral bands occupy very fine wavelengths (Figure 4.9), while the image features and shape features reveal the disparity and association among adjacent pixels from different directions at a wavelength.

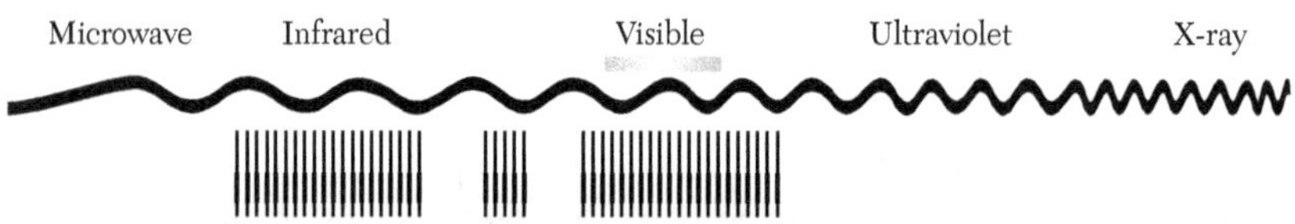

Figure 4.9. Hyperspectral imaging with hundreds of narrow bands.

Over the past decade, hyperspectral image analysis has matured into one of the most powerful and fastest growing technologies in the field of remote sensing. Of the various types of remote sensing, hyperspectral imaging has become one of the most active areas of research. spectral information from hyperspectral images establishes new application domains and poses new technological challenges in data analysis (Vangi et al., 2021). With the available high spectral resolution, subtle objects and materials can be extracted by hyperspectral imaging sensors with very narrow spectral bands for applications such as detection, urban planning, agriculture, identification, surveillance, and quantification. Hyperspectral imagery can be used to map divergent surface targets and topographical features, materials for mineral mapping, and to detect soil properties, including moisture, organic content, and salinity. They are helpful in vegetation studies (species identification, plant stress, productivity, leaf water content, and canopy chemistry), soil type mapping, geology (mineral identification and mapping), hydrology (snow grain size, liquid/solid water differentiation), lake, river, and ocean applications, including biochemical studies (photoplankton mapping activity), and water quality (particulate and sediment mapping; Burke et al., 2001).

NASA successfully launched the Hyperion imaging spectrometer (part of the EO-1 satellite) in 2000, which provides images in 220 spectral bands (0.4–2.5 μm) with 30 m resolution. It can capture 7.5 km × 100 km of land area/image, and provides detailed spectral mapping with high radiometric accuracy to map complex land ecosystems (Birk & McCord, 1994). The AVIRIS (Airborne Visible/Infrared Imaging Spectrometer) is another example of NASA's hyperspectral airborne sensor which collects data in 224 contiguous channels with wavelengths between 0.4–2.5 μm. The CHRIS (Compact High Resolution Imaging Spectrometer) is a fully programmable (i.e., in spatial resolution, total swath, and spectral band settings) spectrometer, which provides five distinct angular views, but does not cover the SWIR range (Chavez et. al., 1994).

Several other hyperspectral missions include the PRISMA(PRecursore IperSpettrale della Missione Applicativa) Italian mission working in a wavelength range 0.40–2.5 μm with a 30 m resolution, the EnMAP (Environmental Mapping and Analysis Program) German mission working between 0.42–2.5 μm at 30 m resolution, and the HISUI (Hyperspectral Imager SUIte) Japanese mission working within 0.40–2.5 μm at 30 m resolution. Many other sensors include, HyMAP CASI (Compact Airborne Spectrographic Imager), DAIS (Digital Airborne Imaging Spectrometer),

ROSIS (Reflective Optics System Imaging Spectrometer), AISA (Airborne Imaging Spectrometer for Applications), HYDICE (Hyperspectral Digital Imagery Collection Experiment), MIVIS (Multispectral Infrared Visible Imaging Spectrometer), and so forth. (Transon et al., 2018). The details of some of the popular hyperspectral sensors are given in Chapter 3.

Processing techniques generally identify the presence of materials/objects through measurement of spectral absorption features. Often the hyperspectral data are post-processed to derive surface reflectance through the use of atmospheric radiative transfer models (Birk & McCord, 1994). Maximum likelihood methods, neural network architectures, support vector machines (SVM), Bayesian approaches, as well as kernel methods have been successfully used in recent years for the identification/classification of hyperspectral data.

Spectral libraries are best described as a collection of representative spectra of a variety of materials. For a "pure" material, these spectral characteristics are called *end-members*. End-members can be measured in the laboratory, in the field, or can be extracted from remotely sensed imagery. Several feature selection, spectral feature extraction, spectral libraries, and classification methods have been developed to deal with the challenging intrinsic nature of hyperspectral data. Some of the traditional approaches for classification include SMA (spectral mixture analysis), and SAM (spectral angle mapper), while advanced machine learning algorithms include SVM (support vector machine), RF (random forests), ANN (artificial neural network), and deep learning CNN (convolutional neural network) architectures.

For analysis of hyperspectral data, such as AVIRIS and MODIS, the available hyperspectral signature libraries could be used (Morisette et al., 2002). Spectral libraries are best described as a collection of representative spectra of a variety of materials, and are crucial for identification of unknown spectra and aid the correction and classification of hyperspectral data by providing end-member spectra. Table 4.11 gives an overview of spectral libraries developed by different organizations. These libraries have been created in lab conditions, and contain a large number of signatures for different types of land use and land cover classes, including minerals and Earth materials. These spectral signatures could be used effectively to match with the reflectance of features/materials on hyperspectral images for their precise and accurate classification with the help of algorithms. However, environmental conditions, topography, climate, and natural variations in features/materials can produce different reflectance patterns

Table 4.11. Comparison of Existing Spectral Libraries

Features	SPECCHIO9 Spectral Input/Output)	USGS Spectral Library	ASTER Spectral Library	DLR Spectral Archive	Concordia Spectrum Research Library	Spectral Analysis
Instruments used	SVC-HR1024	Beckman 5270ASD spectrometer	ASD Spectrometer	ASD spectrometer	Nil	Handheld spectroradiometer
Interface	Java and PHP	—	—	Web	Php	Java
Database	My SQL	—	—	My SQL	My SQL	None
Input File Format	ASD binary, GER, OO, ASCII, Apogee, ENVI SLB, XML, FGI HDF5, SPECPR	ASCII format	ASCII format	ASD binary ASCII	—	ASCII format
Output File Format	CSV, ENVI SLB XML, Direct access from Matlab	ASCII Format Image	—	Metadata zip file Zip file containing ASD binary files	—	—
Available Spectra nos.	80,000 423 (Minerals)	17 (Plants) and man-made Materials	160 (Minerals)	2008	—	—
Development	RSL, University of Zurich, Switzerland	U.S. Geological Survey	California Institute of Technology, U.S.	DLR, Germany	Concordia University, Canada	SAC (ISRO), India

than available in the standard library, and thus can make the interpretation difficult and erroneous (Birk & McCord, 1994).

Hyperspectral images have more advantages than multispectral data for the identification and discrimination of targets/objects because of several reasons (Zhu et al., 2011) including:

1. High spatial resolution.
2. Data frequently collected in a distinct spectral range.
3. Data are contiguous and overlapping, making them more useful to detect information.
4. A contiguous spectrum assists atmospheric windows to be recognized and removal from the radiance signal.
5. Signal-to-noise ratio of the data can be enhanced by comparing pixel spectra.
6. The problem of mixed spectra can be solved by directly deriving the relative abundance of materials.
7. The objects/classes in hyperspectral images can be derived from various spaces, such as the spectral space, image space, and character space.

Drawbacks with hyperspectral images are as follows:

1. It adds a level of complexity to reduce the redundancy from a large number of narrow bands. The increasing data dimensionality and high redundancy between features might cause problems during data analysis.
2. They do not allow regular and synoptic coverages over large areas as compared to multispectral sensors. Moreover, multispectral sensors produce images with lower angular effects due to their much smaller field of view.

Despite technological advances, hyperspectral satellites are still poorly represented in spaceborne missions as compared to multispectral sensors. The classification procedures for hyperspectral data are not as developed as for the analysis of multispectral imagery. As a result, hyperspectral images are still not a very popular tool in remote sensing but are gaining increased use slowly.

4.5 SMALL SATELLITES

The first satellite, Sputnik-1, launched on October 4, 1957, was a small satellite. The first microsatellite was built by enthusiasts of the amateur radio community and launched in the early 1960s. The term "microsatellite" was very probably coined by members of the AMSAT-NA (Radio Amateur Satellite Corporation—North America) community (Sweeting, 1991). In the early 1980s, the engineering of microsatellites, took a radical change in approach to make them smaller and lighter to achieve cost reductions. If a traditional bigger satellite which requires huge investments fails, it leads to loss of many instruments/sensors, costing a large amount of money and human effort. The small satellite segment provides significant benefits in terms of cost, flexibility, speed, and access to space, allowing a high degree of design flexibly for both the platform and its instrument payload (Anderson, 2009). So a larger number of small satellites, each of which is dedicated to a particular mission objective and carries a single instrument/sensor, are being launched. The low weight of small satellites allows them to be launched into space at a much lower cost, which opens up low-cost access to space for countries without a space program, companies, and educational institutions. The classification of small satellites is given in Table 4.12.

Table 4.12. Small Satellite Classification (Gansvind, 2019)

Types of satellites by weight	Weight
Zepto-satellite	0.1 g –1 g
Atto-satellite	1 g –10 g
Femto-satellite	10 g– 100 g
Pico-satellite	100 g –1 kg
Nano-satellite	10 kg
Microsatellite	10 kg –100 kg
Small/Minisatellite	100 kg –500 kg
Medium satellite	500 kg –1000 kg
Large satellite	> 1000 kg

The relatively lower weight and complexity of these small satellites, combined with the standardized small footprint, mean that they can be launched with a much shorter lead time than traditional satellites. These small satellites typically have shorter development cycles, smaller development teams, and, consequently, lower cost, both for the development and for the launch of the satellites (Anderson, 2009). Need for a systematic and continuous survey of the Earth's surface with a minimal time interval between surveys of any area at an acceptable reduced cost gave rise to fundamental design in the structure of small satellites. The advent of small satellite technology in multiple areas, such as micro-propulsion, microelectronics, long distance communications, gathering scientific data, and increased low cost rideshare launch availability, has also led to new mission concepts and missions such as broadband Internet, communications, and Earth observation, which consist of mega constellations of small satellites (Lappas & Kostopoulos, 2020). These small satellites are gaining momentum in almost every area of space, including communication, remote sensing, technology demonstration, and science and exploration. The small satellites provide an important source of remote sensing data. In addition, many of them can be launched simultaneously to operate together in space as a constellation. In all, the benefits of small satellite missions are considered to be: (*a*) more frequent mission opportunities, and therefore faster application of data; (*b*) a larger variety of missions, and therefore greater diversification of potential users; (*c*) more rapid expansion of the technical and/or scientific knowledge base; and (*d*) greater involvement of local and small industry.

A key player in the field of small satellites has been Surrey Satellite Technology Ltd (SSTL) who has built and launched a long list of small satellites (Kramer & Cracknell, 2008). An important development in small satellites was the CubeSats, which owe their origin to Jordi Puig-Suari of California Polytechnic State University and Bob Twiggs of Stanford University, who proposed the CubeSat reference design in 1999 with the aim of enabling students to design, build, test, and operate a spacecraft for low earth orbit (LEO) within the time and financial constraints of a graduate degree program (Cracknell, 2018). The first six CubeSats, weighing about 1 kg each, entered orbit in June 2003. A CubeSat is a small satellite that is made up of multiples of 10 cm × 10 cm × 10 cm cubic units, and has a mass of less than 1.33 kg per unit (Anderson, 2009). The CubeSat became a standard over time by a process of emergence. Generally, the CubeSats have piggybacked on launches of major spacecrafts, the first one launched in June 2003. As on January 1, 2021, more than 1350 CubeSats have been launched, involving national space agencies as well as many commercial entities. Recently, CubeSats even ventured into deep space when two were launched in May 2018 to support the Insight mission to Mars (Gansvind, 2019).

Table 4.13 shows various mega-constellations with microsatellites, which indicates that microsatellites up to a mass of 200–300 kg are the most popular trend in the space industry and that the 200–300 kg mass is the near

Table 4.13. Details of Mega-Constellations of Small Satellites (Lappas & Kostopoulos, 2020)

Constellation	Number	Maker	Weight (kg)	Year	Availability	Orbit (km)	Status
Iridium Next	66 + 9 spares	Thales Alenia+ Orbital ATK	860	2009	2018	780	Complete
Boeing	1396–2956	Boeing Satellite	N/A	2016	N/A	1200	Unknown
LeoSat	78–108	Thales Alenia	1250	2015	2022	1400	First launches in 2021
OneWeb	882–1980	OneWeb Airbus JV	145	2015	2020	1200	6 pilot satellites in February 2019
SpaceX Starlink	4425–11,943	SpaceX	227	2015	2020	550	120 launched in 2020
O3b (SES)	20 O3b 7 O3bm	Thales Alenia (O3b) Boeing (O3bm)	700	2008: O3b 2017: O3bm	2014: O3b 2021: O3bm	8000	O3b complete
Telesat LEO	117–512	TBC	300	2016	2021	1000–1248	Prototypes 2018
CASIC Hongyun	156	CASC	< 500	2017	2022	160–2000	Prototypes 2018
CASIC Hongyun	320	CASC	< 500	2017	2023	1100	Prototypes l 2018

optimum to: (*a*) squeeze as many small satellites into medium size launch vehicles such as Falcon 9, and Ariane 6, (*b*) launch 2–3 small satellites in dedicated small launchers such as VEGA, LauncherOne, Electron, and so on, and (*c*) mass produce in lean, highly automated, automotive style production lines, as OneWeb and SpaceX are currently undertaking (Kramer & Cracknell, 2008).

The ISRO has developed a versatile nano-satellite platform of about 5 kg to accommodate various experimental payloads in demonstration missions lasting from six months to a year. These missions are undertaken by small satellite platforms to reduce cost, time, and associated risks. In order to facilitate systematic development, ISRO has developed a standardized spacecraft bus for small satellite platforms. These fall under broad categories of 400 kg, 100 kg, and 10 kg satellites. In this category, the IMS-1 satellite (100 kg) was launched in 2008. It carried a highly miniaturized multispectral camera MX-T (5 kg). It had a four band imaging system that provided performance at par with the LISS-II camera of IRS-1A/1B. It also carried a compact imaging spectrometer HySI-T. This instrument provided measurement of the reflected the spectrum in the visible and near infrared wavelengths in 64 contiguous bands. The weight of HySI-T was only 3 kg. The Indian INS-1B nano-satellite, with scientific equipment for measuring the amount of neutral atomic hydrogen in the Earth's exosphere, also carries an experimental camera with origami optics (Gansvind, 2019).

Today, India has the world's largest constellation of remote sensing satellites in operation, which is being extensively used for resource management and integrated planning. The development of smaller, cheaper satellite technologies in recent years has led many companies to explore new ways of using low Earth orbit satellites. In the future, a low-cost communications network between low Earth orbit satellites can be established to form a spatial remote sensing network. This network would integrate with a large number of distributed ground sensors to establish ground-space remote sensing. In addition, satellites can easily cover large swaths of territory, thereby supplementing ground-based platforms. Thus, data distribution and sharing would become very easy.

The Dove satellite constellation, operated by Planet Labs Inc., has been deployed from both the International Space Station and conventional rockets to LEO to take daily images of the Earth surface. The Dove-1 nano-satellite was launched on April 21, 2013. Planet Labs had 87 Dove and 5 RapidEye satellites launched into orbit. In 2017, Planet launched an additional 88 Dove satellites. By September 2018, the company had launched nearly 300 satellites. In 2020, Planet Labs launched six additional high-resolution SkySats, and 35 Dove satellites. Following a January 2021 launch of 48 Planet SuperDoves, the company now operates a global constellation of over 200 active nano-satellites. The satellite constellation currently consists of more than 200 nano-satellites, weighing 5 kg each, and providing 3 m multispectral image resolution (Lappas & Kostopoulos, 2020). The constellation is capable of imaging the entire Earth surface each day. The constellation makes up history's largest fleet of Earth imaging satellites and can image up to 250 million km^2 per day, some of which is available under an open data access policy, providing up-to-date information relevant to a variety of mapping applications such as climate monitoring, crop yield prediction, urban planning, disaster response, and monitoring deforestation around the world.

In addition to Dove, there are 13 SkySat satellites, under control of the RapidEye, for a global agricultural monitoring system managed by the Planet company (Lebedev & Gansvind, 2010). The SkySat satellites weigh about 110 kg, and with dimensions of 0.6 m × 0.6 m × 0.95 m have imaging equipment created by the SkyBox Imaging startup (later Terra Bella), which obtains images in the PAN band with a resolution of 90 cm at nadir and in four spectral channels in RGB and NIR with a ground resolution of 2 m. It is also possible to record video with a resolution of 1.1 m in a PAN channel with a frequency of 30 frames per second. The creation of a lightweight carrier reduces the cost of launching a kilogram of mass and meets the growing demand for small satellite launch services.

On August 18, 2020, SpaceX's Falcon 9 rocket launched SkySats 19, 20, and 21 on yet another successful Starlink rideshare mission. Much like SkySats 16–18, which were launched by SpaceX on June 13, 2020, SkySats 19–21 were successfully injected into a drop-off orbit of approximately 207 × 370 km, 53 degree inclination, completing the campaign of 21 satellites originally planned by the SkyBox team in 2009 (Safyan, 2020). These three new SkySats join the 18 others already in orbit and significantly expand the capacity to provide world class, high-resolution images to a variety of commercial, governmental, academic, and nonprofit organizations.

The smallsat (< 500 kg mass satellites) market is going through significant expansion in terms of capabilities and demand. In the last couple of years, numerous companies have produced solutions, largely based around a constellation approach, to better deliver services and reach out to new users. The logic of lower cost constellations is to provide global connectivity from one system (Satcom) or high-frequency change detection (in Earth observation) or connecting devices and vehicles (information) for the Internet of Things (IoT), Machine to Machine (M2M), and traffic monitoring (AIS or ADS-B). It is aided by the advancement of satellite system miniaturization permitted by new technologies and/or advances in related sectors, particularly in computational technology with smallsats now providing operational services that were previously only achievable through heavier satellites.

It is expected that in the next few years, small spacecraft will be a regular and accepted capability for Earth remote sensing measurements. They are going to make new types of improved science measurements that otherwise would not be possible with other normal platforms. There is a steady downward trend in the mass of small satellites. According to the Space Works forecast, about 3,000 satellites with a mass of up to 50 kg will be launched by 2022, which is several times more than the expected number of large launches with a mass exceeding 1,000 kg (Kostev et al., 2016). The global consultancy Euroconsult predicts that although fewer than a thousand smallsats were launched from 2006 to 2015, up to 7,000 smallsats are likely to be launched between 2018 and 2027 for a variety of missions (Behrens & Lal, 2019). The year 2017 represented a 205% increase in nano/microsatellites launched as compared to 2016. The future is likely to see more small satellites, each of which is dedicated to a particular mission objective and carries a single instrument. Through this approach more and more countries around the world are becoming involved in Earth observations from space, not just in using the data from the major established systems but also in constructing their own systems.

4.6 SELECTION OF REMOTE SENSING IMAGES

Remotely sensed data, including both airborne and spaceborne sensor data, vary in spatial, radiometric, spectral, and temporal resolutions. Understanding the strengths and weaknesses of different types of remote sensing data is essential for the selection of suitable remotely sensed data for image classification. Scale, image resolution, and the user's need are the most important factors affecting the selection of remotely sensed data. The user's need determines the nature of classification and the scale of the area, thus affecting the selection of a suitable spatial resolution of remotely sensed data. In general, a detailed classification system is needed for classification at a local level, and thus high spatial resolution data, such as IKONOS, QuickBird, and SPOT 5 HRG data are helpful. At a regional scale, medium spatial resolution data, such as Landsat TM/ETM+, IRS LISS II, and Terra

ASTER are the most frequently used data. At a continental or global scale, coarse spatial resolution data, such as AVHRR, MODIS, IRS WiFS, and SPOT VEGETATION data are preferable (Lu & Weng, 2007).

Another important factor influencing the selection of sensor data is the atmospheric condition. The frequent cloudy conditions in the moist tropical regions are often an obstacle for capturing high-quality optical sensor data. Therefore, in such regions different kinds of radar data serve as an important supplementary data source. Since multiple sources of sensor data are now readily available, image analysts have more choices to select suitable remotely sensed data for a specific study. A combination of multi-sensor data with various image characteristics is usually beneficial to the application (Lefsky & Cohen 2003). In this situation, the economic condition is often an important factor that affects the selection of remotely sensed data and the time and labor that can be spared to the classification procedure, thus affecting the quality of classification results.

The main characteristics of a sensor that are important to the application include many parameters, briefly given as follows.

4.6.1 Spatial Characteristics

Although there are guidelines for selecting an appropriate spatial resolution, most analysts rely on experience and trial and error. The resolution of an image is selected that is a factor of 3 times less than the size of the features to be identified. For example, if you want to visually delineate features with a minimum size (minimum mapping unit) of 1 hectare (100 m × 100 m), a 30 m spatial resolution is probably sufficient, but to identify tree crowns of roughly 3 m × 3 m, finer resolution data at 1 m may be required. The positional accuracy with reference to known locations is a concept correlated with the spatial resolution. Positional or locational accuracy is generally better for ortho-rectified images. For a large area, a high spatial resolution is linked to large data volumes, which increase the time needed to analyze the images. To achieve high spatial resolutions, some sensors have a PAN band, which has good spectral width, including much of the visible and NIR portion of the EMS. For example, the PAN band on the Landsat 7 ETM+ sensor is 15 m, which can be incorporated with other bands to enhance the visual sharpness of the image, valuable for mapping some important land use and land cover types.

Table 4.14 shows the resolution and corresponding accuracy achieved from some of the satellite imagery sources available in the market today.

Table 4.14. Resolution of Images and Accuracy Achieved (Setyawan, 2019)

Spatial resolution (m)	Accuracy (m)
0.31	< 5.0
0.41	3.0
0.55	23
0.82	9
1.50	35
0.40	7.8
0.50	9.5

Figure 4.10 shows a qualitative comparison of the different optical and SAR satellite sensors and LiDAR instruments in terms of their aerial coverage and data resolution/quality with reference to earthquake studies (Rathje & Adams, 2008). In this figure, low-resolution, medium resolution, and high-resolution are taken as relative terms and not defined in terms of absolute values. The low-resolution optical (e.g., Landsat) and SAR (e.g., Envisat) satellites have 15–30 m spatial resolution, but provide the largest aerial coverage (in terms of scene size), such that fewer scenes are required to cover a large area. These satellite images are not detailed enough to generate comprehensive damage data due to an earthquake, yet they can provide generalized estimates of relative damage across a region. The

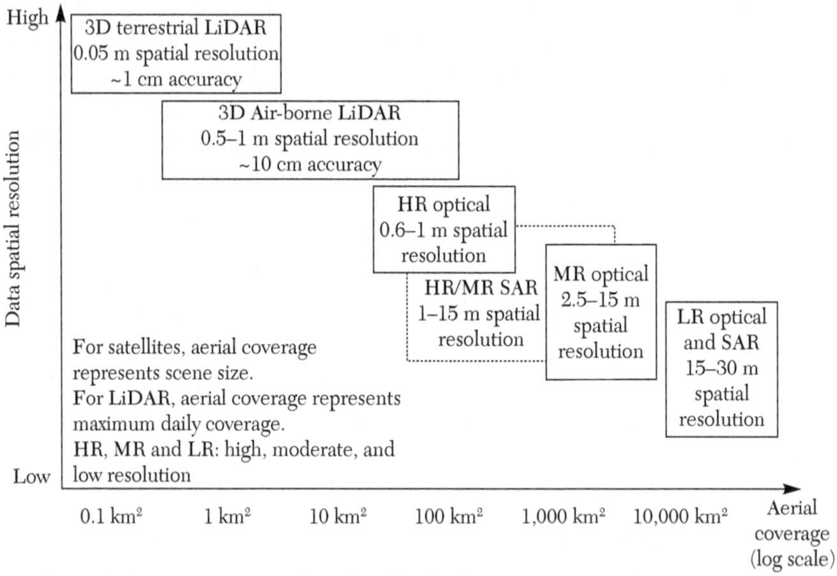

Figure 4.10. Comparison of resolution and aerial coverage of various remote sensing data (Rathje & Adams, 2008).

moderate-resolution optical satellites have a resolution between 2.5 to 15 m, but smaller scene sizes. The high-resolution optical satellites provide the most detailed data, but because of the smaller aerial coverage (up to 100 km^2 per scene) it is difficult to collate and expensive to obtain high-resolution imagery over the entire area affected by an earthquake. Moderate and high resolution SAR images can be used to obtain the details, particularly when clouds may hinder acquisition of moderate to high-resolution optical imagery. Airborne LiDAR can generate 3D digital terrain models over large to moderate areas (1–100 km^2 per day), but at significant cost to acquire the data. Finally, terrestrial LiDAR can provide the most detailed 3D models of damage, but it requires significant time for data collection and analysis if the area is large.

4.6.2 Spectral Characteristics

When evaluating the spectral quality of a particular image, there are three variables that are usually considered: bandwidth, band placement, and the number of bands. The *spectral bandwidth*, which refers to the range of wavelengths detected by a particular sensor, is particularly important when using hyperspectral imagery. However, analysis of hyperspectral imagery is still a new, developing field. *Band placement* defines the portion of the EMS that is used for a particular image band, which is important in order to identify a particular object of interest. This would enable the detection of spectral signatures that are characteristic of certain plant species or land uses. The number of bands is generally less important to carry out visual interpretation, which tends to use only three bands at a time, but is very important when using automated digital classification methods.

4.6.3 Repeat Interval

The repeat interval is the minimum time interval required by a sensor to record a particular feature/area twice. Wide swath widths tend to be associated with low spatial resolution, and are linked to shorter repeat cycles, thus increasing the temporal resolution. Some sensors with a very wide field of view can acquire multiple images of the same area on the same day. The pointable sensors have the advantage that they have the capability to view an area at an angle from a vertical line so they reduce the repeat time for recording an area. Early warning systems for sudden loss of habitat, such as earthquake, fire, and so forth, may require frequent (daily to weekly) acquisition of satellite images.

4.6.4 Radiometric Characteristics

A sensor records the intensity of a given wavelength as a single whole number between the minimum and maximum of a range, known as the radiometric resolution. For example, Landsat TM sensors can store values from 0 to 255, whereas IKONOS sensors can store values from 0 to 2,048. These values define the amount of variation within a wavelength band that can be detected, which is one aspect of sensitivity. Sensitivity is also defined by the sensor's dynamic range. Sensors of particular wavelength bands have extremes of sensitivity above and below which they cannot differentiate change in intensity (Lu & Weng, 2007).

Monitoring or assessment programs may require many categories (such as land use and land cover types) that are represented on the image. Accuracy of assessment is usually higher when mapping broader types of classes. The level of precision can be improved when using data of high spatial, temporal, spectral, and radiometric resolutions. The trade-off between these resolutions needs to be reconciled for the selection of satellite images for a particular application. Each remote sensing application has its own unique resolution requirements which need to be understood.

4.6.5 Image Area

The ground area covered by an image defines the footprint of the image. It depends on the swath width of the sensor, which can vary from as little as 8 m to more than 2,000 km. High spatial resolution images cover less ground area than lower resolution images. Large areas require a large number of scenes at higher spatial resolutions, but only a few scenes at moderate/poor resolution will cover the same area. Mosaicking a large number of scenes can be problematic and time-consuming, especially if the adjacent images are acquired during different seasons. It can also create inconsistencies in mosaicking because each image is taken under specific environmental conditions. It means that the same feature on the ground can produce a different reflectance based on various parameters, including humidity, cloud cover, and time of day. On the other hand, it saves a lot of work if the entire study area is covered on a single image, but the requirement would depend on the objectives and purpose of analysis.

4.6.6 Multi-Angle Images

Some satellite sensors can be pointed (off-nadir) over a particular target area to acquire more frequent images. For a satellite with off-nadir viewing capabilities, users can request a particular feature to be targeted at two different or more occasions of the satellite in the orbit, thereby acquiring more than one image of the same area in a short time. Many sudden and short-lived phenomena, such as earthquakes, requires images at very short intervals. Another advantage of pointable sensors is that they can be used to acquire stereo-images, which can be viewed in 3D to create DEMs.

4.6.7 Image Availability and Its Cost

Of the large number of potential image types, only a few are practical for monitoring over an entire country because these are either costly or have too small a data archive, especially outside the range of the country that launched the satellite. For example, the most practical data sources for biodiversity monitoring are those that record data from IR to visible bands to ultraviolet. Landsat data have been the most heavily used data because of their relatively medium spatial resolution (30 m) and moderate cost (Lu & Weng, 2007). The SPOT HRV images of similar quality to Landsat, having a pixel size of 20 m, can be costly for many for national level mapping. Recent data collected from comparable Indian satellites (IRS) at higher spatial resolutions are also useful because of their low cost, but may have smaller collections of archived data.

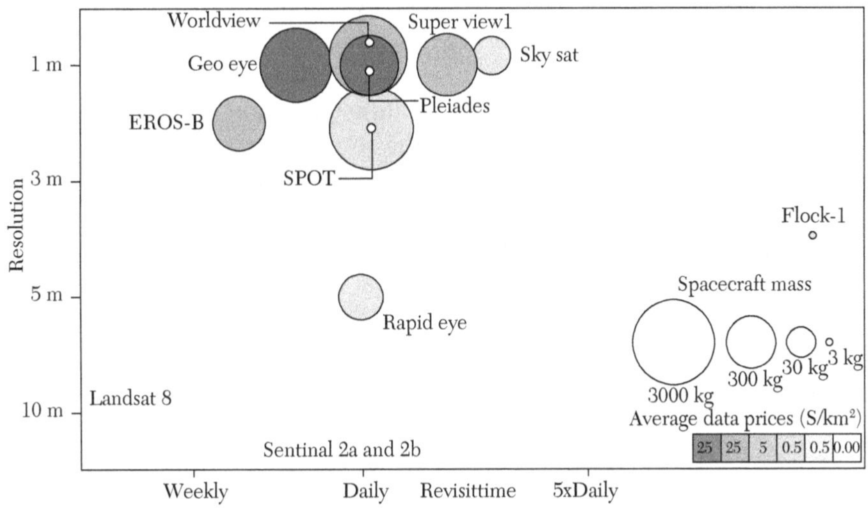

Figure 4.11. Spatial resolution *vs.* revisit time for various satellites.
Source: Satellite Applications Catapult 2017, adapted from EO21 Project

Figure 4.11 shows a relationship between spatial resolution and temporal resolution of various satellites as well as related costs. The coarser spatial resolutions, such as 250 m–1 km, from the MODIS and SPOT VEGETATION sensors, can be used for land cover monitoring. These data are usually free, well archived, and available soon after their acquisitions. However, they may not be optimal for monitoring the habitat extent and its rate of change. In most areas, changes occur in many small patches that are more detectable with higher resolution images. However, coarse resolution images can be processed quickly, and provide a valuable complement to finer resolution data. The availability of both types of data (coarser and higher resolutions) means that countries can maintain a monitoring system in which coarse resolution imagery can be used to estimate the change as part of an early warning system, and higher resolution data can be analyzed less frequently to produce condition and change assessments of greater precision. Very high spatial resolution data are usually very expensive, and it will be uneconomical to conduct monitoring of large areas from high-resolution commercial satellites.

4.6.8 Cost-Effectiveness of Analysis

Using remote sensing in combination with field surveys is likely to be the most effective solution to monitoring many land use and land cover classes at regular intervals. This is especially true for the monitoring of large areas and assessing the accuracy of classification. Since the total cost is likely to be considerable, the needs must be clearly defined by the users for deriving the information. According to Mumby et al. (1999), four types of costs are encountered when undertaking remote sensing activity: (*a*) setup costs, (*b*) field survey costs, (*c*) image acquisition costs, and (*d*) cost of processing the images. The largest of these are the setup costs, such as the acquisition of hardware and software. This may make up 40–72 percent of the total cost of the project, depending on the objectives. However, increase in computational power is cutting down the costs of associated hardware and software. If setup costs are fixed, that is, if the required hardware and software are already available, then field survey costs may dominate the project budget at approximately 80 percent of total costs (or 25 percent with set-up costs), but again it would depend on the objectives and outcome of the project. For example, the field survey is a vital component of any land use and land cover mapping program, and may constitute approximately 70 percent of the time spent on a project. In general, the more time and effort spent on field surveying and ground-truthing, the higher the accuracy of the results analyzed from images.

The third major cost is associated with the acquisition of images. The selection of images is made considering the trade-offs between the map accuracy and the cost of imagery; the latter depends on the size of the study area and choice of sensor. For example, SPOT MS may be a cost-effective satellite sensor for mapping an area of 60 km × 60 km (it would fall within a single SPOT scene), but for larger areas, Landsat TM is a cost-effective data. The ASTER data is also economical but new acquisitions must be requested ahead of time online. The MODIS data is free thus far, but it has a short historical record and is suitable only for coarse scale regional monitoring. The relative cost-effectiveness of digital airborne sensors and aerial photography are more difficult to ascertain because they are case specific. Generally, the acquisition of digital airborne imagery is more expensive than the acquisition of color aerial photography.

4.6.9 Technical Expertise

After the data are acquired, technical knowledge and expertise are needed to process and analyze the imagery, and this is likely to cost more in salaries than the imagery or hardware costs. In addition, a successful monitoring program requires repeated images to carry out the mapping and analysis; all except setup costs are likely to be recurring (Turner et al., 2003). One big challenge to incorporate remote sensing technology into a large project work is to acquire the technical expertise required to handle and interpret the data. Managing even small quantities of satellite imagery requires specialized software, hardware, and trained personnel. However, new software tools are more users-friendly making remote sensing data accessible to everyone; in addition, the possibilities for training in remote sensing are growing rapidly. The number of experts who can work with these platforms and sensors is likely to grow in the future with the increased availability of open-source data and software.

CHAPTER 5

Image Preprocessing Approaches

5.1 INTRODUCTION

Image preprocessing may include the detection and restoration of bad lines, image enhancement, geometric rectification of image, radiometric calibration and atmospheric correction, and topographic correction. If ancillary data are also to be used, quality evaluation of these data are also necessary before they can be incorporated into a classification procedure. Accurate image registration of remotely sensed data is a prerequisite for a combination of different images which are to be used together in a classification process (Lu & Weng, 2007). Preprocessing functions involve those operations that are normally required prior to analysis of the main data and extraction of information.

Preprocessing operations, sometimes referred to as *image restoration and rectification*, are done to correct for sensor- and platform-specific radiometric and geometric distortions of data, which will also lead to enhancement of images. Enhancement of images is done to visualize the imagery and make the interpretation better (Garg, 2021). It improves the interpretability of images through human vision. Radiometric errors may be present in the image due to variations in scene illumination, viewing geometry, atmospheric conditions, and sensor noise and response. Each of these will vary depending on the sensor and platform used to acquire the image and the prevailing atmospheric conditions during the data acquisition. Although some corrections for illumination, atmospheric influences, and sensor characteristics are done prior to distribution of data to the users, still the image may not be fully optimized for interpretation and analysis.

Digital imagery has the advantage that the pixel values (DN) can be manipulated through some mathematical function. It may also be desirable to convert and/or calibrate the data to known (absolute) radiation or reflectance units to facilitate a comparison between the multitemporal data or multi-sensor data. Variations in illumination and viewing geometry between images can be corrected by modeling the geometric relationship. Radiometric and geometric corrections are particularly required if several images collected by different sensors at different dates are to be used and results compared, or used to create a mosaic from these multiple images in order to maintain a uniform illumination condition from one scene to another scene.

5.2 GRAY LEVEL THRESHOLDING

This method is generally used to segment an input image into two distinct classes; one for those pixels which have DN values above a threshold value defined by the analyst, and another for those pixels below the chosen threshold value. Thresholding can be used to prepare a binary image with 0 and 1 values, where 1 indicates the feature which we want to highlight and 0 denotes the features which we want to supress. Figure 5.1 shows gray level thresholding carried out on Landsat-8 OLI panchromatic image to

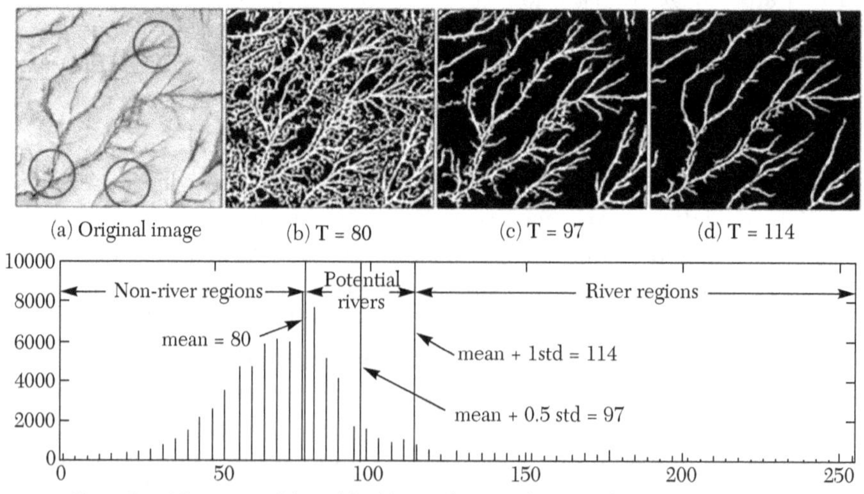

Figure 5.1. Histogram of the original image (bottom figure) and river detection under different global thresholds: (a) original image (top left); (b), (c), and (d) are the images after applying different threshold values (Yang et al., 2015).

create a binary image showing only the river and non-river pixels (Yang et al., 2015). Various threshold values of 80, 97, and 114 produced drainages of various orders, showing a dense pattern with the lower threshold value. Three circles on the original image show the thin rivers considered as potential rivers whose gray values are in the range of mean and mean +1 standard deviation (σ) of the image. The value of mean $+0.5\sigma$ is considered as an optimal threshold to discern thin rivers from the image background.

The gray level slicing method often highlights a certain range of desired gray level imagery. The application includes improving certain features, such as water in satellite imagery and flaws in X-ray imagery. There are two basic methods for level slicing. One is to display a high value for all gray levels within a desired range and a low value for all other gray levels. Another approach is based on transformation, such as brightening the desired gray level range but still maintaining background and gray level tones in an image.

5.3 IMAGE ENHANCEMENT

Image enhancement simply refers to the techniques used to increase the human interpretability and ability to derive meaningful information from remotely sensed images. No generic algorithm exists which could be universally applied to produce the optimum brightness and contrast in an image consisting of several objects, for example, forest, agriculture, roads, rivers, snow, urban, open land, and so forth. Enhancement of an image can be carried out by using different methods, and there is no general theory for determining "good" image enhancement, when it comes to human vision (Castleman, 1996). Thus, each image requires a custom-based approach for its enhancement, depending on the image quality, image features, topography of the area, orientations, shape and size of the features, and so on.

Enhancement is required to identify the features present on the image and make accurate visual interpretation. It is also needed to digitize the features (vector form) from the image (raster form) required for GIS work. Visual inspection of the image plays a very strong role in visual as well as digital interpretation of remote sensing. Through enhancement, the image quality is improved so that the resulting image is better than the original image for a specific application. The main aim of image enhancement is to bring out the details that are hidden in an image or to increase the contrast of a low/poor contrast or dark image. However, it does not add any additional information to the original image but it enhances the visual appearance of existing features.

Image enhancement can be performed by either improving the contrast or removing the noise or blurriness from the image. The contrast enhancement can be better understood with the concept of an image histogram. It is necessary to display the histogram to understand the image characteristics before undertaking the enhancement. Various enhancement techniques will manipulate/change the range of DN values of an image, which can be seen by displaying its histogram, prior and after the enhancement.

Image enhancement is done by increasing the apparent distinction between features in the image, which involves the use of a number of statistical and image manipulation functions (Garg, 2019). Image features are enhanced by three broad operations: (*a*) point operations, where the value of a pixel is enhanced independent of characteristics of neighborhood pixels within each band of the image; (*b*) local (neighborhood) operations, where the value of a pixel under consideration is enhanced based on DN values of neighboring pixels; and (*c*) global operations, where the output value at a specific coordinate is dependent on all the values in the input image. They are shown in Figure 5.2. The types of operations that can be applied to digital images to transform an input image a[*i,j*] into an output image b[*i,j*]. Image enhancement techniques based on point operations are also called as *contrast enhancement techniques*, while the techniques based on neighborhood operations are also known as *spatial enhancement techniques*.

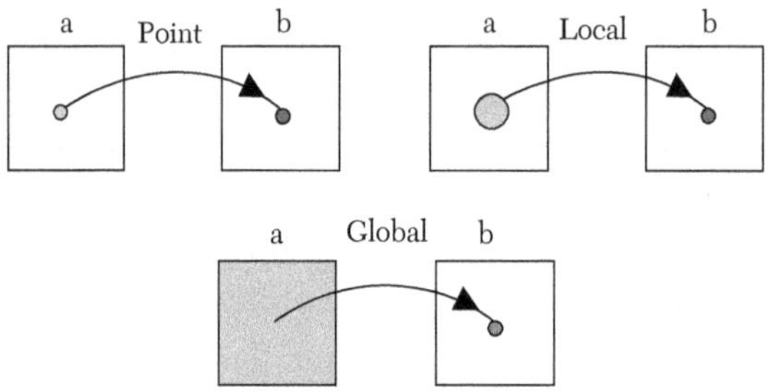

Figure 5.2. Image enhancement for point, local, and global operations; (a) input image and (b) output image.

5.4 CONTRAST ENHANCEMENT

One of the most important quality factors in a satellite image is its contrast. Contrast is related to the color and brightness, and is the difference in DN values of two different objects at the same wavelength. The contrast of an image can be defined as "the ratio of maximum intensity (DNmax) to the minimum intensity (DN_{min}) over an image, that is DN_{max}/DN_{min}." Contrast is created by the difference in luminance reflected from two adjacent surfaces (Lillesand et al., 2004).

Sometimes during image acquisition, low contrast may result due to poor illumination, lack of dynamic range in the image sensor, and so forth. The unfavorable environmental conditions reduce the contrast and the details in captured remote sensing images. If the contrast of an image is too low, it might be impossible to distinguish between two objects and they would be perceived as the same object to human vision. Our visual system is more sensitive to contrast than absolute luminance; therefore, we can perceive the world similarly regardless of the considerable changes in illumination conditions. If the contrast of an image is highly concentrated on a specific DN values range, information may be lost in those areas which are excessively and uniformly concentrated. Therefore, there is a need to optimize the contrast of an image in order to represent all features.

Contrast enhancement is frequently referred to as one of the most important operations in image processing. The aim of contrast stretching is to increase the dynamic range of gray levels in the image being processed. Since contrast is an important quality factor in remote sensing images, contrast enhancement techniques are required for better information representation and visual perception. The demand of high-quality remote sensing images is rapidly increasing due to their widespread use in various applications, such as military, geosciences, water, object detection, agriculture, and infrastructure.

The intensity or brightness of the pixels in a digital image can be graphically depicted through a histogram. A typical grayscale digital image and its corresponding histogram are presented in Figure 5.3. In this histogram, grayscale values are plotted on an x-axis for an 8-bit image with

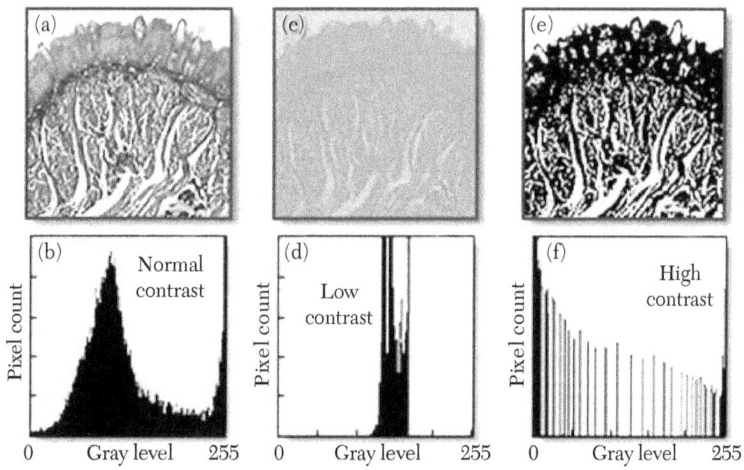

Figure 5.3. Contrasts and histograms of three types of images.

range from 0 to 255 (a total of 256 gray levels), and the number of pixels comprising each gray level (frequency) is plotted on a y-axis (Richards & Jia, 2013). Three types of images along with their histogram are shown here: normal contrast, low contrast, and high contrast. The first histogram occupies the entire range of DN values and shows almost a normal distribution, the second histogram shows mid-range DN values with a higher frequency, and the third type of histogram presents the higher frequency with DN in the lower range.

The histogram provides a convenient representation of a digital image by indicating the relative number of pixels at each brightness level and the overall intensity distribution of the image in general. Statistics derived from the histogram can be employed to compare contrast and intensity between images. In addition, the number of pixels in the histogram can be used to determine area measurements of a specific feature. Histograms can be altered by image processing algorithms to produce corresponding changes in the image quality.

Contrast enhancement involves altering the narrow range of original DN values so that a greater range of DN values are occupied, thereby increasing the contrast between objects as well as their backgrounds. It basically improves the interpretability for human views, and provides "better" input for image processing techniques (Richards & Jia, 2013). There are many different techniques and methods of enhancing the contrast and details in an image. Two broad categories of enhancement techniques are; linear and nonlinear techniques, and each of them has further categorization, as presented in Table 5.1. Some popular techniques are briefly explained as follows:

Table 5.1. Categories of Image Enhancement

Linear enhancement	Nonlinear enhancement
Minimum-maximum	Histogram equalization
Percentage linear	Histogram matching
Piecewise linear	Logarithmic
	Inverse log
	Exponential
	Arc tangent
	Power law

5.4.1 Linear Contrast Enhancement

Linear contrast enhancement, also known as *contrast stretching*, expands the original DN values of the image linearly into a new range. It is the simplest form of enhancement which involves identifying the minimum and maximum brightness values (DN) in the image and applying a linear transformation to stretch this range to occupy the full range (e.g., 0–255 in an 8-bit image; Garg, 2019). It is best applied to remotely sensed images with Gaussian type histograms, where all the brightness values fall within a narrow range of the histogram. In imagery, the raw digital data often occupies a small portion of the available range of DN values. Linear enhancement involves changing these original values so that a greater range is achieved, thereby increasing the contrast between the targets/objects and their backgrounds.

There are three methods of linear contrast enhancement: (*a*) minimum-maximum linear contrast stretch, (*b*) percentage linear contrast stretch and, and (*c*) piecewise linear contrast stretch.

5.4.1.1 Minimum-maximum linear contrast stretch

In the min-max method, spectral differences can be detected by the minimum and maximum DN values and stretched to full gray levels. For example, Figure 5.4 (*a*) shows a minimum value of 84 and a maximum

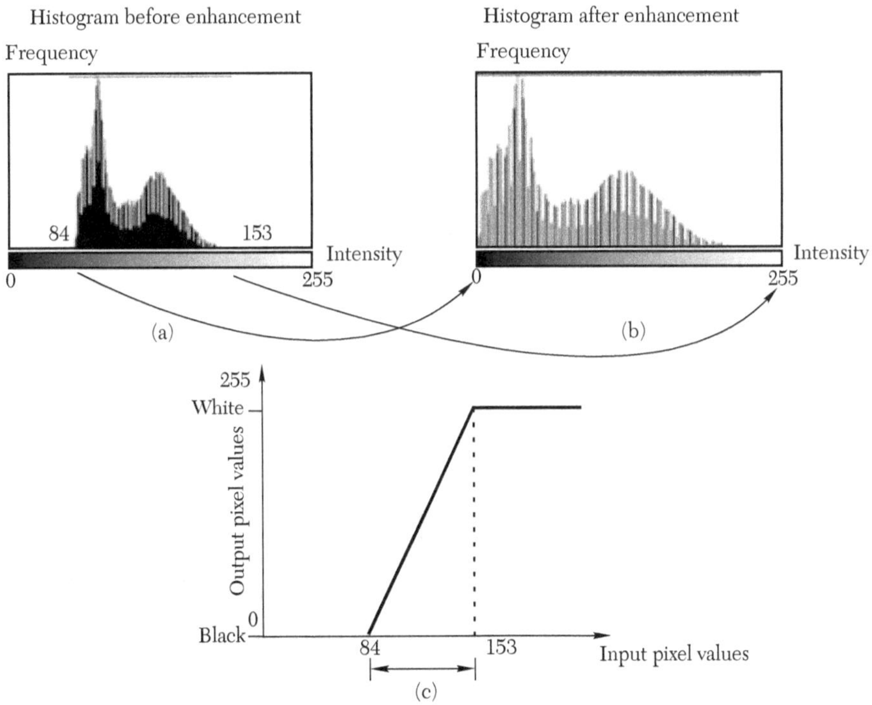

Figure 5.4. Histogram (a) original image and (b) after enhancement, and (c) DN values input versus output images.

value of 153 in an 8-bit image. The minimum brightness value of 84 can be brought to 0, and the maximum value of 153–255. These 70 levels (84–153) are occupying less than one-third of the full 256 levels available. When such an image is viewed without enhancement, the values of 0 to 63 and 154 to 255 are absent. The relationship used is given as follows (Jain, 1989):

$$DN' = [(DN - DN_{min})/(DN_{max} - DN_{min})] \times \text{No. of gray levels} \tag{5.1}$$

Where DN′ represents the output digital values in the new image, while DN represents the digital value at that pixel location from the input image. In this equation, the "DN_{min}" and "DN_{max}" are the minimum and maximum DN value, respectively, in the input image. Here "No. of gray levels" represents the total number of intensity values that can be assigned to an image to be enhanced (e.g., it is 255 in an 8-bit image).

A linear stretch will uniformly expand this small range so that it covers the full range of values from 0 to 255 (Figure 5.4 [*b*]). The intermediate gray level values of the original image would be similarly and proportionately set to new values. These types of enhancements are best applied to remotely sensed images with normal or near-normal distribution, that is all the brightness values fall within a narrow range of the image histogram and only one mode is apparent. It enhances the contrast in the image so that light toned areas appear lighter and dark areas appear darker, making visual identification and interpretation of features much easier (Castleman, 1996). This approach of linear contrast stretching is known as a minimum-maximum linear contrast stretch, and the distribution of DN in the values in input image and output image is shown in Figure 5.4 (c). Figure 5.5 shows the results of linear contrast enhancement on a Landsat ETM+ band 4 image (a, b, c, and d) where many details in the land area, including habitation, are much clearer, as compared to the original image.

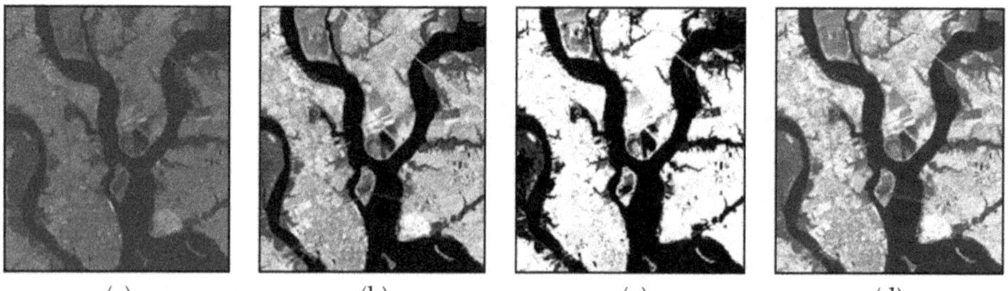

(a) (b) (c) (d)

Figure 5.5. (a) Original Landsat ETM+ image, (b) linear contrast enhancement, (c) piecewise linear enhancement, and (d) histogram equalized image.

5.4.1.2 *Percentage linear contrast stretch*

The percentage linear contrast stretch is similar to the minimum-maximum linear contrast stretch except that this method uses a specified minimum and maximum values that lie in a certain percentage of pixels from the mean of the histogram. In this approach, a standard deviation (σ) from the mean value is often used to push the tails of the histogram beyond the original minimum and maximum values (Jain, 1989). It is not necessary that the same percentage be applied to each tail of the histogram distribution. As an example, water pixels in an image lying between 0 and 12 DN values, when stretched between 0 and 255, may reveal variations in water (polluted and clear). Similarly, a percentage stretch of the same image between values 25–45 may yield detailed vegetation information, useful for the delineation of healthy vegetation.

5.4.1.3 *Piecewise linear contrast stretch*

It is used when the distribution of a histogram in an image is bi- or trimodal. In such cases, it is required to stretch only certain DN values of the histogram for enhancement of selected areas. A piecewise linear contrast enhancement involves the identification of a number of linear enhancement steps that expands the brightness ranges in the modes of the histogram (Colwell, 1983). It can be expressed by:

$$f(x,y)=\begin{cases} ax, & 0\le x\le x_1 \\ b(x-x_1), & x_1\le x\le x_2 \\ c(x-x_2) & x_2\le x\le B \end{cases} \tag{5.2}$$

where $f(x, y)$ is the piecewise linear contrast stretch in the image; a, b and c are appropriate constants, which are the slopes in the respective regions; and B is the maximum DN value.

A piecewise linear contrast stretch normally assumes that the data values are continuous and there is no break in the values between high, middle, and low range (Garg, 2019). It is a very powerful enhancement procedure, and therefore it requires a greater understanding with the modes of the histogram and the features they actually represent on the image. Figure 5.6 shows the concept used in a piecewise linear contrast stretch. In this method, a series of small min-max stretches are established within a single histogram. In the histogram, several breakpoints are introduced that increase or decrease the contrast of the image for a given range of values. The higher the slope, the narrower the range of values, which results in a wider output spread for those same values, and thus increases the contrast for that range of values. A low sloping line results in a lower contrast for the same range of values. Figure 5.5 (c) presents the results after piecewise linear enhancement (e.g., water pixels are enhanced).

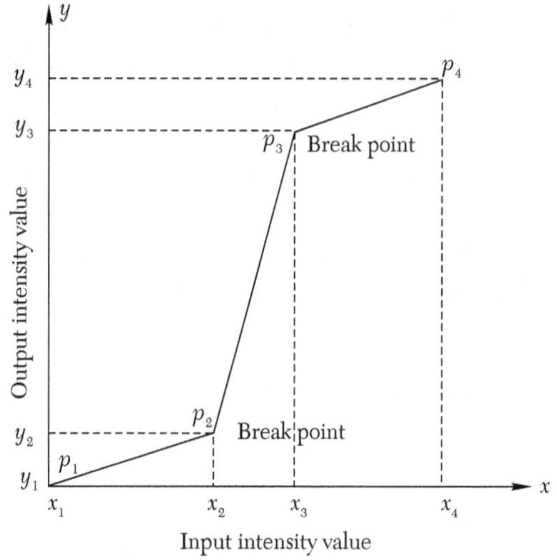

Figure 5.6. Concept used in piecewise linear enhancement.

5.4.2 Nonlinear Contrast Enhancement

In these methods, the input and output DN values follow a nonlinear transformation. One major disadvantage of a linear stretch is that it assigns an equal number of DN levels for infrequently occurring values as it does for frequently occurring values. This can result in some of the features being enhanced, while other features, which may be of greater importance, may not be sufficiently enhanced. Under such circumstances, nonlinear contrast enhancement techniques perform better. They effectively preserve the edges and details of images. Nonlinear contrast enhancement includes histogram equalization, logarithmic method, exponential method, power-law transformation method, and so on for enhancement of images (Castleman, 1996).

5.4.2.1 Histogram equalization

Histogram equalization is one of the most useful forms of nonlinear contrast enhancement for adjusting image contrast. A histogram simply plots the frequency at which each gray level occurs. In histogram equalization, all pixel values of the image are redistributed so there are approximately an equal number of pixels to each of the user-specified output gray scale classes, as shown in Figure 5.7 (Jain, 1989). A uniform distribution of the input range of values across may not always be an ideal enhancement. In such cases, a histogram equalization approach may give better results.

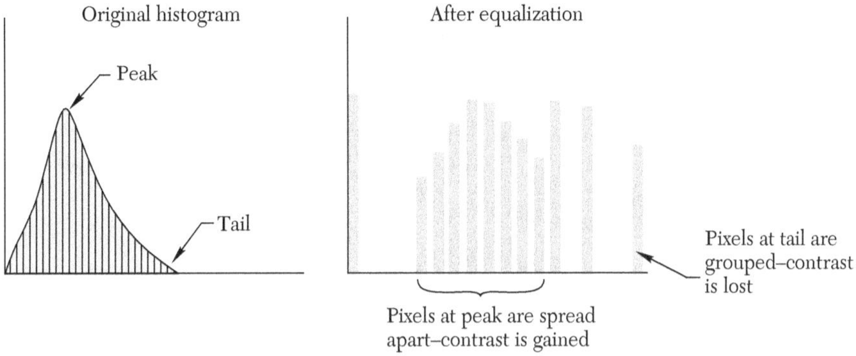

Figure 5.7. Original histogram (left) and histogram equalization (right).

The histogram equalization approach assigns more display values (range) to the frequently occurring portions of the histogram. In this way, the detail in such areas will be better enhanced relative to those areas of the original histogram where values occur less frequently. It accomplishes this by effectively spreading out the most frequent intensity values, that is stretching out the intensity range of the image. The histogram equalization automatically reduces the contrast in very light or very dark parts of the image associated with the tails of a normal distributed histogram. It is effective only when the original image has poor contrast to start with (Figure 5.5 [*d*]), otherwise histogram equalization may degrade the image quality. It can also separate pixels into distinct groups, if there are very few output values over a wide range.

The histogram equalization method in general will flatten out the probability distribution of an image and increase its dynamic range. However, its strength depends on the contrast of the original image. The lower the contrast, the greater the strength is. Since its simplicity makes it easy to implement, histogram equalization is usually applied in many areas, including texture synthesis, medical image processing, and speech recognition. An advantage of this method is that it is a fairly straightforward technique and an invertible operator. So theoretically, if the histogram equalization function is known then the original histogram can be recovered. A disadvantage of the method is that it increases the contrast of background noise, while decreasing the usable signal (Richards & Jia, 2013). In addition, the generated result is over-enhanced if there are several peaks in the histogram, which leads to a saturation of artifacts and harsh appearance in the enhanced image.

5.4.2.2 Logarithmic stretching

In logarithmic stretching, the logarithmic function is used for scaling the original DN levels into the output range. It is a two-step process. In the first step, the *log* values of the input DN values are computed. In the second step, the *log* values are linearly stretched to fill the complete range of DN values (e.g., 0–255 in 8-bit image). The general form of logarithmic stretching uses the following form (Lillesand & Kiefer, 1994).

$$DN_{out} = a \log_e (b \times DN_{in}) \tag{5.3}$$

where a and b are constants, DN_{in} is the input DN value from the original image, and DN_{out} is the output value after logarithmic enhancement.

The *log* transformation maps a narrow range of low input gray level values into a wider range of output values. It has the greatest impact on the brightness values found in the darker part of the histogram, as seen from Figure 5.8. A logarithmic stretch compresses the higher brightness values within an image and disproportionately expands the darker values. In this approach, the original darker values are given a greater relative contrast than that of the original values at the higher end of the brightness scale. This allows better visual discrimination among the darker features of the image, while maintaining some limited recognition of brighter features. For example, water features, which are typically darker in tone, would benefit from logarithmic stretches, while the land features, which are brighter,

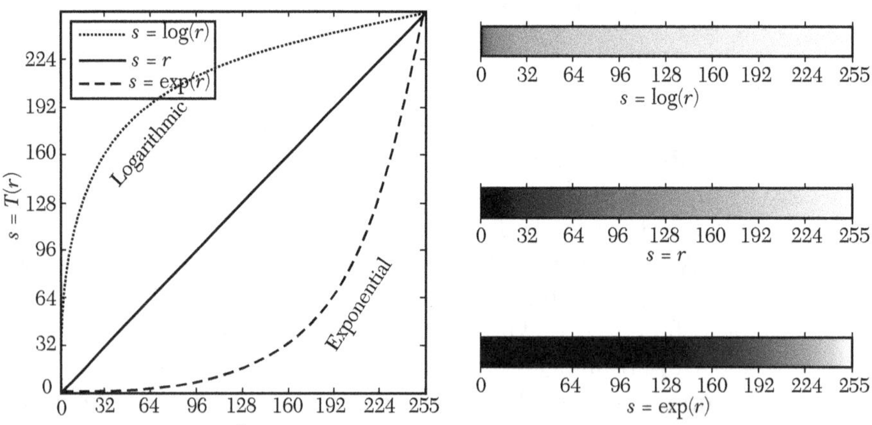

Figure 5.8. Effect of logarithmic enhancement and exponential enhancement.

would still be somewhat recognizable (Jenson, 2007). Figure 5.9 (middle image) shows the results of logarithmic enhancement. As expected, the darker areas are enhanced by the logarithmic enhancement.

(a) (b) (c)

Figure 5.9. (a) Original image, (b) logarithmic enhancement image, and (c) exponential enhancement image.

5.4.2.3 Exponential stretching

Exponential function stretches the image in an opposite way to logarithmic enhancement. It enhances the brighter pixels, while decreasing the dynamic range in low DN value regions (dark), as is clear from Figure 5.8. The equation used could be as follows (Richards & Jia, 2013).

$$\mathrm{DN}_{\mathrm{out}} = ae^{b\times \mathrm{DN}_{\mathrm{in}}} \tag{5.4}$$

where a and b are constants, $\mathrm{DN}_{\mathrm{in}}$ is the input DN value from the original image, and DNout is the output value after exponential enhancement.

Figure 5.9 (right image) shows the results of exponential enhancement applied to an image. The brighter areas are enhanced by the exponential enhancement.

5.4.2.4 Power-law transformations

It is an alternative to both the logarithmic and exponential transforms. It is the "raise to a power" transform in which each input pixel value is raised to a fixed power (Jain, 1989):

$$\mathrm{DN}_{\mathrm{out}} = a\,(\mathrm{DN}_{\mathrm{in}})^{b} \tag{5.5}$$

where a and b are constants; if the value of $b > 1$, it enhances the contrast of high-value portions of the image at the expense of low-value regions, while for $b < 1$, the effect is reversed. This gives the power-law transform properties similar to both the logarithmic ($g < 1$) and exponential ($g > 1$) transforms.

5.5 SPATIAL FILTERING

A digital image can be viewed as a two-dimensional function $f(x, y)$, and the x–y plane indicates spatial position information, called the *spatial domain*. The filtering operation based on the x–y space neighborhood is called *spatial domain filtering*. The spatial filtering process can be represented mathematically as (Richards & Jia, 2013):

$$g(x, y) = T\,[f\,(x, y)] \tag{5.6}$$

where $g\,(x, y)$ is the output image, T is an operator, and $f\,(x, y)$ is the input image.

A matrix of coefficients is used to average the value of each image pixel with the neighborhood of pixels surrounding it, which is shown as:

$$\overline{g}(x,y) = \frac{1}{K}\sum_{i=1}^{K} g_i(x,y) \tag{5.7}$$

Spatial filtering methods mainly focus on point to point (pixel to pixel) manipulation of the image, as they process every single pixel intensity of the image. Figure 5.10 shows the process of spatial filtering with a 3 × 3 filter (with 3 rows and 3 columns; a total of 9 pixels). The filter is also known as a *template, kernel,* or *window*). For ease of computation, the filters can have odd kernel dimensions (3 × 3, 5 × 5, 7 × 7, and so on). Convolution

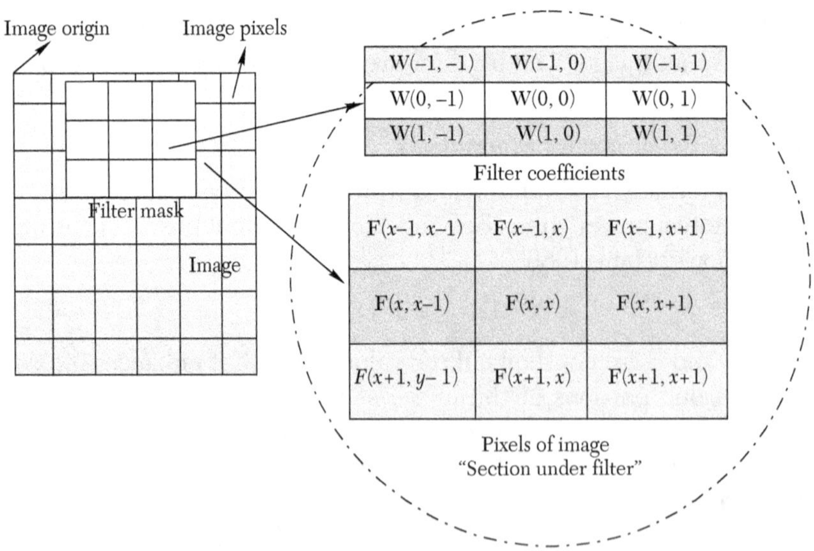

Figure 5.10. Applying a 3 × 3 filter to an image at a central pixel.

filtering is a common mathematical method of implementing spatial filters. In this, each pixel value is replaced by the mathematical operator applied over an area on the periphery of that pixel (Garg, 2019). Generally, spatially filtered images must be contrast stretched to use the full range of image display. Each pixel of the filter in spatial filtering is assigned a weight (w). For example, the new value R at the center of the filter is computed as:

$$R = (1/9) \times [w(-1, -1) f(x - 1, y - 1) + w(-1, 0) f(x - 1, y) + \ldots + w(0, 0) f(x, y) + \ldots + w(1, 0) f(x + 1, y) + w(1, 1) f(x + 1, y + 1)] \quad (5.8)$$

In the filtering process, the filter is moved pixel by pixel in the image function $f(x, y)$ so that the center of the filter coincides with the pixel (x, y). At each pixel (x, y), the filter's value is computed based on the specific elements of the filter. Spatial filtering is used for land use studies, highlighting textures of urban features, road systems, and agricultural areas. It is also useful for enhancing the lineaments that may represent significant geological structures, such as faults, veins, or dykes (Sabins, 1996). It can also enhance image texture for discrimination of lithology and drainage patterns.

Filters usually suppress (de-emphasize) certain frequencies and pass (emphasize) others. Filters that pass high frequencies and, hence, emphasize fine detail and edges, are called *high-pass filters*; *low-pass filters* suppress high frequencies, and are useful in smoothening of an image, and may reduce or eliminate "salt and pepper" type noise (Garg, 2019). Low-pass filtering tends to reduce deviations from local averages, and thus smoothens the image. The difference between the input image and the low-pass filtered image is the high-pass filtered output. The low-pass and high-pass filtered images derived from Landsat TM band 5 are shown in Figure 5.11.

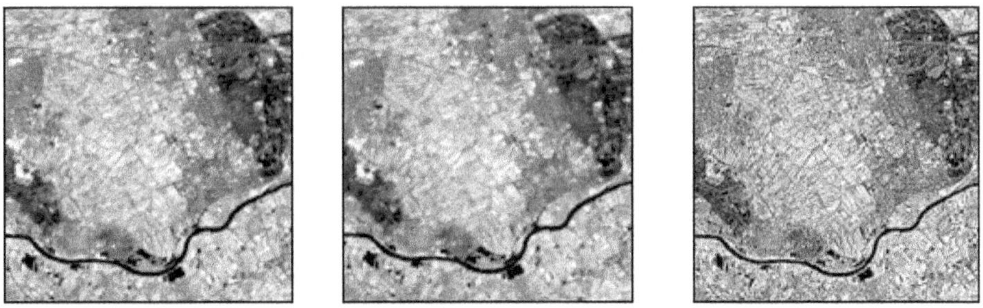

Figure 5.11. Landsat TM Band 5 of New York area (left), results of low-pass filter (middle), and results of high-pass filter (right) superimposed on original image.

In general, images of practical interest consist of several dominant *spatial frequencies*. Fine details in an image involve a larger number of changes per unit distance than the image features having broader details. Figure 5.12 presents an example of a low spatial frequency image and a high spatial frequency image. The mathematical technique for separating an image into its various spatial frequency components is called *Fourier analysis* (Jain, 1989). After an image is separated into its components by "Fourier Transform," it is possible to emphasize certain groups (or bands) of frequencies relative to others and recombine the spatial frequencies into an enhanced image.

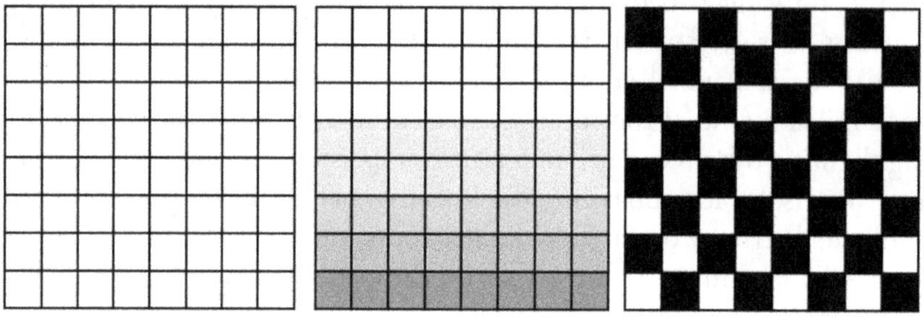

Figure 5.12. Spatial frequency images.

5.5.1 Low-Pass Filters

The low-pass filter is also known as a *smoothing* or *blurring filter* because it removes high spatial frequency details and preserves the low frequency components. These filters are mainly used to smoothen image features and to remove noise but often at the cost of degrading an image's spatial resolution (blurring; see Figure 5.11, middle image). The filter elements contain equal weights, replacing the central pixel value with a weighted average of the surrounding DN values. Low-pass filters are employed to emphasize the broader changes present within the image, but are very useful for noise reduction, as an intermediate step to image enhancement. High frequency initiates noise into an image that decreases image quality (Richards & Jia, 2013). Low-pass filters are briefly explained as follows.

1. **Average (Mean) filter:** It replaces the data value by the average of the given data point. The filter is given as follows:

1/9	1/9	1/9
1/9	8/9	1/9
1/9	1/9	1/9

2. **Median filter:** It replaces each central pixel with the median value within the kernel size. The median filter arranges all the values within the kernel in increasing or decreasing order, and searches for the "middle" value (i.e., median pixel value) within the kernel size. Therefore, it smoothens an image while preserving the edges larger than the kernel dimensions. A median filter is considered to be superior to the average filter because of two reasons. Firstly, the median is a set of n numbers (e.g., 9 in the case of a 3 × 3 kernel) where n is an odd integer which is always one of the data standards existing in the set so it does not compute a new set of values. Secondly, the median is less convoluted to error or to detect data consequence (Garg, 2019).

3. **Gaussian filters:** A Gaussian filter is a smoothening operator quite similar to the mean filter, but based on a Gaussian shape, which smoothens, preserving the edges more than a mean filter. The Gaussian convolution function is represented as (Jain, 1989):

$$G(x,y) = \frac{1}{2\pi\sigma^2} e^{\left(\frac{x^2+y^2}{2\sigma^2}\right)} \tag{5.9}$$

It is assumed that the distribution has a zero mean. Actually, it is possible to use the Gaussian operator for both smoothening and enhancement purposes even using a corresponding template. The standard deviation of the Gaussian filter determines the degree of smoothening. The equivalent convolution kernel that approximates a Gaussian low-pass filter is given as follows.

0.0007	0.0256	0.0007
0.0256	0.8948	0.0256
0.0007	0.0256	0.0007

5.5.2 High-Pass Filters

High-pass filters remove low spatial frequency details and preserve the high frequency components linked to local variations. This operation can be performed by subtracting the low-pass filter image from the original image to obtain a new image with enhanced local contrast (Richards & Jia, 2006). The high-pass filter enhances the spatial frequencies that are less than the size of the kernel matrix used in the operation. Therefore, the selection of kernel size produces a strong control on the level of detail generated from the process. For example, very small features (e.g., roads, canals) may require small kernel sizes (i.e., 3 × 3), whereas very large features, such as buildings, reservoirs, and so forth, may require much larger kernels (i.e., 7 × 7 or bigger) in order to effectively enhance them.

High-pass filters enhance the differences between values of neighboring pixels; these changes in values are represented by edges and lines. They are used for image sharpening or edge enhancement. The image of a high-pass filter is shown in Figure 5.13 (middle image). The edges are the border between two types of surfaces (forest–field), and lines are rivers, streams, and roads. High-pass filters enhance such objects, which are smaller than a half of a filter window. Bigger objects are usually suppressed. The elements of a high-pass moving kernel matrix characterized by a high central value, typically surrounded by negative weights, are given as follows (Jain, 1989).

−1	−1	−1
−1	8	−1
−1	−1	−1

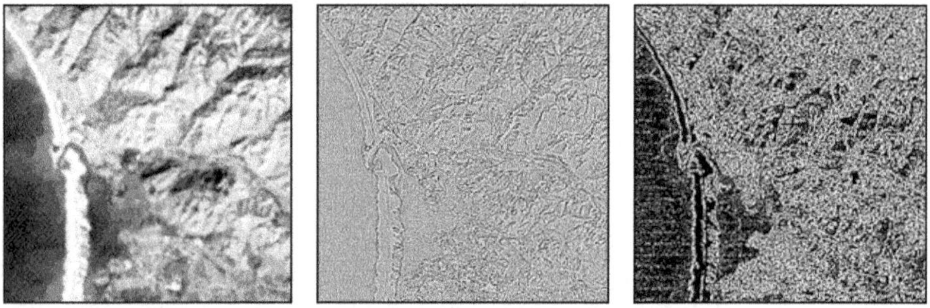

Figure 5.13. Landsat TM Band 2 (left), high-pass filtered image (middle), and results after Sobel filter (right).

5.5.2.1 Edge detection filters

Edge detection is a fundamental tool in image processing, machine vision, and computer vision, particularly in the areas of feature detection and feature extraction (Richards & Jia, 2013). The edges for an image are always the important characteristics that offer an indication for a higher frequency. The points at which image brightness changes sharply are typically organized into a set of curved line segments, known as *edges* (Garg, 2019). Edges are boundaries between different textures. Detection of edges for an image may help for image segmentation, data compression, as well as image matching such as image reconstruction, and so on. It involves a set of mathematical methods which aim at identifying points at which the image brightness changes sharply or, more formally, has discontinuities. A classical method of edge detection involves convolving an image with an operator (a 2D filter), that is constructed to be sensitive to large gradients in an image while returning values of 0 in uniform regions. The output image from edge detection shows a binary image with the edges shown in a white tone.

The geometry of the operator determines the characteristic direction in which it is most sensitive to edges. Operators can be optimized to horizontal, vertical, or diagonal edges. Edge detection is difficult in noisy images, since both the noise and images contains high frequency content. Operators used in noisy images are typically larger in scope, so they can average enough data to discount localized noisy pixels. Edge detection filters are also used to reduce the noise resulting in blurred and distorted images (Jenson, 2007).

Consider a continuous brightness function instead of a discrete brightness in an image, as shown in Figure 5.14.

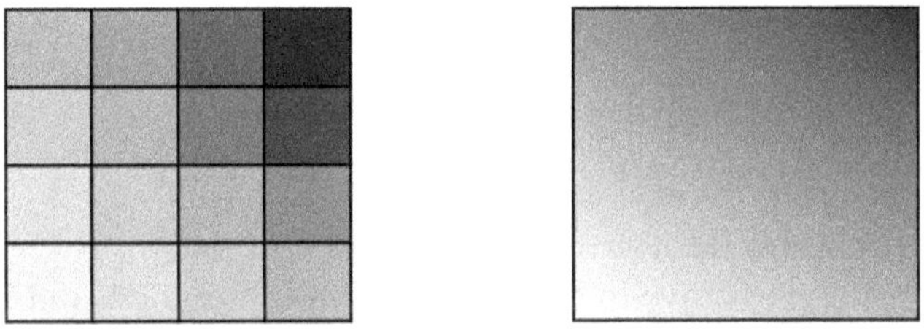

Figure 5.14. Discrete brightness and continuous brightness image.

The spatial gradient (∇) can be defined as (Jain, 1989):

$$\nabla\phi(x,y) = \frac{\delta}{\delta x}\phi(x,y)i + \frac{\delta}{\delta y}\phi(x,y)j \tag{5.10}$$

where $\phi(x, y)$ is the brightness value at pixel location x, y and i and j are unit vectors in the x and y directions, respectively.

For edge detection, the magnitude and direction of change is given, respectively, by:

$$|\nabla| = \sqrt{\nabla_1^2 + \nabla_2^2} \quad \text{and} \quad \angle\nabla = \tan^{-1}(\nabla_2/\nabla_1)$$

where $$\nabla_1 = \frac{\delta}{\delta x}\phi(x,y) \qquad \nabla_2 = \frac{\delta}{\delta y}\phi(x,y) \tag{5.11}$$

In other words, the magnitude of the vector is the vector (Pythagorean) sum of the gradient in the x-direction and the gradient in the y-direction.

Equation 5.11 is for continuous gradients. For discrete gradients (i.e., across pixels in imagery), the derivatives are replaced with differences. Two difference-based spatial operators are the Roberts operator and the Sobel operators (Gupta, 1991). Three general approaches to edge detection are used: (*a*) using an edge-detection template, (*b*) subtracting a smoothed image from its original, and (*c*) calculating spatial derivatives (spatial gradients). There are many filters used for edge detection, such as Robert, Prewitt, Sobel, Laplacian of Gaussian (LOG) operators, Canny operators, Hilbert

Transform, and so on. The most common method for edge detection is to calculate the differentiation of an image. The first-order derivatives of an image are computed using the gradient, and the second-order derivatives are obtained using the Laplacian.

In *Prewitt gradient filters*, the filters are specified by (Colwell, 1989):

$$
\begin{aligned}
[h_x] &= \frac{1}{3}\begin{bmatrix} 1 & 0 & -1 \\ 1 & 0 & -1 \\ 1 & 0 & -1 \end{bmatrix} = \frac{1}{3}\begin{bmatrix} 1 \\ 1 \\ 1 \end{bmatrix} \times [1 \quad 0 \quad -1] \\
[h_y] &= \frac{1}{3}\begin{bmatrix} 1 & 1 & 1 \\ 0 & 0 & 0 \\ -1 & - & -1 \end{bmatrix} = \frac{1}{3}\begin{bmatrix} 1 \\ 0 \\ -1 \end{bmatrix} \times [1 \quad 1 \quad 1]
\end{aligned}
\tag{5.12}
$$

Both h_x and h_y are derivatives in the x and y direction, respectively. Each filter takes the derivative in one direction and smoothens in the orthogonal direction using a one-dimensional version of a uniform filter.

A *Sobel filter* is based on convolving the image with a small, separable, and integer-valued filter in the horizontal and vertical direction (Jain, 1989). Technically, it is a discrete differentiation operator, computing an approximation of the gradient of the image intensity function. At each point in the image, the result of the Sobel operator is either the corresponding gradient vector or the norm of this vector. The gradient approximation that it produces is relatively crude, in particular for high frequency variations in the image. Mathematically, the operator uses two 3x3 kernels which are convolved with the original image to calculate approximations of the derivatives; one for horizontal changes, and another for vertical. If G_x and G_y are two-dimensional convolution operations which at each point contain the horizontal and vertical derivatives, the resulting gradient approximations can be combined to give the gradient magnitude, using (Colwell, 1989):

$G_x =$

−1	0	1
−2	0	2
−1	0	1

$G_y =$

−1	−2	−1
0	0	0
1	2	1

$$G = \sqrt{G_x^2 + G_y^2} \tag{5.13}$$

we can compute the gradient direction using the formula:

$$\theta = \tan^{-1}\left(\frac{G_y}{G_x}\right) \tag{5.14}$$

where, for example, θ is 0 for a vertical edge which is darker on the right side.

The Sobel filter, being nondirectional, detects edges in all directions, as shown in Figure 5.13 (right image). In this example, the Sobel edge enhancement algorithm is applied that finds an overabundance of discontinuities to emphasize the sharp boundaries. This filter highlights abrupt discontinuities, such as rock joints and faults. It gives the brightness value as well as the direction of edges. The Sobel is a product of averaging and a differential kernel, hence it computes the gradient after smoothing the noisy pixels.

5.5.2.2 *Directional filters*

A directional filter forms the basis for edge detection methods. An edge within an image is detected when a steep gradient occurs between adjacent pixel values. This change in values is measured by the first derivatives (often referred to as slopes) of an image. Directional filters can be used to compute the first derivatives of an image. Directional filters can be designed for any direction within a given space. Figure 5.15 shows various direction filters in a 3 × 3 window. For images, x- and y-directional filters are commonly used to compute derivatives in their respective directions. The following array is an example of a 3 × 3 kernel for an x-directional filter (the kernel for the y-direction is the transpose of this kernel). The array is an example of one possible kernel for an x-directional filter (Jain, 1989). The resulting image shows some edge information. This information represents the slopes in the x-direction of the image. The filtered image can then be scaled to highlight these slopes.

–1	0	+1
–1	0	+1
–1	0	+1

Vertical

–1	–1	–1
0	0	0
+1	+1	+1

Horizontal

0	+1	+1
–1	0	+1
–1	–1	0

Diagonal NW/SE Edge

–1	+1	0
+1	0	–1
0	–1	–1

Diagonal NW/SW Edge

Figure 5.15. Directional filters.

In the first filter, the central value is the horizontally accumulated difference; in the second filter, the central value is the vertically accumulated difference; in the third filter, the central value is the accumulated difference across a NW/SE line; and in the fourth filter, the central value is the accumulated difference across a NE/SW line. The results of directional filters are shown in Figure 5.16.

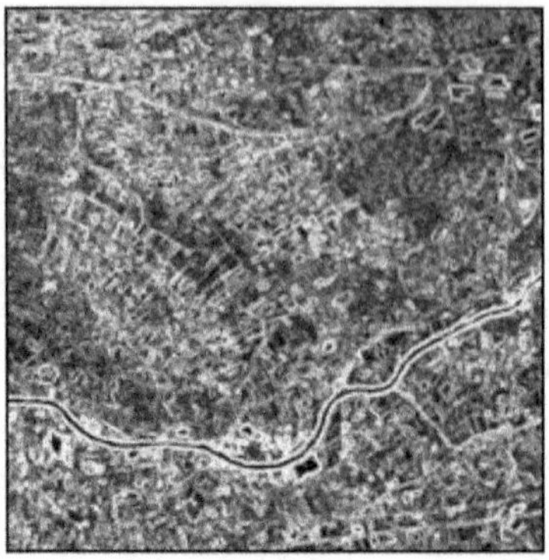

Figure 5.16. Results of directional filters applied to original image shown in Figure 5.11 (left).

5.5.2.3 Sharpening filters

There are a couple of filters that can be used for sharpening the image. For example, the Laplace operator, which is based on the second order differential, is one of the most popular sharpening filters (Jain, 1989). The corresponding filter templates are shown in Figure 5.17. With the sharpening enhancement, two numbers with the same absolute value represent the same response, so w_1 is equivalent to the following template w_2: Laplacian edge enhancement weights within the filter window are distributed in such way that the sum of the central pixel and all other pixels is equal to zero. The sharper results are obtained if the original image is added to the Laplace filtered image, as shown in Figure 5.18.

0	1	0
1	–4	1
0	1	0

1	1	1
1	–8	1
1	1	1

0	–1	0
–1	4	–1
0	–1	0

–1	–1	–1
–1	8	–1
–1	–1	–1

Figure 5.17. Filters for sharpening the image.

Figure 5.18. The sharp image is the superimposition of the original image shown in Figure 5.11 (left) with the results of the Laplace filter.

5.6 NOISE REMOVAL

Noise in an image may be due to irregularities or errors that occur in the sensor response and/ or data recording and transmission. Common forms of noise include *systematic striping* or *banding and dropped lines*. Both of these effects are to be corrected before enhancement or classification is performed (Colwell, 1983). Striping was common in early Landsat MSS data due to variations and drift in the response over time of the six MSS detectors. The "drift" was different for each of the six detectors, causing the same brightness to be represented differently by each detector. The overall appearance was thus a "striped" effect. The process made a relative correction among the six detectors to bring their apparent values in line with each other. Dropped lines occur when there are system errors which result in missing or defective data along a scan line. Dropped lines are normally corrected by replacing the line with the pixel values in the line above or below, or with the average of the two values.

5.7 CLOUD REMOVAL

Clouds affect all visible and IR bands, hiding features twice: once with the cloud, another with its shadow. Clouds cannot be completely eliminated, although images can be corrected for cloud shadows by applying some advanced algorithms. Shadows due to topography of the terrain can be removed by Ratio methods. If there are clouds in the image, it can be removed by creating a mosaic with another overlapping image. Various studies have used CNN, the inpainting algorithm, the cloning method, and so forth, for removal of clouds from satellite images.

5.8 RADIOMETRIC CORRECTIONS

The scattering may reduce, or attenuate, some of the energy illuminating the Earth's surface as well as energy traveling from the target to the sensor, affecting radiometric accuracy. Radiometric correction is a preprocessing method to reconstruct physically calibrated values by correcting the spectral errors and distortions caused by sensors, sun angle, topography, and the atmosphere (Castleman, 1996). With large variations in spectral response from a diverse number of objects (e.g., urban, agriculture, forest, open land, snow, water, etc.), no generic radiometric correction algorithm could be applied to display the optimum brightness range and contrast for all the objects. Thus, for each image, a custom adjustment of the range and distribution of brightness values is usually done. Various methods of atmospheric correction can be applied ranging from detailed modeling of the atmospheric conditions during data acquisition to simple calculations based solely on the image data.

Radiometric corrections involve removal of sensor or atmospheric "noise," to more accurately represent the ground conditions. It may be required to (*a*) improve the variations within an image, (*b*) correct for data loss, (*c*) remove the haze, (*d*) create a mosaic, and (*e*) compare multibands and temporal images. It is done to modify the DN values to account for noise, that is, contributions to the DN values that are a result of the intervening atmosphere and sun-sensor geometry (Eastman, 1999). Figure 5.19 shows the image mosaic created after applying radiometric corrections to the original image.

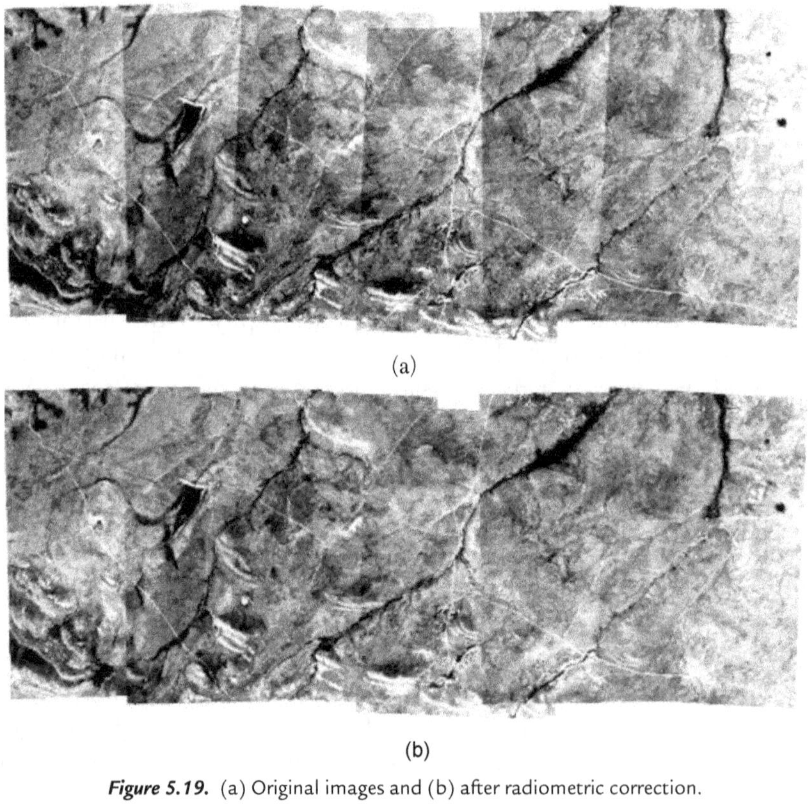

(a)

(b)

Figure 5.19. (a) Original images and (b) after radiometric correction.

5.8.1 Reflectance to Radiance Conversion

For many quantitative applications of remote sensing data, it is necessary to convert the DN values to measurements in units which represent the actual reflectance or emittance from the surface/object. This may be done based on detailed knowledge of the sensor response and the way in which the analog signals (i.e., the reflected or emitted radiation) are converted to digital numbers, called *analog-to-digital (A-to-D)* conversion (Colwell, 1983). By solving this relationship in the reverse direction, the absolute radiance can be calculated for each pixel, so that comparisons can be made accurately over time and between different sensors. The DN reflectance values can be converted to absolute radiance values (L), given as follows. This is useful when comparing the actual reflectance from different sensors, for example, TM and SPOT, or TM and ETM (Landsat-5 and -7), as follows:

$$\mathrm{DN} = a\,\mathrm{L} + b \tag{5.15}$$

where, a = gain and b = offset

The radiance value (L) can be calculated as:

$$\mathrm{L} = [\mathrm{L}_{max} - \mathrm{L}_{min}] \times (\mathrm{DN}/255) + \mathrm{L}_{min} \tag{5.16}$$

where L_{max} and L_{min} are known values from the sensor calibration (given in the handbook of that satellite or supplied by the agency who has designed the sensor).

5.8.2 Atmospheric Correction Models

Atmospheric models can be used to account for the effects of scattering and absorption in the atmosphere. A number of parameters are required to accurately apply the atmospheric correction, such as the amount of water vapor and distribution of aerosols (Richards & Jia, 2013). Sometimes this data can be collected by field instruments that measure atmospheric gases and aerosols, but this is often an expensive and time-consuming task. Other satellite data can also be used to help estimate the amount and distribution of atmospheric aerosols. Many software packages include special atmospheric correction modules that use atmospheric radiation transfer models to produce an estimate of the true surface reflectance.

5.8.3 Atmospheric Haze Correction

The simplest method for haze correction is known as the *dark object subtraction method* which assumes that in case of no haze, the pixel will have a zero DN value; for example, deep water in NIR will have complete absorption, and therefore those pixels should have zero DN values. This method examines the observed brightness values in an area of shadow or for a very dark object (such as a large clear lake) and determines the minimum DN value (Lillesand et al., 2004). The correction is applied to all DN values of each band by subtracting the minimum observed DN value, determined for each band, from all pixel values in each respective band. Since scattering is wavelength dependent, the minimum value will vary from band to band. This method is based on the assumption that the reflectance from these dark features, if the atmosphere is clear, should be very small, if not zero. If their values are much greater than zero, then they are considered to be affected by atmospheric scattering. Lower wavelengths are more subject to haze, which falsely increases the DN values.

This method is used when there is no data available on atmospheric conditions and aerosol properties at the time the image was acquired. The accuracy of these techniques are generally lower than the physically based correction methods, but they are the simplest and most used atmospheric correction methods, as they don't require any atmospheric data. However, these methods do have problems if dark pixels are not present in the image.

Figure 5.20 shows the process involved which includes identifying one dark (x_1) and one bright (x_2) object on the affected image which can also be identified on the ground. Their ground reflectances are also measured as Rx_1 and Rx_2, respectively. Radiance at the sensor of x_1 and x_2 objects can be computed from the image as Lx_1 and Lx, respectively. The two points x_1 and x_2 plotted on the graph are shown by a line whose equation is $R = (L - a) \times S$ where a is the intercept and S is the slope of the line $= (Rx_1 - Rx_2)/(Lx_1 - Lx_2)$.

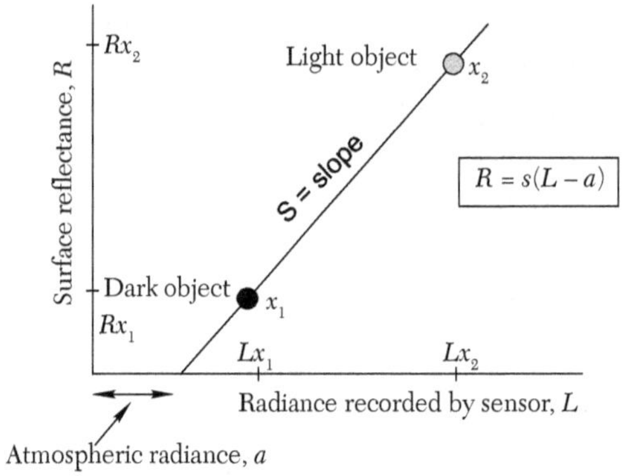

Figure 5.20. Graph between dark pixel and white pixel reflectance and radiance values.

5.9 GEOMETRIC CORRECTIONS

Geometric corrections include correcting for geometric distortions due to sensor-Earth geometry variations, and conversion of the data to real-world coordinates (e.g., latitude and longitude) on the Earth's surface. Geometric corrections are required when several images are to be mosaicked and/or compared together (e.g., change assessment) for quantitative measurements (Garg, 2019). Raw remote sensing data contain geometric distortions without any real geographic coordinates, and therefore these are not as such useful for overlay with other map layers, or a comparison between image scenes.

There are two types of corrections affecting the geometry of the image, that is, *systematic* and *non-systematic errors*. Systematic errors, also known as *predictable errors*, are introduced by the remote sensing system itself or in combination with the Earth's rotation or curvature characteristics. These systematic distortions are often identified and corrected using prelaunch or in-flight platform ephemeris (i.e., the geometric characteristics of the sensor system and the Earth at the time of data acquisition).

Non-systematic errors or *random errors* are usually introduced by phenomena that vary in nature through space and time. The most important external variables that can cause geometric errors in remote sensor data are random movements by the aircraft or spacecraft at the time of data collection, which usually involve *altitude* changes and/or *attitude* changes (roll, pitch, and yaw; Colwell, 1983). Geometric distortions due to these changes can be corrected using ground control points (GCPs) and appropriate mathematical models. A GCP defines a location of a specific point on the ground surface (e.g., a road intersection, corner of a building) that can be identified on the imagery and located accurately on a map and whose real coordinates are known.

To geometrically rectify a remotely sensed image to a map coordinate system, two operations are to be performed: *spatial interpolation* and *intensity interpolation*. In spatial interpolation, the geometric relationship between the input pixel coordinates (column and row; referred to as x, y) and the associated map coordinates of this same point (X, Y) is established through a process known as *georeferencing*. The intensity interpolation involves the determination of pixel brightness values of the output image using the brightness values from the input image where the pixels of output images have different locations and orientations due to transformation of input image into output image. It is done through a process known as *resampling* (Richards & Jia, 2013).

5.9.1 Georeferencing of Images

Georeferencing is the process of geospatially referencing data and information objects—datasets, maps, photographs, and imagery—to their proper locations on Earth. It means that the internal coordinate system of a digital map or aerial photo or satellite image can be related to a ground system of geographic coordinates. Georeferencing can be seen as an umbrella term for techniques which are concerned with the unique identification of geographical objects, such as roads, places, bridges, buildings, or agricultural areas, with their geographical locations.

Georeferencing usually takes place using a *projection system* and *geodetic reference system*, which comprise a standard coordinate system for the Earth (Garg, 2021). A map projection is the means by which rgw 3D surface of the Earth is drawn on a 2D surface (map, image). An appropriate map projection is selected for this purpose. If a relatively small area is to be mapped, the map projection may not be as important as when mapping large areas (countries, continents) because in small areas the amount of distortion in a particular projection may be almost negligible. For selecting a proper projection system, the general guidelines are: (*a*) if the country to be mapped is in the tropics, a cylindrical projection is used; (*b*) if it is important to maintain proper proportions, data should be displayed using an equal area projection (however, shape may not be preserved); (*c*) equal area projections are well suited to thematic data; and (*d*) where shape is important, a conformal projection is used (Hackeloeer et al., 2014).

Having selected a map projection, one needs an estimation of the Earth's shape. Earth is assumed as a sphere, however, the Earth more closely resembles an ellipsoid (also called an oblate spheroid, hence the term map spheroid). Several ellipsoids/spheroids are available for use, as given in Table 5.2. Other reference systems may also be adopted depending on the region of the globe and specific application fields; the option is available in image processing software.

Table 5.2. Examples of Some of the Spheroids in Use around the World

Spheroid	Area
Clarke 1866	North and Central America, Caribbean, and the Philippines
Clarke 1880	France and Africa
Bessel 1841	Indonesia
Everest 1830	India
Australian National	Australia
Helmert	Egypt

Georeferencing is also required as the directions of satellite motion and onboard sensors, while taking images, are not exactly north-south or west-east, respectively, so images are not a perfect square shape but they have some skewed shape. The georeferencing process orients the image to the Earth's reference system so it can be viewed, compared, and analyzed with other geographic data (Richards & Jia, 2013). It is fundamental to GIS work as it refers to the process of assigning locations to geographical objects within a geographic frame of reference.

To georeference an image, the positions of GCPs are determined. The GCPs are normally selected as prominent objects whose geographical locations can be accurately determined either from the maps or GPS. A number of GCP pairs are used to establish the nature of the geometric coordinate transformation that must be applied to rectify every pixel in the output image (X, with a value from a pixel in the unrectified input image (x, y). The GCPs ideally should be well distributed throughout the image, in each of the corners and middle of the image. At least four GCPs are required for georeferencing an image using a first order transformation model, but additional points help increase the accuracy. Once the image is georeferenced, each pixel on it has real coordinates associated.

The entire process of georeferencing involves establishing the GCPs, inputting the known geographic coordinates of these points, selecting the coordinate system and other projection parameters, running the transformation model, and then minimizing the *rms* (root mean square) error. The *rms* error is the difference between the actual coordinates of the GCPs and the coordinates predicted by the transformation model (polynomial), which should be within ± one pixel, to accept the results of georeferencing process. If this value is higher, the GCPs selected which give a higher *rms* error are to be modified/deleted/replaced. Again *rms* error is computed as above, until the threshold of ± one pixel error is achieved. The accuracy of georeferencing depends on the number, accuracy, and distribution of the GCPs and the choice of polynomial model applied. A first order polynomial is given as follows (Jain, 1989).

$$\begin{aligned} X &= ax + by + c \\ Y &= dx + ey + f \end{aligned} \tag{5.17}$$

where X and Y are the coordinates of the output image/map, x and y are the coordinates from the input image, and a to f are the coefficients which are to be determined.

There are six unknowns (a to f) in the previous equations, which require a minimum of three GCPs to compute the value of these coefficients.

Once these are known, any image coordinate can then be substituted in the equations to get the corresponding ground position on the used map coordinate system. In general, a higher order of polynomial with more numbers of GCPs is preferred. The higher the order, the more complex the distortion that can be corrected, as shown in Figure 5.21. However, transformations higher than the third order are rarely needed. Higher order transformations require more GCPs, as shown in Table 5.3, and, thus, will involve progressively more processing time.

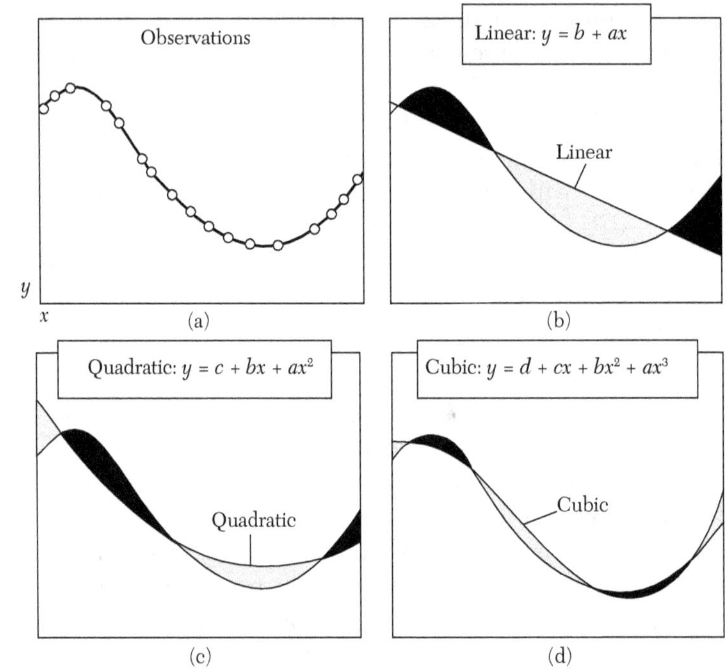

Figure 5.21. Various polynomials and associated distortions.

Table 5.3. Polynomial Order and GCPs Required

Polynomial order	Minimum GCPs required
1	3
2	6
3	10
4	15
5	21
6	28
7	36

Figure 5.22 (*a*) shows the original image where corners of the wall are skewed due to the satellite view angle. Figure 5.22 (*b*) presents the geometrically corrected image in which the rectangularity of the wall corners is maintained.

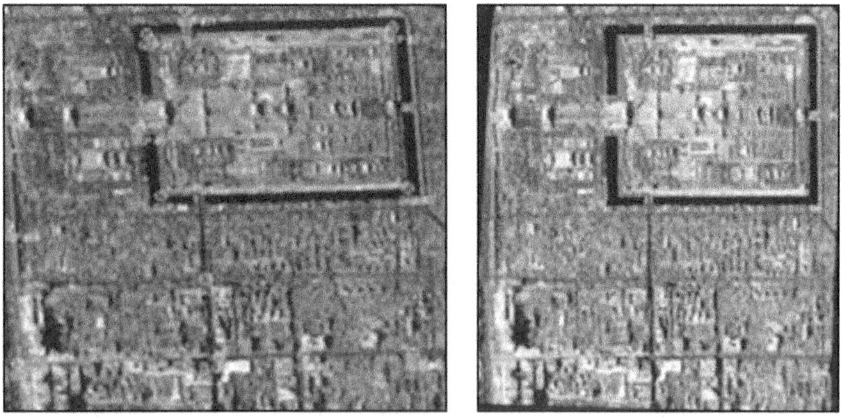

(a) (b)

Figure 5.22. (a) The original image (left) and (b) geometrically corrected image (right).

5.9.2 Resampling Methods

After the coordinate transformation, the pixels in the image may have been oriented differently than the way they were originally present in the raw image. The attribute value of the pixel location is to be interpolated for the cells oriented to the new coordinate system. In order to actually geometrically correct the original image, a procedure called *resampling* is used to determine the digital values to place in the new pixel locations of the corrected output image (Garg, 2019). The resampling process calculates a new DN value of the pixel from the original digital pixel values in the raw image. There are three commonly used resampling methods: *nearest neighbor*, *bilinear interpolation*, and *cubic convolution*.

5.9.2.1 Nearest neighbor resampling

It uses the DN value from the pixel in the original image which is nearest to the new pixel location in the corrected image. Once the location of the cell's center on the output image dataset is located on the input image, the nearest neighbor will determine the location of the closest cell center on the input image and assign the value of that cell to the cell on the output image. This is the simplest and most popular method used in resampling. This technique is a good choice for discrete (categorical) data since it does not alter the value of the input cells. Since the output cell values remain the same, nearest neighbor should be used for nominal or ordinal data, where each value represents a class, member, or classification (categorical data such as a land-use, soil, or forest type). The nearest neighbor can be performed before radiometric correction and classification (Castleman, 1996).

The nearest neighbor method, however, may result in some pixel values being duplicated while others are lost. It also tends to result in a disjointed or blocky image appearance. When this method is used to resample from a larger to a smaller grid size, there is usually a "stair stepped" effect around diagonal lines and curves. On linear thematic data (e.g., roads, streams), the method may result in breaks or gaps in a network of linear data.

The process has been explained through Figure 5.23(*a*). Consider an output-corrected image created from an input (original) image that has been rotated by some degree in a georeferencing operation, and thus will require resampling. For each output cell, a value needs to be derived from the input image. The output cell is light gray which has to be processed. In the nearest neighbor approach, the cell of the input image which is closest to the output cell is the dark gray cell. The DN value of this dark gray cell is selected and replaced in the light gray shade of the output image by this approach. This process is repeated for each cell in the output image.

5.9.2.2 Bilinear interpolation resampling

It takes a weighted average of the surrounding four pixels in the original image nearest to the new pixel location of the output image. It applies weights based on the distance of the four nearest cell centers smoothening the output pixel. The weighted averaging process alters the original pixel DN values and creates entirely new DN values in the output image. This may be undesirable if further processing and analysis, such as classification based on spectral response, is to be done. If this is the case, resampling may best be done after the classification process. Since the values for the output cells are calculated according to the relative position and the value of the input cells, bilinear interpolation is preferred for data where the location from a known point or phenomenon determines the value assigned to the cell (that is, continuous surfaces). Elevation, slope, temperature gradients, annual precipitation, intensity of noise from an airport, and salinity of the groundwater near an estuary are some examples represented as continuous surfaces, and are most appropriately resampled using bilinear interpolation. This interpolation method results in a smoother-looking surface than can be obtained using the nearest neighbor method (Campbell, 2011).

The process is shown in Figure 5.23(b). The output cell is shaded in light gray which has to be processed by taking the four nearest cells of the input image shown in dark gray shade. For bilinear interpolation, the four input cell centers (dark gray points) nearest to the processing cell center (light gray) are identified, the weighted average is calculated as per the distance, and the resulting value is assigned as the output value for the processing cell (light gray). This method is often used when changing the cell size of the data, such as in resolution merging SPOT and TM scenes, and produces less "blocky" images, and is more accurate than the nearest neighbor. Since pixels values are averaged, bilinear interpolation has the effect of a low-frequency convolution. Edges are smoothed, and some extremes of the data file values are lost.

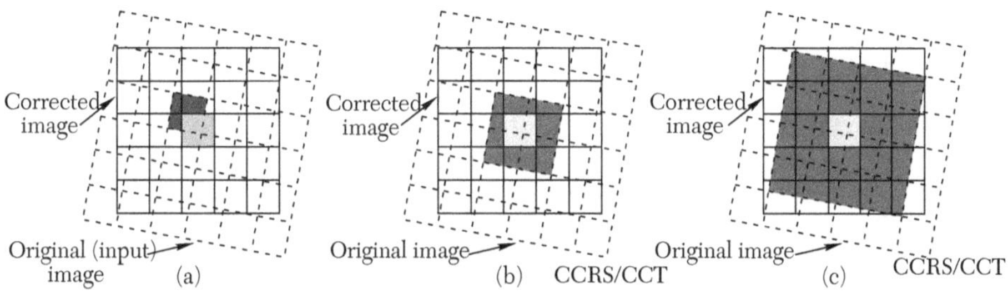

Figure 5.23. Resampling methods (a) nearest neighbor, (b) bilinear and (c) cubic convolution. ***Source:*** Antrop and De Maeyer, 2005.

5.9.2.3 Cubic convolution resampling

It computes a distance weighted average of a block of sixteen pixels from the original image which surrounds the new pixel location of the output image. Cubic convolution is similar to bilinear interpolation, except that the weighted average is calculated from the 16 nearest input cell centers and their DN values. As with bilinear interpolation, this method also results in completely new pixel DN values, so it is not recommended to use output image for radiometric analysis. However, these two methods (i.e., bilinear and cubic convolution) produce images which have a much sharper appearance with no blocky appearance as produced by the nearest neighbor method (Campbell, 2011).

Figure 5.23(c) demonstrates how the output DN value is calculated for cubic convolution. The 16 input cells (dark shades) nearest to the processing cell (light gray shade) are identified, the weighted average of DN values is calculated, and the resulting DN value is assigned as the output value for the processed cell (light gray shade). Cubic convolution has a tendency to sharpen the edges of the data more than the bilinear interpolation since more cells are involved in the calculation of the output DN value. Cubic convolution is recommended when the cell size of the data is being dramatically reduced, such as merging aerial photographs and TM data.

Bilinear interpolation or cubic convolution should not be used on categorical data since the categories will not be maintained in the output raster dataset. However, all three techniques can be applied to continuous data, with nearest neighbor producing a blocky output, bilinear interpolation producing smoother results, and cubic convolution producing the sharpest. The nearest neighbor is easiest of the three methods to compute and the fastest to use. Figure 5.24 shows the results of all three resampling methods.

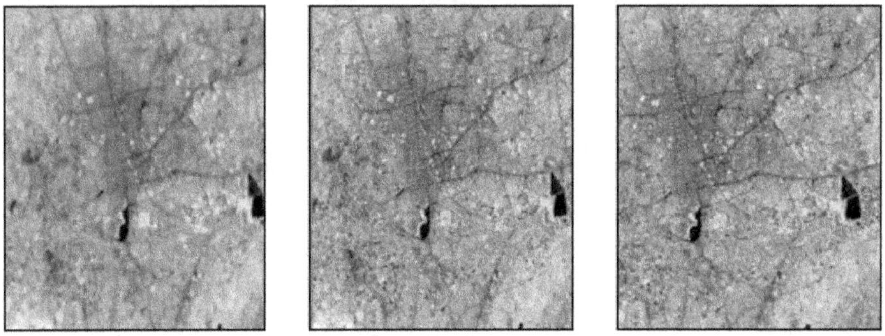

(a) (b) (c)

Figure 5.24. Comparison of three methods: (a) nearest neighbor, (b) bilinear convolution, and (c) cubic convolution.

5.10 IMAGE TRANSFORMATIONS

Image transformations typically involve the manipulation of multiple bands of data, whether from a single multispectral image or from two or more images of the same area acquired at different times (i.e., multi-temporal image data) to get the additional synthetic images (Jain, 1989). Because of the multispectral character of remote sensing images, the new images can be created through transformations that generate new pixel descriptions from the old, and thereby create synthetic image components, or bands, as shown in Figure 5.25. The new components are related to the old brightness values of the pixels in the original set of spectral bands using a mathematical operation, which is usually linear.

Some examples of transformations are simple arithmetic operations, such as image subtraction, image addition, image division, and image multiplication methods. Others are based on the mathematical concept of the derivative, such as vegetation indices (VI), normalized difference vegetation index (NDVI), principal component analysis (PCA), tasselled cap transformations (TCT) and so forth, which could be used for creation of additional images, and used along with the original raw images for analysis and classification (Richards & Jia, 2013). With the availability of images from various sensors at different resolutions, it has become necessary to combine the images from two or more sensors to get a new image having optimum information content. This transformation, called *image fusion*, is also being used frequently in remote sensing data analysis. All these transformations generate new images from two or more sources which highlight a particular feature of interest better than the original input images. Alternatively, it might be possible to preserve most of the essential information content of the original image using a reduced number of the transformed dimensions. The transformed images can be displayed in the three primary colors to create a color composite and for compressed transmission and storage of data (Campbell, 2011).

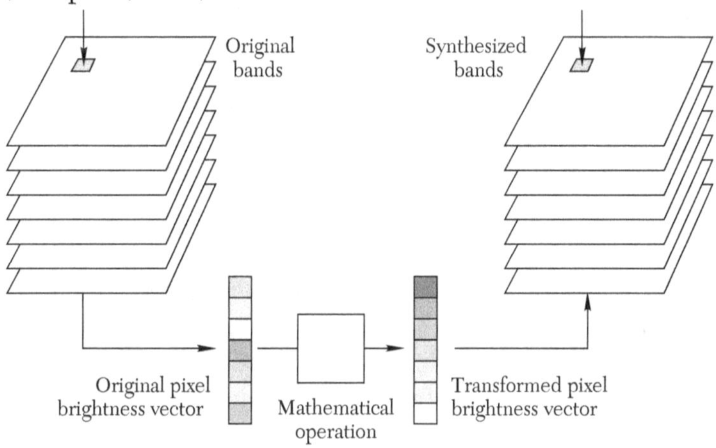

Figure 5.25. Transformation of images.

5.10.1 Arithmetic Operations

One of the image transformation techniques is to apply simple arithmetic/logical operations to the image data. Basic arithmetic operations, like addition, subtraction, multiplication, and division, are performed on two or more images of the same geographical area. The prerequisite for this operation is that all input images should be georeferenced together. These input images may either be individual bands from images acquired at different dates or separate image bands from a single multispectral image. A subtraction operation between images is frequently used to determine the changes that have occurred between the dates of these images. It is the most popular operation used in change detection studies, such as mapping changes in urban development around cities and for identifying areas where deforestation is occurring.

Division (ratio) of images is another popular operation which is used to create output images useful for vegetation studies as well as eliminating topographic effects. Several ratio images can be derived from multispectral bands of a sensor and used to generate ratio color composites in an RGB display.

5.10.2 Ratio Index

Band ratio is an image transformation technique that is applied on a multispectral image to enhance the contrast between the features by dividing pixel by pixel the DN values in one band image by the DN values of another band image of the same area (Garg, 2019). It is an effective technique for suppressing the topographic shadows which are predominant in hilly regions. For a given incident angle of solar radiation, the radiation energy received from a land surface depends on the angle between the land surface and the incident radiation. Therefore, solar illumination on a land surface varies with the terrain slope and aspect, which results in topographic shadows. In remotely sensed images, the reflectance values in NIR and red bands are used to derive ratio images which are almost free from the topographic shadow effect.

A ratio image of Landsat MSS band 7 (NIR 0.8–1.1 μm) divided by band 5 (Red 0.6 to 0.7 μm) would result in a ratio much greater than 1.0 for vegetation, and a ratio of around 1.0 for soil and water. Thus the discrimination of vegetation from other surface cover types is significantly enhanced. Also, the ratio images are able to identify areas of unhealthy or stressed vegetation, which show low near-infrared reflectance, as the ratio values would be lower than for healthy green vegetation. Another benefit of spectral ratioing is that the image displays the relative values (i.e., ratios) instead of absolute brightness values, as variations in scene illumination as a result of topographic effects are reduced (Lillesand & Kiefer, 1994). Thus, although the absolute reflectances for forest covered slopes may vary depending on their orientation relative to the sun's illumination, the ratio of their reflectances between the two bands should always be very similar, as shown conceptually in Figure 5.26. The ratio values for deciduous and conifer forests each in sunlight and shadow portions are 0.93 and 0.92, and 0.70 and 0.71, respectively, eliminating the effect of shadow while interpreting such images.

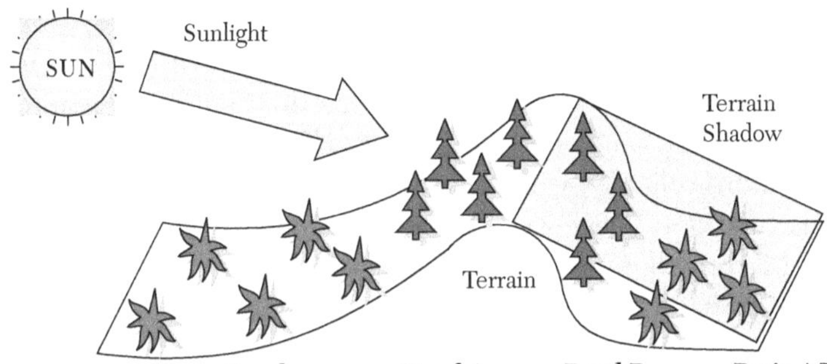

	Land cover	Band A	Band B	Ratio A/B
Deciduous Forest	Sunlit	140	150	0.93
	Shadow	56	61	0.92
Conifier Forest	Sunlit	102	145	0.70
	Shadow	41	58	0.71

Figure 5.26. Supressing the effect of terrain shadow by ratio images.

Many VIs (vegetation indices), including the ratio VI, have been developed based on several arithmetic operations. The ratio VI is the ratio of the near-infrared band (NIR) to the red band of a multispectral image, which may be used to highlight the vegetated areas. It is expressed as:

$$VI = NIR/Red \tag{5.18}$$

If both the red and NIR bands (or the VIS and NIR) have similar reflectance, the value of the ratio is 1 or close to 1. Ratio values for bare soils generally are near 1; as the amount of green vegetation increases, the ratio increases. The value of the ratio can increase far beyond 1 up to 30. Figure 5.27 shows the Landsat TM Band 3 and its ratio image (band 4 [NIR]/band 3 [Red]), where the ratio image shows a sharp difference between land and water.

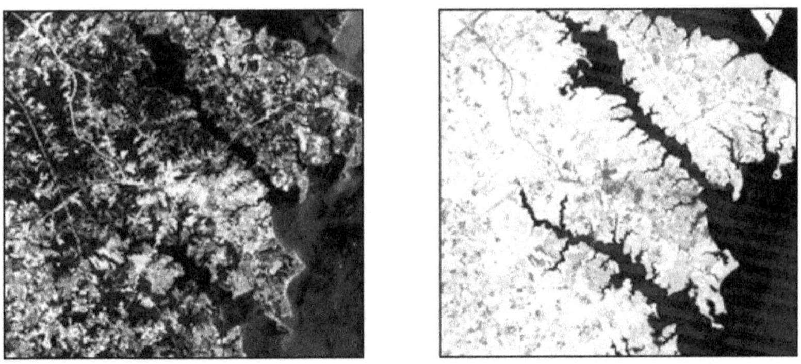

(a) (b)

Figure 5.27. (a) Landsat TM Band 3 and (b) ratio image band 4 /band 3.

5.10.3 NDVI Vegetation Index

Vegetation indices (VIs) are spectral combinations of two or more bands devised to enhance the spectral signature of vegetation, mainly derived from red and NIR bands (Richards & Jia, 2003). For this reason, they are used for a more reliable evaluation of photosynthetic activity and structural variations of vegetation cover. Vegetation indices operate by contrasting intense chlorophyll pigment absorption in the red with the high NIR reflectance of leaf mesophyll. Crop types may be suitably identified by using bands in these vegetation indices that are spectrally uncorrelated. These can also provide quantitative measures, based on vegetation spectral properties that can be used to quantify biomass or vegetative vigor (Castleman, 1996). The rationale for spectral vegetation indices is to use the spectral signatures of green and healthy vegetation as compared to those of stressed/unhealthy vegetation. Earth materials such as bare soil, sand, exposed rock, concrete, and asphalt, generally exhibit a steady rise in reflectance, without strong variations in the visible to NIR regions, whereas, green vegetation exhibits an increasing spectral reflectance from visible to NIR bands. Vegetation reflectance is very low in the blue and red regions, slightly higher in the green band, and finally shows the greatest spectral response in the NIR.

More than 200 VIs have been mentioned in scientific literature, however, the NDVI (normalized difference vegetation index) is very frequently used worldwide for the study of forest cover, vegetation, and crop classifications (Gibson, 2000; Richards & Jia, 2013). It is defined as the ratio involving the sums of and differences between spectral bands for various bands of the same sensor. The NDVI algorithm subtracts red reflectance values from NIR and divides it by the sum of NIR and red bands, and is computed as follows:

$$\text{NDVI} = (\text{NIR} - \text{Red})/(\text{NIR} + \text{Red}) \tag{5.19}$$

The arithmetic combination of the NDVI exploits the different spectral responses of vegetation cover in the visible spectral (red) and NIR bands. It provides a dimensionless numerical value. The formula is designed as a ratio in order to normalize its variability field between – 1 and +1. The normalization of the NDVI reduces the effects of variations caused by atmospheric contaminations. Clear water has a negative NDVI value close to –1. Slightly higher than 0 for soils and between 0.4 and 0.7 for vegetation, dense vegetation such as that found in temperate and tropical forests or crops at their peak growth stage can exceed 0.8 and is close to 1.0 is for the rainforest. The NDVI value is an indicative of plant photosynthetic activity, and has been found to be related to the green LAI (leaf area index) and the fPAR (fraction of photosynthetically active radiation) absorbed by vegetation. High or low values of the vegetation index identify pixels covered by substantial proportions of healthy or stressed or dead vegetation, as shown in Figure 5.28, where the NDVI value for healthy vegetation would be higher, whereas for stressed vegetation or dead vegetation it would be much lower, obtained using red and NIR wavebands. When the plant becomes dehydrated or stressed, the leaves reflect less NIR light, but the same amount in the visible range (Garg, 2019). Thus, mathematically combining these two signals can help differentiate a healthy plant from a sick plant.

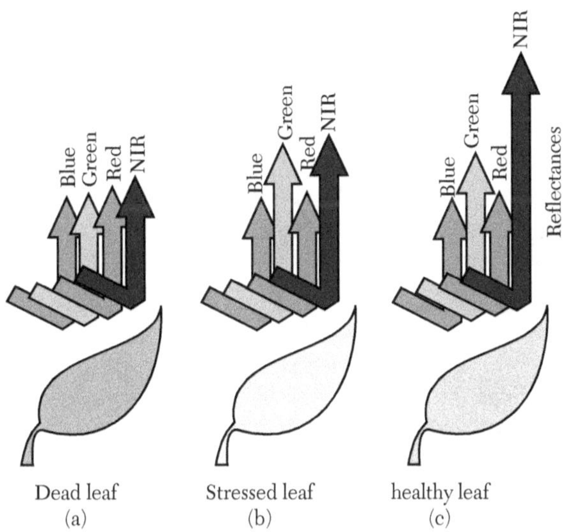

Figure 5.28. NDVI values for (a) dead vegetation, (b) stressed vegetation, and (c) healthy vegetation.

NDVI has found a wide application in vegetative studies, such as estimating crop yields, pasture land, rangeland, vegetation health, changes in vegetation over time, and so forth. It is often directly related to other ground parameters, such as percent of ground cover, photosynthetic activity of the plant, surface water, LAI, and the amount of biomass. Many factors affect NDVI values, like plant photosynthetic activity, total plant cover, biomass, plant and soil moisture, and plant stress. Because of this, NDVI is correlated with many ecosystem attributes (e.g., net primary productivity, canopy cover, bare ground cover). Due to the ratio of two bands, NDVI helps compensate for differences both in illumination within an image due to slope and aspect and differences between images due to time of day or season when the images were acquired. Thus, NDVI makes it possible to compare images over time to study ecological changes. It has been used to monitor vegetation condition, vegetation health, as well as cover and phenology (life cycle stage) over large areas, and therefore can provide early warning on droughts and famines. The AVHRR sensor onboard the NOAA series of satellites has been used for monitoring vegetation conditions at a continental and global scale.

The NDVI map generated from MODIS is shown in Figure 5.29. It has been found that there is a significant relationship between NDVI values and vegetation density. The higher the value of NDVI, the denser the vegetation and lesser the erosion.

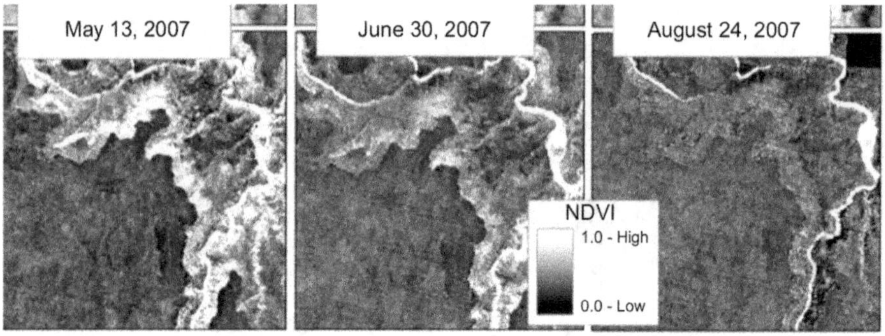

Figure 5.29. NDVI images generated from Landsat TM band 4 and band 5 data.

5.10.4 Other Indices

The SAVI (Soil adjusted vegetation index) is similar to NDVI but is used in areas where vegetative cover is low (< 40%). When a significant amount of the soil surface is exposed, the soil reflectance can influence the NDVI values. Light reflected from the soil can have a significant effect on NDVI values (changing the values by up to 20%).

$$\text{SAVI} = \frac{(\text{NIR} - \text{Red})}{(\text{NIR} + \text{Red} + \text{L})} \times (1 + \text{L}) \tag{5.20}$$

where *L* is a correction factor which ranges from 0 for very high vegetation cover to 1 for very low vegetation cover. An *L* value of 0.5 is typically used for intermediate vegetation cover, and 0.25 for higher density vegetation cover. When *L* is equal to zero, SAVI becomes the same equation as NDVI. The adjustment factor *L* can be found by trial and error until a factor that gives equal vegetation index results for the dark and light soils is found.

There are a large number of other indices (Bannari et al., 1995) which have been developed for the enhancement of vegetation and other associated features. These are given in Table 5.4.

Table 5.4. Various Vegetation Indices Used in Remote Sensing

Index	Wavebands
Difference Vegetation Index (DVI)	NIR-Red
Weighted Difference Vegetation Index (WDVI)	NIR-aRed Where a is the slope of the soil line
Normalized Difference Snow Index (NDSI)	(Visible-SWIR)/(Visible+ SWIR)
Normalized Green (NG)	Green/(NIR + Red + Green)
Chlorophyll Vegetation Index (CVI)	CVI = (NIR × Red)/(Green2)
Perpendicular Vegetation Index (PVI)	1/√a2+1 (NIR – aRed – b) Where a and b are the coefficients derived from the soil line: NIR soil = a Red soil + b.
Transformed Vegetation Index (TVI)	0.5 × [120 × (NIR–Green)– 200 × (Red–Green)]
LAI	LAI = 0.57 × exp(2.33 × NDVI). or LAI = $(((1+NDVI)/(1–NDVI)) \times NDVI)^{0.5}$
Transformed Adjusted Vegetation Index (TASI)	a(NIR – aRed – b)/[aNIR + Red – a b + χ (1 + a^2)] Where a and b are slope and intercept of the soil line, and χ has been adjusted so as to minimize background effect, and its value is 0.08.
Modified Soil Adjusted Vegetation Index (MSAVI)	NIR – Red/[(NIR+ Red+ L) (1+L)] Where L = 1 – 2 a × NDVI × WDVI
Generalized Soil Adjusted Vegetation Index (GESAVI)	NIR – Red b – a/Red + Z Where Z is the soil adjustment coefficient. To normalize soil effects, Z value is found at 0.35.
Enhanced Vegetation Index (EVI)	2.5(NIR – Red)/(NIR + 6Red – 7.5Blue + 1)
Normalized Burn Ratio (NBR)	(NIR – SWIR)/(NIR + SWIR)
Infrared Percentage Vegetation Index (IPVI)	NIR/(NIR + Red)
Atmospherically Resistant Vegetation Index (ARVI)	(NIR – (Red –1 × (Blue – Red)))/(NIR + (Red –1 × (Blue – Red)))
Normalized Pigment Chlorophyll Index (NPCI)	(Red – Blue)/(Red + Blue)
Optimized Soil-Adjusted Vegetation Index (OSAVI)	[1.5 × (NIR – Red)]/[(NIR + Red + 0.16)]
Stressed Related Vegetation Index (STVI)	MIR × Red/NIR
Vegetation Condition Index (VCI)	$[(NVDI_j - NDVI_{min})/(NDVI_{max} - NDVI_{min})] \times 100$

(Contd.)

Index	Wavebands
Enhanced Normalized Difference Vegetation Index (ENDVI)	((NIR + Green) – (2 × Blue))/((NIR + Green) + (2 × Blue))
Atmospherically Resistant Vegetation Index (AVRI)	(NIR – (2Red – Blue))/(NIR + (2Red – Blue))

5.10.5 Tasseled Cap Transformation

The TCT (tasseled cap transformation), also known as the *Kauth-Thomas transformation*, was developed for enhancing the spectral information contents of satellite images. The TCT was originally derived by Kauth and Thomas (1976) for Landsat MSS four bands based on a spectral analysis of the growth of wheat in fields (Castleman, 1996). The transformation got its name from the triangular shape of the graph by the distribution of data in 2D space when the red band values of pixels are plotted against the near infrared pixel values. This shape further looked like a cap shape in *n*-dimensional (bands) data space. The TCT is an agriculture-specific transformation designed to highlight the most important, spectrally observable phenomena of crop development, in such a way that discrimination among crops and crops from other vegetative cover is maximized. Its basis lies in the behavior of crop trajectories with time in infrared versus red, and red versus green band images. Thus, the plot of the data for a wheat field resembles a cap as shown in Figure 5.30, which is the basis of the work of Kauth and Thomas. The TCT coefficients are defined to maximize the separation of the different growth stages of wheat.

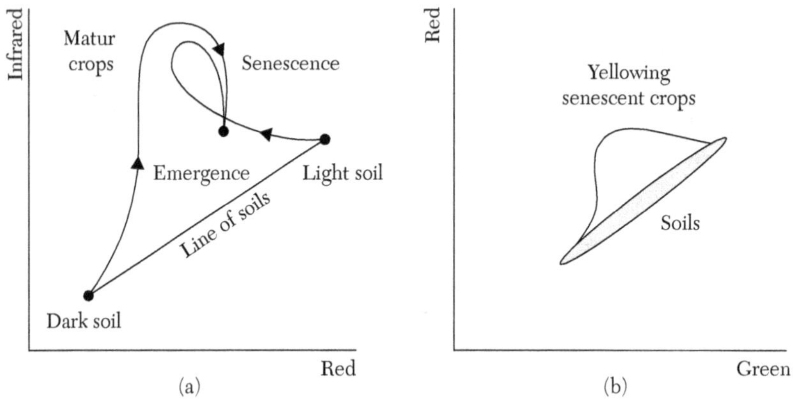

Figure 5.30. The graph between red and infrared bands to depict the tasseled cap transformation (Kauth & Thomas, 1976)

It is a linear affine transformation based on the conversion of a given input image data set (or Cartesian reference system) in a new image data set (or

Cartesian reference system) of composite values. The values of coefficients are obtained through weighted sums of the input image data where the weights are statistically derived (Campbell, 2011). A TCT is performed by taking the "linear combination" of the original image bands—similar in concept to the PCA (principal components analysis).

The TCT is a conversion of the original bands of an image into a new set of bands that are useful for vegetation mapping, and identifying key forest attributes, including species, age, and structure (Cohen et al., 1995). It is performed on a pixel by pixel basis to better highlight the underlying structure of the image, given as follows:

$$\text{TCT} = \text{T (DN)} + \text{B} \tag{5.21}$$

where TCT is the transformed output, T is the transformation coefficient, DN is the value of the pixel, and B is Bias. The transformation coefficients depend on the sensor, because different sensors have different bands, which in turn have different spectral responses. The four indices are computed from four input MSS (bands 4–7), as given below:

$$\begin{aligned} \text{SBI} &= +0.433\text{MSS4} - 0.632\text{MSS5} + 0.586\text{MSS6} + 0.264\text{MSS7} \\ \text{GVI} &= -0.290\text{MSS4} - 0.562\text{MSS5} + 0.600\text{MSS6} + 0.491\text{MSS7} \\ \text{YVI} &= -0.829\text{MSS4} - 0.522\text{MSS5} + 0.039\text{MSS6} + 0.149\text{MSS7} \\ \text{NSI} &= 0.223\text{MSS4} + 0.012\text{MSS5} - 0.542\text{MSS6} + 0.810\text{MSS7} \end{aligned} \tag{5.22}$$

Each TCT index is created by the summation of the DN value of image band 1 times a coefficient plus the DN value of image band 2 times another coefficient, and so on. These coefficients are like weights to create the TCT indices, which are derived statistically from images and empirical observations and are specific to each imaging sensor (Garg, 2019). The first component is called as the SBI (soil brightness index). Brightness is defined in the direction of soil reflectance variation. It is obtained from a weighted sum of all bands, that is, urbanized and bare soil areas. This index measures the characteristics of soil types. The second component, called as the GVI (greenness vegetation index), is perpendicular to the first component and its axis passes through the point of maturity of the plants. Greenness is defined in the direction of vegetation reflectance variation. It is obtained from the contrast of the visible bands (high absorption) with the IR bands (high reflectance); that is, the greater the biomass, the brighter the pixel value in the image (Cohen et al., 1995). This component is typically used as an index of photosynthetically active vegetation, and is a measure of vigor of vegetation. The third component corresponds to an axis perpendicular to the first and second axis and passing through the point which represents the YVI (yellowness vegetation index or wetness index; e.g., soil or surface moisture, amount of dead/dried vegetation). Wetness concerns the moisture status of the environment (soil and plant moisture). It is obtained from the contrast of the sum of the visible and NIR with the sum of longer infrared bands, that is, water bodies are very bright. The greater the moisture content, the brighter the response. These three components are shown in Figure 5.31. The fourth component represents the projection onto an axis perpendicular to other three, and is called as the NSI (non-such index). This component is generally not used and is treated as random noise. A TCT produces the same number of output components as input bands, but all of the components may not be useful. Generally, the first three tasseled-cap components (SBI, GVI, YVI) contain useful information. The first component contains most of the information from the image, and each subsequent component contains less of the previous component information, and the fourth component contains much of the image "noise," and therefore is not used.

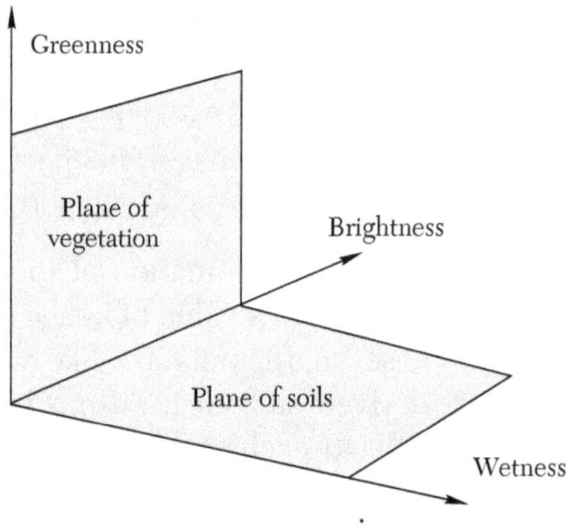

Figure 5.31. Representation of greenness, brightness, and wetness components on three axes.

The output images from the first three TCT components along with the original image of Landsat, Band 7, are shown in Figure 5.32. Visualizing the success of TCT on MSS bands, it was later extended to the Landsat TM (Crist & Cicone, 1984), ETM (now available in a routine of PCI Geomatics software), IKONOS sensor (Horne, 2003), and many more. All these transformations uncorrelate the multispectral data by the weighted sums of the input bands to extract a greater amount of information allowing an easier identification of distinct surface features. Crist and Cicone (1984) gave the relationship, as follows, to compute the tasseled cap transformation to six channels of Landsat TM data. The weights are different and the third

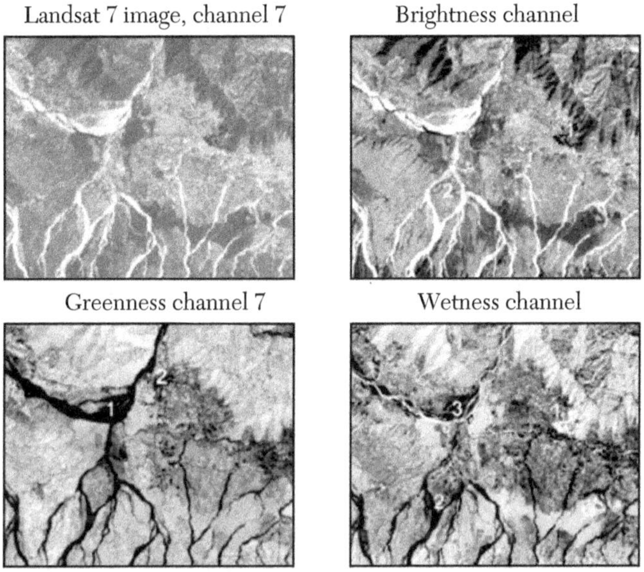

Figure 5.32. Original image and three tasseled cap transformations.

component was taken to represent soil wetness rather than yellowness as in Kauth and Thoma's original formulation. The weighted sums adapted from the TM input bands (1, 2, 3, 4, 5 and 7) input channels are:

$$\begin{aligned}
SBI &= 0.3037TM1 + 0.2793TM2 + 0.4343TM3 + 0.5585TM4 + 0.5802TM5 \\
&\quad + 0.1863TM7 \\
GVI &= -0.2848TM1 - 0.2435TM2 - 0.5436TM3 + 0.7243TM4 + \\
&\quad 0.0840TM5 - 0.1800TM7 \\
YVI &= 0.1509TM1 + 0.1793TM2 + 0.3299TM3 + 0.3406TM4 - 0.7112TM5 - \\
&\quad 0.4572TM7
\end{aligned} \tag{5.23}$$

The weighted sums adapted from the ETM four input channels are given as follows:

$$\begin{aligned}
SBI &= 0.1544\ Blue + 0.2552\ Green + 0.3592\ Red + 0.5494\ NIR \\
GVI &= -0.1099\ Blue - 0.1255\ Green - 0.2866\ Red + 0.8226\ NIR \\
YVI &= 0.3191\ Blue + 0.5061\ Green + 0.5534\ Red + 0.0301\ NIR
\end{aligned} \tag{5.24}$$

The weighted sums developed by Horne (2003) for the IKONOS four channels are as follows:

$$\begin{aligned}
SBI &= 0.326\ Blue + 0.509\ Green + 0.560\ Red + 0.567\ NIR \\
GVI &= -0.311\ Blue - 0.356\ Green - 0.325\ Red + 0.819\ NIR \\
YVI &= -0.612\ Blue - 0.312\ Green + 0.722\ Red - 0.081\ NIR \\
NSI &= -0.650\ Blue + 0.719\ Green - 0.243\ Red - 0.031\ NIR
\end{aligned} \tag{5.25}$$

Jackson et al. (1980) proposed the ASBI (adjust soil brightness index) and AGVI (adjust green degree vegetation index), which can be expressed as follows:

$$ASBI = 2.0 \times YVI \quad (5.26)$$

$$AGVI = GVI - (1 + 0.018 \times GVI)\, YVI - NSI^2 \quad (5.27)$$

5.10.6 Principal Component Analysis

Principal components analysis (PCA) is a linear transformation technique which uncorrelates multivariate data by translating and/or rotating the axes of the original feature space, so that the data can be uncorrelated in a new component space (Campbell, 2011). The data of two or more bands from a sensor may be similar and convey essentially the same spectral information. PCA is used as an important transformation technique in digital image processing of satellite images where a number of correlated bands of the image data are reduced to few uncorrelated bands that represent most of the information present in the original dataset.

PCA is a popular dimensionality reduction technique (i.e., it compress the information contained in an original n-channel data set into fewer than n "new" channels or components) called *principal components* (PCs; Figure 5.33). The first principal component (PC-1) accounts for the maximum proportion of variance from the original dataset. All subsequent orthogonal components (i.e., PC_2, PC_3, and so on) account for the maximum proportion of the remaining variance (Garg, 2019). The purpose of PCA is to translate and/or rotate original axes so that original DN values on axes X_1 and X_2 are redistributed (reprojected) onto new axes (Figure 5.33). Usually, the first three principal components account for the vast majority of the variance found within the dataset, which can be used for image enhancement or classification. The first three principal components can

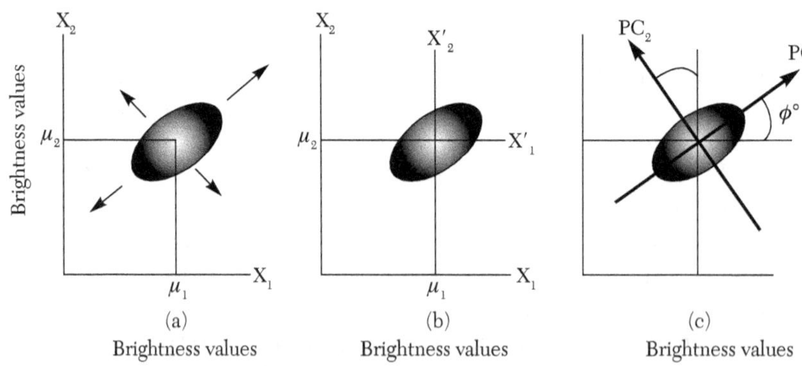

Figure 5.33. PC1 and PC2 components.

also be used to create an FCC where the ground features are enhanced for visual interpretation. The FCC image may effectively be used as a base for ground truth collection in supervised classification of land use and land cover, geologic interpretation, and so forth.

PCA can be applied to multispectral and hyperspectral remotely sensed data. The minimum noise fraction (MNF) method can be used with hyperspectral data for noise reduction. PCA has been extensively used in studies like change detection, image enhancement, and so on (Jensen, 1986). It is also widely used in pattern recognition, image classification to improve the accuracy of classification, and the remote sensing application, which mathematically establishes a new set of variables describing the variance in the original data set. PCA is also useful in providing maximum visual separability of image features.

Transformation of original data on X_1 and X_2 axes onto PC_1 and PC_2 axes requires transformation coefficients that can be applied linearly to the original pixel values. These new axes are called PC_1 and PC_2, and so on, and contain decreasing amounts of the variance found in the dataset in the same order. PCA is equivalent to transforming the data to a new coordinate system with a new set of orthogonal axes, like the tasseled cap transformation. It compresses the data by eliminating the redundancy in the data. Many times, the first three or four resulting images from the principal components will describe more than 95% of the variance. The remaining individual bands can be dropped as they have very little variance (Richard & Jia, 2013). Since the new set contains fewer bands and more than 95% of the variance of the original images, the computations will be faster, and the accuracy is maintained. The result of this transformation will have the same number of bands as the specified number of principal components (one band per axis).

PCA creates new variables as weighted sums of the different bands of data. The weights used in principal component analysis are determined statistically and mathematically from the data. PCA translates the original axes to a new set of axes, with each axis orthogonal to the other's 1st axis or PC associated with the maximum amount of variance (the ellipsoid's major axis), the 2nd axis (orthogonal to the 1st) contains the next highest amount of variation, and so on.

By computing the correlation between each band and each PC, it is possible to determine how each band "loads" or is associated with each PC. In addition to PC images, the PCA also produces the following outputs:

eigenvalues and eigenvectors. Eigenvalues contain information about the percent of total variance explained by each PC. Eigenvectors indicate degree of correlation between each band and each principal component. Figure 5.34 shows seven PCA images created from Landsat TM images.

Principal component 1. Principal component 2. Principal component 3.

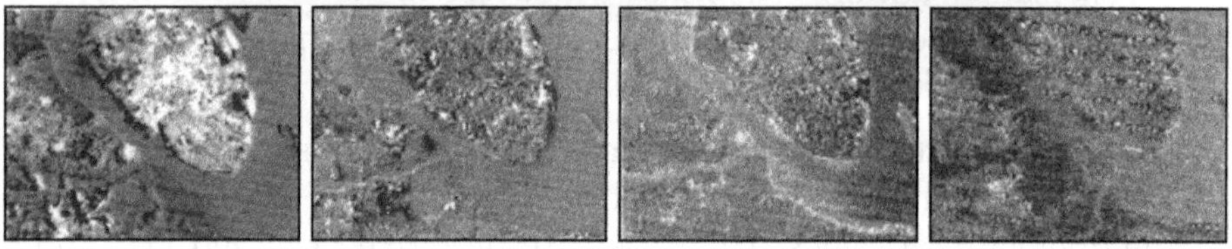

Principal component 4. Principal component 5. Principal component 6. Principal component 7.

Figure 5.34. Seven PCA images derived from Landsat TM imagery (Jensen, 2005).

If one is utilizing the famous scikit-learn library, the PCA function available there can be used. However, in general, the steps used for PCA are as follows:

1. Compute the covariance matrix of the data set in vector space.
2. Calculate the eigenvalues of the covariance matrix.
3. The diagonal matrix with the eigenvalues along the diagonal will be the covariance matrix of the transformed axes (principal component axes).
4. Find the matrix of eigenvectors (Di) for each individual l of interest by solving for $[\Sigma x - \lambda_i \mathrm{I}]g_i = 0$. for that λ.
5. Transpose the Matrix D to produce principal component transformation matrix (g). The number of rows in g will equal the number of spectral dimensions from which the eigenvalues and eigenvectors were calculated.
6. For each g matrix (derived from a given l) the original data values (in the original x coordinate system) are multiplied by the rows in g (g_1, g_2, … g_n where n is the number of dimensions in vector space) to produce coordinates in the transformed dimension (new y coordinate system). Each axis in the original spectral space will be multiplied by its

corresponding row in the g matrix to produce the transformed coordinate system (principal component).

7. Steps 4–6 are repeated until the desired number of principal component transformations have been executed.

5.11 IMAGE FUSION APPROACHES

Since the beginning of the 1990s, the availability of satellite image data at finer resolutions has dramatically increased, ranging from less than a meter to few kilometers spatial resolution. In order to derive maximum information from these images, it is necessary that the data is combined in most efficient way for the specific application. Fusion techniques have been successfully applied in the space and Earth observation domains, computer vision, medical image analysis, defence security, and so forth (Zhang, 2010). These are useful for a variety of applications, ranging from object detection, recognition, identification, and classification to object tracking, change detection, decision making, and so on. Research on data fusion has a long history in the remote sensing community because fusion products are the basis for many applications. Fusion of SAR and optical imagery has been applied to topography mapping, fused interferometric SAR and optical imagery to DEM extraction, and fused optical and polarimetric SAR images for digital line graphics production. Fusion of LiDAR data and imagery has also been widely adopted for a variety of applications ranging from DEM generation, 3D object extraction and modeling, and land cover mapping (Santurri et al., 2010).

Data fusion is an effective approach oriented to information extraction based on the synergetic exploitation of data originating from different sources. It aims to produce better results than those obtained by a single image of the same sources. The multispectral sensors usually have high spectral content but lower spatial resolution as compared to PAN sensors which have a higher spatial resolution and a wide spectral bandwidth. A PAN sensor gives a high spatial resolution image, but it lacks spectral resolution, which does not contain any color information. Therefore, a fusion based enhancement technique is used to get the spatial and spectral information from the image. To optimally benefit from the advantages of multispectral images (i.e., high spectral resolution) and PAN images (i.e., high spatial resolution), the two products are often combined or fused for improved visual image interpretation and information retrieval. It produces a high-quality visible representation of the images, and therefore it may be considered as an image enhancement technique (Welch & Ehlers, 1987).

Multi-sensor image fusion techniques combine two or more geometrically registered images of the same scene into a single image that is more easily interpreted than any of the originals. PAN and multispectral images can be obtained by several satellites such as SPOT, QuickBird, IKONOS, Landsat, WorldView, GeoEye, OrbView, IRS, Leica ADS40, and Pléiades. With appropriate combination of these data, an output image can be created with the best characteristics of both, namely high spatial and high spectral resolution. The fused images may improve interpretation capabilities and provide more accurate results. A general procedure for image fusion is presented in Figure 5.35. Data fusion involves two major steps: (*a*) geometric registration of two datasets, (*b*) feature extraction, (*c*) radiometric calibration, and (*d*) merging of spectral and spatial information contents to generate a new dataset that contains the enhanced information from both datasets (Lu & Weng, 2007).

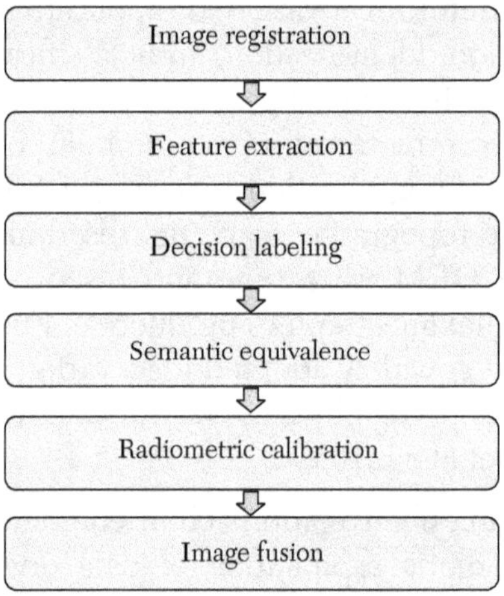

Figure 5.35. A general procedure for image fusion.

Data fusion is a general and popular multidisciplinary approach, which combines data from multiple sources to improve the potential values and interpretation performances of the source data, as well as to produce high-quality visual data (Santurri et al., 2010). Remote sensing data fusion, as one of the most commonly used techniques, aims to integrate the information acquired with different spatial and spectral resolutions from sensors mounted on satellites, aircraft, and ground platforms to produce the fused data that contains more detailed information than each of the sources. Such an increase in remote sensing and ancillary datasets opens up the possibility of utilizing multimodal datasets jointly to further improve the performance of the processing approaches for various applications. The sharp and recent increase in the availability of data captured by different sensors combined with their considerably heterogeneous natures poses a serious challenge for the effective and efficient processing of remotely sensed data. Many methods have been developed for that purpose and produce multispectral images having the highest spatial resolution available within the dataset.

Since optical satellites have trade-offs in spatial, spectral, and temporal resolutions, several data fusion techniques have been developed for reconstructing synthetic data that have the advantages of different sensors (Pohl & van Genderen, 1998). Depending on which pair of resolutions has a trade-off, these technologies can be broadly divided into two categories: (*a*) spatiospectral fusion to merge fine spatial and fine spectral resolutions; and (*b*) spatiotemporal fusion to blend fine-spatial and fine-temporal resolutions.

5.11.1 Spatio-Spectral Fusion

Many satellite sensors observe the Earth's surface at different spatial resolutions in different wavelengths. For example, the spatio-spectral fusion technique will fuse fine spatial resolution images (e.g., 0.5 m WorldView PAN image) with coarse spatial resolution images (e.g., 2 m WorldView multispectral image) to create fine spatial resolution images for all bands. Spatio-spectral fusion is also termed *pan-sharpening* when the available fine spatial resolution image is a single PAN image. When multiple fine spatial resolution bands are available, spatiospectral fusion is referred to as *multiband image fusion*, where two optical images with a tradeoff between spatial and spectral resolutions are fused to reconstruct the fine spatial and fine spectral resolution imagery. Multiband image fusion tasks include multiresolution image fusion of single satellite multispectral data (e.g., MODIS and Sentinel-2) as well as hyperspectral and multispectral data fusion.

5.11.2 Spatiotemporal Fusion

In this approach, the amount of information can be maximized through integrating the spatial, spectral, and temporal attributes that can lead to accurate stable predictions and enhance the final output (Zhang, 2010). Spatiotemporal fusion is a technique to blend the fine spatial resolution but coarse temporal resolution (e.g., Landsat) data with fine temporal resolution but coarse spatial resolution data (e.g., MODIS) to create fine spatiotemporal resolution data. Its implementation is performed based on the availability of at least one coarse-fine spatial resolution image pair (e.g., MODIS-Landsat image pair acquired on the same day) or one fine spatial resolution land cover map that is temporally close to the prediction day. Over the past decade, several spatiotemporal fusion methods have been developed. Spatio-temporal fusion can be applied within local and global fusion frameworks, as well as on various data processing levels depending on the desired techniques and applications. Spatiotemporal fusion has been an active area of research over the last few decades.

For remote sensing-based global monitoring, there always exists a trade-off between spatial resolution and temporal revisit frequency (i.e., temporal resolution). For example, the MODIS satellite can provide data on a daily basis, but the spatial resolution (250 m to 1000 m) is often too coarse to provide explicit land cover information, as such information may exist at a finer spatial scale than the sensor resolution. The Landsat TM sensor can acquire images at a much finer spatial resolution of 30 m, but has a limited revisit capability of 16 days. Fine spatial and temporal resolution data may be crucial for timely monitoring of highly dynamic environmental, agricultural, or ecological phenomena on the Earth's surface. The recent development in UAV/drone technology also provides a huge amount of multisource data with very high spatial and temporal resolutions.

In general, remote sensing fusion techniques can be classified into three different levels: pixel-level, feature-level, and decision-level (Pohl & van Genderen, 1998), as shown in Figure 5.36. The pixel-level image fusion methods are well suited for images acquired from the same sensor, since they undergo similar calibration processes and minimum spectral differences in terms of having the same number of bands and bandwidth ranges in the spectrum, whereas feature- or decision-level fusion is more flexible and able to handle heterogeneous data, such as combing elevation data (e.g., LiDAR) with satellite images.

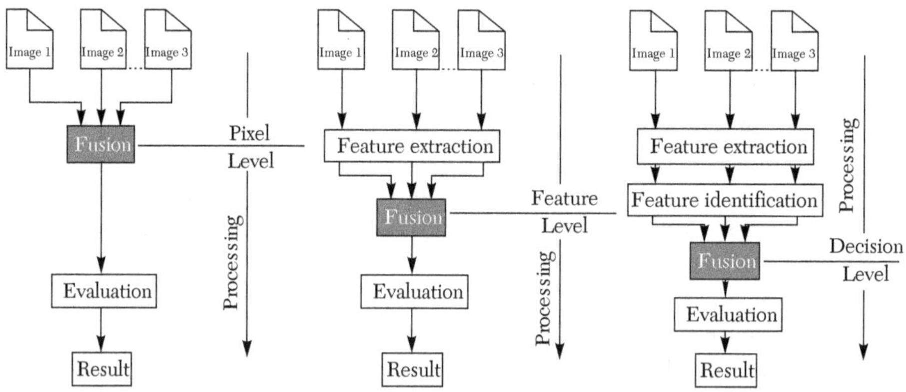

Figure 5.36. Various levels of fusion approaches.

Pixel-level image fusion: It is a direct low-level fusion approach. It involves a pixel-to-pixel operation, where the physical information (e.g., DN values, elevation, thermal values, etc.) associated with each pixel within two or more images is integrated into a single value which is expected to be more informative and synthetic than either of the input data. The purpose of pixel-level fusion of optical images is mainly to improve the spatial resolution, enhance structural and textural details, and retain the spectral fidelity of the original multispectral data. For this reason, it is also called *pan-sharpening*. This procedure is extremely useful in object-based image analysis, where very high resolution images are required to extract the objects of interest (Santurri et al., 2010). In agricultural applications, for example, farm boundaries are often extracted from pan-sharpened high resolution multispectral images using image segmentation approaches. For pan-sharpening, for example, the HIS (intensity-hue-saturation), PCA, BT (Brovey transform), and GS (Gram-Schmidt) methods can be used.

Pan-sharpening follows several algorithms, but the common one is known as component substitution, which involves different steps, as follows:

1. *Up-sampling*: This is the up-sampling of the color bands to match the resolution of the PAN band.
2. *Alignment*: The up-sampled PAN and color bands are aligned to decrease artifacts caused by the misregistration.
3. *Forward transform*: The up-sampled PAN and color bands are then converted to alternative color-space.
4. *Intensity matching*: The color band intensity is matched to the intensity of the PAN band in transformed space.
5. *Component substitution*: There is a direct substitution of the PAN band for the intensity component transformed.
6. *Reverse transformation*: This is performed by substituting component intensity to transform it back to the original color.

For multi-temporal data, the purpose of pixel-level fusion is to highlight the changes between different times, using either the same or different sensors. The algorithms for pixel-level fusion of remote sensing images include the CS (component substitution) fusion technique (Aiazzi et al., 2007), modulation-based fusion techniques (Gangkofner et al., 2008), MRA (multi-resolution analysis)-based fusion techniques (Amolins et al., 2007), STARFM (spatial and temporal adaptive reflectance fusion model), and STRUM (spatial and temporal reflectance unmixing model).

Feature-level image fusion: Feature level fusion extracts various features, for example, edges, corners, lines, texture parameters, and so on, from different data sources, and then combines them into one or more feature maps that may be used instead of the original data for further processing. This is particularly important when the number of available spectral bands becomes so large that it is impossible to analyze each band separately.

Methods applied to extract features usually depend on the characteristics of the individual source data, and therefore may be different if the data sets used are heterogeneous. It involves extracting and matching distinctive features from two or more overlapping images using methods such as dimensionality reduction, like the PCA, LDA (linear discriminant analysis), SIFT (scale invariant feature transform), SURF (speeded-up robust features), and so forth. Fusion is then performed using the extracted features and the coefficients corresponding to them. Some other common methods that include spatiotemporal fusion on the feature level are sparse representation and deep learning algorithms. Typically, in image processing, such fusion requires a precise (pixel-level) registration of the available images. Feature maps thus obtained are then used as input to preprocessing for image segmentation or change detection (Santurri et al., 2010).

Feature-level fusion methods deal with data at higher processing levels than pixel-level methods. Firstly, the feature extraction procedures are applied, and then the fusion process using advanced techniques takes place. For example, the extraction of objects using segmentation procedures is required in fusion at feature level. Features corresponding to characteristics, which are dependent on their environment, such as shape, extent, and neighborhood, are extracted from the original images. The similar objects from multiple sources are assigned to each other and then fused for further assessment.

Decision-level image fusion: It is a high-level fusion method that requires each image to be processed individually until an output (e.g., classification map) is obtained. The outputs are then post-processed using decision-level fusion techniques. In other words, once the input images are processed individually, the decision-level rules are used to combine extracted information to reinforce common interpretation and resolve differences and furnish a better understanding of the observed objects (Gangkofner et al., 2008). This level of fusion can include the previous two levels of fusion within its operation. Decision-level fusion combines the results from multiple algorithms to yield a final fused decision. When the results from different algorithms are expressed as confidences (or scores) rather than decisions, it is called soft fusion; otherwise, it is called hard fusion. Methods of decision-level fusion include voting methods, statistical methods, and fuzzy logic-based methods.

Fusion methods are applied in varied applications, such as remote sensing image classification, fingerprint verification, and so forth. Decision-level fusion has great uses in distributed and parallel processing systems. Some useful decision-fusion methods, applied in different applications, are voting, rank-based, Bayesian inference, and Dempster-Shafer methods. However, it is a challenge for improvement of fingerprint verification, especially when the quality of images is low. In such cases, high-level methods based on statistical and neural network methods and the combination of those approaches are useful.

High-level fusion methods, such as at feature-level and decision-level, are essential in order to use multifeatures, including spectral context, structural context and texture characteristics comprehensively. They require multisource data, such as SAR, optical images, LiDAR, and ground data. Combinations of multifeatures can improve the accuracy of image classification and information extraction. The existing mathematical tools of high-level fusion methods include probability theory, evidence theory, fuzzy and possibility theory, neural networks, and so on. One trend is the application of contemporary machine-learning techniques. SVM and ensemble learning, and other new developments in machine learning, can be applied to high-level fusion.

A disadvantage of pixel-level fusion techniques to the fusion of SAR and optical images is that either spectral features of the optical imagery or the microwave backscattering information is destroyed, or both of them simultaneously. Therefore, it is necessary to develop a specifically tailored SAR-optical image fusion technique which can fully utilize those two types of image sources. So far, classifier combination has been an effective measure, which not only chooses the basic classifier corresponding to the SAR and optical imagery, but also integrates fusion results from different basic classifiers.

The previous categorization does not encompass all possible fusion methods, since input and output of data fusion may be different at different levels of processing. Pohl and Van Genderen (1988) provided a good review of literature on various methods of multi-sensor data fusion, including color/related techniques (e.g., color composite, intensity/hue/saturation, and luminance/chrominance), and statistical/numerical methods (e.g., arithmetic combination, PCA, high pass filtering, regression variable substitution, canonical variable substitution, component substitution, wavelets, Laplacian pyramid, curvelets, and contourlets).

5.11.3 Multi-Resolution Approach

The multi-resolution approach (MRA) merges the spatial information from a high-resolution image with the radiometric information from a low-resolution image. The process is to sharpen the low-resolution image. In recent years, powerful MRA techniques such as wavelets, curvelets, and others have become popular because of an increase in computational power and the availability of algorithms in commercial remote sensing software. The MRA-based approaches decompose images into a number of channels depending on the local frequency content (Nunez et al., 1999). A pyramid is used to represent the multi-scale models for the original image. With each increasing level, the original image is approximated at coarser spatial resolution. In between the individual pyramid levels, the transform is performed using wavelet and curvelet transforms.

Garzelli and Nencini (2005) have proposed a fusion based on MRA to describe how high-pass information is modeled from a PAN image. Pradhan et al. (2006) have proposed an MRA which is extended to discrete functions. The space is conserved and determines the best possible number of decomposition levels required for merging images with a particular resolution ratio. If the resolution ratio is high, the decomposition levels are higher to produce better results; however, the computational complexity is higher due to more decomposition levels.

5.11.4 Wavelet Transformation

Recently, wavelet-based data analysis has become one of the fastest growing research areas being used for a number of different applications. Wavelets are mathematical functions which were initially developed independently in diverse application fields like mathematics, electrical engineering, quantum physics, and seismic geology (Zhang, 2010). In the last few years, wavelet have also found new application in satellite data analysis. Signal wavelet decomposition using DWT (discrete wavelet transform) using multi-resolution time-scale analysis provides better results than DFT (discrete fourier transform). The wavelet transform offers a number of advantages as compared to traditional Fourier methods, especially in searching for discontinuities and sharp spikes in analyzing the signals. The shape of the wavelet function reflects the type of features present in the signal/image under processing. As an example, for abrupt changes, the Harr function is the most recommended, whereas in the case of smooth variations, a damped cosine function is more adequate (Pohl & van Genderen, 1998).

Wavelets split the data into different frequency components, and then study each component with a resolution matched to its scale. In the field of image processing, DWT acts in this way: an image is decomposed, with each level corresponding to a coarser resolution band. There are two basic types of wavelet transforms: (*a*) the first type is widely used for image compression and cleaning (noise and blur reduction), designed to be easily reversible (invertible); and (*b*) the second type is designed for signal analysis, for example, to detect noise, study signals, and determine how the frequency content of a signal evolves. Some techniques only focus on image enhancement without any reduction of noise, such as the histogram equalization method in the wavelet domain, whereas other methods mainly focus on noise reduction.

5.11.5 Brovey Transformation

Brovey transformation is a straightforward technique for merging data from more than one sensor. It takes into account the factor of the chromaticity transform. For applications where images are to be fused into a single composite image by taking the sequence of images through diverse sensors, the Brovey transformation is considered to be the simplest and most applicable approach. It standardizes the three multispectral bands used for RGB to append the intensity and brightness into the image. It adds the component of brightness and intensity in the resultant fused image by performing multiplication of the result with the anticipated data. It includes an RGB color transform technique that is known as color normalization transform to avoid disadvantages of the multiplicative technique. It is helpful for visual interpretation but generates spectral distortion.

5.11.6 IHS Transformation

A simple and commonly used approach to fuse multi-pectral and panchromatic images is the RGB (red-green-blue) and IHS (intensity-hue-saturation) transformation of three-dimensional color spaces. In its simplest form, the RGB-IHS transformation approach first transforms an RGB image color composite made from three multispectral bands into an IHS color space, which results in three images; intensity, hue and saturation.

This transformation enables a replacement of the "intensity" image of the IHS transform (which is derived from the multispectral image) to be replaced by the high spatial resolution panchromatic image. This new intensity image, together with the original hue and saturation images (from the multispectral image) are then transformed back into an RGB color space for visualization. Variants to this simple technique exist and include the pixel addition method (Chavez et al., 1994), in which the high spatial resolution panchromatic image is added in equal amounts to each of the (three) multispectral bands.

The RGB color space represents the amount of red, green, and blue components in a color, and therefore it is not easy to distinguish one color from another only using the RGB color coordinates. The use of the IHS transform, which converts three images from the RGB space into the IHS color space, can overcome this problem. An alternate approach to color is the IHS, which is useful because it presents colors nearly as human beings perceive them. The IHS system is based on the color sphere where the vertical axis represents intensity (I), the radius is saturation (S), and the circumference is hue (H). The intensity axis represents brightness variations and ranges from black (0) to white (255); no color is associated with this axis. The hue represents the dominant wavelength of color, and its values commence with 0 at the midpoint of red tones and increase counterclockwise around the circumference of the sphere to conclude with 255 adjacent to 0. The saturation represents the purity of color and ranges from 0 at the center of the color sphere to 255 at the circumference. A saturation of 0 represents a completely impure color, in which all wavelengths are equally represented and which the eye will perceive a shade of gray that ranges from white to black depending on intensity. Intermediate values of saturation represent pastel shades, whereas high values represent purer and more intense colors.

Similar to TCT and PCA, IHS also performs a rotation from the RGB axis to a new orthogonal IHS. This axis rotation can be obtained by using a different formula to separate spatial (I) and spectral (H, S) information from a standard RGB image. The IHS space provides information closer to human color perception, in particular: (*a*) intensity is the overall brightness and conveys about how close the image is to black or white, (*b*) hue provides information about the actual perceived color (i.e., the dominant or average wavelength of light), and (*c*) saturation gives information about the purity of the color (Pohl & van Genderen, 1998).

The value of I, H, and S can be computed as:

$$
\begin{aligned}
I &= R + G + B \\
H &= \frac{G - B}{I - 3B} \\
S &= \frac{I - 3B}{I}
\end{aligned}
\tag{5.28}
$$

The transformation from the RGB color space to a given IHS space is characterized by the following properties: (*a*) nonlinearity, (*b*) reversibility, and (*c*) independence of each component from the others; that is, it may be varied without affecting the others. IHS is considered an ideal tool for image processing ranging from contrast enhancement to image fusion aimed at (*a*) enhancing the spatial content and (*b*) preserving the spectral properties of the data set to be processed. The IHS transformation is identified to be the most frequently used method for improving the visual display of multisensor data (Welch & Ehlers, 1987), but the IHS approach can only employ three image bands, and the resultant image may not be suitable for further quantitative analysis such as classification.

CHAPTER 6

IMAGE CLASSIFICATION

6.1 INTRODUCTION

The interpretation of image data is different from that of maps. There are various approaches and quantitative methods for extracting useful information from a variety of remote sensing data products. In order to take advantage of and make optimum use of remote sensing data, meaningful information can be extracted from the imagery to create various thematic maps such as roads, rivers, urban, agricultural crops, forests, water, and so forth. Interpretation and analysis of remote sensing images involve identification and/or measurement of various targets/objects present on an image (Garg, 2019). Objects/targets in remote sensing images may be any feature which can be represented by a point, line, or area feature. These objects/targets can easily be distinguished if they have different contrast with other features around it in the image.

Image classification methods are broadly divided into two methods, manual interpretation and digital interpretation methods. Visual interpretation of remote sensing images requires the identification of various objects on the images by the human eye. However, it does not allow for full exploitation of the data characteristics, as human eyes cannot interpret all spectral differences in imagery. Much interpretation and identification of targets/objects in remote sensing imagery is performed manually or visually, that is by a human interpreter. However, in many cases, this is done by using digital imagery displayed in a pictorial form on a computer screen and using the capabilities of image processing software for its interpretation and analysis (e.g., on-screen digitization).

On the other hand, digital interpretation methods allow for quantitative analysis of all spectral bands in imagery simultaneously, using image processing software. These digital methods are able to detect subtle differences in DN values on the image that human eyes fail to detect. Several features show a change in appearance and shape with time, and this can affect the interpretations of the features carried out on temporal images. Some examples include: urbanizations, soil moisture/soil color as it dries up, crops growing and maturing, flooded regions, forest types, and seasonal rivers. In addition, several features/objects would appear in different tones and textures on images that are often collected under different atmospheric and sun angle conditions across a season. Therefore, it is important that the atmospheric and radiometric corrections are applied before taking up interpretation of these images (Garg, 2021).

Interpretation and analysis of remote sensing imagery involves the identification and/or measurement of various objects in order to extract useful information. Both the methods (visual and digital) have, however, their own advantages and disadvantages. In some cases, a mix of both methods may be employed when analyzing the images. In fact, the ultimate decision about the selection of the manual method or digital method of interpretation depends on the utility and relevance of the information derived for the practical application at hand. The various steps involved in interpretation of images are presented in Figure 6.1.

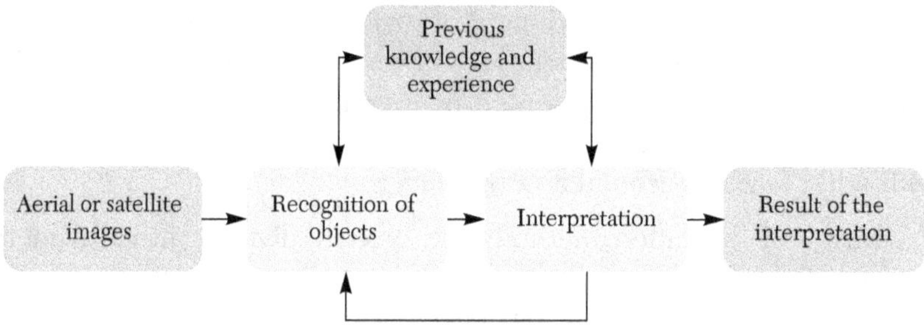

Figure 6.1. Steps in interpretation of an image.

6.2 MANUAL (VISUAL) INTERPRETATION METHODS

Interpretation of aerial photographs and satellite imagery is a method of obtaining information about the objects and landscape. *Photointerpretation* is defined as the act of examining aerial photographs/images for the purpose of identifying objects and judging their significance. It is a process of studying the geographical features from images based on the detection, identification, and spatial localization of individual objects and terrain shapes. The interpretation of imagery is more difficult than the everyday visual interpretation of our surroundings, as we perceive difficulty in depth assessment when viewing a 2D image. For this purpose, we can view the images stereoscopically so as to create a 3D model of objects/ground from 2D images in the overlap region (Colwell, 1983).

Indeed, interpretation benefits greatly in many applications when images are viewed in stereo, as visualization (and therefore, recognition) of targets/objects is enhanced dramatically. Viewing objects on 2D images provides a very different perspective, as in normal circumstances we are familiar with 3D viewing of the terrain. Combining the 2D perspective at a different scale and lack of recognizable details can make even the most familiar objects difficult to interpret in an image. In addition, we can see the objects only in visible wavelengths, and therefore the imaging of features outside this wavelength is more difficult to comprehend.

Earlier, when digital images and computerized classification were not readily available, images were mostly examined and interpreted by visual interpretation methods (Reddy, 2012). The level of collected information depends on multiple things, such as training and experience of the analyst, scale of the photograph/image, geographic location of an area, associated maps, and ground observation data. The knowledge of the individual analyst further influences the visual interpretation of the result. Partly the interpretation is dependent on the experience of the analyst, as his knowledge and skills are applied in identification of objects. The basic task in the interpretation of aerial and satellite images is systematically "reading" their contents, which includes correct recognition and classification of objects, determining their properties, quantitative and qualitative characteristics, accurate spatial (positional) location of objects, examination and evaluation of interrelationships between the objects and phenomena, and analyzing these linkages and identifying patterns characterizing the attributes.

6.2.1 Visual Interpretation Elements

There are seven fundamental parameters or elements that are used in the interpretation of airborne photographs or remote sensing images. These are tone or color, texture, pattern, shape, size, shadow, site, and association. In some cases, a single element alone is sufficient for successful identification; in others, the use of several elements will be required. Observing the differences between targets/objects and their backgrounds involves comparing different targets/objects based on the visual interpretation elements (Garg, 2019). Figure 6.2 shows these elements which have been arranged on the basis of their spatial arrangement and degree of complexity, considering the tone/color variations to be the primary element of visual image interpretation. The nature of each of these interpretation elements is described as follows:

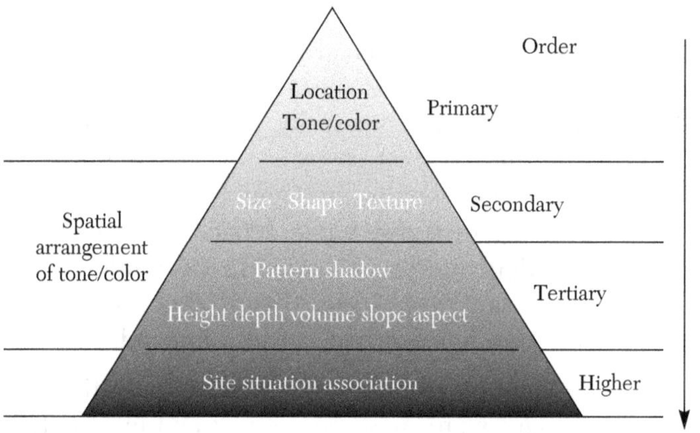

Figure 6.2. Order of importance of visual interpretation elements (Colwell, 1983).

1. **Tone or color:** Tone is the relative brightness of gray levels on a black and white image or color/FCC image (Reddy, 2012). It is the measure of the intensity of the reflected or emitted radiation of the objects of the terrain. An Earth surface feature/object often reflects or emits different proportions of energy referred to as the spectral signature of the object in the EMS. Lower reflected objects appear relatively dark and higher reflected objects appear bright. Generally, tone is the fundamental element for distinguishing between different targets or features. Variation in tone also allows the elements of shape, texture, and pattern of objects to be distinguished.

   Human eyes can discriminate only 16–20 gray levels in a B&W photograph/image, while more colors can be distinguished in a color/FCC photograph/image. Therefore, color images are preferred for visual interpretation. For example, vegetation reflects much in the NIR region of the EMS, therefore in standard FCC vegetation appears red, which is more identifiable. Figure 6.3(*a*) represents an image of a part of the Shivalik region with Dehradun city in the center, where various tones represent different features,such as forest, sand, river, urban, and so forth.

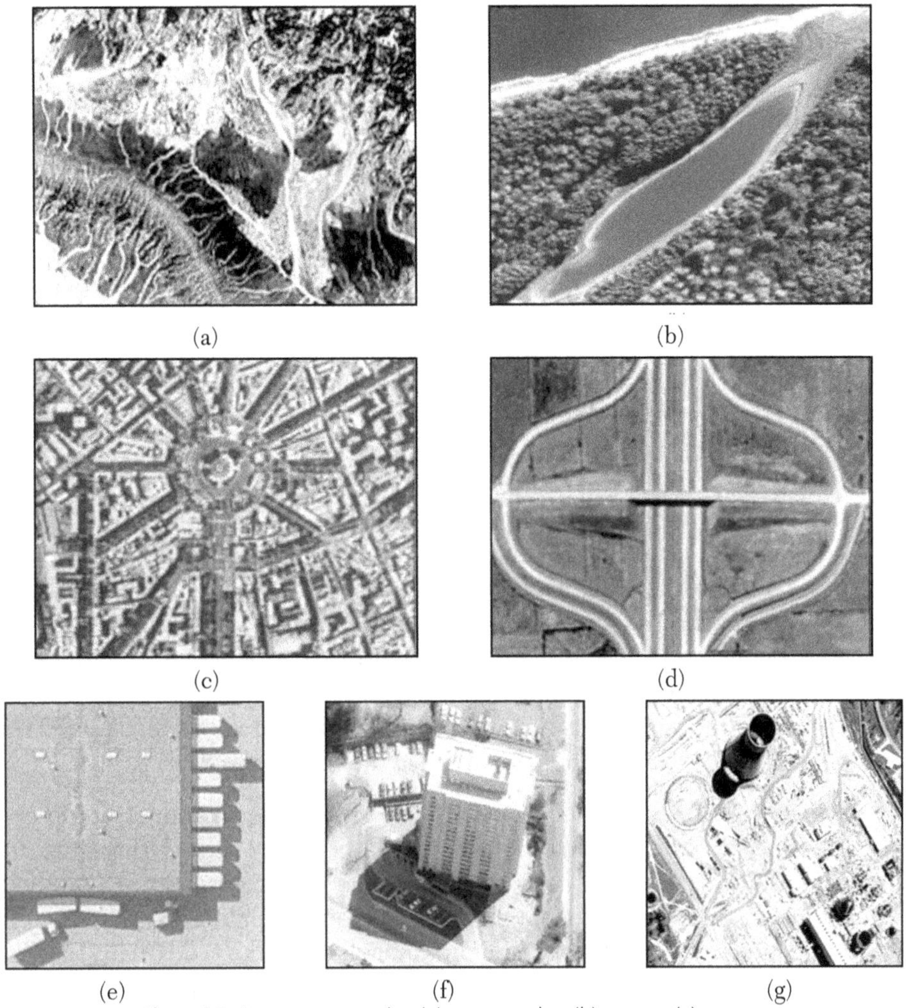

Figure 6.3. Images representing (a) tone or color, (b) texture, (c) pattern, (d) shape, (e) size, (f) shadow, and (g) site and association.

2. **Texture:** It refers to the arrangement and frequency of tonal variation in a particular area of an image. Texture is produced by an aggregate unit of features which may be too small to be clearly discerned individually on the image (Lillesand & Kiefer, 1994). It depends on shape, size, pattern, and shadow of terrain features. Texture is always scale or resolution dependent. Some objects may have similar tone, but difference in texture helps their identification. As an example, in a high-resolution image grassland and tree crowns have a similar tone, but grassland will have a smooth texture as compared to trees. Smooth texture refers to

less tonal variation and rough texture refers to abrupt tonal variation in an image or photograph. Smooth textures are most often the result of uniform, even surfaces, such as fields, asphalt, or grasslands. Rough textures would consist of a mottled tone where the gray levels change abruptly in a small area, whereas smooth textures would have very little tonal variation. A target with a rough surface and irregular structure, such as a forest canopy, results in a rough-textured appearance. Texture is one of the most important elements for distinguishing features in radar imagery. Figure 6.3(*b*) shows smooth texture exhibited by water in the center and rough texture by forests in the surroundings.

3. **Pattern:** The repetition of certain general form or relationship in tones and texture creates a pattern, which is characteristic of this element in image interpretation (Lillesand & Kiefer, 1994). The pattern can be observed as a spatial arrangement of the features or objects which may be natural and artificial or arranged randomly and systematically on the landscape. Objects both natural and man-made have some kind of pattern which aids in their recognition. For example, orchards with evenly spaced trees and a new urban area set up with regularly spaced houses are good examples of a pattern. If we further want to analyze the temporal pattern of urban expansion in an area, we can observe from multiple images that the city grows either in radial or elongated form. Figure 6.3(c) shows a radial pattern of the road network.

4. **Shape:** It refers to the general form, its configuration, structure, or outline of an individual object which helps in recognizing features from an image (Reddy, 2012). The shape of certain features can be a very distinctive clue for their interpretation. Straight edge shapes (e.g., rectangular) typically represent urban or agricultural fields, while natural features, such as forests and lakes, are generally more irregular in shape. Generally, regular shapes, square, rectangles, circles, and ovals are a sign of man-made objects, for example, buildings, roads, and cultivated fields, whereas irregular shapes with no distinct geometrical shape are a sign of natural objects, for example, a river or forest. Figure 6.3(d) shows a peculiar symmetrical shape which indicates the design of intersections of roads for the smooth flow of traffic.

5. **Size:** The size of objects in an image is a function of scale. It is one of the important elements and has the most distinguishable characteristics of an object (Sabins, 1996). The size of objects on images must be considered in the context of the image scale or resolution. It is important to assess the

size of an object relative to other objects in a scene, as well as its absolute size, to aid in the interpretation. The most measured parameters are length, width, perimeter, area, and occasionally volume. Measuring the size of an object allows the interpreter to draw many possible alternatives. For example, the size may be a distinguishable characteristic between land uses, for example, large buildings such as a factory or warehouse can be identified as a commercial property, whereas smaller buildings may indicate residential properties. Figure 6.3(e) shows various sizes of vehicles parked in parking space, which can differentiate between a normal car and a truck.

6. **Shadow:** Shadow may provide an idea of the profile and relative height of targets/objects without any stereoscopic imagery, which may make identification easier (Reddy, 2012). This element is helpful in image interpretation, and gives a real clue of an object's identity, but sometimes creates difficulties in identification. For example, the shadow of a building provides valuable information about the relative height of the building on the ground that may be a multiplestory building, and so forth. Knowing the time of photography and the solar elevation/illumination, the height of objects can be estimated. Shadow is also useful for enhancing or identifying the topography and landforms, particularly in radar imagery. On the other hand, the shadows can reduce the interpretation accuracy in their area of influence, since targets within shadows (due to darker tones) are much less (or not at all) discernible compared to their surroundings. Figure 6.3(f) shows a tall building which is casting its shadow on the ground, obscuring ground information.

7. **Site and Association**: These are higher order complex key elements and are very important to identify the objects in the image when objects are not clearly identified using the previous elements. Site refers to topographic or geographic location; for example, sewage treatment facilities are positioned at low topographic sites near streams or rivers to collect waste flowing through the system from higher locations. The site has unique characteristics of the landscape. Aspect, topography, geology, soil, vegetation, and cultural features on the landscape are distinctive factors that the interpreter should be aware of when examining a site. The relative importance of each of these factors will vary with local conditions, but all are important. For example, some vegetation grows in swamps while others grow on sandy areas or on the sunny side of hills.

Association takes into account the relationship between other recognizable objects or features in proximity to the target of interest. The relationship of the feature to the surrounding features provides clues towards its identity. Often, many of the rock types have distinct topographic expressions; for example, some kinds of sedimentary rocks are typically exposed in the form of alternating ridge and valley topography. A very high reflectance feature in the Himalaya valley may be either snow or cloud. The identification of features that one would expect to associate with other features may provide information to facilitate the identification. For example, in urban settings, smooth green space generally refers to grassland or a playground, and not agriculture land. Site and association are utilized in combination of each other to draw a meaningful and logical conclusion and are rarely used independently. For example, commercial properties may be associated with proximity to major transportation routes. Nuclear power plants are associated with a source of cooling water, weather patterns can be associated with pollution sources, and so forth. Figure 6.3(g) shows a big cooling tower which provides a clue of the presence of a thermal power plant.

6.2.2 Visual Interpretation Keys

Interpretation keys are a set of guidelines used to assist the interpreters in rapidly identifying the features (Colwell, 1983). These keys aid in identifying thematic classes from image feature characteristics. Basically, the interpretation keys can be one of two generic types: selective and elimination keys. Selective keys are arranged in such a way that interpreters simply select the example that most closely corresponds to the object they are trying to identify, for example, industries, landforms, and so on. Elimination keys are arranged so that the interpreters follow a precise stepwise process that leads to the elimination of all items except the one(s) that they are trying to identify.

Image interpretation depends on the interpretation keys established by an experienced analyst from prior knowledge and the study of images. Generally, standardized keys are established to eliminate the differences between different interpreters. Determination of the type of key and the method of presentation to be employed depend upon: (*a*) the number of objects or conditions to be identified, and (*b*) the variability typically encountered within each class of features or objects within the key. Interpretation keys are more easily constructed and used for the identification of man-made objects and features than for natural vegetation and landforms. For typical land cover types, these keys are presented in Table 6.1.

Table 6.1. Typical Image Interpretation Keys for Interpreting Land Cover Types

Classes	Band 4	Band 5	Band 6	Band 7	Bands 4, 5, 7 (BGR)	Bands 4, 5, 7 (RGB)
Snow	PW	PW	PW	PW	PW	PW
Cloud	W	W	W	W	W	W
Haze	W	W	–	–	W	W
Forest	DGR	BL	W	W	R	G
Grass	GR	DG	W	W	P	BY
Bare land	GR	W	W	W	W	W
Wetland	GR	W	GR	DGR	LB	RP
Urban	GR	W	GR	DGR	LB	RP
Water	DGR	BL	BL	BL	B	BP
Shadow	BL	BL	BL	BL	BL	BL

Source: http://stlab.iis.u-tokyo.ac.jp/~wataru /lecture/rsgis/rsnote/cp7/t7-5-2.gif
Note: PW–Pure White, W–White, DRG–Dark Gray, R–Red, B–Blue, P–Pink, G–Green, LB–Light Blue, RP– Reddish Purple, GR–Gray, BL–Black, BY–Brandish Yellow and BP–Bluish Purple.

Most interpreters prefer to use elimination keys in their analyses (Colwell, 1997). Keys can be used in conjunction with any type of remotely sensed data. Basically, the interpretation keys help the interpreter organize the information present in image form and guide the analyst to correctly identify the unknown objects. For analysis of natural features, training and field experience are often essential to achieve consistent results. As an example, a thematic map (right) prepared from visual interpretation of a satellite image (left) is shown in Figure 6.4.

Figure 6.4. Satellite image (left) and its mapping through visual interpretation (right).

6.2.3 Visual Interpretation Aids

On the basis of tasks to be carried out, visual interpretation aids are required for viewing, including enlargement, transfer of detail, and measurement. Interpretation aids can also include equipment for monoscopic viewing and stereo-viewing. Some of the projection devices can be used for both viewing and transfer of details from an image to a map at the same scale or a different scale. These devices, such as enlargers and light tables, provide the means for simultaneous viewing of both the image and the map. Enlargers can be used for paper prints or film negatives to make the enlarged view and carry out mapping/interpretation at a different scale. Light tables, in addition to the instruments used for optical projection, are used for delineation of features from hardcopy paper prints at the same scale as the paper print. Handling of images for interpretation is relatively easier using light tables. Sometimes, a handheld magnifying lens is also used to see the details on an image in magnified form. A magnified view can remove certain doubts, particularly linear features.

6.2.4 Field Data Collection and Verification

Field verification can be considered as a form of collateral data which is typically conducted to assist in the analysis process and testing the reliability of generated thematic maps. Collateral data is also called "ancillary data," and it may be available in the form of text, tables, maps, graphs, or image metadata. Essentially, this is the process of familiarizing the analysts with the area or type of features present therein. This type of verification is done prior to the interpretation to develop a visual "signature" of how the feature(s) of interest appear on the ground as well as on the image. So, field data is an important data source both for visual interpretation and digital interpretation methods.

The nature, amount, timing, method of data acquisition, and data integration procedures should be carefully thought out for the field verification. For better interpretation, field visits to the area may be done twice, prior to analysis and after analysis of the image. After an analysis is made, it is conducted to verify the accuracy of the interpretation. Fieldwork is sometimes considered as being three times as expensive as the lab analysis. The amount and type of field work required for a given application may vary greatly, and is generally dependent upon the (*a*) type of analysis involved; (*b*) image quality, including scale resolution and information to be interpreted; (*c*) accuracy requirements for classification; (*d*) experience and knowledge of the interpreter about the sensor, area, and subject; (*e*) terrain conditions; (*f*) access to ancillary material; and (*g*) cost considerations.

6.3 DIGITAL INTERPRETATION

The image classification process involves a categorization of all pixels of a digital image into one of several land use and land cover classes, or "themes." In a broad sense, *image classification* is defined as the process of categorizing all pixels in an image to obtain a given set of labels or land use and land cover themes (Lillesand & Keifer, 1994). Normally, multispectral data are used to perform the digital classification and, indeed, the spectral value present for each pixel is used as the numerical basis for categorization of image pixels.

According to Lu and Weng (2007), major steps of image classification may include: (*a*) selection of a suitable classification system; (*b*) deciding image classes, such as urban, agriculture, water, forest, and so on; (*c*) conducting field surveys and collecting ground information; (*d*) image preprocessing for enhancement of geometric and radiometric qualities of satellite imagery; (*e*) feature extraction and selection; (*f*) image classification; (*g*) post-processing: filtering and classification; and (*h*) comparing classification results with the field (ground truth) data for accuracy assessment. Various steps followed in digital image classification of remotely sensed data to create typical thematic resource maps are shown in Figure 6.5.

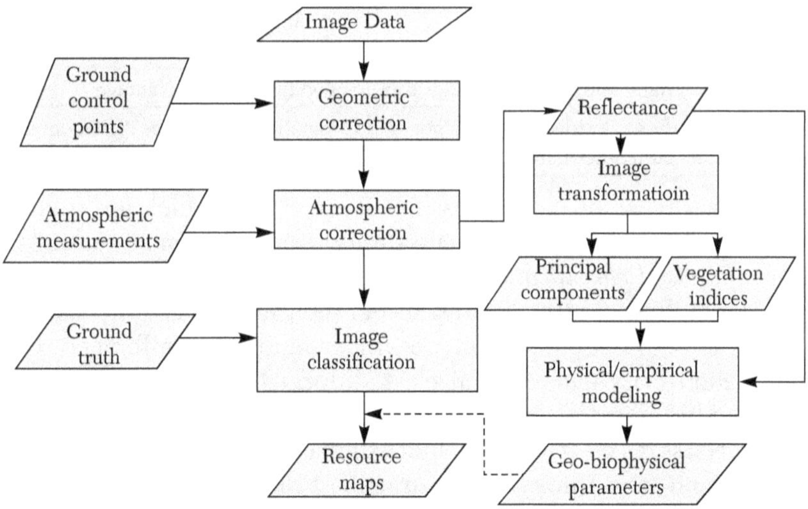

Figure 6.5. Various steps involved in digital image classification of remotely sensed data.

A comparison between visual and digital methods of interpretation is given in Table 6.2.

Table 6.2. Visual Interpretation versus Digital Interpretation of Remotely Sensed Images

S.No.	Manual interpretation	Digital interpretation
1	Hardcopy data are required.	Digital data are required which may be expensive.
2	Requires simple and inexpensive equipment.	Requires specialized, and often expensive equipment and software. So start-up cost may be high.
3	Limited to analyzing only a single channel of data or a single image at a time.	Simultaneous interpretations of images with several wavelengths is possible.
4	Uses brightness of image, so very good for extraction of spatial information.	Uses DN values so very good for extraction of spectral information.
5.	It is not possible to change the characteristics of hardcopy images.	Radiometric and geometric corrections can be applied through preprocessing of image.
6.	It involves subjectivity, that is, interpretation results may vary with the time and person depending upon the experience and knowledge of the interpreter.	It is an objective process which utilizes the spectral signature of objects through a computer, so results can be quickly reproduced any number of times.
7	It is a qualitative method.	Extraction of quantitative information is possible and easier.
8	Cost-effective for small areas and for one-time interpretation.	Cost-effective for large geographic areas, and for repetitive interpretations.
9	No algorithm is used.	Possibility to explore alternatives with complex interpretation algorithms.
10	Time-consuming approach for a large area.	Speed may be an advantage.
11	Output product is hardcopy, so it is not compatible with other data.	Compatible with other digital data (e.g., GIS).

Two broad types of digital classification methods which are used by the majority of the image processing systems to analyze remote sensing images are: *supervised classification* and *unsupervised classification* (Jain, 1989). In order to illustrate the differences between supervised and unsupervised classification, it is important to understand the difference between information class and spectral class. Information class is a class that is specified by an image analyst. It refers to the information to be extracted. Spectral class is a class which includes similar gray levels in the multispectral space. In an ideal information extraction task we can directly associate a spectral class in the multispectral space with an information class. There are many limitations of both supervised and unsupervised classification methods, and therefore several new classification approaches, including a "hybrid classification method," have been developed. These are shown in Table 6.3 and are further discussed in Chapter 7.

Table 6.3. Summary of Some Classifiers Used for Image Classification

Criteria	Categories	Example of classifiers
Whether training samples are used or not	Supervised classification approaches	Maximum likelihood, Minimum distance, Artificial Neural Network (ANN), Decision Tree (DT)
	Unsupervised classification approaches	ISODATA, K-means clustering
	Hybrid approaches	Expert systems, Artificial intelligence Bayesian formalism, Evidential reasoning, Multiple ANN, Voting rules
Whether parameters such as mean vector and covariance matrix are used or not	Parametric classifiers	Bayesian classifier, Maximum likelihood, Linear Discriminant Analysis (LDA)
	Nonparametric classifiers	ANN, DT, Evidential reasoning, Support Vector Machine (SVM), Expert system, K-Nearest Neighbor (K-NN), Logical regression, and Multilayer Perceptron (MLP)
Which kind of pixel information is used	Per-pixel classifiers	Most of the classifiers, such as Maximum likelihood, Minimum distance, ANN, DT, Genetic Algorithm (GA), Random Forest (RF), and SVM
	Sub-pixel classifiers	ANN, Fuzzy-set classifiers, Fuzzy expert system, Possibilistic c-Means, Regression modeling, Regression tree analysis, Noise clustering, Spectral Mixture Analysis (SMA)
Which kind of pixel information is used	Object-oriented classifiers	Image segmentation, E-cognition, and ArcGIS Feature analyst
	Per-field classifiers	GIS-based classification approaches
	Rule based	Knowledge-based algorithm
Whether output is a definitive decision about land cover class or not	Hard classifiers (CRISP)	Most of the classifiers, such as Maximum likelihood, Minimum distance, ISODATA, Parallelepiped, Centroid (k-means), ANN, DT, SVM
	Soft classifiers	Fuzzy-set classifiers, Fuzzy c-means, Sub-pixel classifier, Spectral mixture analysis, ANN, GA, DT, Possibilistic c-Means (PCM), Linear mixer models

(Contd.)

Criteria	Categories	Example of classifiers
Whether spatial information is used or not	Spectral classifiers	Maximum likelihood, Minimum distance, ANN
	Contextual classifiers	Iterated conditional modes, Marko Random Field (MRF), Point-to-point contextual correction, Frequency-based contextual classifier
	Spectral-contextual classifiers	Extraction and classification of homogeneous objects (ECHO), MRF, Combination of parametric or nonparametric and contextual algorithms

6.3.1 Supervised Classification

Supervised classification requires selection of an appropriate classification scheme, and then identification of training areas in the image that best represents each class. These techniques are mostly used for quantitative analysis, and are broadly divided into *soft classification methods* and *hard classification methods*. The difference in these two methods is that they identify and describe the regions in spectral space. Some methods seek a segmentation based on probability distributions while others use simple geometric partition. Hard classification generates firm boundaries between classes, whereas soft classification sometimes leads to spatially overlapped regions.

In a supervised classification, the analyst identifies in the image homogeneous representative areas of different land use and land cover types (information classes) of interest. These representative areas are also called as *training areas* or *training samples*. The purpose of generating training samples is to assemble a set of statistics that describes the spectral response patterns for each land use and land cover class to be classified in the image (Lillesand & Kiefer, 2004). A sufficient number of training sites of each class is selected to represent the variation present within each class in the image. The process of defining the training areas of a class is usually carried out by drawing the polygons of known classes over the image, interactively. These polygons are labeled as such and used for computation of statistics by the algorithm. The DN values in all spectral bands for the pixels comprising these areas are used to "train" the algorithm and to identify spectrally similar areas for each class. The flowchart of supervised classification is presented in Figure 6.6(*a*).

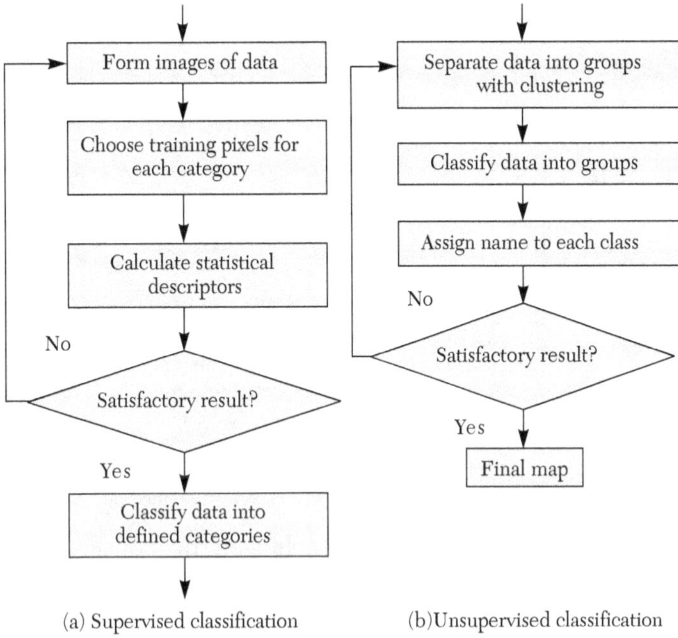

Figure 6.6. Flow diagram of (a) supervised, and (b) unsupervised classification techniques.

Supervised classification has three distinct stages, namely the training, classification, and output stage, as shown in Figure 6.7. Training is the identification of a sample of pixels of known classes gathered from the reference data, such as field visits, existing maps, reports, and aerial photographs. In supervised classification, an analyst is assumed to have a priori knowledge of the area to locate training samples, which represent homogeneous areas of known land use and land cover classes (Garg, 2021). The training process helps in developing statistical parameters which describe the spectral response pattern for the information classes that are being considered for image classification. The variation in spectral reflectances of a training data set per class in multispectral bands is used to generate the statistics, such as the mean values and covariance (the extent to which two variables vary together) from different wavelength bands for each class. Thus, a sufficient number of homogeneous training samples for each class to represent the tonal variation present within each class in the image are selected. It is important that the training samples are well distributed throughout the image, as far possible, as they act as representative for the entire image, and also incorporate the site specific variations of that class in the training data.

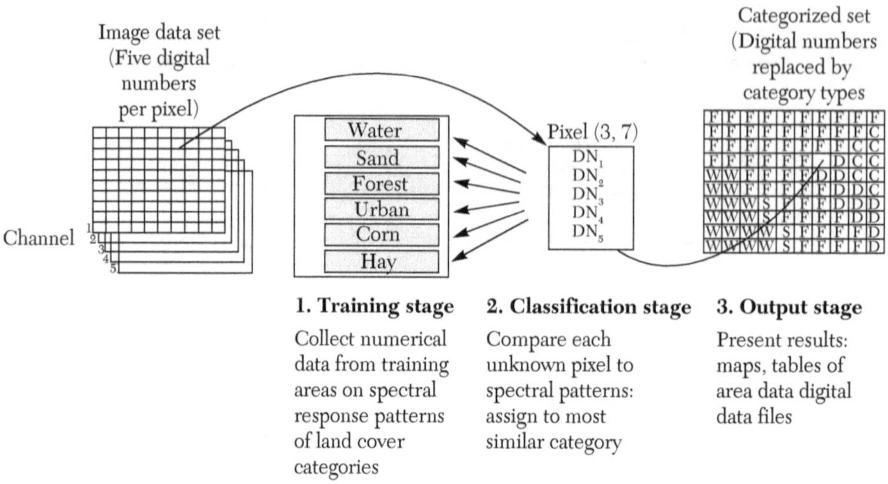

Figure 6.7. Stages in the supervised classification method.

An improved classification can be achieved by considering a suitable classification algorithm with a sufficient number of training samples. Training samples are often prepared either by field work or from other data sources, such as aerial photographs and satellite imagery of fine spatial resolution based on a single pixel or polygon. In the case of coarse resolution data, the selection of training samples is often tedious as it contains large regions of mixed pixels. Mixed pixels are due to the existence of different classes in the same pixel.

The selection of appropriate training samples is also based on the analyst's familiarity with the geographical area and knowledge of the land use and land cover types present in the image. It is important to understand that the training samples are selected judiciously and scientifically. The training sample size can vary from a minimum of 10N–100N per class, where N is the number of bands used for classification (Jenson, 2007). A histogram for each band of the training areas/samples can be drawn to check the homogeneity of the class. The normal histogram with a single peak would indicate a single class, but a bimodal response could point to a subclass present in the class. In such cases, the training sample locations can be reselected or modified. Thus, the classification accuracy may be improved considerably if each training class is found to have only one single peak in the histogram of the training sample.

The training areas developed in one scene may or may not be transferable to an entire study area consisting of many images. If ground conditions, lighting conditions, or atmospheric effects change from one image to another, then training areas must be developed independently for each scene. Furthermore, training areas may not be transferable across time, as the real changes in the land use and land cover occurring at a training site location over time will cause inaccurate classification results in the temporal image.

The key characteristics of selecting the training areas are as follows:

1. *Shape:* Shapes of training areas are not so important provided that shapes do not prohibit accurate delineating and positioning of outlines of regions on digital images. Usually, it is easy to define rectangular or polygonal areas and minimize the number of vertices.

2. *Location:* Location is important as each informational category should be represented by several training areas well-distributed in the image. Training areas must be positioned in such locations that favor accurate and convenient transfer of their shapes from maps and aerial photographs to the digital image. Since the training data represents the entire area within the image, it must not be clustered in one region of the image. As most of the time the analyst uses direct field observations for the selection of training data on an image, the requirement of an even distribution of training data often has practical constraints, as only those sites which are easily accessible must have been visited on the ground. Often, the use of good maps and aerial photographs can provide the basis for accurate delineation of training samples that could not be inspected in the field due to some constraint.

3. *Number:* The optimum number of training areas depends on the number of classes to be identified, their diversity, and the resources available in delineating the training areas. Ideally, each information category, or each spectral subclass, should be represented by a number (perhaps 5 to 10 at minimum) of training areas to ensure that spectral properties of each class are represented. It is usually better to define many small training samples than to use only a few large samples.

4. *Uniformity:* Perhaps the most important property of a good training sample is its uniformity and homogeneity. Data within each training area should exhibit a unimodel frequency distribution for each spectral band to be used.

Another important criterion for selection of training samples is to avoid overlap in the spectra for each class. The spectral separability between the classes is important to determine while finalizing the purity of the training sample (Congalton, 1991). It is a statistical measure of distance between the two signatures. The separability can be calculated for any combination of bands that is used in the classification; thus the bands that are not useful in the classification can be discarded. There are several processes to perform that can determine whether the spectral signature data are a true representation of the pixels to be classified for each class.

The spectral signatures that are created for a training a sample can be evaluated by a number of methods (Colwell, 1983): (*a*) graphical method, (*b*) Euclidean distance, (*c*) divergence, (*d*) transformed divergence, and (*e*) Jeffries-Matusita distance (JM). These are described as follows:

(a) Graphical method: It is drawing and viewing ellipse diagrams as well as scatterplots of data values for every pair of bands. For example, Figure 6.8 shows a scatterplot of two classes in a two-dimensional space (scatterplot) which are not spectrally separable between band A and band B, while the same classes are spectrally separable between band C and band D.

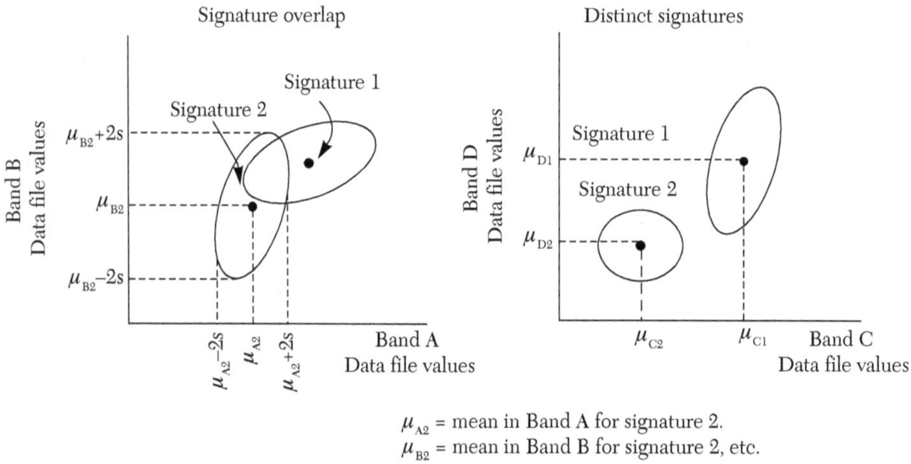

Figure 6.8. Scatterplot of two classes spectrally not separable (left), and spectrally separable (right).

(b) Euclidean distance: It is a statistical measure of the distance between two spectral signatures. For Euclidean distance evaluation, the spectral distance between the mean vectors of each pair of signatures is computed. If the spectral distance between two samples is not significant for any pair of bands, then they may not be distinct enough to produce a successful classification. Euclidian distance is computed as follows:

$$\text{Euclidean distance (ED)} = \sqrt{\sum_{i=1}^{n}(d_i - e_i)^2} \tag{6.1}$$

where n = number of bands, i = particular band, d_i = data file value of pixel d in band i, and e_i = data file value of pixel e in band i.

(c) Divergence: The divergence (D) is computed as:

$$D_{ij} = \frac{1}{2} tr((C_i - C_j)(C_i^{-1} - C_j^{-1})) + \frac{1}{2} tr(C_i^{-1} - C_j^{-1})(\mu_i - \mu_j)(\mu_i - \mu_j)^T) \quad (6.2)$$

where i and j are the signatures of two classes being compared, C_i = covariance matrix of signature I, μ_i = mean vector of signature i, tr = trace function, T = transposition function and between class i and j.

(d) Transformed Divergence (TD): The parametric TD separability measure is a commonly used measure. It provides a covariance weighted distance between class means to determine whether class signatures are separable. It is computed as:

$$TD_{ij} = 2000\left(1 - \exp\left(\frac{-D_{ij}}{8}\right)\right) \quad (6.3)$$

where D_{ij} is computed using equation 6.2. The rationale of the TD is to define the separation between two normal distributions (characterized by their mean vectors and covariance matrices), but most metrics used do not follow a Gaussian distribution. In this respect, the TD is only indicative in cases of normality, or when classifiers based on the assumption of normality (e.g., maximum likelihood) are used. The scale of the TD values can range from 0 (no separability) to 2,000 (completely separable). As a general rule, if it is greater than 1,900, then the classes are considered to be separable. Between 1,700 and 1,900, the separation is fairly good, and below 1,700, the separation is poor (Jensen, 1996). The larger the separability values are, the better the final classification results will be.

(e) Jeffries-Matusita Distance (JM): The JM distance has a saturating behavior with increasing class separation, like the TD. However, it is not as computationally efficient as the TD (Jensen, 1996). Both transformed divergence and the Jeffries-Matusita distance have upper and lower bounds. The JM is between 0 and 1,414. The formula for computing the JM distance is as follows:

$$\alpha = \frac{1}{2}(\mu_i - \mu_j)^T \left(\frac{C_i + C_j}{1}\right)^{-1} (\mu_i - \mu_j) + \frac{1}{2}\ln\left(\frac{|(C_i + C_j)/2|}{\sqrt{|C_i| \times |C_j|}}\right) \quad (6.4)$$

$$\mathrm{JM}_{ij} = \sqrt{2(1 - e^{-\alpha})} \quad (6.5)$$

where: i and j = the two signatures (classes) being compared, C_i = the covariance matrix of signature i, μ_i = the mean vector of signature i, ln = the natural logarithm function, and $|C_i|$ = the determinant of C_i (matrix algebra).

In the second stage of supervised classification (see Figure 6.7), the training pixels are used to derive various statistics (DN values) for each class, and are correspondingly assigned signatures. In the third stage, each pixel of the image is allocated to the class with which they show greatest similarity based on the established signatures. Thus, based on statistics of these training sites, the entire image is then assigned pixel-wise a user-defined land use type (urban, corn, hay, etc.) or land cover type (water, sand, forest, etc; Richards & Jia, 2013). The algorithm determines the unique "signatures" (range of DN values) for each training class, and uses the spectral characteristics of the training areas to classify the remainder of the image. Each pixel in the image is then labeled as the class it most closely "resembles" by comparing the signatures of training areas. Since the analyst is "supervising" the categorization of a set of specific classes, the method is called supervised classification. The flow diagram in Figure shows the procedure used (left diagram) in supervised classification.

Accuracy of supervised classification results mainly depends on the analyst's ability to collect a sufficient number of training areas which are homogeneous and in sufficient number (Hord, 1992). At the end of classification, if a particular class has not been picked up properly for classification by the algorithm, the image is again classified by refining the training areas/samples for that class. The process allows refining the training areas/samples many times, until acceptable accuracy of classification is achieved. It is therefore in a way a process, where the analyst supervises the classification of images into a defined, set of classes. Thus, in a supervised classification, the information classes are identified which are then used to determine the spectral classes representing them.

Once the statistical characteristics are computed for each information class, the image is classified using any supervised method like the minimum distance classifier, the maximum likelihood classifier, and the parallelepiped classifier. These classifiers are explained as follows. Figure 6.9(a to c) shows the original image, selection of training sample, and results of supervised classification, respectively.

6.3.1.1 Minimum distance classification

Minimum distance classifies image data on a database file using a set of possible class signature segments as specified by the signature parameter (Reddy, 2012). Each segment specified in the signature, for example, stores spectral signature data pertaining to a particular class. Only the mean

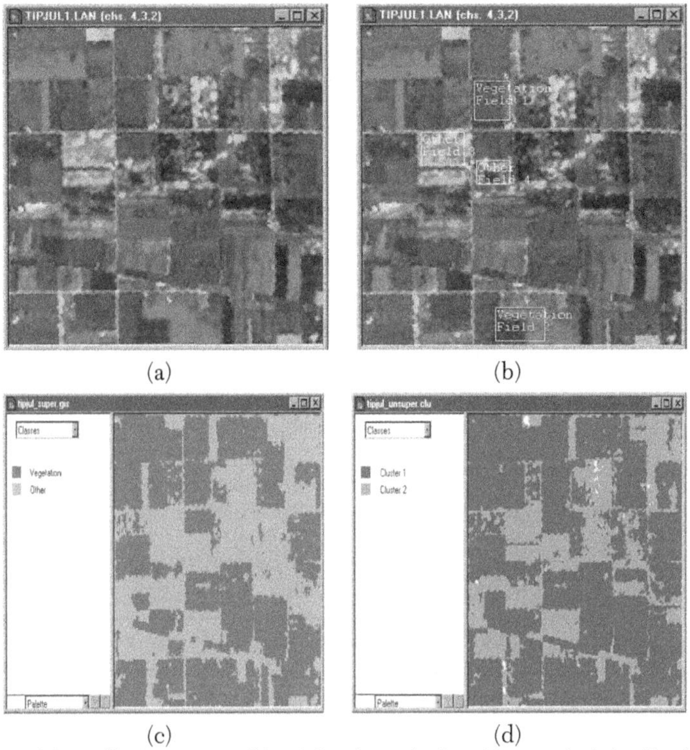

Figure 6.9. (a) Satellite image FCC, (b) training data selection, (c) supervised classification, and (d) unsupervised classification (please see color figures in companion files).

vector in each class signature segment is used. It is the simplest classifier where the (*a*) analyst first computes the mean of each training class, (*b*) the distance (Euclidian or Mahalanobis) of each pixel from the mean is calculated, and (*c*) each new pixel is assigned to that class whose distance is nearest to the mean of the training classes (as shown in Figure 6.10). This process is used to classify unknown image data to classes which minimize the distance between the image data and the class in multifeature space.

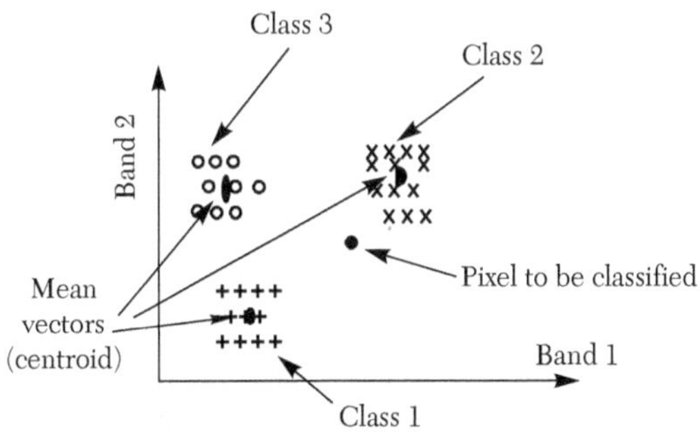

Figure 6.10. Minimum distance classifier.

For supervised classification, the cluster groups are formed by DN values of pixels within the training samples defined by the analyst. Each cluster can be represented by its centroid, often defined as its mean value. As each pixel is considered for assignment to one of the several classes, the multidimensional distance to each cluster centroid is calculated, and the pixel is then assigned to the closest cluster. Thus, the classification proceeds by using the "minimum distance" from a given pixel to a cluster centroid defined by the training data (Garg, 2019). In its simplest form, minimum distance classification may not always be accurate; there is no provision for accommodating differences in variability of classes, and some classes may overlap at their edges. Minimum distance classifiers are direct in concept and in implementation but are not widely used in remote sensing work. The advantages and disadvantages of the minimum distance classifier are summarized in Table 6.4.

Table 6.4. Advantages and Disadvantages of the Minimum Distance Classifier

Advantages	Disadvantages
Fast and simple to process.	
Since every pixel is spectrally closer to either one sample mean or another, there are no unclassified pixels.	Pixels that should be ideally unclassified (i.e., they are not spectrally close to the mean of any sample, within reasonable limits) become classified. However, this problem is alleviated by thresholding out the pixels that are farthest from the means of their classes.
The fastest decision rule to compute, except for then parallelepiped classifier.	Considers only mean value. Does not consider class variability. For example, a class like an urban land cover class is made up of pixels with a high variance, which may tend to be farther from the mean of the signature. Using this decision rule, outlying urban pixels may be improperly classified. Inversely, a class with less variance, like water, may tend to overclassify (that is, classify more pixels than are appropriate to the class), because the pixels that belong to the class are usually spectrally closer to their mean than those of other classes to their means.

Euclidean distance is the most common distance unit used in low dimensional data sets. Distance is usually measured in the real world. While the Euclidean distance unit is useful in low dimensions, it doesn't work well in high dimensions and for categorical variables. In this sense, Mahalanobis distance tends to be more robust to noisy data.

The Mahalanobis Distance (MD) is similar to Euclidean distance, except that the covariance matrix is used in the equation. Variance and covariance are figured in so that clusters that are highly varied lead to similarly varied classes and vice versa. It is a well-known statistical distance function, where a measure of variability can be incorporated into the distance unit directly. It is a distance measure between two points in the space defined by two or more correlated variables. It takes the correlations within a dataset between the variables into consideration. If there are two non-correlated variables, the MD between the points of the variable in a 2D scatterplot is the same as the Euclidean distance when the covariance matrix is the unit matrix. This is exactly the case if the two columns of the standardized data matrix are orthogonal. The MD depends on the covariance matrix of the attribute and adequately accounts for the correlations. Here, the covariance matrix is utilized to correct the effects of cross-covariance between two components of a random variable (Colwell, 1983). The equation for the MD classifier is as follows:

$$\text{Mahalanobis distance (MD)} = (X - M_c)^T\,(Cov_c)^{-1}(X - M_c) \qquad (6.6)$$

where: c = a particular class, X = the measurement vector of the candidate pixel, M_c = the mean vector of the signature of class c, Cov_c = the covariance matrix of the pixels in the signature of class c, Cov_c^{-1} = inverse of Cov_c^{-1} and T = transposition function.

Mahalanobis distance is fast and no pixels are left unclassified. It requires the least time for computation among all the other supervised classification techniques. The disadvantage of this technique is that it takes into account only the mean value, and does not incorporate variability of signatures, so it is less efficient than the maximum likelihood technique.

6.3.1.2 Maximum likelihood classifier

The maximum likelihood classifier (MLC) is one of the most popular methods of classification in remote sensing, in which a pixel with the maximum likelihood is classified into the corresponding class (Garg, 2019). In MLC, the probability value of pixels is taken into consideration for classifying the pixels. It assumes that the statistics for each class in each band are normally distributed. It calculates the probability that a given pixel belongs to a specific class and then compares, and the pixel is assigned to the class where the probability value is highest (Figure 6.11).

This method is highly efficient when classifying multispectral images. It accounts for the variability of signatures, and is considered to be most accurate. Though this is an efficient technique, it requires larger computational time. It relies heavily on normal distributed signatures of the classes. It is considered to give more accurate results than the parallelepiped classification, however it is much slower due to extra computations. If a

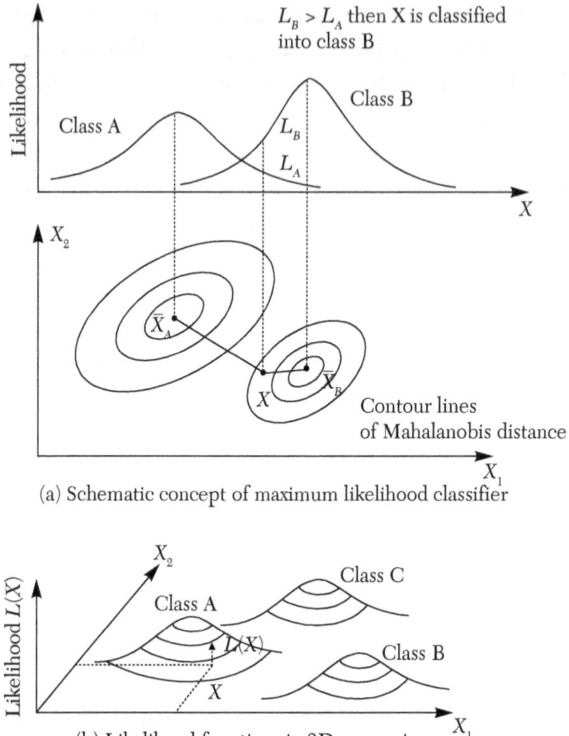

(a) Schematic concept of maximum likelihood classifier

(b) Likelihood functions in 3D expression

Figure 6.11. Concept of maximum likelihood classifiers.

priori knowledge exists that the probabilities are not equal for all classes, the weights can be specified for particular classes. This variation of the maximum likelihood decision rule is known as the Bayesian decision rule (Hord, 1982). Unless a priori knowledge of the probabilities exists, it is recommended that they are not be specified.

In Gaussian distribution, each class is assumed to be multivariate normally distributed; that means each class has a mean m_i that has the highest likelihood of occurrence. The likelihood function decreases exponentially as the feature vector x deviates from the mean vector m_i. The rate of decrease is governed by the class variance; the smaller the variance, the steeper the decrease will be, and the larger the variance, the slower the decrease will be.

The algorithm assumes that classes in the input data have a Gaussian distribution and that signatures are well selected, but this is not always true. It uses a probability density function based on the Bayes decision rule, meaning that each pixel is assigned to the class that has the highest probability (maximum likelihood). Each pixel is labeled according to the decision rule (Colwell, 1983):

$$x \in \omega_i \quad \text{if} \quad p(x|\omega_i)p(\omega_i) > p(x|\omega_j)p(\omega_j) \qquad \text{for all} \qquad j \neq i \tag{6.7}$$

The classes are represented by ω_i, $i = 1...m$ where m is the number of classes. x is a column vector with the brightness values for the pixel in each measurement band; therefore $p(x|\omega_i)$ describes the chance of finding a pixel at position x in spectral space from each class ω_i. $p(\omega_i)$ is referred to as prior probability and describes the probability that pixels from class ω_i appear anywhere in the image. $p(x)$ can be removed, since it is not class dependant. This leads to the *discriminant function* as follows:

$$g_i(x) = \ln p(x|\omega_i)p(\omega_i) = \ln p(x|\omega_i) + \ln p(\omega_i) \tag{6.8}$$

Substitution of equation (6.7) into equation (6.8) gives a new decision rule, as follows:

$$x \in \omega_i \quad \text{if} \quad g_i(x) > g_j(x) \qquad \text{for all} \qquad j = i \tag{6.9}$$

Classes in spectral space are considered to be normally distributed (Gaussian distribution) which leads to the Gaussian maximum likelihood classifier, as:

$$g_i(x) = -\ln |C_i| = (x - m_i)^T C_i^{-1} (x - m_i) \tag{6.10}$$

where m_i and C_i are the mean vector and covariance matrix of the data in class ω_i, respectively. If the highest probability of a pixel falls below a specified value, a threshold can be set and the pixel won't be classified in any of the available classes, which will yield a high accuracy.

The MLC is the most common classification method used for remotely sensed data. With the MLC, it is guaranteed that the error of misclassification is minimal, if $p(x|\omega_i)$ is normally distributed. Unfortunately, the normal distribution cannot always be achieved. In order to make the best use of the MLC method, it is important that the training samples will generate distributions as close to the normal distribution as possible. The MLC is relatively robust but it has a limitation when handling data at nominal or ordinal scales. The computational cost increases considerably as the image dimensionality increases. In two-dimensional space, the class boundary cannot be easily determined. Therefore, in the MLC algorithm, boundaries are not used but probabilities are compared.

While using the MLC algorithm, several issues are to be taken into consideration:

1. Sufficient ground truth data should be sampled to allow accurate estimation of the mean vector and the variance-covariance matrix;

2. The inverse matrix of the variance-covariance matrix becomes unstable in the case of high correlation between two image bands; and

3. When the population is not normally distributed, the MLC algorithm cannot be applied (Hord, 1982). The advantages and disadvantage of the MLC are summarized in Table 6.5.

Table 6.5. Advantages and Disadvantages of the MLC

Advantages	Disadvantages
The most popular classifier as well as most accurate (if the input samples/clusters have a normal distribution).	The algorithm takes a long time to compute. The computation time increases with the number of input bands.
Takes the variability of classes into account by using the covariance matrix, as does the Mahalanobis distance.	Insufficient ground truth data and/or correlated bands can affect the results. A thresholding condition can be introduced into the classification rule to separately handle ambiguous feature vectors.
Normally classifies all the pixels.	It relies heavily on a normal distribution of the data in each input band which may not always be true. Tends to overclassify signatures with relatively large values in the covariance matrix. If there is a large dispersion of the pixels in a cluster or training sample, then the co-variance matrix of that signature contains large values.

6.3.1.3 Parallelepiped classification

Parallelepiped classification, sometimes also known as *box decision rule* or *multilevel slicing*, is based on the ranges of values within the training data to define regions within a multidimensional data space. This is a widely used decision rule based on simple Boolean and/or logic (Jain, 1989). Training data in n spectral bands may be used in performing the classification. Brightness values from each pixel of the multispectral imagery may be used to produce an n-dimensional mean vector.

In this method, a parallelepiped-like (i.e., hyper-rectangle) subspace is defined for each class. Using the training data for each class, the limits of the parallelepiped subspace can be defined either by the minimum and maximum pixel DN values in the given class, or by a certain number of standard deviations (σ) on either side of the mean of the training data for the given class. The accuracy of classification depends on the selection of the lowest and highest values in the statistics of each class. In the 3D feature space, this forms a rectangular box; one for each class (as shown in Figure 6.12[a]). If the pixel falls inside the parallelepiped, it is assigned to the class, however, if the pixel falls within more than one class, it is put in the overlap class. If the pixel does not fall inside any class, it is assigned to the null class or unclassified pixels. For correlated data, some parallelepipeds can overlap, as illustrated in Figure 6.12(b).

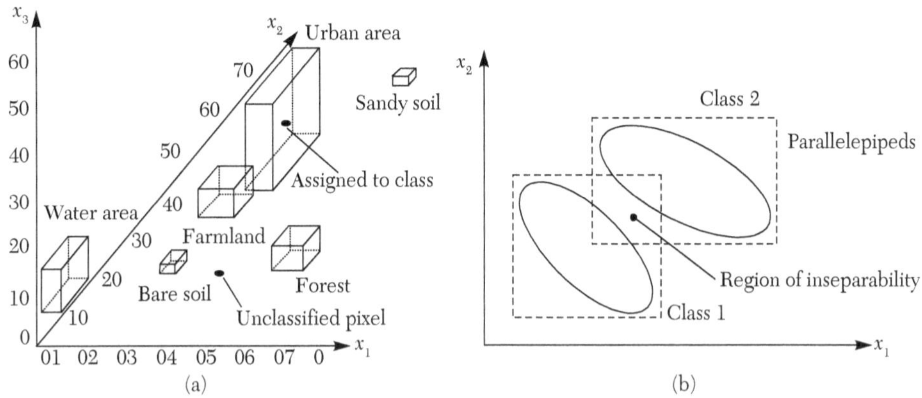

Figure 6.12. (a) Parallelepiped classifier, and (b) correlated data showing regions of inseparability.

The advantage of this method (Jenson, 2007) is that (*a*) it is a simple, fast, and robust classifier, (*b*) the speed of this classifier is high, and (*c*) it is good for non-normal distributions. It has some disadvantages due to which it is rarely used, for example: (*a*) there can be significant gaps between the parallelepipeds, and the pixels within these regions will remain unclassified; (*b*) it does not incorporate variability; (*c*) it allows class overlap, and (*d*) classes can include pixels spectrally distant from the signature mean. Other advantages and disadvantages are given in Table 6.6.

Table 6.6. Advantages and Disadvantages of Parallelepiped Classifier

Advantages	**Disadvantages**
Fast and simple, since the data file values are compared to limits that remain constant for each band in each signature, thus cutting processing time.	Since parallelepipeds have corners, pixels that are actually quite far, spectrally, from the mean of the signature may be unclassified.
Not dependent on normal distributions. Often useful for a first-pass, broad classification, this decision rule quickly narrows down the number of possible classes to which each pixel can be assigned, thus cutting processing time.	There are problems when class ranges overlap—rules must be developed to deal with overlap areas as it may reduce the accuracy of the results.
Good for helping decide if additional classes are needed (if there are many unclassified pixels).	The areas with gaps between the parallelepipeds are designated as unclassified.

6.3.2 Unsupervised Classification

Unsupervised classification in essence reverses the process of supervised classification. It requires less input from the analyst to process the image, as the analyst does not predefine the land use and land cover types (Garg, 2021). The analyst can define the number of output classes required and spectral variance (such as the standard deviation, σ) within each class. The algorithm divides the image into those defined number of classes, based entirely on the spectral data and with no prior knowledge of what land use and land cover types may be present in the image.

The classification algorithm searches and analyzes the image, and groups the pixels into natural groupings of clusters which are deemed to be uniquely representative of the image. The clustering algorithms are used to determine the natural (statistical) groupings or structures in the data (Figure 6.13[a]). Several iterations are required to be performed to determine the optimal number of classes used for initialization of the algorithm. The final result of this iterative clustering process may result in some clusters that the analyst must combine or some clusters that must be broken down, and that will require a further iteration (Figure 6.13[b]). The results may be checked with ground truth data or some reference data. If there is a mismatch between the two, the complete process can be restarted by changing the input parameters into the algorithm, until satisfactory results are achieved.

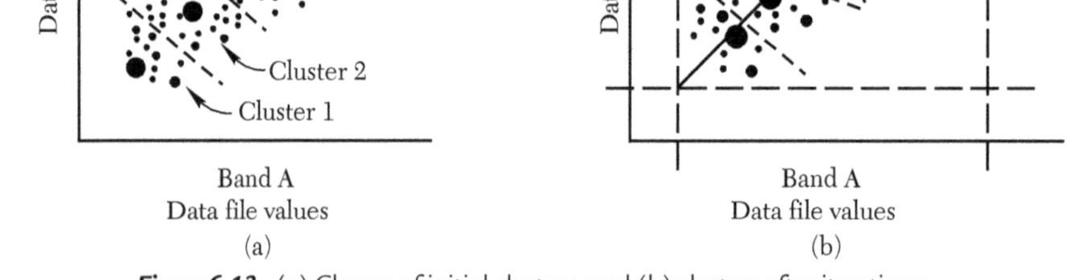

Figure 6.13. (a) Classes of initial clusters, and (b) clusters after iterations.

After classification, the image analyst will determine if these arbitrary classes have some meaning. These classes are known as spectral classes, the identity of which will not be initially known. The analyst compares the classified data with some form of reference data, such as large scale images/

photos, maps, or site visits, to determine the identity and assign the names to these spectral classes. The basic assumption is that values within a given land use and land cover type should be close together in the measurement space (i.e., have similar gray levels), whereas data in different classes should be comparatively well separated (i.e., have very different gray levels; Eastman, 1999). The flow diagram in Figure 6.6(b) shows the procedure used in unsupervised classification, and Figure 6.9(d) presents the results of unsupervised classification.

There are two types of clustering methods, hierarchical clustering and partitioning clustering. The former groups the data with a sequence of partitions, while the latter one divides the data into a prespecified number of clusters. The histogram of the image is plotted, and the most prominent number of peaks in the histogram can also be considered as the number of clusters present in the image. A significant amount of time is required to be spent while trying to determine the physical meaning of a class identified by the algorithm. Time spent trying to optimize and interpret the results by the unsupervised method may far exceed the time an analyst would have spent in selecting the training samples for supervised classification. Thus, unsupervised classification is not completely without human intervention. However, it does not start with a predetermined set of classes as in a supervised classification (Hord, 1992).

Unsupervised classification is becoming increasingly popular in long-term maintenance of a GIS database. There are now systems that use extremely fast clustering procedures, requiring less operational parameters. Thus, it is becoming possible now to train GIS analysis to undertake the automated classification that meets typical map accuracy standards. With suitable ground truth accuracy assessment procedures, this tool can provide a rapid means of producing quality land use and land cover data in GIS on a continuing basis that can be uploaded on a website for the utilisation of various users. However, the classes identified in one image may not be the same classes found on a second image. Since it is impossible to ensure consistency in class identification from one image to the next, unsupervised classification is not very useful for change detection studies.

In both supervised and unsupervised approaches, several iterations are usually conducted before a classification is completed. In supervised classification, this usually involves modifying the training data, while in unsupervised classification, it usually requires selecting any ambiguous classes and redoing the analysis either to split the classes or merge the classes that are then labeled separately with known cover types. Table 6.7

lists the advantages and disadvantages of supervised and unsupervised classification techniques.

Table 6.7. Advantages and Disadvantages of Supervised and Unsupervised Classification Techniques

Techniques	Advantages	Disadvantages
Supervised	Generates informational classes that represent features on the ground.	Information classes may not match spectral classes.
	Training areas may be reusable (assuming they do not change with time; e.g., roads).	Homogeneity of information classes varies.
	Information categories are distinct first, and then their spectral separability is examined.	Difficulty in selecting well distributed training sites. Training areas may not encompass unique spectral classes.
	Errors can be easily identified and solved.	Large time required for training phase of algorithm.
	It is easy to apply, and results can be improved at any stage.	The classification process has to be repeated if new data (samples) are added.
Unsupervised	No prior knowledge of the area is required.	Spectral classes do not necessarily represent features (informational classes) on the ground.
	Human error is minimized.	Does not consider the spatial relationships in the data.
	Unique spectral classes are produced.	Interpretation of spectral classes, and their merging or splitting can be very time-consuming.
	Relatively fast and easy to perform.	Spectral properties may vary with time, across images.
	Automatically classifies raw image data into useful information.	The algorithm determines spectrally separable classes, and then defines their information value.

The two most frequent clustering methods used for unsupervised classification are K-means and the Iterative Self-Organizing Data Analysis Technique (ISODATA). These two methods rely purely on spectrally pixel-based statistics and incorporate no prior knowledge of the characteristics of themes being studied. These are described as follows.

6.3.2.1 K-means method

K-means is one of the simplest unsupervised learning algorithms that solves the well-known clustering problem. K-means is the most popular one due to its simplicity, efficiency, and empirical success in recognizing multivariate data (Jain, 1989). The goal of K-means is to reduce the variability within the cluster. In this approach, the classes are determined statistically by assigning pixels to the nearest cluster mean based on all available bands. It assigns first an arbitrary initial cluster vector. The second step classifies each pixel to the closest cluster. In the third step, the new cluster mean vectors are calculated based on all the pixels in one cluster. The second and third steps are repeated until either the distances the mean cluster vector that have changed from one iteration to another or the percentage of pixels that have changed between the iterations, is small. A number of methods for finding distances in multidimensional data space is available. One of the simplest is Euclidean distance. Because the K-means approach is iterative, it is computationally intensive. Therefore, it is often applied only to image sub-areas rather than full scenes.

In this technique, the summation of squares distances (SS distances), defined as errors between each pixel and its assigned cluster center, is minimized using the following relationship.

$$SS_{\text{distances}} = \Sigma_{\forall x}[x - C(x)]^2 \tag{6.11}$$

wherein $C(x)$ pixel x is assigned is to the mean of the cluster. Minimizing the $SS\text{d}_{\text{istances}}$ is equivalent to minimizing the mean squared error (MSE). The MSE is a measure of the within cluster variability and represented as:

$$MSE = \frac{\sum_{\forall x}[x - C(x)]^2}{(N - c)^b} = \frac{SS_{\text{distances}}}{(N - c)^b} \tag{6.12}$$

where b represents the number of spectral bands, N is the number of pixels, and c indicates the number of clusters. The K-means implementation steps in ENVI 5.0 are shown in Figure 6.14.

The K-means algorithm does not necessarily find the most optimal configuration corresponding to the global objective function minimum. The algorithm is also significantly sensitive to the initial randomly selected cluster centers. The K-means algorithm can be run multiple times to reduce this effect.

6.3.2.2 Iterative Self-organizing Data Analysis Technique (ISODATA)

The ISODATA represents a comprehensive set of heuristic (rules of thumb) procedures that have been incorporated into an iterative classification algorithm (Jenson, 1986). Self-Organizing refers to the way in which it locates clusters with minimum user input. The ISODATA makes a large

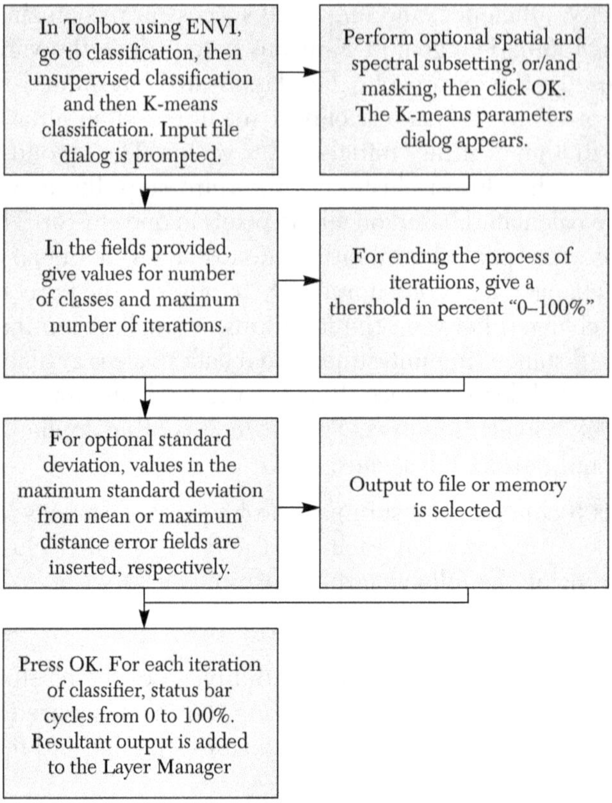

Figure 6.14. The K-means implementation work flow in ENVI 5.0.

number of passes through the remote sensing dataset until specified results are obtained. It is a modification of the K-means clustering algorithm, which includes (*a*) merging clusters if their separation distance in multispectral feature space is below a user's specified threshold, and (*b*) rules for splitting a single cluster into two clusters.

The method starts by arbitrarily establishing n cluster means based on the means and standard deviations (σ) of the bands in the input file ("n" number of classes to be classified and specified by the user). Each cluster represents a group of pixels which have similar spectral characteristics in the input bands. A minimum distance criterion is then used to assign each pixel to the 'nearest' cluster. The cluster means are then recalculated and each individual pixel is again compared to the new cluster means and assigned to the nearest cluster; this process is iterated a specified number of times.

The clusters are merged if either the number of members (pixels) in a cluster is less than a certain threshold or if the centers of two clusters are closer than a certain threshold (Richards & Jia, 2013). Clusters are split into two different clusters if the cluster standard deviation exceeds a predefined value and the number of members (pixels) is twice the threshold for the minimum number of members. The ISODATA algorithm allows the number of clusters to be adjusted automatically during the iteration by merging similar clusters and splitting clusters with large standard deviations. After each iteration, the normalized percentage of pixels whose assignments are unchanged since the last iteration is displayed. When this number reaches the convergence threshold, the program terminates. It is possible for the percentage of unchanged pixels to never converge or reach the convergence threshold. Therefore, it may be beneficial to monitor the percentage, or specify a reasonable maximum number of iterations, so that the program does not run indefinitely.

Implementation steps of ISODATA are shown in Figure 6.15. A typical ISODATA algorithm normally requires the analyst to specify the following parameters:

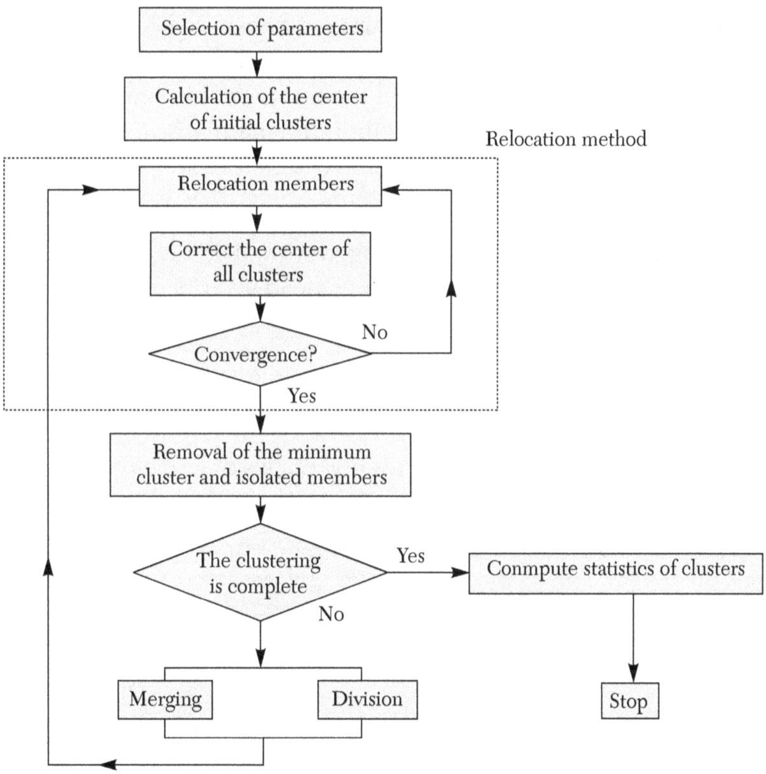

Figure 6.15. ISODATA implementation work flow.

1. The maximum number of clusters to be identified by the algorithm (e.g., 15 clusters). However, only fewer classes may be found in the final classification map after splitting and merging take place.
2. The maximum percentage of pixels whose class values are not to be changed between iterations. When this number is reached, the ISODATA algorithm terminates automatically. Some datasets may never reach the desired percentage unchanged, and if this happens, it is necessary to interrupt processing and edit the input parameters.
3. The maximum number of iterations ISODATA is to classify pixels and recalculate cluster mean vectors. The algorithm terminates when this number is reached.
4. If a cluster contains less than the minimum percentage of members, it is deleted and the members are assigned to an alternative cluster. This also affects whether a class is going to be split. The default minimum percentage of members is often set to 0.01.
5. Maximum standard deviation values (σ) between 4.5 and 7 are typically chosen. When the standard deviation for a cluster exceeds the specified maximum standard deviation and the number of members in the class is greater than twice the specified minimum members in a class, the cluster is split into two clusters. The mean vectors for the two new clusters are the old class centers $\pm 1\sigma$.
6. If the split separation value is changed from 0.0, it is replaced by the σ in determining the locations of the new mean vectors ± the split separation value.
7. Minimum distance between cluster means is assigned (e.g., normally 3.0) and the clusters with a weighted distance less than this value are merged.

Table 6.8 lists the advantages and disadvantages of K-means and ISODATA clustering techniques.

Table 6.8. Advantages and Disadvantages of K-means and ISODATA Clustering Techniques

Classifiers	Advantages	Disadvantages
K-means	Fast and simple to process	Could be influenced by the number and position of the initial cluster centers specified by the analyst, the geometric properties of the data, and clustering parameters.

Classifiers	Advantages	Disadvantages
ISODATA	Fast and simple to process	Needs several parameters.
	Being iterative, clustering is not geographically biased to the top or bottom pixels of the data file.	The clustering process is time consuming, because it can repeat many times.
	It is highly successful at finding the spectral clusters that are inherent in the data, as long as enough iterations are allowed.	Does not account for pixel spatial homogeneity.

6.4 POST CLASSIFICATION

Post-classification comparisons have been successfully applied to the process of mapping changes over large areas. The key to conducting accurate post-classification comparisons is consistent interpretation of the images used. This is especially important if there are many images to analyze and more than one analyst is involved. Care must be taken to ensure that the areas that have not changed over time are being interpreted consistently. An error assessment of the final product is therefore important and necessary.

6.4.1 Smoothening Filters

Any classification procedure usually leaves a small number of isolated, generally poorly classified or unclassified pixels that are often located at the boundaries between two distinct areas. Such data often presents a "salt-and-pepper" appearance due to the inherent spectral variability encountered by a classification when applied on a pixel-by-pixel basis (Lillesand & Kiefer, 1994). In such a case, homogenizing the classification by reassigning the pixels to one or the other class is desirable.

Since the resulting image after classification is often speckled, with individual, isolated pixels of one class surrounded by pixels of another class, filters are usually applied to the classification to smooth or generalize the final classification result. A majority filter is normally used where a moving window of 3 × 3 or 5 × 5 is passed through the classified image. In the operation, the central pixel of the window is changed to the majority class within the window. If there is no majority class in the window, the identity of the center pixel is kept unchanged. As the windows progresses through the dataset, the original class values are continually used, and not the modified ones from the previous window operation (Eastman, 1999). The majority filter eliminates very small clusters of pixels of the same class, while it smoothens edges between groups of pixels of differing classes (Figure 6.16). These filters can also incorporate some spatial weighting function, if required. Output can also be smoothed more than once, if there are large numbers of isolated pixels.

4	4	4	4	3	7
4	4	7	7	7	7
5	5	7	7	6	7
5	5	5	5	5	6
7	7	5	5	5	5
7	0	5	2	1	1

(a)

=

4	4	7	7	7	7
4	4	7	7	7	7
5	5	7	7	7	7
5	5	5	5	5	6
7	7	5	5	5	5
7	7	5	5	5	5

(b)

Figure 6.16. (a) Original data (b) results after applying majority filter.

There are several other filters which act as sieves where small patches of pixels of the same class are reclassified, but large ones are not. This eliminates small patches without modifying the edges between the larger and unfiltered patches. The analyst has to make a decision as to the type and level of filtering applied based on knowledge and desired result, which has consequences on the quality, accuracy, and applicability of the final product.

Change in land cover may be assessed by direct analysis of the raw satellite data from different dates or in a post-classification comparison in which two classified maps from different dates are compared to each other. Estimates of change from the technique tend to have greater errors because of differences in interpretation during the classification process for each image date. The better approach is to directly map changes from the raw satellite data collected from the different times. Precise georeferencing of images (making sure the same pixels from different acquisition dates overlay each other precisely) is necessary for accurate estimation of change when using this technique. It is usually possible to attain a precision of less than one pixel's width.

6.4.2 Accuracy Assessment

Before using a thematic map derived from remote sensing data, it is important to assess the classification accuracy of the map. Image classification is considered to be incomplete without estimating the accuracy of output (Garg, 2021). In the process of accuracy assessment, it is commonly assumed that the difference between an image classification result and the reference data is due to classification error. In addition to errors from the classification itself, other sources of errors, such as positional errors resulting from georeferencing, number of classes, and poor quality of training samples, all affect classification accuracy. However, in order to derive a reliable classification accuracy, errors in the reference data also need to be examined, especially when the reference data are not obtained from a field survey. A traditional approach for accuracy assessment of a remote sensing-derived map is by overlaying the reference map on top of the map obtained from remote sensing (Figure 6.17), and finding out the difference quantitatively. In remote sensing, the accuracy is assessed by two popular methods, the error matrix and the kappa coefficient, which are explained as follows.

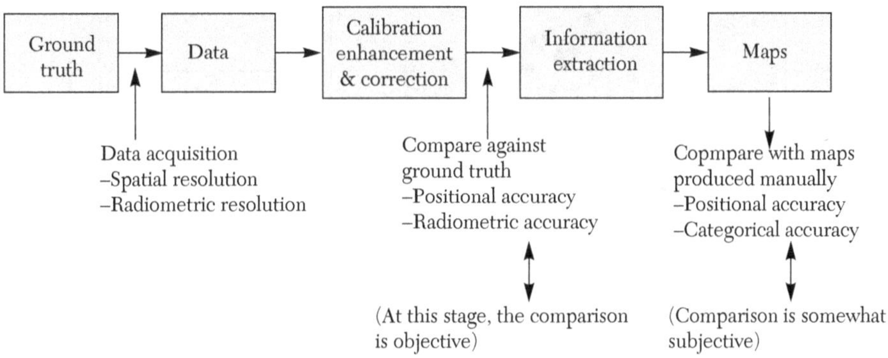

Figure 6.17. Ground truth data for accuracy assessment.

6.4.2.1 Error matrix

The error matrix approach is the most widely used in accuracy assessment (Foody, 2002). For evaluation of classification errors, a classification error matrix is typically formed, which is sometimes called a *confusion matrix* or *contingency table*. In order to properly generate an error matrix, the factors considered are: (*a*) quality of reference data, (*b*) classification scheme, (*c*) sampling scheme, (*d*) spatial autocorrelation, and (*e*) sample size and sample unit (Congalton & Green, 1999). Selection of a suitable sampling strategy is a critical step (Congalton, 1991) in accuracy assessment. The major components of a sampling strategy include sampling unit (pixels or polygons), sampling design, and sample size (Muller et al., 1998). Possible sampling design methods include random, systematic, stratified random, and cluster sampling, as shown in Figure 6.18.

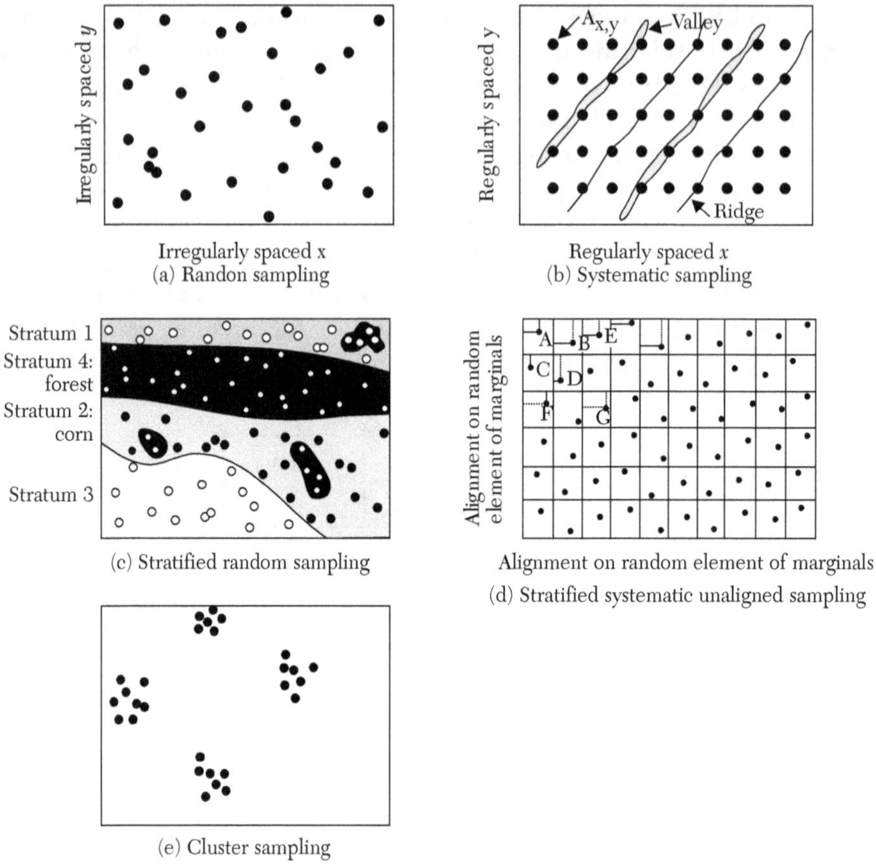

Figure 6.18. Important sampling techniques.

1. *Random sampling*: It is done using a completely random process. Observations are randomly taken and no rules are used.
2. *Systematic block sampling:* It is done by taking the observations at regular intervals according to a planned strategy.
3. *Stratified random sampling*: It is done by taking a minimum number of observations randomly placed in each class. Sampling points are generated proportionate to the distribution of classes in the image.
4. *Random sampling with grid*: It is done by dividing the area into grids and observations are taken randomly; one in each grid.
5. *Cluster sampling*: It is done by a randomly selected cluster of several nearby observations. These nearby observations can be randomly selected, systematically selected, and so on.

An error matrix is usually a quantitative method of characterizing image classification accuracy, although uncertainty and confidence analysis of classification results has also gained some attention recently, and spatially explicit data on mapping confidence is regarded as an important aspect in effectively employing classification results for decision making (Liu & Zhou, 2004). The error matrix compares the pixels or polygons in a classified image against the ground reference data (Jensen, 2005). It is a table that shows the correspondence between the classification results with respect to the reference image. The error matrix is created with the help of reference data which includes thematic maps, aerial photos, ground surveys, GPS surveys, and high-resolution satellite data. Digital aerial photography and videography are also very valuable, especially when the photos and video frames can be automatically linked geographically with the image classification results. These data provide an excellent source to assist satellite image interpretation and validation of final classified maps.

Since it is practically impossible to test each and every pixel in the classification image, a representative sample of reference points in the image with known class values is used. Ground reference pixels (training samples) used to train the classification algorithm are not used for the assessment of classification accuracy to remove unbiased reference information (Gibson, 2000). Normally half of the training samples are used for classification, and the remaining half for estimating the accuracy after classification. Based on binomial probability theory, the number of samples required (N) can be computed.

$$N = \frac{Z^2(p)(q)}{E^2} \tag{6.13}$$

where p is expected percent accuracy

$q = 100 - p$

E = allowable error

Z = 2 (from the standard normal deviate of 1.96 for the 95% two-sided confidence level).

For example, a sample for which the expected accuracy (p) is 90% at an allowable error (E) of 5% (i.e., it is 95% accurate), the number of points necessary for reliable results is:

$$N = [2^2\,(90)(10)]/5^2$$

So a minimum of 144 points would be required.

Important accuracy assessment elements, such as *overall accuracy*, *omission error*, *commission error*, and *kappa coefficient*, can be derived directly from the error matrix. The accuracy assessment based on a normalized error matrix is regarded as a better presentation than the conventional error matrix (Stehman, 2014). The error matrix approach is only suitable for "hard" classification, assuming that the map categories are mutually exclusive and exhaustive and that each location belongs to a single category. This assumption is often violated, especially for classifications with coarse spatial resolution imagery. The traditional error matrix approach is not appropriate for evaluating the classification results, such as from fuzzy logic. Accordingly, many new measures, such as conditional entropy and mutual information (Maselli et al., 1994), fuzzy-set approaches (Woodcock & Gopal, 2000), symmetric index of information closeness (Foody, 1996), Renyi generalized entropy function (Ricotta & Avena, 2002), and parametric generalization of Morisita's index (Ricotta, 2004) have been developed. A modified kappa coefficient and tau coefficient have been developed as improved measures of classification accuracy (Ma & Redmond, 1995). One critical issue in assessing fuzzy classification is the difficulty of collecting reference data. More research is thus needed to find a suitable approach for evaluating fuzzy classification results. Some of the classification accuracy assessment algorithms can be found in Rosenfield and Fitzpatrick-Lins (1986).

The *overall accuracy* of the classification map is determined by dividing the total correct pixels (sum of the major diagonal) by the total number of pixels in the error matrix. The overall accuracy of classification itself can be misleading, as it does not reveal whether the error is evenly distributed between classes or some classes are really bad and some are really good. Therefore, it is always better to compute the values of commission and omission errors. It is likely that the overall accuracy of an image classification might be quite high, whereas an individual class may have a considerably lower accuracy. Figure 6.19 presents an example of accuracy assessment.

If the accuracy of individual classes is most important to users, the classification results cannot be accepted as such, even if the overall accuracy is higher. The *producer's accuracy* of the classification is the total number of correct pixels in a category divided by the total number of pixels of that category as derived from the reference data (i.e., the column total). This statistic indicates the probability of a reference pixel being correctly classified, as the analyst might be interested to know how well a certain area is classified. *Omission error* is computed as (100-producer accuracy [%]).

	Reference Data					
	Residential	**Commercial**	**Wetland**	**Forest**	**Water**	**Row total**
Residential	**70**	5	0	13	0	88
Commercial	3	**55**	0	0	0	58
Wetland	0	0	**99**	0	0	99
Forest	0	0	4	**37**	0	41
Water	0	0	0	0	**121**	121
Column total	73	60	103	50	121	**407**

Overall accuracy = 382/407= 93.86%

Producer's Accuracy (omission error)		**User's Accuracy (commission error)**	
Residential = 70/73 = 90%	4% omission error	Residential = 70/88 = 80%	20% omission error
Commercial = 55/60 = 92%	8% omission error	Commercial = 55/58 = 95%	5% omission error
Wetland = 99/103 = 96	4% omission error	Wetland = 99/99 =100%	0% omission error
Forest = 37/50 = 74%	26% omission error	Forest = 37/41 = 90%	10% omission error
Water = 20/22 = 100%	0% omission error	Water = 121/121 =	100% 0% omission error

Computation of Khat Coefficient of Agreement

$$\hat{K} = \frac{N\sum_{i=t}^{k} x_{ii} - \sum_{i=t}^{k}(x_{i+} \times x_{+i})}{N^2 - \sum_{i=t}^{k}(x_{i+} \times x_{+i})}$$

where N = 407

$$\sum_{i=t}^{k} x_{ii} = (70 + 55 + 99 + 37 + 121) = 382$$

$$\sum_{i=t}^{k}(x_{i+} \times x_{+i}) = (88 \times 73) + (58 \times 60) + (99 \times 103) + (41 \times 50) + (121v121) = 36{,}792$$

$$\text{therefore } \hat{K} = \frac{407(382) - 36792}{407^2 - 36792} = \frac{155474 - 36792}{165649 - 36792} = \frac{118682}{128857} = 92.1\%$$

Figure 6.19. Computation of various accuracy measures.

User's accuracy is the total number of correct pixels in a category divided by the total number of pixels in that class on the map (i.e., the row total). User's accuracy is the probability that a pixel classified on the map actually represents that category on the ground. *Commission error* is computed as (100-user's accuracy [%]). Additionally, the producer's and user's accuracy cannot be separately used as an indicator of the classification accuracy, as these values do not provide complete details (Richards & Jia, 2013).

6.4.2.2 Kappa coefficient

Another accuracy indicator is the kappa coefficient (K). It is a measure of overall statistical agreement of an error matrix, which takes non-diagonal elements of the error matrix into account. Kappa analysis is recognized as a powerful method for analyzing a single error matrix and for comparing the differences between various error matrices (Foody, 2002). It is a measure of agreement or accuracy between the remote sensing–derived classification map and the reference data as indicated in the error matrix by (*a*) the major diagonal, and (*b*) the chance agreement, which is indicated by the row and column totals. The kappa coefficient is computed as shown in Figure 6.19.

The kappa coefficient varies between 0–(Congalton & Green, 1999). The value of 0 indicates that there is a mismatch between the reference data and the classified image, and the classification is considered to be random. If the value is 1, the classified image and the reference data both have a true agreement with 100% accuracy. Thus, the higher the value of the kappa coefficient, the more accurate the classification from a remotely sensed image. Negative values can occur but they are spurious. Thus, if K > 0.80, it represents strong agreement and good accuracy with > 80%. If it is between 0.40–0.80, it is moderate accuracy, and K < 0.40 is poor accuracy. Figure 6.20 present an example of kappa coefficient accuracy using various IRS sensors at varying spatial resolutions. It is observed that as the spatial resolution increases (e.g., WiFS), the value of the kappa coefficient decreases.

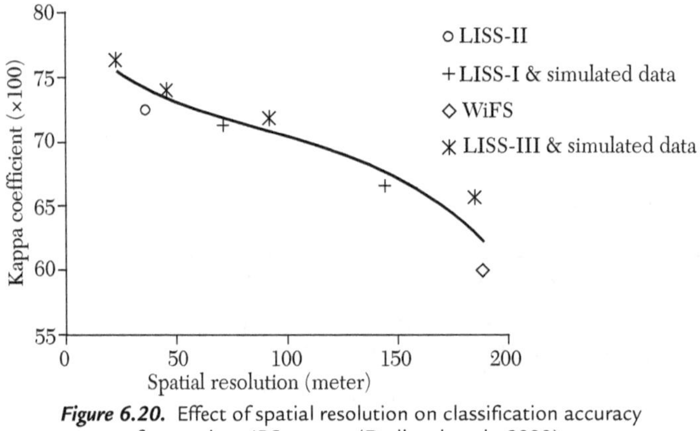

Figure 6.20. Effect of spatial resolution on classification accuracy from various IRS sensors (Dadhwal et al., 2002).

The kappa coefficient reflects the difference between actual agreement and the agreement expected by chance. The classification accuracy for each class indicates if there is any overestimation or underestimation in classification based on the input from the reference data (Conglaton & Green, 1999). The kappa coefficient uses all elements in the error matrix and not just the diagonal ones. The estimate of kappa is the proportion of agreement after chance agreement is removed from consideration.

The kappa coefficient is not as sensitive to differences in sample sizes between classes, and is therefore considered a more reliable measure of accuracy. Another advantage is that two classification products can be statistically compared. For example, two classification maps can be created using different algorithms, but the same reference data can be used to verify them. For example, two coefficients can be derived, as $\hat{K}_1, \hat{K}_2$, For each $\hat{K}$ the variance $\hat{V}$ can also be calculated by using the following equation:

$$\hat{V} = \frac{1}{N(1-p_c)^4}\{p_{ii}[(1-p_c)-(p_i+p_{+i})(1-p_o)]^2 + (1-p_o)^2\sum_{i=1}^{nb}\cdot\sum_{i=1}^{nb}p_j(p_t+p_{jt})^2-(p_o p_c-2p_2+2p_c+p_o)^2\} \tag{6.14}$$

It has been suggested that a z-score be calculated by:

$$Z = \frac{\hat{K}_1 - \hat{K}_2}{\sqrt{\hat{V}_1 + \hat{V}_2}} \tag{6.15}$$

A normal distribution table can be used to determine where the two $\hat{K}$s are significantly different from Z. For example, if Z > 1.96, then the difference is said to be significant at the 0.95 probability level. Given the previous procedures, we need to know how many samples are required to be collected and where they should be placed.

1. The larger the sample size, the more representative an estimate can be obtained; therefore, more confidence can be achieved.
2. In order to give each class a proper evaluation, a minimum sample size should be applied to every class.

The goal of accuracy assessment is to quantitatively determine how effectively pixels are grouped into the correct feature classes in the area under investigation. The accuracy of a thematic map can be determined by constructing the matrix along with kappa statistics in order to test whether any difference exists in the interpretation work. Kappa statistics consider a measure of overall accuracy of image classification and individual category accuracy as a means of actual agreement between classification and observations. Kappa statistics have been shown to be a statistically more sophisticated measure of classifier agreement, and thus, give better interclass discrimination than the overall accuracy.

CHAPTER 7

STATE-OF-ART CLASSIFICATION TECHNIQUES

7.1 INTRODUCTION

Digital image classification uses the quantitative spectral information contained in an image, which is related to the composition or condition of the objects on the Earth's surface. There are various classification techniques that have been developed and widely used to produce various thematic maps (Jenson, 2005), as already given in Table 6.3 of the previous chapter. In general, spatial resolution is one of the most important factors that affects classification details and influences the selection of a suitable classification algorithm. In high-resolution images, objects are made up of several pixels, so object-based classification is considered superior to traditional pixel-based classification. In medium/low resolution images, pixels and objects are similar in scale, so both pixel-based and object-based image classifications can perform well (Mather, 2004).

Classification methods can be grouped as parametric and nonparametric, or hard and soft classifiers, or per-pixel, sub-pixel, per-field. Parametric and nonparametric image classification techniques are based on supervised learning. A serious drawback of the hard classifier is that a large quantity of spectral information is lost in determining the pixel membership (Gibson, 2000). When the pixel contains a single land cover class, the pixel records spectral characteristics of that class, but when the pixel contains a mixture of land cover classes, the pixel value is a function of the reflectance from the mixture of classes. Generally, the more cover classes a pixel contains, the more spectral characteristics of that class it has. As the proportions of land cover classes change from pixel to pixel, the spectral characteristics also change. Thus, a mixed or heterogeneous pixel has spectral characteristics that differ from those of a homogeneous pixel. In addition, changes in condition within a given land cover class also cause variation in spectral characteristics.

In remote sensing, pixel measurement vectors are often considered as points in a spectral space. Most image classifications are based on remotely sensed spectral responses. Spectral characteristics of a pixel measurement vector determine its position in a spectral space, and thus help in classification of that pixel. Due to the heterogeneity of landscapes and the limitations in spatial resolution of remote-sensing imagery, mixed pixels are common in medium and coarse spatial resolution data. The presence of mixed pixels has been recognized as a major problem, affecting the effective use of remotely sensed data in per-pixel classifications (Cracknell, 2018).

Traditionally, information classes are implicitly represented as classical sets. Thus, a partition of spectral space is based on the principle of classical set theory. Decision surfaces are defined precisely by some decision rules (for example, the decision rule of a conventional maximum likelihood classifier) to separate the regions. Such a partition is usually called a hard partition. Due to the complexity of biophysical environments, spectral confusion may be common among various land cover classes (Maxwell et al., 2018). Thus, ancillary data are often used to modify the classification image based on some expert rules. For example, forest distribution in mountainous areas may be related to elevation, slope, and aspects. Data describing terrain characteristics can therefore be used to modify the classification results based on the knowledge of specific vegetation classes and topographic factors.

Various classification algorithms which have been used in remote sensing image analysis are described here. All these classification algorithms are currently not available as part of most commercial image processing software. Therefore, some open-source remote sensing software have been presented in the chapter which might be useful for applying these modern classification approaches. Prasad et al. (2015b) presented a good summary of several of these classification approaches. The selection of a suitable classifier may require consideration of many factors, such as classification accuracy, algorithm performance, spatial scale, the variance of classes, and computational resources.

7.2 ADVANCED CLASSIFICATION TECHNIQUES

Classifiers such as ISODATA, k-means, k-nearest neighbor (k-NN), minimum distance, Mahalanobis distance, parallelepiped, MLC, and spectral angle mapper (SAM) are based on assumptions of data distribution. The performance of such classifiers depends largely on the accuracy of the estimated model parameters. These classifiers suffer from the high dimensionality of satellite imagery. As a result, it might be difficult to select a significant number of training samples. Another drawback of these classifiers is the difficulty of combining spectral data with the ancillary data.

The hard classifiers, for example, ISODATA, parallelepiped, k-means, and MLC, develop classes by combining the spectra of all the pixels in a training set from a given feature (Mather, 2004). The classes contain the contribution of all the pixels in the training set. The hard classifiers assume that the pixels are pure, and categorize them into one and only one class. In such classifications, each pixel belongs to the class it most closely resembles. This makes the hard classifiers inefficient in handling the problem of mixed pixels. It treats the mixed pixels as noise, uncertainty, or error. These classifiers work well in homogenous areas (e.g., forests, water bodies), and generally produce low classification accuracy in heterogeneous areas.

Per-pixel algorithms such as decision trees (DT), neural network (NN), and support vector machines (SVM); sub-pixel algorithms such as the fuzzy classifier, and the fuzzy expert system; contextual-based approaches (e.g., ECHO); knowledge-based algorithms (rule-based); and object oriented classifiers (e.g., eCognition) have been used for various applications and generation of specific thematic maps from remote sensing images (Jian & Zhao, 2009). Automated classification approaches such as object-based classification, the standardized object oriented automatic classification (SOOAC) method based on fuzzy logic, fuzzy sets, expert systems, knowledge-based stratified classification, artificial neural networks (ANN), DT, and Bayesian and hybrid classifiers are frequently used for digital image classification (Haralick, 1979). Parametric classifiers such as the Bayesian classifier use algebraic possibility for allocation to each class. The input parameters required, like mean and covariance, are derived from the training data and used in these classifiers. In per-pixel based advanced methods, only the spectral data for each individual pixel is used to classify it, whereas in contextual methods the data from neighboring cells are included to assist the classification of each pixel. However, some contextual classifiers use information about the texture (spectral variance) around a given pixel (Jian & Zhao, 2009).

The pixel of an image may have two or more classes, such as trees with rivers, or boundaries of more than two land cover types. Sub-pixel classification approaches provide a more appropriate representation and accurate area estimation of land covers than per-pixel approaches, especially when coarse spatial resolution data are used. Different approaches have been used to derive a soft classifier, including fuzzy-set theory, Dempster–Shafer theory, certainty factor, softening, and neural networks (Mather, 2004). The fuzzy-set technique and spectral mixture analysis (SMA) classification are the popular approaches used to overcome the mixed pixel problem (Foody & Cox, 1994).

Advanced classification algorithms such as ANN, SVM, DT algorithms, and so on are highly applied for image classification, and have generally outperformed conventional classifiers. They are especially suitable for incorporation of nonspectral data into the classification process. There are many techniques that can be applied for feature extraction from remotely sensed data, such as PCA, minimum noise fraction (MNF), transform discriminant analysis (TDA), decision boundary (DB), feature extraction (FE), nonparametric weighted feature extraction (NPWFE), wavelet transform (WT), and SMA (Kamavisdar et al., 2013).

Nonparametric classifiers such as k-NN, logical regression and multilayer perceptron (MLP) are used when no density function is available. They approximate the probability density function for further use. Advanced nonparametric classifiers, such as NN, DT, and knowledge-based classifiers, may be more suited to deal with multisource data, such as a combination of spectral signatures, textures, context information, and ancillary data, and therefore these techniques have gained increased attention in recent years (Fauvel, 2007).

Many of the advanced classification algorithms are currently available as part of most commercial image processing software. Table 7.1 provides a comparison of some common geospatial and remote sensing software packages and statistical and data analysis packages having advanced remote sensing data classification methods. eCognition, a commercial software package (Trimble, USA), provides automated methods to optimize or tune the user-defined parameters. In addition, all of the statistical software packages offer automated means to perform parameter tuning. For example, both the e1071 and caret packages in R offer methods to tune a variety of learning algorithms (Maxwell et al., 2018). Statistical and data analysis software, such as R (*https://cran.r-project.org/*), MATLAB (*https://www.mathworks.com/products/matlab.html*), Scikit-learn for Python (*http://Scikit-learn.org/stable/*), and Weka (*http://www.cs.waikato.ac.nz/ml/weka/*), currently offer the best environment to do so. An added benefit of R, Scikit-learn, and Weka is that they are all currently free (please see Section 7.3). Additionally, there is some integration between these free tools; for example, an R interface to Weka is available using the RWeka package and Scikit-learn can be implemented within Weka using wekaPython. The open-source and free geospatial software tool QGIS currently has implementations of all of the modern algorithms (Halounova, 2007). There are many online resources available for learning how to implement ML classification using these tools.

Table 7.1. Some Remote Sensing and Statistical Software with Advanced Classification Methods (Modified from Maxwell et al., 2018)

Software	**Algorithm**					
	k-NN	**ANNs**	**DTs**	**Boosted DTs**	**RF**	**SVM**
ArcGIS 10.8/ArcPro 2		✓	✓		✓	✓
eCognition Developer 10	✓	✓	✓	✓	✓	✓
ENVI 5.6 TerrSet/IDRISI 18.3	✓	✓	✓		✓	✓
ERDAS Imagine 2020	✓	✓	✓	✓	✓	✓
QGIS 3.18	✓	✓	✓	✓	✓	✓
MATLAB R 2021a	✓	✓	✓	✓	✓	✓
R 4.1	✓	✓	✓	✓	✓	✓
Python Scikit-learn 0.24	✓	✓	✓	✓	✓	✓
Python Weka 3.9.5	✓	✓	✓	✓	✓	✓

Some advanced classification approaches used for improved classification of remotely sensed data are presented in the following sections.

7.2.1 Artificial Neural Network (ANN)

Neural Networks (NN), despite having been around for decades, have received much attention only in the last few years in computer vision and machine learning (ML). *Artificial Neural Network* (ANN) is defined as a computing system made up of number of simple, highly interconnected processing elements, which process information by their dynamic state response to external inputs. Neural networks can be thought of as classifiers that extract hierarchical features from raw data (i.e., pixel values) and learn models for various vision related tasks, such as object recognition, semantic segmentation, and so forth. ANNs has a normal tendency for storing experimential knowledge. They are nonparametric classifiers, and are generally based on data mining techniques known as artificial intelligence (AI) methods, which are used to solve many real-world problems (Garg, 2021).

The architecture of an ANN is shown in Figure 7.1. The elements of an ANN are neurons (equivalent to biological axons), which are organized in layers; each layer consists of a set of neurons. The principle behind the ANN is defined on the behavior of the human brain in which learning algorithms are used for classification and prediction. The ANN is a group of multiple perceptrons or neurons at each layer. The number of nodes at the output layer is equal to the number of information classes, whereas the number of nodes at the input is equal to the dimensionality of each pixel. In addition, ANNs typically have hidden nodes arranged in one or more additional layers (Lu & Weng, 2007). The network may or may not have hidden node layers, making their functioning more interpretable. It is also known as a feed-forward neural network because inputs are processed only in the forward direction. These types of neural networks are one of the simplest variants of neural networks as they pass information in one direction, through various input nodes, until it makes it to the output node.

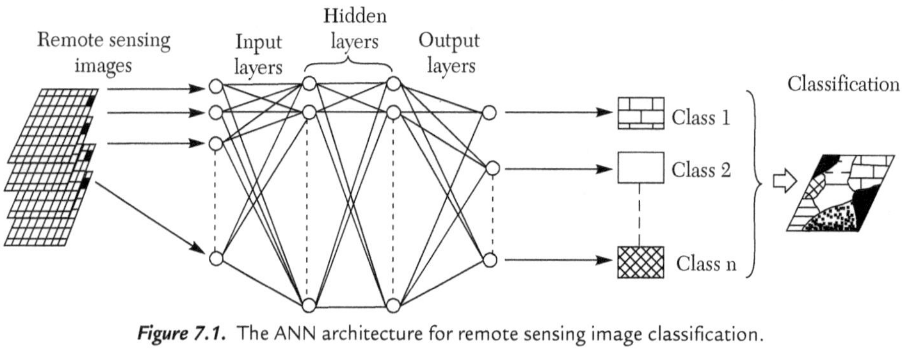

Figure 7.1. The ANN architecture for remote sensing image classification.

The key characteristic of an ANN is that all neurons in one layer are connected to all neurons in adjacent layers, and these connections have respective weights. It may use the average of the weighted sum of numerous sigmoids to define a decision function. The weights on the connections, in combination with the typically nonlinear activation function that further modifies values at each neuron, determine how input values are mapped to values on the output nodes. Each artificial neuron generates an output based on certain activation functions, such as piecewise linear, sigmoid,

Gaussian, and so forth. (Kamavisdar et al., 2013). As the number of neurons in the hidden layer are increased, and especially yet more hidden layers are added, it rapidly increases the potential for describing very complex decision boundaries.

In ANN modeling, the back-propagation technique typically uses error distribution criteria to train the neural networks because of its flexible and adaptive ability. During the training phase, the ANN learns about the regularities present in the training data and then constructs rules that can be extended to unknown data (Foody, 1996). In the learning phase, the network learns the connection weights iteratively from a set of training samples. The network gives an output, corresponding to each input, which is compared to the desired known output. The error between these two is used to modify the weights of the ANN. The training procedure ends when the error is below a predefined threshold value. Thereafter, all the testing data are fed into the classifier to perform the classification. In an ANN, it is easy to incorporate supplementary data in the classification process in order to improve classification accuracy. The ANN algorithms are extremely efficient (Figure 7.26[b]) when the classification process is not a simple one. A well trained network is capable of classifying highly complex data.

There are several ANN algorithms that can be used to classify remotely sensed images, such as multilayer perceptron (MLP), fuzzy ARTMAP classification, self-urganized feature map (SOM), and radial basis function network (RBFN). The MLP is the most widely used type of ANN, which is a supervised learning technique called back-propagation for training the network. It is a feed-forward ANN model that maps input data sets onto a set of appropriate outputs. The advantages and disadvantages of ANN are briefly given in Table 7.2.

Table 7.2. Advantages and Disadvantages of Different Classifiers (Kamavisdar et al., 2013)

S.No.	Technique	Advantages	Disadvantages
1.	ANN	It is a nonparametric classifier. It is a universal functional approximator with arbitrary accuracy. Capable to present functions such as OR, AND, NOT. It is a data-driven selfadaptive technique. Efficiently handles noisy inputs. Computation rate is high.	It is semantically poor. The training of ANN is time-consuming. Can produce nonoptimal classifications. Encounters the problem of over-fitting. Many user-determined parameters to specify. Difficult in choosing the type of network architecture.

(Contd.)

S.No.	Technique	Advantages	Disadvantages
2.	CNN	It automatically detects the important features without any human supervision. It is computationally efficient. It is quite robust even under challenging conditions, such as illumination, complex background, size and orientation of the images, and different resolutions.	Sometimes takes much longer to train. However, after training, their testing time efficiency is much faster. Need large data sets, and their proper annotation, which must be performed by domain experts. Poor data labeling could create significant reduction in performance and precision. Problems with pretrained models on similar and smaller data sets. Optimization issues because of the models' complexity, as well as hardware restrictions.
3.	OBIA	Very high accuracies can be obtained. Fuzzy logic rationale applicable. Additional object features (beside spectral information) can be used for classification. Very useful if spectral variance within a class is higher than spectral variance between classes (e.g., urban).	Iterative, very time-consuming process. Quality is highly dependent on competence of user. Rules set are often only partly transferable. Good for smaller areas, but not so much for mass data throughout.
4.	Decision Tree	Can handle nonparametric training data. Does not require an extensive design and training. Provides hierarchical associations between input variables to forecast class membership and provides a set of rules that are easy to interpret. Simple and computationally efficient, quite flexible and versatile. Only little additional effort is required once the decision tree is constructed. Decisions are based on physical properties. Implemented in all commercial remote sensing software packages.	A priori knowledge required. The usage of hyperplane decision boundaries parallel to the feature axes may restrict their use in which classes are clearly distinguishable. Becomes complex calculation when various values are undecided and/or when various outcomes are correlated. Construction of tree is labor-and time-intensive. Decision rules are only transferable with limitations (variation in time and space).
5.	Fuzzy Measure	Efficiently handles uncertainty. Properties are described by identifying various stochastic relationships.	Without a priori knowledge, output is not good. Precise solution depends upon direction of decision.

S.No.	Technique	Advantages	Disadvantages
6.	Fuzzy C-Means (FCM)	Gives best results for overlapped datasets. Data point is assigned membership to each cluster center as a result of which data point may belong to more than one cluster center.	It gives high membership values for the outlier points, and therefore it has difficulty in handling outlier points. In a cluster, membership of a data point depends directly on the membership values of other cluster centers, which may lead to undesirable results. It faces problems in handling high dimensional data sets and a large number of prototypes. It is sensitive to initialization and is easily trapped in local optima.
7.	Possibilistic C-Means Algorithm (PCM)	It can find meaningful clusters as defined by dense regions. It overcomes the need to specify the number of clusters and it is highly robust in the presence of noise and outliers.	Extremely sensitive to good initialization. It may lead to generation of coincident clusters since the columns and rows of the typicality matrix are independent of each other. If the initialization of each row is not sufficiently distinct, it may lead to coincident clusters.
8.	k-NN	It requires the tuning of only one free parameter. It is nonparametric (i.e., it does not assume any particular statistical distribution of the training data). It converges to the optimal Bayes bound under certain conditions.	Computationally expensive when the training dataset is large. The majority voting decision strategy generally is adopted to perform the classification task. It overlooks the geometric configuration of the K-nearest training samples in the decision process, there by resulting in an under-exploitation of their discrimination potential.
9.	AI-Based	Reduction in human error. Helps in repetitive work. Faster decisions and efficient communication. Rational decision-maker.	Costs overrun. Lack of trained personnel. Lack of practical products. Potential for misuse.

(Contd.)

S.No.	Technique	Advantages	Disadvantages
10.	Machine Learning Based	Compatibility to high-performance computing architectures. Accuracy in prediction, forecasting, and classification. Addresses optimization, scalability, and portability.	It is lacking in representation of large and complex models for the systems not linear in nature. Unable to handle large volumes of data. Difficult to learn hierarchical features and generalize.
11.	RF	Reduces the risk of overfitting and the required training time. Exhibits superior performance in cases with noise and weak discrimination data, and is insensitive to the initialization of parameters. Runs fast and efficiently when handling large datasets. Superior method for working with missing data.	By having many trees, the ability to visualize the trees is effectively lost. The sensitivity of RF classification to the sampling design needs to be taken into account in order to reduce misclassification.
12.	SVM	Flexible in the choice of threshold. Makes use of a nonlinear transformation. Provides a good generalization capability. The problem of over-fitting is eliminated. Reduction in computational complexity. Simple to manage decision rule complexity and error frequency. Good generalization (performs well for heterogeneous classes). Good results with only a small number of training data. Works well also with high number of dimensions (multi/hyperspectral).	Result transparency is low. Training is time-consuming as new training data required for each input data set (even after, e.g., atmospheric correction differences in class variance). Structure of algorithm is difficult to understand. Determination of optimal parameters is not easy when there is non-linearly separable training data. Nonlinear discriminable data require additional computing time.
13.	SAM	Fast and easy approach. Comprehensible for the user. Relatively robust against illumination differences (topography, light source, sensor, etc.). Comparability of image spectra with lab spectra.	Illumination tolerance is accompanied by insensitivity for detecting certain physiologic changes. Similar spectra, which only significantly differ in their albedo (e.g., needle leaf and broad leaf forests), are wrongly classified.

S.No.	Technique	Advantages	Disadvantages
14.	SMA	Objective method without subjective influence of the producer. Quantitative results on sub-pixel basis. Also detects objects that are not perceived visually. Especially suitable for chemical, geological, and mineralogical investigations. Suitable for hyperspectral data.	Preprocessing of the images and spectra is very complex. Difficult in topographically heterogeneous areas (DTM quality). Minimum number of recording channels is required. Software packages with programming interface are needed, e.g., ENVI (IDL, C).

7.2.2 Convolutional Neural Network (CNN)

Computer vision is an interdisciplinary field that has been gaining huge attention in recent years for object detection. Object detection aids in pose estimation, vehicle detection, autonomous cars, security and surveillance, and many industrial applications. The convolutional neural network (CNN) is a modern deep learning (DL) method which is being widely used for image analysis tasks, such as image classification, and object detection and segmentation, by using a sequence of feed-forward layers (Khan et al., 2017). The CNNs take advantage of the underlying structure in images, that is topological information (spatial information, such as adjacency and rotations). In satellite images, scale and orientation changes are the main characteristics of targets/objects. Moreover, the edge information of targets/objects in satellite imagery contains very prominent and concise attributes. The major challenges in target/object detection in satellite imagery include presence of targets/objects in different sizes, different orientations, and at very close locations. The object detection algorithms try to draw a bounding box around the object of interest to locate it within the image. Also, there could be many bounding boxes representing different objects of interest within the image.

The CNN is similar to the traditional neural network, and it is made by neurons that have learnable weights and biases. This model uses a variation of multilayer perceptrons and contains several layers, convolutional, pooling, and activation layers, followed by a fully connected layer that produces the output. Unlike an ANN, the neurons in a CNN layer are not connected to all other neurons, but connected only to a small region of neurons in the previous layer. The major difference between a traditional ANN and CNN is that only the last layer of a CNN is fully connected, whereas in an ANN each neuron is connected to every other neuron. The CNN automatically learns complex features present in the image using different layers which has "learnable" filters, and combines the results of these filters to predict the class or label probabilities of the input image (Khan et al., 2020).

As shown in Figure 7.2, a convolutional layer is the basic layer of a CNN that may consist of one or more convolutional layers used for extracting features (Song et al., 2019). The first layer might detect the lowest level features, such as lines, corners, and edges in the image. The next layers might detect middle level features, such as shapes and textures, and finally higher level features, such as structure of the plant, will be detected by higher layers in the network. The unique technique of building up from lower level features to higher level features in an image makes CNN most useful in many applications. The convolution operation works on a small local area of the image with the size of a convolutional kernel. The convolutional kernel is a learnable weight matrix whose output goes through an activation function, and a convolved feature map is generated. The feature map can be the input for the subsequent convolutional layers. These convolutional layers will create feature maps that record a region of the image which is ultimately broken into rectangles and sent out for nonlinear processing. This way, more sophisticated features can be extracted after several convolutional layers are stacked layer by layer. The hidden layers in CNN learn to detect different features in an image. The weights and biases for all neurons in a hidden layer are the same. Thus, all hidden neurons detect the same features in different regions of an image. The neurons in each feature map also share the weight of a convolutional kernel in the operation, which ensures that the parameters in the network will not increase significantly, even if the convolutional layers continue to increase, thereby reducing the storage requirement for the CNN model (Ren et al., 2017).

Pooling layers in the CNN model aim to capture features, which ensure that the network can still learn effective features, even if a small

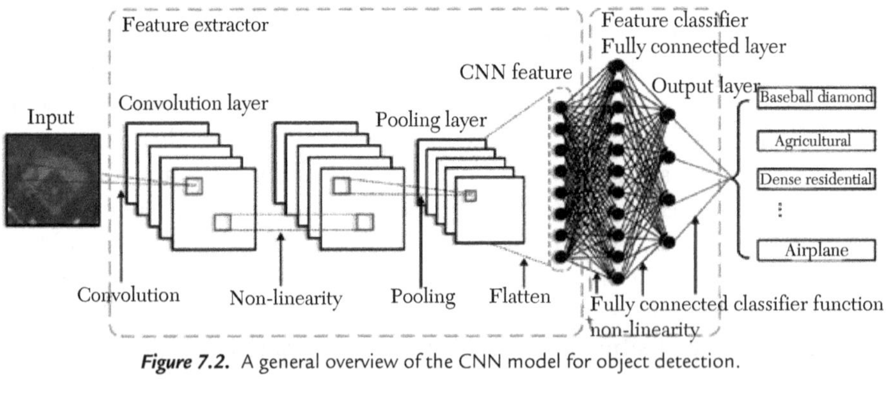

Figure 7.2. A general overview of the CNN model for object detection.

amount of input data shift occurs. These layers are not sensitive to their precise locations. Pooling reduces the dimensionality of the feature map by condensing the output of small regions of neurons into a single output, thus simplifying the following layers and reducing the number of parameters to learn. It also preserves the important information in the feature maps, thereby further reducing the computation of network training (Salah, 2017).

A fully connected layer is composed of multiple hidden layers where each hidden layer contains multiple neurons, and each neuron is fully interconnected with the neurons of the subsequent layer. One-dimensional feature vectors obtained by flattening feature maps, after operations in the convolutional and pooling layers, are used as the input for a fully connected layer. The objective of a fully connected layer is to map these features into a linearly separable space and coordinate with the output layer in classification. The output layer uses a function such as the softmax function (Liu et al., 2017) and SVMs to output the classification results.

The activation function is generally nonlinear, which enables the network to be capable of learning on layer-wise mapping. Common activation functions include sigmoid, rectified linear unit (ReLU; Hara et al., 2015), tanh, and maxout functions (Goodfellow et al., 2013). The ReLU can map the output of a neuron to the highest positive value, or if the output is negative, ReLU maps it to zero. The final layer connects the neurons from the last hidden layer to the output neurons which produce the final output. The class probabilities are determined by the value of each node in the final layer. A loss function, which is also referred to as a cost function or an objective function, is used to represent the extent of previous inconsistencies between the value predicted by the model and the actual value.

The CNN architecture uses the images as input which allows the encoding of certain properties into the architecture. Some popular CNN architectures include AlexNet, Visual Geometry Group (VGG) Networks, ResNet, Fully Convolutional Network (F-CNN), and also a recent development for GoogleNet called Inception-v4. AlexNet uses ReLu for the nonlinear part, instead of a tanh or sigmoid function, which was the earlier standard for traditional neural networks (Song et al., 2019). The advantage of ReLu over the sigmoid function is that it trains much faster than the latter because the derivative of sigmoid becomes very small in the saturating region, and therefore there are almost no updates to the weights.

While ANN is ideal for fast processing of the data, such as image data, text data, and tabular data, by forward facing algorithms and solving problems, the CNN requires many more data inputs to achieve its high accuracy rate. In many cases, multiple methods of data augmentation are required to be used for data processing in CNN in order to get the same accuracy as ANN. The CNN doesn't process data in a forward-facing way, but rather refers to the same data multiple times when creating maps (Khan et al., 2020). This process will achieve the same accuracy for data classification problems. The ANN has the ability to implicitly detect complex nonlinear relationships between dependent and independent variables. The image classification problems using ANN become difficult because 2D images need to be converted to 1D vectors. This increases the number of trainable parameters exponentially, and ultimately increases the storage and processing capability, and thus would be expensive. Figure 7.3 illustrates the proposed model of satellite image classification that is based on CNN.

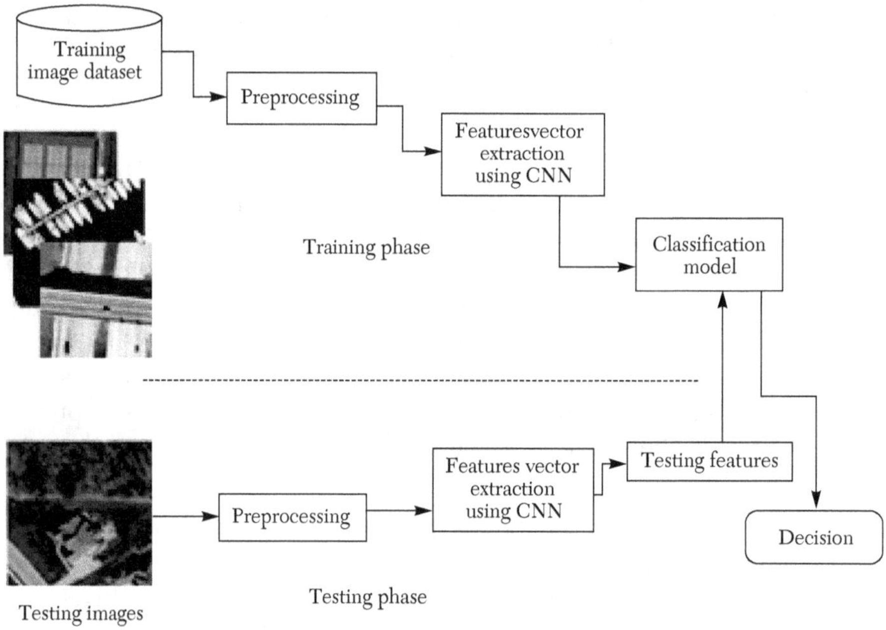

Figure 7.3. Use of CNN for image classification.

The ANN tends to be a less popular choice when analyzing images because of the reliance on valid data inputs, while the CNN works in a compatible way with images as input data. The ANN is a comparatively lightweight way of solving data classification problems. The main advantage of CNN as compared to ANN is that it automatically detects the important features without any human supervision. This is why CNN is an ideal solution to computer vision, image classification, and image-dependent machine learning problems. The advantages and disadvantages of CNNs are given in Table 7.2. In general, CNN tends to be a more powerful and accurate way of solving classification problems. However, ANN is still dominant for problems where datasets are limited and image inputs are not necessary.

7.2.3 Recurrent Neural Network (RNN)

Recurrent neural networks (RNNs) have traditionally been used for a discrete time-sequence analysis. In an RNN, the input and output data can be of varying length, therefore, certain tasks that involve sequential inputs, such as speech and language processing, and time-series satellite images, often benefit more from RNNs (Mou et al., 2017). They save the output of processing nodes and feed the result back into the model (they don't pass the information in one direction only). This is how the RNN model is said to learn to predict the outcome of a layer. Each node in the RNN model acts as a memory cell, continuing the computation and implementation of operations. If the network's prediction is incorrect, then the system self-learns and continues working toward the correct prediction during back-propagation.

The most exciting application of back-propagation is to train the RNNs. With the time of the computation involved in forward computation, the RNNs will generate deep feed-forward networks to learn long-term dependencies, and thus, it is difficult to learn and store information for very long (Maa et al., 2019). To address this problem, several specialized memory units have been developed, for example, the long short-term memory cell and gated recurrent unit, to augment the networks. Therefore, with the development of their architecture and ways of training, the RNNs have been successfully and extensively applied in predicting the next character in the text, or the next word in a sequence, as well as also have been extended to other more complex tasks of remote sensing images. The RNNs have been used for the vehicle classification task, which has application to illegal vehicle type recognition, traffic surveillance, and autonomous navigation, among others (Lakhal et al., 2018). The basic difference between the ANN, CNN, and RNN is given in Table 7.3.

Table 7.3. A Comparison of ANN, CNN, and RNN Approaches (Lakhal et al., 2018)

	ANN	CNN	RNN
Type of data	Tabular data, Text data	Image data	Sequence data
Parameter sharing	No	Yes	Yes
Fixed length input	Yes	Yes	No

(Contd.)

	ANN	CNN	RNN
Recurrent connections	No	No	Yes
Spatial relationship	No	Yes	No
Performance	ANN is considered to be less powerful than CNN, RNN.	CNN is considered to be more powerful than ANN, RNN.	RNN includes less feature compatibility when compared to CNN.
Application	Facial recognition and computer vision.	Facial recognition, text digitization, and natural language processing.	Text-to-speech conversions, time-series satellite data analysis.
Advantages	Storing information on the entire network. Ability to work with incomplete knowledge. Has fault tolerance. Has a distributed memory.	High accuracy in image recognition problems, weight sharing. Automatically detects the important features without any human supervision.	Remembers each and every piece of information. Time series prediction is even used with convolutional layers to extend the effective pixel neighborhood.
Disadvantages	Hardware dependence. Unpredicted behavior of the network.	Large training data needed. Does not encode the position and orientation of an object. Lack of ability to be spatially invariant to the input data.	Training is difficult. It cannot process very long sequences if using tanh or ReLU as an activation function.

7.2.4 Region-Based CNN

Object detection in remote sensing images becomes increasingly important due to its use in a wide range of practical applications, such as urban planning and traffic monitoring. In the field of target/object detection, region based CNNs (R-CNNs) can be used to extract the convolution features of regions and make localization and classification by region more robust, leading to better segmentation (Ren et al., 2017). Girshick et al. (2014) proposed a method, called R-CNN, where selective search is utilized to extract just 2,000 regions from the image, which are called region proposals. Therefore, now, instead of trying to classify a huge number of regions, one can just work with 2,000 regions (Gandhi, 2018). Many real-life satellite scenes cover a large area with relatively sparsely distributed targets/objects, but the existing algorithms work well on large-scale satellite scenes containing large numbers of small targets in crowded neighboring environments.

The process of target/object detection includes three key modules: region recommendation, feature extraction, and region classification. First, region recommendation is used to generate approximately 2,000 region recommendations for the input image to form the candidate detection set. Second, feature extraction is used to extract the fixed length 4,096 feature vectors from each region recommendation. Last, the third module classifies regions to score and filter each recommended region.

The R-CNN takes a huge amount of time to train the network, as an image would be classified into 2,000 region proposals (Gandhi, 2018). It cannot be implemented in real time, as it takes around 47–49 seconds for each test image. The selective search algorithm is a fixed algorithm. Therefore, no learning is happening at that stage. This could lead to the generation of bad candidate region proposals. In addition to storing the feature map of the region proposal, lots of disk space is required. Considering the slow computation speed of R-CNNs, the common region proposal methods used in R-CNNs include selective search and edge boxes.

Compared to R-CNN, which is primarily used in the field of target/object detection, the fully convolutional neural network (F-CNN) is more commonly used in the field of semantic segmentation. After a target feature is extracted, the F-CNN is used for feature reconstruction, which improves the computational efficiency and eliminates the shortcomings of constraining local features. The SegNet model is a common F-CNN that comprises of a group of symmetric encoders and decoders. Its structure is simple, and model detection is fast. The U_net is also a common F-CNN that is typically used in medical images that are typically large and, thus, similar in size to the remote sensing images.

7.2.5 Fast R-CNN

Some of the drawbacks of R-CNNs were incorporated to build a faster object detection algorithm, called *Fast R-CNN*. The Fast R-CNN algorithm structure is simple, and continuously extracts the feature map through a series of convolution and pooling operations, and finally uses the fully connected layer for target/object detection and judgment (Ren et al., 2017). The approach is similar to the R-CNN algorithm, but, instead of feeding the region proposals to the CNN, the input image is fed to the CNN just once to generate a convolutional feature map. From the convolutional feature map, the regions of proposals generated by selective search are identified and warped into squares. These region proposals are then projected on to the feature maps generated by the CNN. This process is called *RoI* (*region of interest*) *projection*.

Before starting with RoI projection, one needs to know about the subsampling ratio. It is the ratio of the feature map size to the original size of the image (Gandhi, 2018). By using an RoI pooling layer, squares are reshaped into a fixed size so that they can be fed into a fully connected layer.

Fast R-CNN is faster than R-CNN because 2,000 region proposals are not fed to the CNN every time. Instead, the convolution operation is done only once per image, and a feature map is generated from it. It also combines different parts of architecture (such as ConvNet, RoI pooling, and the classification layer) in one complete architecture. It also uses the softmax layer instead of SVM in its classification of the region proposal, which proves to be faster and generates better accuracy than the SVM. Most of the time taken by Fast R-CNN during detection is a selective search region proposal generation algorithm. Hence, it is the bottleneck of this architecture which can be dealt with in Faster R-CNN.

7.2.6 Faster R-CNN

Both of the algorithms (R-CNN and Fast R-CNN) use selective search to find out the region proposals. Selective search is a slow and time-consuming process affecting the performance of the network. Therefore, Ren et al. (2017) proposed an object detection algorithm called *Faster R-CNN* that eliminates the selective search algorithm and allows the network ro learn the region proposals (Song et al., 2019). In a Faster R-CNN, a region proposal network is used to generate candidate regions and an internal deep network is used to replace the candidate region proposals. This technique gives less than 2,000 region proposals, is faster than the selective search, as accurate as selective search or better, and is able to propose overlapping RoIs with different aspect ratios and scales.

Similar to Fast R-CNN, in Faster R-CNN the image is provided as an input to a convolutional network which provides a convolutional feature map. Instead of using selective search algorithm on the feature map to identify the region proposals, a separate network is used to predict the region proposals. The predicted region proposals are then reshaped using an RoI pooling layer which is later used to classify the image within the proposed region and predict the offset values for the bounding boxes, as shown in Figure 7.4. Therefore, it is considered a better technique for real-time object detection (Gandhi, 2018).

Table 7.4 compares the candidate region-based target detection methods (sliding window, R-CNN, and Faster R-CNN) in terms of input images for the CNN as well as their advantages and disadvantages.

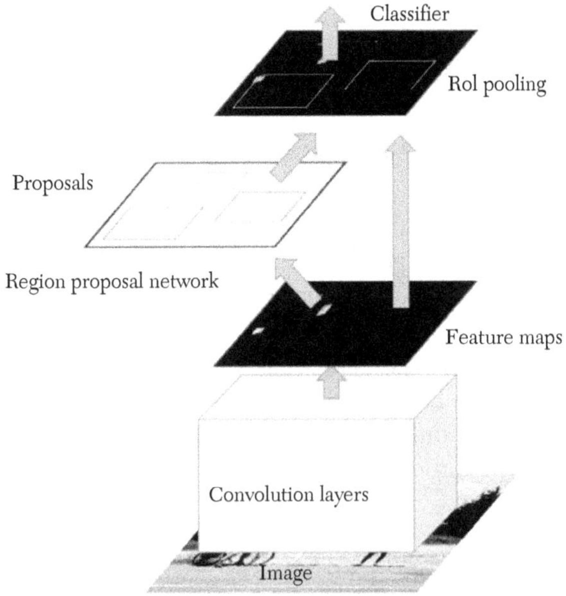

Figure 7.4. The architecture of Faster R-CNN.

Table 7.4. A Comparison of Object Detection Methods (Song et al., 2019)

Method	**Sliding window method for detection**	**Regional proposal method for detection**	
		R-CNN	**Faster R-CNN**
Input image for the CNN	Image area	Image area	Entire image (no fixed size is required)
Advantages	Relatively high object identification capability.	Relatively high object identification capability. Relatively few candidate regions which can be controlled within thousands or hundreds of regions.	Relatively high object identification capability. End-to-end object detection. Independent of a region proposal method. Very high detection speed.
Disadvantages	The method of exhaustion generates a large number of candidate regions, resulting in a significant volume of repeated calculations in the subsequent operations, and therefore has relatively low efficiency when used to process large remote sensing images.	Relatively long time to calculate candidate regions. Feature extraction will result in repeated calculations.	Relatively complex training process.

There are a lot of object detection algorithms, such as R-CNN, Fast R-CNN, spatial pyramid pooling (SPP), YOLO (you only look once), SSD (single shot detector), and so forth. The selection of the appropriate algorithm is a complex problem. Figure 7.5(*a*) presents an understanding of the speed versus accuracy trade-off of some algorithms. It shows that for achieving higher accuracy, Faster R-CNN is the better choice. However, if high computation is not required, the SSD is the best. If accuracy is not too much of a concern, but speed is a concern, the YOLO will be selected. The SSD seems to be a good choice as it can be run on a video and the accuracy tradeoff is very little. Figure 7.5(*b*) presents the performance of the SSD, YOLO, and Faster R-CNN on objects of various sizes. At large sizes, the SSD seems to perform similar to Faster R-CNN. However, the accuracy decreases as the object size becomes smaller. In addition, the choice of a right object detection method is crucial and depends on the problem at hand.

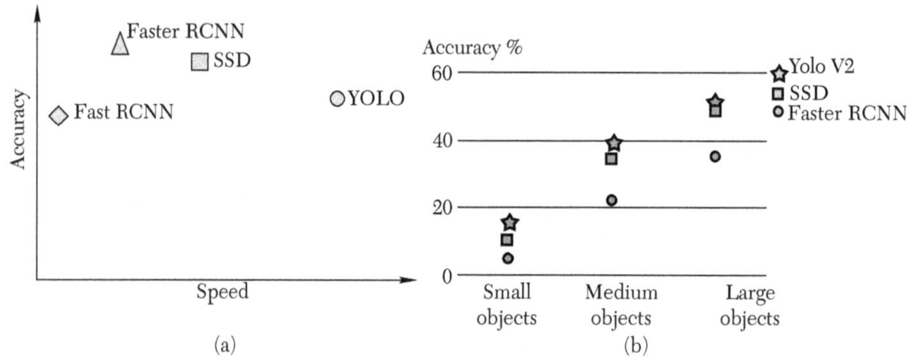

Figure 7.5. Object detection algorithms: (a) speed vs. accuracy and (b) object size vs. accuracy.

7.2.7 Object-Based Image Analysis (OBIA)

Pixel-based analysis is not successful in performing on high-resolution satellite data. While pixel-based image analysis is based on the information in each pixel, object-based image analysis (OBIA) is based on information from a set of similar pixels, called objects or image objects (Burnett & Blaschke, 2003). Several analysis techniques have been developed for medium-resolution satellite images (10–100 m), and these were less effective and more time-consuming when used on high-resolution data. It can be better understood with Figure 7.6, which depicts the relationship between the spatial resolutions and the objects to be detected. The three situations shown in the Figure 7.6 require completely different techniques to extract objects from the image datasets. Situation (*a*) requires sub-pixel techniques, (*b*) needs pixel-by-pixel techniques, and (*c*) requires the use of the OBIA approach when the task is to identify, classify, and characterize a given object. In a low-resolution image, pixels are significantly larger than objects; in medium-resolution images, pixels, and object sizes are of the same order; and in high-resolution images, pixels are significantly smaller than the size of the objects.

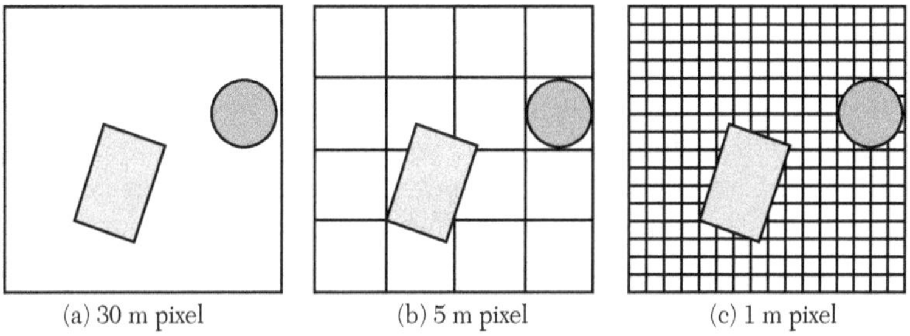

Figure 7.6. Relationship between objects and spatial resolutions in (a) low-resolution, (b) medium-resolution, and (c) high-resolution images.

The key motives behind the transition from pixel-based to object-based analysis of remote sensing data are (Kettig & Landgrebe, 1976): (*a*) the demand for improved interpretative values of remote sensing data in different applications (mostly in time comparison studies, and requirements of remote sensing data used for planning); (*b*) increased availability of high-resolution satellite data, on which one can observe surface objects in greater detail (including the increased interest in the contextual validation of the image content); and (*c*) higher level of development of technological equipment and algorithms used for processing remote sensing data (accessibility to a wider user society; transfer of GIS object-based spatial analysis toward the field of raster remote sensing data and their particularities).

The OBIA provides more meaningful information than pixel-based image analysis by allowing for less well-defined edges or borders between different classes. On maps, divisions between different types of vegetation, for example, where a shrubland meets a grassland, are generally represented by a single line. In nature, no such smooth change occurs. Instead, the area where the shrubland meets the open grassland is a transition area, called an *ecotone*, containing characteristic species of each community, and sometimes species unique to the ecotone itself. The OBIA allows for this area of transition by using fuzzy logic. That is, the objects that occur within the ecotone belong to, and are thus considered members of, both the shrubland and grassland classes. The membership value of a pixel to a class varies from 0.0 (no membership) to 1.0 (100% complete membership to a class, and thus no ambiguity). An object in an ecotone might have 70%

membership within the shrubland class, and 30% membership within the grassland class. This is a more realistic approach than the objects belonging strictly in one class or another, but not both.

The object-oriented classification of a segmented image is substantially different from performing a per-pixel classification. The OBIA classifier groups the pixels of homogenous image objects into representative shapes and sizes. The algorithm incorporates both spectral and spatial information in the image segmentation phase. The result is the creation of image objects defined as individual areas with shape and spectral homogeneity which may be recognized as segments or patches in the landscape. The analysis is not constrained to using just spectral information, but the mean spectral information along with various spatially distributed variables (e.g., elevation, slope, aspect, population density) associated with each object (polygon) in the dataset may also be used. This introduces greater flexibility and robustness for the use of this classifier. Once selected, the spectral and spatial attributes of homogeneous image objects are then analyzed using traditional classification algorithms (e.g., nearest-neighborhood, minimum distance, maximum likelihood) or knowledge-based and fuzzy logic classification approaches.

With the availability of the eCognition software in 2000, the OBIA received a steep rise (*https://geospatial.trimble.com*). This software was the first commercial product that could be used to perform a quality OBIA of multispectral data. Later, the OBIA was followed by software modules ERDAS Imagine (module Objective), ArcGIS (Feature Analyst), and ENVI (Feature Extraction within ENVI Zoom). Nowadays, object-based analysis of satellite data is a well-established technique that it is based on the concepts of segmentation, and edge detection, as well as object detection and classification (Johansen et al., 2010). The process flowchart involved in classification using OBIA is shown in Figure 7.7. It includes multi-resolution segmentation, selection of training areas, computation of statistics, and classification. The two most common segmentation algorithms to be used are: Multi-Resolution Segmentation in eCognition and Segment Mean Shift in ArcGIS.

In OBIA, different methods can be used to classify objects. In addition, one can use:

1. *Shape*–for classification of buildings, shape statistics, such as "rectangular fit" may be used. This tests an object's geometry to the shape of a rectangle.

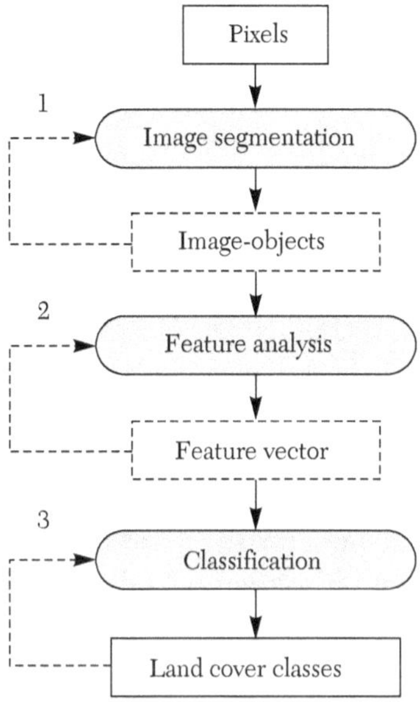

Figure 7.7. Generalized flow chart for OBIA.

2. *Texture*–the homogeneity of an object can be used; for example, water is mostly homogeneous with a smooth texture, but forests have shadows and give a rough texture.
3. *Spectral*–the mean value of spectral properties such as NIR, SWIR, red, green, or blue can be used.
4. *Geographic context*–objects having proximity and distance relationships between neighbors.

The OBIA algorithm is an interactive, multiphase process. It enables the intermediate results to be checked, and immediate improvements can be made through immediate parameter fine-tuning. The spatial relationship information contained in image objects allows for more than one level of analysis. This is critical because image analysis at the landscape scale requires multiple, related levels of segmentation, or scale levels. The objects in OBIA provide complex information on various scales (through multiple segmentations with different parameter settings), and thus OBIA is more suited to landscape scale analyses. The classification uses clear semantic rules that can also be used to enhance or omit certain typical object characteristics (e.g., linearity, length, width, rectangularity of buildings) or enhance their key differences (e.g., typical size in nature). Object-based classification is a relatively new method of remote sensing, and therefore there is no general rule that defines the relations between the objects obtained within the segmentation process and the geographical objects (Burnett & Blaschke, 2003). The levels and hierarchical relations between objects that are obtained at different spatial scales are relatively poorly studied. Table 7.2 presents the advantages and disadvantages of OBIA classifiers used in image classification.

Several studies compared pixel-and object-based classification, and showed that satellite data of medium (e.g., Landsat TM/ETM+ and SPOT-5) and high resolutions (IKONOS, QuickBird, WorldView) yield better results with the object-based approach (Yan et al., 2006). The OBIA offers advantages as compared to other methods. For example, it uses a vast variety of remote sensing data characteristics (e.g., spectral, spatial, temporal) and can be combined with GIS functionalities (Figure 7.8). The classification uses all characteristics for their classification (e.g., shape, texture, relations with other segments). Objects can be classified using their spatial relationships with adjacent or nearby objects. The identified objects are generally in vector form, which is easier for post-processing than the pixel-based classification results. The generalization can also be carried out

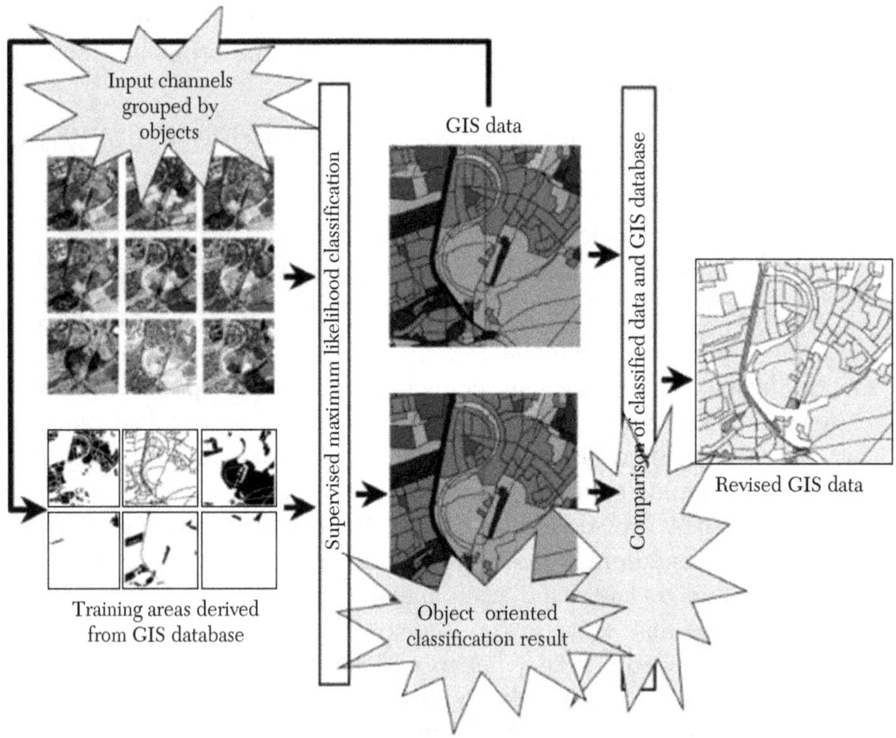

Figure 7.8. The OBIA using remote sensing data characteristics for integrating results in GIS.

during the processing phase (e.g., elimination of small objects based on their shapes or sizes). Since the basic computation entities are objects (and not pixels), it reduces the demand on computer algorithms, and at the same time enables the users to utilize more complex computation techniques and a wider set of data characteristics (Ren et al., 2017). The OBIA is able to filter out the undesired information and assimilates other pieces of information into a single object. This is analogous to how the human eye filters information that is then translated by the brain into an image that makes sense. For example, the pixels in an image are filtered and grouped to reveal a pattern, like that of an orchard or tree plantation.

There are several limitations of OBIA. When processing extensive databases (large area of interest, high spatial resolution, or both), powerful processing hardware is needed, since numerous pixels are processed simultaneously during the multispectral image segmentation of object-based classification. Segmentation sometimes may not provide a uniform solution, as even with a minimal change in the radiometric resolution, the segmentation parameters or the preprocessing procedures yield different results. The problem of using thresholds for segmentation is that they are data dependent. For example, the percentage of vegetation pixels varies significantly between satellite images of summer or winter due to phenological changes, illuminating light, weather conditions, soil type, or time of day. Therefore, the same thresholds can't be used for different datasets. The processes that enable multi-scale object-based analysis do not have enough support in topological linking, querying through scale, and object tracing with practical implementations. Due to the complex geographical reality and the diversity of satellite images, the processes are not designed to be fully automatic, but rather to suit a broad spectrum of different applications. So, automization remains limited merely to highly specialized tasks and to identify particular objects. From the point of view of discrimination/preservation of the basic geometrical object properties, object determination is not strong enough. The repeatability of the discrimination process in various natural and technological conditions is also poor.

7.2.8 Decision Tree (DT)

The decision tree (DT) classifiers are nonparametric, and more efficient than the single-stage classifiers (Quinlan, 1987). With a DT classifier, decisions are made at multiple levels, and therefore they are also known as *multilevel classifiers*. In designing a DT classifier, it is desirable to construct an optimum tree so as to achieve the highest possible classification accuracy with the minimum computation. A tree is termed as *univariate* if it splits the node using a single attribute, or a multivariate, if it uses several attributes. The binary tree classifier is considered as a special case of a DT classifier (Maxwell et al., 2018).

The basic concerns in a DT classifier are the separation of groups at each nonterminal node and the choice of features that are most effective in separating the group of classes. This method consists of the root node (the starting node), partitioning the nodes, and the terminal node that represents the group of pixels allocated to the same class. The DT is an iterative and progressive method of pattern recognition based on a hierarchical rule approach (Salah, 2017). If the DT characterizes too much detail or noise in the training data, an over-fitting process may occur, reducing the classification accuracy.

The DT calculates class membership by repeatedly partitioning a dataset into homogeneous subsets using a variety of binary splitting rules (Tso & Mather, 2009). A DT is built from a training set which consists of objects, each of which is completely described by a set of attributes and a class label. Attributes are a collection of properties containing all the information about one object. Unlike class, each attribute may have either ordered (integer or a real value) or unordered values (Boolean value). These rules are derived from training data using statistical methods and based on the "impurity." If all pixels contained by a given node belong to the same category, the node is pure and the impurity is zero. A hierarchical classifier permits the acceptation and rejection of class labels at each intermediate stage.

Several statistical algorithms for building decision trees are available, including CART (classification and regression trees), C4.5, CHAID (chi-squared automatic interaction detection), and QUEST (quick, unbiased, efficient, statistical tree) (Pal & Mather, 2003). These algorithms generally use the recursive-partitioning algorithm, whose input requires a set of training examples, a splitting rule, and a stopping rule. Partitioning of the tree is determined by the splitting rule, and the stopping rule determines if the class in the training set can be split (branches) further. If a split is still possible, the classes in the training set are divided into subsets by performing a set of statistical tests defined by the splitting rules. The test that results in the best split is selected and applied to the training set, which divides the training set into subsets. This procedure is recursively repeated for each subset until no more splitting is possible.

Figure 7.9 presents an example of the DT classification rules developed for land use types based on the DN values of SPO-5 multispectral data and

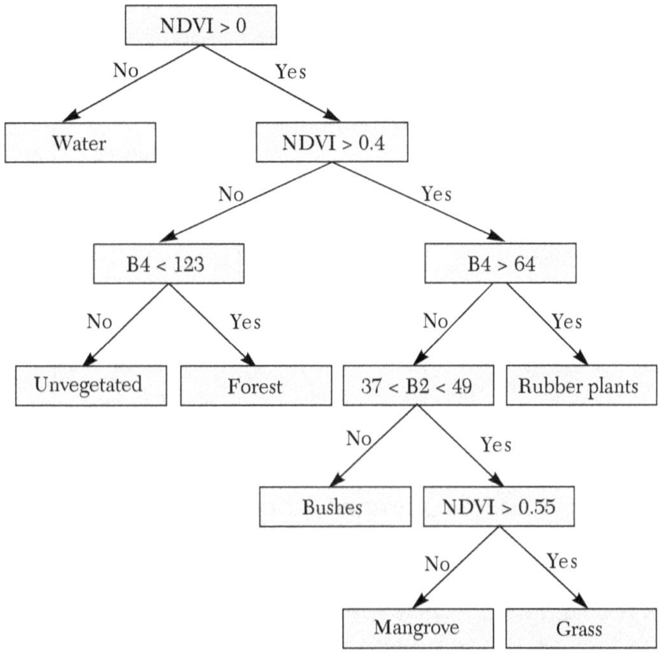

Figure 7.9. An example of a decision tree approach.

NDVI of land use types. If the logical condition at a given node is fulfilled, the right branch is chosen; otherwise the left branch is followed. The numbers indicate the variables and their values that are used as thresholds for each node condition. The process continues until the node becomes pure, and is assigned as a possible terminal node.

Table 7.5 provides a brief comparison of the four most widely used DT methods (Song & Lu, 2015). The DTs based on these algorithms can be constructed using data mining software that is included in widely available statistical software packages.

Table 7.5. Comparison of Different Decision Tree Algorithms

Methods	CART	C4.5	CHAID	QUEST
Measure used to select input variable	Gini index; Towing criteria	Entropy info-gain	Chi-square	Chi-square for categorical variables; J-way ANOVA for continuous/ ordinal variables

(Contd.)

Methods	CART	C4.5	CHAID	QUEST
Pruning	Pre-pruning using a single-pass algorithm	Pre-pruning using a single-pass algorithm	Pre-pruning using chi-square test for independence	Post-pruning
Dependent variable	Categorical/ Continuous	Categorical/ Continuous	Categorical	Categorical
Input variables	Categorical/ Continuous	Categorical/ Continuous	Categorical/ Continuous	Categorical/ Continuous
Split at each node	Binary; Split on linear combinations	Multiple	Multiple	Binary; Split on linear combinations

The most widely used splitting rules are: (*a*) entropy; (*b*) the gain ratio or information gain (IG); and (*c*) Gini models. Entropy measures homogeneity, and aims to decrease the entropy until a pure terminal node, with zero entropy, is reached. The IG is a measure of reduction in entropy that would result from using the splitting node rule (Quinlan, 1987). The variable that achieves the highest IG value will be chosen to split the data at that node. One drawback of this approach is that the variables with relatively high variances are generally selected. This would lead to a bias toward a large number of splits. In order to overcome this problem, the IG can be adjusted by the entropy of the partitioning. The Gini index measures the impurity of the node and separates the largest homogeneous group from the remaining training data (Breiman, 2001). The Gini index of all parts is summed for each split rule. The split rule with the maximum reduction in impurity and minimum Gini index is selected.

DTs have many advantages. The model logic can be visualized as a set of “if–then” rules. These can utilize categorical data, and once the model has been developed, classification is extremely rapid because no further complex mathematics is required. Problems with DT is to include the possibility of generating a nonoptimal solution and overfitting. The latter is normally addressed by pruning the tree, removing one or more layers of splits (i.e., branches). Pruning, however, reduces the accuracy of classifying the training data but generally increases the accuracy of dealing with unknowns (Pal & Mather, 2003). The advantages and disadvantages of DT are given in Table 7.2. Figure 7.10 shows an original image (FCC) and the corresponding image after classification using a DT approach.

Figure 7.10. FCC (IRS LISS III Image, Oct. 2004) of Mayurakshi Reservoir, Jharkhand, India, Red (R), Green (G), Near Infrared (NIR) bands (left image), and land use classes from a decision tree approach (right image; Ghose et al., 2010) (please see color figures in companion files).

7.2.9 Extraction and Classification of Homogeneous Objects (ECHO)

One classification algorithm that utilizes both the spectral and spatial (i.e., textural) characteristics of the data in the decision process is referred to as the ECHO (extraction and classification of homogeneous objects) classifier. The ECHO is a multistage spatial-spectral classifier developed by Kettig and Landgrebe (1976) that has elements of a parametric per-pixel classifier as well as elements related to texture classification; hence, it is a hybrid technique in nature (Fauvel, 2007). Currently ECHO is supported in both Windows and Mac formats in the Purdue/NASA MultiSpec software package that is available from the MultiSpec website (*http://dynamo.ecn.purdue.edu/~biehl/MultiSpec/*).

The ECHO technique defines the boundary around an area having generally similar spectral and textural characteristics, and then the entire area within the boundary is classified as a unit into one of the classes for which training statistics have been developed. It relies on a two-stage procedure whereby homogeneous objects are recognized in local areas, and each object as a whole is spectrally classified using a non-contextual algorithm (Kettig & Landgrebe, 1976). The ECHO classifier substantially reduces the number of classifications, resulting in a potential increase in speed (and thus decrease in cost). It simplifies complex mixtures of pixels and can extract the essence of a mass of seemingly complex spectral responses, and often accurately extract the dominant class.

Use of both spectral and spatial characteristics in the data results in map outputs which have a smooth appearance on maps, whereas classification algorithms that are based only on spectral data and involve per-pixel classifiers often result in output maps with a "salt and pepper" (noisy) appearance. Based on the spectral signatures of the classes, many advanced pixel-based classifiers such as SVM, and NN, have been used in the literature, but these do not use the spatial content of the image, resulting in noisy thematic maps (Karakahya et al., 2003). Approaches like MRF and Monte Carlo optimization use contextual information, but the main drawback of these algorithms is the computing time, which can be extended even for small datasets. The morphological profile (MP) can be used as an alternative way of exploiting the spatial information (Fauvel, 2007). Compared to the MRF-based classifiers, the MP offers the possibility of making use of geometrical information (shape, size, etc.) and performs well on many types of data (panchromatic, multispectral and hyperspectral data).

The classification of images from ECHO involves four stages (Lu et al., 2003): (*a*) partitioning the feature space into cells (e.g. 2 × 2, 3 × 3, 4 × 4, etc.); (*b*) determining the homogeneity of pixels within each cell by the user's defined thresholds, and each cell is either considered a single entity or individual pixels within the cell remain as single pixels; (*c*) aggregating the cells and individual pixels based on spectral statistical associations between them; and (*d*) processing the aggregate of cells of pixels and single pixels by a classifier to get the final results.

7.2.10 Fuzzy Classifiers

A pixel corresponds to an area on the ground. Quite often, such an area contains a mixture of land cover classes, for example, grass and underlying soil. The mixed classes may also take place when the size of the pixel is larger than the size of the features about which information is to be extracted. In addition, different conditions may exist within a cover class. For example, within a pixel, vegetation may be in different conditions that are caused by factors such as plant health, age, and water content.

Fuzziness is a type of imprecision that characterizes classes without crisply defined boundaries. These defined classes are called *fuzzy sets*. Zadeh (1965) initially introduced the idea of "fuzzy set" to deal with the fuzziness in a definable way, and since then the theory of fuzzy sets has been developed and applied to many disciplines, including handling the uncertainty of quantification in human geography, and in GIS. Fuzzy set theory can provide a better representation of geographical information created by a mixture of classes. In fuzzy representation for remote sensing image analysis, land cover classes can be defined as fuzzy sets and pixels as set elements.

Pixels with mixed classes or in intermediate conditions can be described by their membership values, which can vary continuously between 0 and 1. For example, if a pixel contains both soil and vegetation it may have two membership values indicating the extent to which it is associated with these two classes, and it is termed its *fuzzy membership value*. Each pixel is attached with a group of membership value that indicates the extent to which the pixel belongs to certain classes. To some extent, fuzzy membership values reflect the land cover composition of a mixed pixel, which enables a more accurate and realistic representation of land cover, overcoming some of assumptions of the conventional classifications (Foody & Cox, 1994).

Fuzzy representation of geographical information enables a new method for spectral space partition. Such a partition is referred to as a *fuzzy partition of spectral space*. Figure 7.11 illustrates the partition of pixels by a hard classifier (left diagram) and membership values of a pixel in a fuzzy partition in 2D space (right diagram) within three classes; soil, water, and vegetation. Thus, a pixel will have a fuzzy membership to every informational class. Membership values close to one signify a high degree of similarity between a pixel and fuzzy class, while membership values close to zero imply little similarity between the pixel and that class (Islam & Metternicht, 2005).

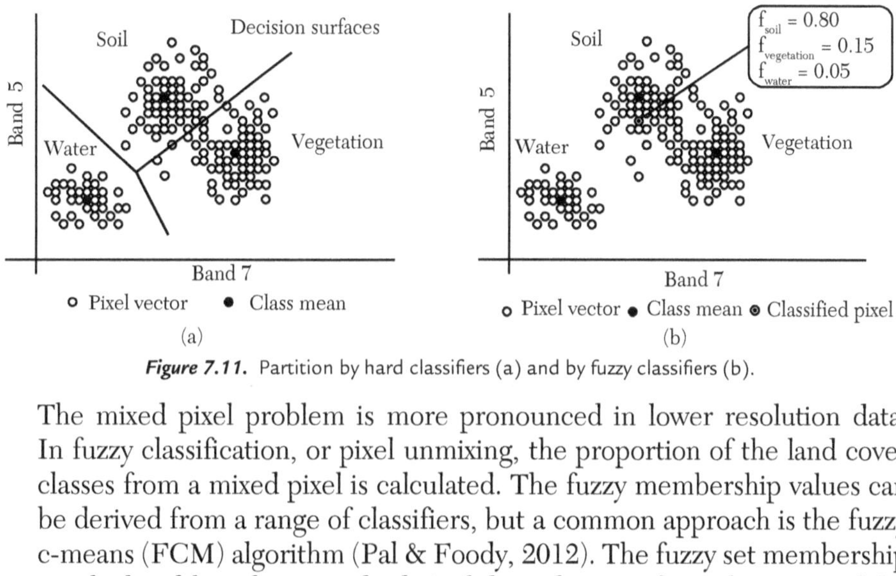

Figure 7.11. Partition by hard classifiers (a) and by fuzzy classifiers (b).

The mixed pixel problem is more pronounced in lower resolution data. In fuzzy classification, or pixel unmixing, the proportion of the land cover classes from a mixed pixel is calculated. The fuzzy membership values can be derived from a range of classifiers, but a common approach is the fuzzy c-means (FCM) algorithm (Pal & Foody, 2012). The fuzzy set membership is calculated based on standard Euclidean distance from the mean of the spectral signature. The underlying logic is that the mean of a signature represents the ideal point for the class, where fuzzy set membership is 1. When distance increases, fuzzy set membership decreases, until it reaches the user-defined distance where fuzzy set membership decreases to 0 (Abdu et al., 2004).

For multispectral satellite images, fuzzy membership values associated to predetermined informational classes can be generated for each band. Fuzzy operators based on fuzzy set theory can be used to integrate the fuzzy memberships of selected bands, as these operators allow manipulating and processing incomplete and/or imprecise information to obtain the most reasonable output (Islam & Metternicht, 2005). The conceptual model devised for the generation of fuzzy classification maps consists of four main steps, as shown in Figure 7.12: (*a*) selection of the best set of satellite images; (*b*) computation of fuzzy membership values for each class on the multiband data set; (*c*) integration of fuzzy membership values for classes using different fuzzy operators; and (*d*) accuracy assessment, which depends upon the threshold selection and fuzzy integral.

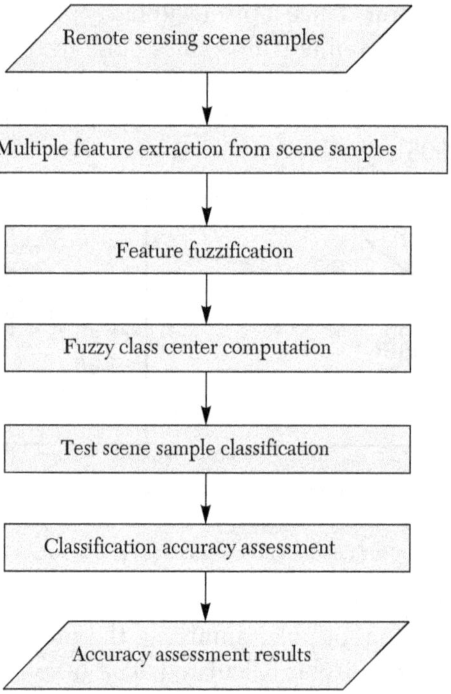

Figure 7.12. Fuzzy classification steps to analyze satellite images.

7.2.11 Fuzzy C-Means (FCM)

The fuzzy c-means (FCM) clustering algorithm was introduced by Ruspini (1969). It is a method of clustering by assigning the membership value into one or more cluster. The membership function in FCM clustering is generated for each pixel showing similarity of pixel membership to each predetermined class. The membership value generated using FCM gives the "degree of sharing" (to which a pixel will be shared among the clusters of the pixel) in various clusters in the feature space.

The FCM algorithm and its derivatives have been used successfully in many applications such as pattern recognition, remote sensing image classification, data mining, and image segmentation (Lei et al., 2013). FCM in image processing works by partitioning the image pixels into a set of fuzzy clusters through an iterative optimization of the objective function, with the update of membership and the cluster center. This is regardless of the actual number of clusters present in the dataset. It is mainly due to the probabilistic constraint on the membership value. In this algorithm, centroids are defined for each of the clusters and the quality of results depends on the initial set of clusters, the value of the centroid, and the number of iterations. The process is shown in Figure 7.13.

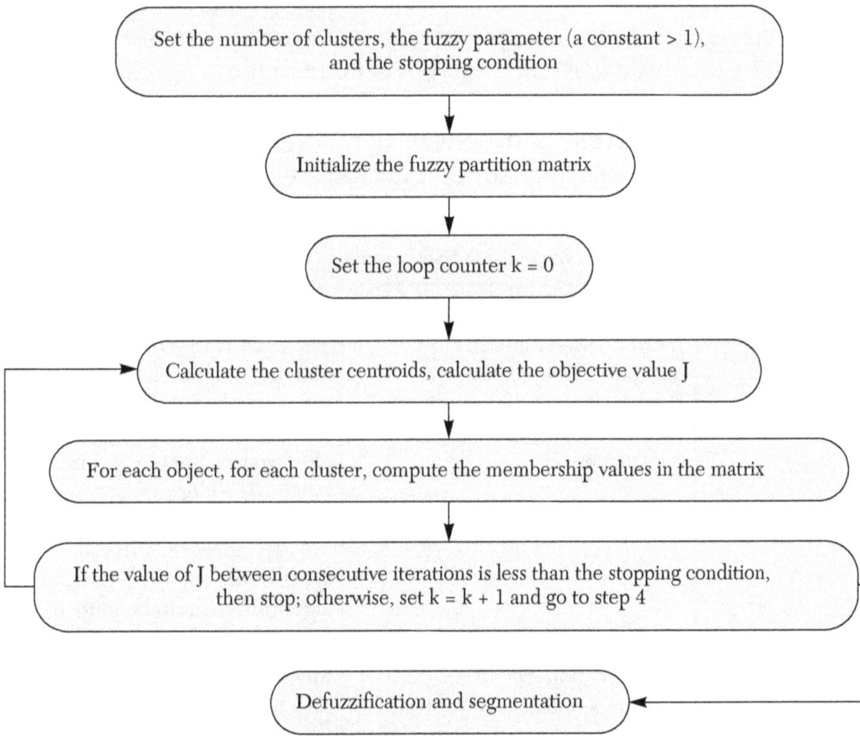

Figure 7.13. The FCM clustering algorithm flowchart.

FCM is the fuzzy variant of k-means, which allows pixels to belong to multiple clusters with varying degrees of membership, as compared to CRISP or hard clustering methods, which force pixels to belong exclusively to one class. In fact, the FCM is an improvement over k-means when the data set can't clearly subdivide into underlying partitions (Abdu et al., 2014). Due to the iterative nature, it is important for an FCM algorithm to select a good set of initial cluster centers randomly so that the algorithm may take fewer iterations to find the actual cluster centers. The pixels on an image are highly correlated, that is the pixels in the immediate neighborhood possess nearly the same feature data. Therefore, the spatial relation of neighboring pixels is an important characteristic which can be very useful in image classification. However, the standard FCM only considers the pixel's spectral information, ignoring the spatial information in the image context, which makes it very sensitive to noise, outliers, and other imaging artifacts.

Abdu et al. (2014) described various fuzzy clustering methods, such as the spatially weighted FCM (SWFCM) clustering algorithm, the membership connectedness method, size-weighted fuzzy clustering, fuzzy statistics-based affinity propagation (FS-AP), and ant colony optimized FCM clustering. Table 7.6 presents a comparison emphasizing their advantages, disadvantages, and the types of input they can use. It is observed that threshold level-based fuzzy clustering is the only technique that has the ability to handle intensity variations in the image. Figure 7.14 shows the results of the FCM classifier sorted into three classes: water, vegetation, and urban regions.

Table 7.6. Advantages and Disadvantages of Various FCM Methods

Technique	Advantages	Disadvantages
FCM	It can be used for either unsupervised or supervised classification. It has advantages over others for preserving image information.	The clustering center locations and the number of clustering influence the validity of clustering results, yet the number of clustering can not be priori to obtain. The clustering center location and characteristics are dependent on artificial speculation, and the FCM algorithm tends to be immersed in local optimum as well.
Spatially weighted FCM clustering	It clusters non-Euclidean structures in data. It is robust of the original clustering algorithms to noise and outliers while retaining computational simplicity.	Single level thresholding. Heavily affected by its parameters.

Technique	Advantages	Disadvantages
Membership connectedness	It has reduced sensitivity to noise, and reduced false segmentation results. It can utilize expert knowledge in the form of selected seeds and segments.	It is faced with intraclass redundancies. It also leads to spatial dependency.
Size-weighted fuzzy clustering method	It detects small regions that often appear in remotely sensed images. It can detect overlapped regions in satellite images.	It does not provide detailed information about the clusters.
Fuzzy-statistics-based affinity propagation technique	It is fast and robust to noise. It reduces dependence among pixels. It avoids poor solutions caused by hard decisions during clustering.	It works well with application of multispectral remote sensing images. It does not handle variation in intensity.
Ant colony optimized FCM clustering	It is robust and versatile in handling a wide range of combinatorial optimization problems. Can be used in dynamic applications (adapts to changes, such as new distances, etc). It provides improved visual interpretation of segmentation results. It reduces computation time.	Theoretical analysis is difficult. It includes sequences of random decisions (not independent). Probability distribution changes by iteration. Research is experimental rather than theoretical.

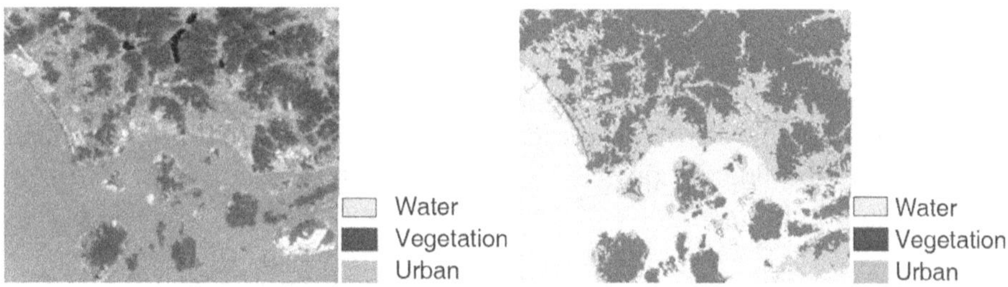

(a) (b)

Figure 7.14. (a) Satellite image and (b) FCM classified image.

7.2.12 The Possibilistic C-Means

Soft classifiers such as the FCM, have a probabilistic constraint, and possibilistic c-means (PCM) is based on a modified version of FCM. The PCM classifier assigns the pixel fractions according to the area it represents inside a pixel, which makes it efficient in handling mixed pixels. The possibilistic approach simply means that the membership value of a point in a cluster (or class) represents the typicality of the point in the class, or the possibility of the point belonging to the class. The PCM clustering gets over the problem by relaxing the probabilistic constraint to produce absolute class memberships, which can be interpreted and used directly to indicate the class proportions of pixels to produce a sub-pixel classification. This may result in a more accurate sub-pixel classification than that produced from the FCM (Lu & Weng, 2007).

Unlike FCM, the membership value generated by the PCM can be interpreted as *degree of belongingness* or *compatibility* or *atypicality*. The *degree of belongingness* implies the degree to which the pixel belongs to a class. The *Degree of compatibility* is the degree by which the pixel is compatible to the other pixels in the cluster and the cluster mean. *Degree of typicality* helps to differentiate between a highly atypical member of the cluster versus a moderately atypical member of the cluster. All three give the possibility of a point or pixel to belong to a class. This is contrary to that of FCM, where it is the *degree of sharing*. The previous difference in the interpretation of membership values in the case of FCM and PCM is observed as PCM relaxes the constraint on FCM. By overcoming this constraint, a higher accuracy of supervised classification using PCM is obtained as compared to that of FCM. The PCM can also handle noise and outliers which affect the prototype parameters, that is the cluster means (Ibrahim et. al., 2013).

The PCM, as a supervised classifier, works better in the case of untrained classes when compared with FCM (as a supervised classifier). Untrained classes are those classes which are present in the image but are not known to the analyst; hence, the classifier is not trained with that unknown class. Comparing this with the constraint relaxed by PCM, that is if the membership value of a pixel in all the clusters is maximum, then it will give the *degree of belongingness* of a pixel to different clusters (Krishnapuram & Keller, 1996). The FCM is primarily a partitioning algorithm, whereas the PCM is primarily a mode-seeking algorithm. The power of the PCM lies in finding meaningful clusters as defined by dense regions. When the data is not severely contaminated, the FCM can provide a reasonable initialization and a scale estimate. Thus, with a proper choice of the scale and "fuzzifier" parameters, the PCM can be used to improve the results of the FCM.

Similar to FCM, sub-pixel classifications can be generated from these class membership values. FCM and PCM clustering, in essence, are unsupervised classification algorithms, but these may be applied in supervised modes (Ibrahim et al., 2013). This can be done by direct input of the class means (i.e., cluster centers) obtained from the training datasets in a supervised classification process. Further, in the formulation of these methods, a weighting factor m that describes the *degree of fuzziness* is provided.

The fuzzifier determines the rate of decay of the membership value. Figure 7.15 shows the membership function generated by the PCM as a function of m. When $m = 1$ (no fuzziness), the memberships are crisp. When m is reaching ∞, the membership function does not decay to zero at all (complete fuzziness). In the FCM also, $m = 1$ corresponds to the crisp case and m reaching ∞ corresponds to the maximally fuzzy case. However, the interpretation of m is different in the FCM and the PCM. In the FCM, increasing values of m represent increased sharing of points among all clusters, whereas in the PCM, increasing values of m represent the increased possibility of all points in the dataset completely belonging to a given cluster. Thus, the value of m that provides satisfactory performance is different in the two algorithms. The earlier studies have shown that there is no optimal value of m, but a value in the range 1.5–3 can generally be adopted (Krishnapuram & Keller, 1996). A value of $m = 2$ is known to give good results with the FCM. However, as can be seen in Figure 7.15, for this value of m the membership function decays too slowly in the case of the PCM.

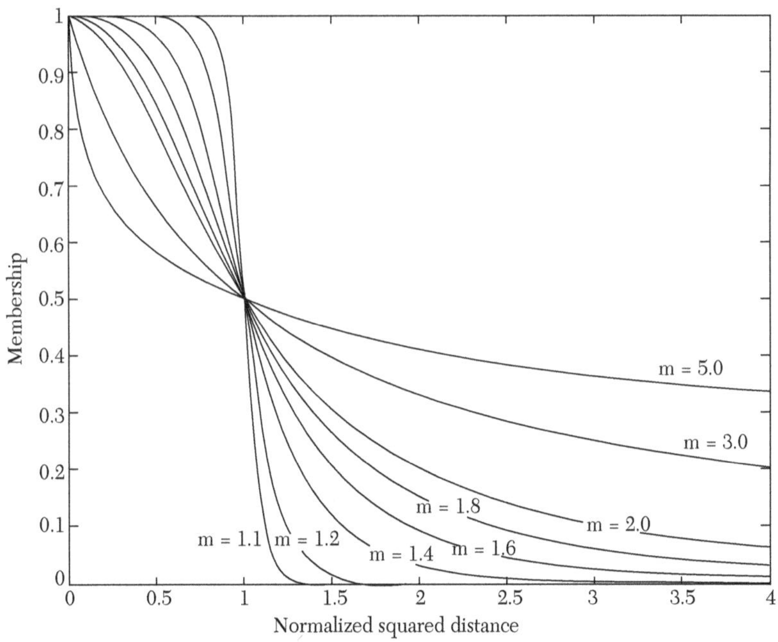

Figure 7.15. Plot of PCM membership function for various values of fuzzifier parameter *m*. (Krishnapuram & Keller, 1996).

7.2.13 K-Nearest Neighbor

The k-nearest neighbo (k-NN) method is used to generalize information from field plots to pixels for map production. The method is highly appreciated because it allows to the prediction of all the measured variables at the same time for each pixel or area of interest. The k-NN classifier is unlike the other classifiers, as each unknown sample is directly compared against the original training data (Maselli et al., 2005). The unknown sample is assigned to the most common class of the k training samples that are nearest in the feature space to the unknown sample. A low k value will therefore produce a very complex decision boundary; a higher k value will result in a greater generalization. Because a trained model is not produced, k-NN classification is expected to require greater resources as the number of training samples increases.

The k-NN technique is implemented in most commercial statistical packages, such as SPSS, and in GIS, such as IDRISI. The source codes implementing the k-NN algorithm are also available online and distributed within open-source frameworks in C++, Java, Visual Basic, and even on the basis of Excel (Chirici et al., 2012). The k-NN algorithm is applied in Weka open-source, a collection of machine learning algorithms for data mining tasks. Tilburg Memory Based Learner (TiMBL) is also an open-source software package implementing k-NN and several other memory-based learning algorithms (Daelemans et al., 2010). In addition, open-source R packages offer nearest neighbor search and a number of different tools facilitating the comparison among different nearest neighbor search algorithms (Crookston & Finley, 2008).

The algorithm performs the assignment by consulting a reference set of labeled patterns (training samples) defined in a n-dimensional input feature space. In particular, the classification of an unlabeled sample carried out by the k-NN algorithm starts with the identification of the subset of $k$training samples that are the closest to it. Various decision strategies can then be adopted to classify the unlabelled sample; the most widely used strategy assigns it to the class that appears most frequently within this subset (Blanzieri & Melgani, 2008). The k-NN algorithm assigns each unknown pixel to the field attributes of the most similar reference pixels for which field data exists. Similarity is expressed as per the distance between the query point and the neighboring ones. Different types of distance metrics can be selected such as Euclidean distance, Manhattan distance, and Mahalanobis distance (Gagliano et al., 2007). The flow diagram of s k-NN classifier used in image classification is shown in Figure 7.16.

The advantages and disadvantages of l-NN are given in Table 7.2. As it overlooks the geometric configuration of the training samples, different approaches have been proposed in the literature to overcome it. For instance, the distance between the unlabeled sample and each of its k-nearest training samples is computed and then used as a weight to influence the global decision process. Other approaches make use of fuzzy sets and Dempster–Shafer theories to better exploit information conveyed by the set of k-nearest training samples.

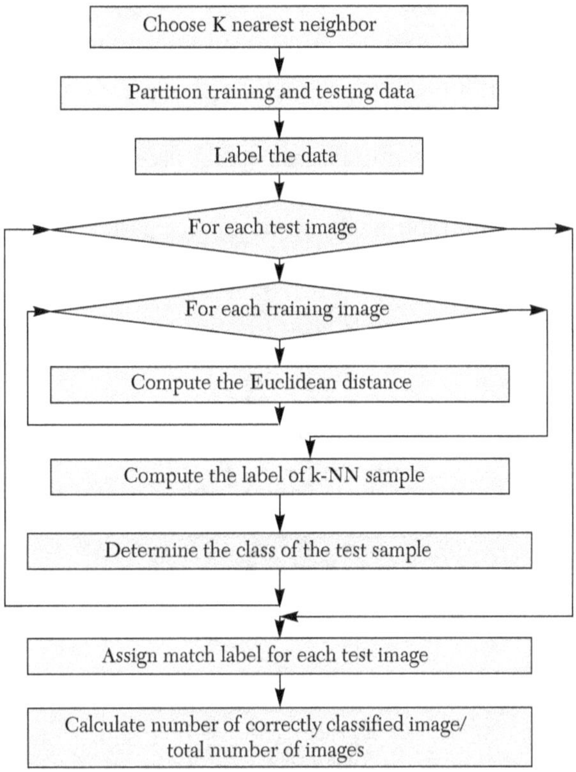

Figure 7.16. Flowchart of k-NN classifier in image classification.

7.2.14 Genetic Algorithm

The genetic algorithm (GA) introduced by John Holland in 1975 is a search optimization algorithm based on the mechanics of the natural selection process (Holland, 1992). The GA is a population-based approach in which members of the population are ranked based on their solutions' fitness. In GA, a new population is formed using specific genetic operators, such as crossover, reproduction, and mutation. It is primarily used in optimization, and is good at refining irrelevant and noisy features selected for classification. GAs have also been used for many machine learning applications, including classification and prediction tasks such as prediction of weather (Mitchell, 1995). These can be used to evolve aspects of particular machine learning systems, such as weights for neural networks, rules for learning classifier systems or symbolic production systems, and sensors for robots. It can be used in feature classification and feature selection, and can handle large, complex, non-differentiable, and multimodal spaces.

A basic GA has five main components: a random number generator, a fitness evaluation unit, a reproduction process, a crossover process, and a mutation operation, as shown in Figure 7.17. Reproduction selects the fittest candidates of the population, while crossover is the procedure of combining the fittest chromosomes and passing superior genes to the next generation, and mutation alters some of the genes in the chromosomes. Different chromosomes constitute a group. The GA searches through a

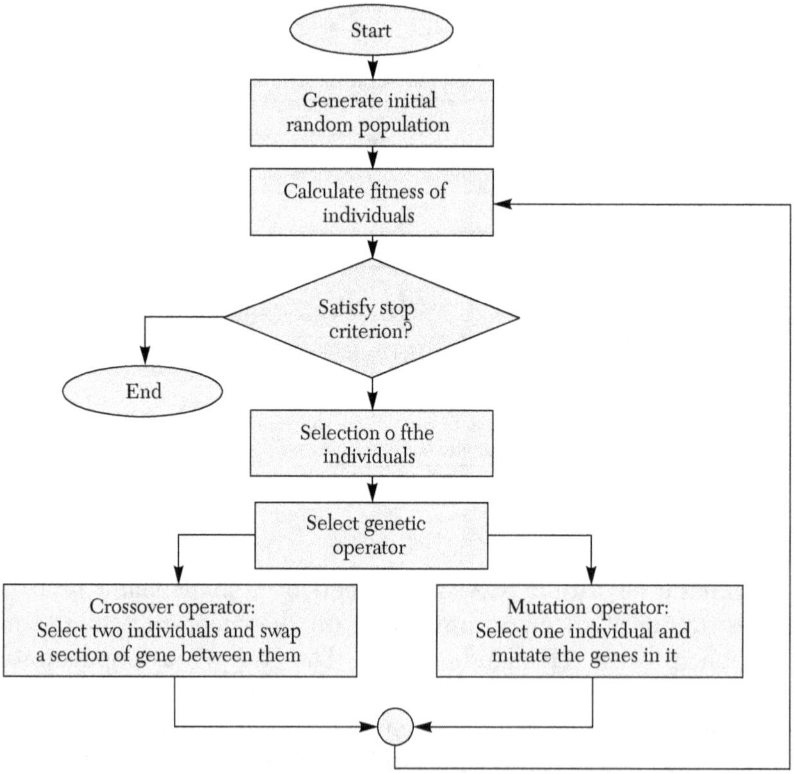

Figure 7.17. General flowchart of the genetic algorithm.

space of "chromosomes," each of which represents a candidate solution to a given problem (Mitchell, 1995). Before searching, an object is assigned a string (chromosome) composed of specific symbols and arranged in a certain order through a "coding mechanism" (Lin & Zhang, 2020). The fitness of the chromosome depends on how well that chromosome solves the problem at hand. Common coding mechanisms include binary, gray, and real number coding mechanisms. Binary coding is one of the most commonly used coding mechanisms, which solves the problem with a string consisting of binary symbols 0 and 1 (Zhou & Sun, 1999). This encoding mechanism is simple and easy to operate, and follows the principle of symbol minimum character set encoding.

The GA is based on the assumptions that computation or development of a scoring function is nontrivial. It maintains a set of candidate solutions in each iteration and selects better individuals from the population according to certain criteria. Thereafter, it generates a new generation of candidate populations through selection, crossover, and mutation. It repeats this process until a certain convergence condition is satisfied (Jian & Zhao, 2009). The *selection operation* refers to the operation of selecting excellent (i.e., high fitness) individuals from the population for the next generation group according to a certain probability. The *crossover operation* refers to the operation of exchanging partial genes between two paired chromosomes in a certain manner, according to the crossover probability to generate two new individuals (Mitchell, 1995). The *mutation operation* refers to the operation of replacing certain gene values of an individual's coding string with other gene values, according to the mutation probability to generate a new individual.

The GA most often requires a "fitness function" that assigns a score (fitness) to each chromosome in the current population (Lin & Zhang, 2020). This function is used to measure the advantages and disadvantages of an individual; the greater the value, the better the individual's quality. The design of a suitable fitness function can reduce optimization time and improve optimization efficiency. The fitness function is the driving force of the GA process, and must be designed according to the requirements of solving the problem. The solution (i.e., model parameters) is represented by an encoding method. Multiple initial solutions are randomly generated and the search begins. The fitness function guides the search direction and a solution is selected, exchanged, and mutated to realize the evolution of the solution. Thus, the optimal solution of the problem is obtained (Holland, 1992).

Image classification techniques ranging from maximum likelihood to neural networks depend on the feature vectors formed by the DN values in each spectral band for each pixel. The features of its neighborhood, like texture, or the average value of nearby pixels, are also necessary to get good spectral information of a pixel. Hence, the hybrid GA may be used to choose these features automatically (Zhou & Sun, 1999).

7.2.15 Artificial Intelligence

Artificial Intelligence (AI) is a term coined in 1950 by Alan Turing, a mathematician who has laid the foundations of computers for the modern age (Turing, 1950). AI is a branch of computer science where machines are programmed and given a cognitive ability to think and mimic actions like humans and animals. The benchmark for AI is human intelligence regarding reasoning, speech, learning, vision, and problem-solving (Garg, 2021). This simulation of human intelligence is measured by the machine's ability to predict, classify, learn, plan, reason, and/or perceive. AI-based works aim to take human approaches to tasks and build upon them, so that the resulting machine can perform these tasks or even do better than humans (Jung et al., 2021).

The main difference between an AI algorithm and a traditional computer algorithm is the way the rules are created in both. In a traditional algorithm, a developer sets the specific rules that define an output for each type of input to the software, whereas AI algorithms are designed to build out their own system of rules, rather than having those rules defined by a developer. Through this approach, it has become possible for computers to start doing AI-based tasks that are traditionally dependent on humans, such as customer service, playing chess, face recognition, video search, textual analysis, or driving cars (Sisodiya et al., 2020).

Machine learning (ML) is a subset of AI, while deep learning (DL) is a subset of ML, and neural networks make up the backbone of DL algorithms (Garg, 2021). Essentially, AI is an umbrella term for any technology that can take up complex problems, like human beings. The interlink between AI, ML, and DL is an important one, and it is built on the context of increasing complexity, as shown in Figure 7.18. AI covers all the works within the field of machine intelligence and describes how ML results are exploited for further use. Typically, it includes recognizing and getting aware of typical situations, making decisions based on the recognized high-level parameters, and predicting future developments. AI examples include: an automated assistant, Google search, recommendations on Netflix or Amazon, or finding a great deal. Every major company and service today is incorporating AI into their system in some form or other. The AI research is most prevalent in areas where machines were previously incapable to make progress, such as communication, reasoning, or predicting an event.

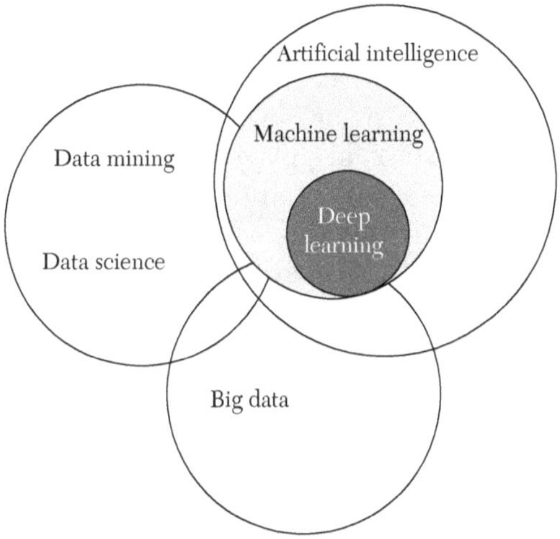

Figure 7.18. Relationship between AI, ML, and DL.

AI has three different levels as follows and as shown in Figure 7.19:

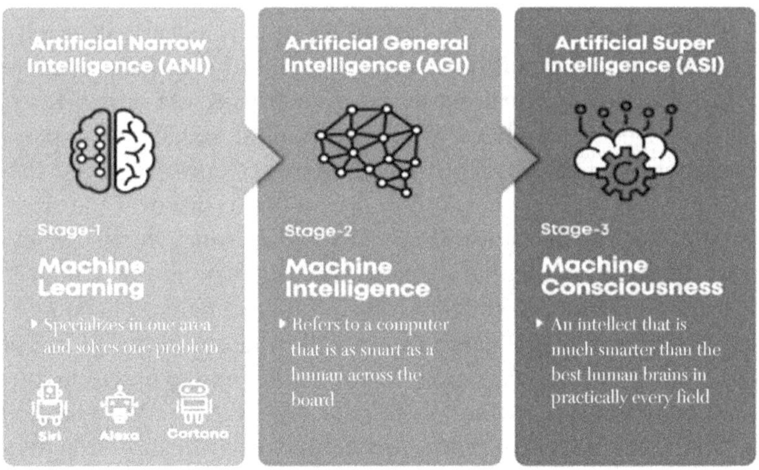

Figure 7.19. Three levels of AI.

1. *Artificial Narrow Intelligence (ANI):* It is said to be narrow AI when the machine can perform a specific task better than a human.

2. *Artificial General Intelligence (AGI):* The AI reaches the general state when it can perform any intellectual task with the same accuracy level as a human being would do.

3. *Artificial Super Intelligence (ASI):* An AI is active when it can beat humans in many tasks.

ANI is considered to be "weak" AI, whereas the other two types (AGI and ASI) are classified as "strong" AI. Weak AI is defined by its ability to complete a very specific task, like playing a video game or face recognition. Almost all AI applications today can be classified as weak AI. This means that the algorithm is only learning what it has been told to learn within an extremely specific field. For example, a virtual assistant can identify verbal commands with ease and can also learn to recognize new, previously unheard variations of these commands. In AGI and ASI, the incorporation of more human behaviors becomes prominent, such as the ability to interpret tone and emotions. Chatbots and virtual assistants, like Siri, are being used frequently, but they are still examples of ANI. The AGI would perform at par with human being, while ASI would surpass a human's intelligence and ability. The ultimate end goal of AI research is to build similar machines, often referred to as strong AI, which would theoretically be able to learn new tasks independent of humans' instructions.

Satellite images are really benefiting from applications of AI and ML to train algorithms to interpret the images. With Big Data processing and analysis, the machines can easily be trained using AI and ML to analyze satellite images at a global scale (Sisodiya et al., 2020). Examples include the generation of radiometrically and geometrically calibrated data cubes or using an automated process of data mining to discover the desired data and also detect artifacts or outliers in the data sets, which are beyond the capabilities of human beings, due to the large volume of satellite images. The discovered and selected datasets are further analyzed in detail by an AI-based algorithm for extracting the specific characteristics of the observed objects. Thus, the results of analysis may contribute to updating the existing models or building new models for the observations. Visualization of the model parameters or extracted information is a verification step to deal with the large complex data volumes. Specific evaluation paradigms are needed to build trust in the obtained results to be used to make predictions. The process is iterative, and when new data are acquired, they will be analyzed further using the capabilities of AI. Various processes associated with AI are shown in Figure 7.20.

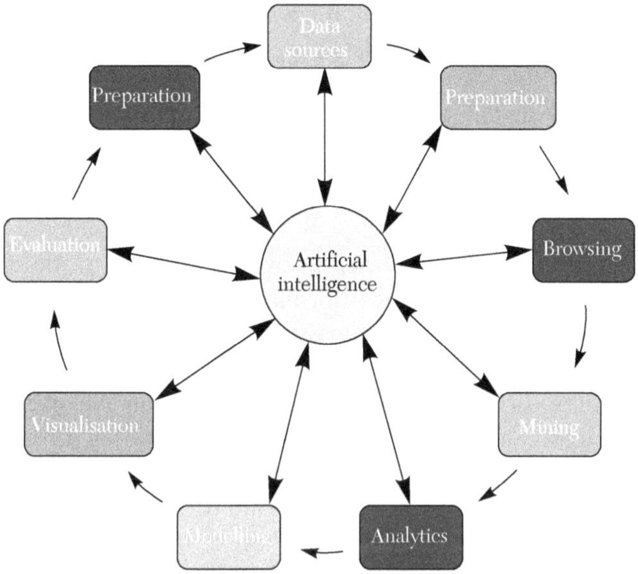

Figure 7.20. AI and various associated processes.

Today remote sensing satellites are capable of sending terabytes of images of the Earth's surface on daily basis, supporting AI-driven applications that provide previously inaccessible insights on global-scale economic, social, and industrial processes. The amount of data generated from satellites is growing exponentially, but the current open-source AI tools are not optimized for satellite imagery. The heavy expense of imagery analytics and interpretation is still often performed by human analysts. The increasing amount of data being produced from satellites, improvements in camera technology, and improvements in data storage and transfer capabilities are garnering increased attention in the geomatics community (Koga et al., 2020). The computer vision technique applied to satellite imagery is one of the most promising applications of AI in space technology today. In the Earth observation industry, DigitalGlobe is a good example of AI's satellite image applications. Companies such as Orbital Insight, Spaceknow, and Rezatec are delivering a wide-range of AI-driven solutions on top of the satellite imagery.

In the field of satellite image analysis, AI-driven satellite data applications are in demand for a number of important reasons (Garg, 2021):

1. The rapid advancements in machine vision in last few years have made challenging tasks easy (such as identifying cars, buildings, or changes over time) that can be rapidly done by machines.

2. With increasing use of satellite images, camera technology enhancement, improved data storage, and data transfer, there is an exponential increase in amount of data being produced from various satellites that can easily be handled and analyzed by AI and cloud based technologies.
3. Satellite image analysis and interpretation performed by human analysts is time-consuming and costly.

TAI applications for satellite imagery are expanded into multiple levels for ML training to extract useful information for developing AI models. AI models are developed through the use of ML or DL technology monitored by satellite images. They need a huge amount of training data for detecting various objects for mapping and monitoring. The AI applications using satellite images can learn to predict various scenarios after analyzing the situation from satellite imagery datasets that are specially created for AI models. These satellite images can be used to in AI to automatically detect various objects, like agricultural fields, forest, buildings, water, and various other man-made structures (Jung et al., 2021).

The AI using satellite images can be used in two broad ways; "single-level" applications and "multilevel" applications (Komissarov, 2019). The first level of application of satellite data uses computer vision to detect various objects, like buildings, roads, open land, and forest land, all of which are important in many applications. Such applications are mainly developed for change detection, such as crop, land use, urban infrastructure, and so on. Objects can easily be monitored from high-resolution satellite images and detected through AI-based algorithms, accomplished by a number of integrated CNNs, such as Faster R-CNN, and YOLO (Koga et al., 2020) for detecting the cars parked in the parking area, crop mapping, yield prediction, and monitoring the urbanization activity of any region of the globe. Currently, many CNN-based solutions exist that have pretrained models. Several CNNs, such as ResNet, VGG (Visual Geometry Group), and Inception, have been trained previously in publicly available generic labeled image datasets such as the image sets in ImageNet. Through transfer learning techniques, these pretrained CNNs are extended, and then several layers are trained further with the specific new labeled dataset for deriving the new solution. Thus, the AI-based new models are tuned for the new object detection. Several objects in satellite imagery are often very small in size, while input images are large in number (several megapixels), and also there is a relative shortage of training data.

Several methods are not optimized to detect very small objects in large satellite images and often produce inaccurate results. The algorithm required for change detection also needs extensive ground truth data, which is a complex and time-consuming task to collect. It becomes more complex when data are to be analyzed at a global scale. At such a scale, normalizing the satellite images is another big challenge related to contrast difference due to weather and natural changes in lighting conditions.

Figure 7.21. AI-based automated car detection in a parking lot (Komissarov, 2019).

In multilevel applications, information extracted from satellite imagery in single-level applications is supported with data from other sources for further application, for example to count the cars in the parking zone which have already been detected in a single-level application (Koga et al., 2020). These images are annotated and used as training data to develop an AI-based algorithm to accurately count the cars on the satellite imagery at parking lots across many cities or the globe. The number of cars in the parking lot may be construed as an indirect indication for retailers' revenue (more cars means more customers and thus more revenue). This data when used with other sources (e.g., socioeconomic) can predict retailers' profits. Another application of use of satellite image data with AI-based models in multilevel applications is crop yield and price prediction. The NDVI indices are critical for agriculture-related models that can be obtained using open-source satellite images, such as Landsat and Sentinel. The AI models based on vegetation indices and information such as weather and soil-related data have proven very successful for the estimation of yields (Jung et al., 2021). The crop yield data supported with supply-demand analysis and other information can be used to create a model to forecast the price, which is useful to commodity traders, agro-based industry, insurers, government, market research firms, farmers, and so forth. The high-value crops may justify a higher technology investment from AI-related products in the near future. In multilevel applications, information extracted from satellite imagery from single-level methods could be one input to a more complex ML system supported by AI.

Agriculture-based applications can benefit from AI technologies because of their potential to leverage big data, which is now becoming easily accessible through the use of UAVs/drones (Jung et al., 2021). The UAVs provide ample opportunity to apply advanced analytics to their data for managing agricultural systems, thus improving the resiliency and efficiency of crop production systems. In addition to the quality of data, when using a large dataset to train the AI models, good accuracy can be achieved even when noisy data is involved, suggesting that the volume of training data is essential in developing robust AI models.

Table 7.7 summarizes the key parameters for some AI based-applications: (*a*) required imagery; (*b*) features, extracted from imagery; (*c*) data, augmenting imagery; (*d*) customers; and (*e*) companies, delivering one or another application. As is seen from this table, very-high resolution imagery (1 m) is required for most of the commercially viable applications. At present, there are only few satellites providing such high-resolution data, but these data are expensive, so it limits the new start-ups comeing up. Table 7.2 lists some advantages and disadvantages of AI-based approaches.

Table 7.7. Key Parameters for Some AI-Based Applications (Komissarov, 2019)

Applications	**Resolution (m) of images**	**Objects detected**	**Other data required**	**Companies/ AI system used**
Retailers' profits prediction	0.3–1.0	Number of cars in parking lots	Socio-economic data	Orbital Insight SPACE_KNOW RS Metrics
Crop yield prediction	5–25	NDVI	Soil related data, Weather data	telluslabs Descartes Labs
Estimation of economic activity	0.3–1	Construction rate, number of cars, aircraft, etc., nighttime luminosity	–	Orbital Insight SPACE_KNOW
Poverty prediction	0.3–1	nighttime luminosity	–	Orbital Insight SPACE_KNOW

7.2.16 Machine Learning Classifier

The roots of machine learning (ML) in remote sensing date back to the 1990s. The major focus of ML is to extract information from data automatically by computational and statistical methods, and develop techniques that allow computers to "learn." It was initially introduced as a way to automate knowledge-base building for remote sensing. Thereafter, ML was soon adopted as an important tool in remote sensing. Now it is being used in all types of works, ranging from an unsupervised satellite image scene classification (Li et al., 2016) to the classification of agricultural crops (Chaturvedi, 2018). For example, Huang and Jensen (1997) described how a knowledge base can be built using minimal input from human experts, and then decision trees are created to infer the rules from the human input for the expert system. Major leaps and bounds in machine vision in the last 5–10 years have made challenging tasks, such as identifying cars, buildings, or changes in landscape over time, easier by machines.

The ML classification has become a major focus in remote sensing applications (Maxwell et al., 2018) and is widely accepted, particularly in operational land cover mapping. The algorithms have proved to be a powerful tool for analyzing satellite imagery of any resolution and any size. The ML algorithms are generally able to model complex class signatures and do not make assumptions about the data distribution (i.e., are nonparametric). It is generally found that these methods tend to produce higher accuracy as compared to traditional parametric classifiers, especially for complex data with a high-dimensional feature space, that is many predictor variables (Pal & Foody, 2012). Despite the increasing acceptance of ML classifiers, parametric methods, particularly maximum likelihood, still appear to be commonly used methods and remain popular tools for image classification.

There are three broad classes of ML algorithms: supervised, unsupervised, and reinforced machine learning. The difference between supervised and unsupervised is that while using supervised algorithms, there is an output column in a dataset, whereas while using the unsupervised algorithms, a huge dataset is used, and it is the function of the algorithm to cluster the dataset into various different classes based on the relation it has identified between different records. Analysts have to look at the data and then divide it based on the algorithms without having any training. There is no target or outcome variable to predict or estimate. Supervised learning is further divided into two groups: classification and regression (Huang & Jensen, 1997). With supervised training, the training data contains the input and target values. The algorithm picks up a pattern that maps the input values to the output and uses this pattern to predict values in the future. Unsupervised learning, on the other hand, uses training data that does not contain the output values. The algorithm figures out the desired output over multiple iterations of training.

The new models generally developed for predicting/classifying the images require a big dataset for training and testing the results of the algorithm. Normally, about 75% of the data is used for training and 25% is used for evaluating the performance of the model after it has been trained; 75:25 is not a fixed ratio, and it can be changed depending on the dataset being used (Breiman, 1996). However, it has to be ensured that the training segment of the dataset have an unbiased representation of the whole dataset and that it is not too small as compared to the testing segment of the dataset. The dataset should not have only one class present but have almost every possible class so that the model is trained over every possible kind of input. The training dataset should not be too small, as it may not give reliable predictions if the model has not been trained over every different type of input. Overfitting the model generally results in making an overly complex model. If the same type of data (the type of data on which it has been trained) is used for evaluating the model performance, a very high prediction/classification accuracy may be achieved. However, if the inputs are modified (different from what the model has used before), the prediction/classification accuracy is low. Overfitting can therefore be done by using a bigger dataset and segmenting the dataset properly. Additionally, it is beneficial to reduce the complexity of the model so that all the extreme cases are not classified.

The third class of the ML algorithm is reinforcement learning, which has a slightly different concept, as the algorithm takes decisions in that environment and the user provides the algorithm. It keeps on improving itself with each decision based on the feedback it gets for its last decision. The algorithm of reinforcement learning is rewarded for every right decision made, and using this as feedback, the algorithm can build stronger strategies. Reinforcement learning is a learning algorithm that is applied to intelligent agents so that they can adjust to the conditions in their environment, and it is achieved by maximizing the value of the "reward" points that can be achieved. This type of algorithm teaches how to deal with a problem, an action that has an impact. The most common example is a self-driving car; the machine is asked to avoid collisions or violations of traffic rules (Chaturvedi, 2018). If an accident or violation occurs, then the machine will be given a negative reward; if the machine does it right, it will be given a positive value. From there, the machine will learn to drive a car and improve further.

There are multiple famous ML algorithms in use today: SVM, NN, RF, k-NN, ANN, DT, k-means, PCA , and so on. (Cross & Jain, 1983). Table 7.8 lists the default parameters whose optimum values are needed to ensure confidence that the best possible classification has been produced. The relative difficulty of running parameter optimization for different classifiers is often cited as a major consideration in selecting an algorithm.

Table 7.8. User-Defined Parameter Examples (Maxwell et al., 2018)

Algorithm	User-defined parameters
k-nearest neighbor	Number of neighbors considered (k)
MLP ANN with back-propagation	Activation function, number of hidden layers, number of nodes in each layer, learning rate, momentum factor, weight initialization, number of iterations
Decision trees	Pruning parameters (cp)
Boosted decision trees (Adaboost method)	Number of iterations or trees (n)
Random forests	Number of trees (n), number of variables randomly sampled as candidates at each split (m)
Support vector machines using polynomial kernel	Cost or slack parameter (C), kernel type, kernel-dependent parameters: polynomial order (p), radial basis gamma (γ)

The growth in ML algorithms proves that learning features from datasets are more efficient and practical than defining the features. Due to the character of "learning by itself," various datasets are being used in several applications such as microwave, LiDAR, DEM, satellite remote sensing, and the fusion of multisource and multi-resolution data. Mostly, these algorithms work well to construct the relationships between inputs and outputs. The ML algorithm does not depend on statistical models; rather, it depends on the relationship between the response data and the learning classifier. Figure 7.22 shows a flowchart for using an ML algorithm.

Selecting an ML classifier for a particular task is challenging, not only because of the wide range of machine-learning methods available but also because the literature appears contradictory, making it difficult to generalize regarding the relative classification accuracy of individual ML algorithms (Maxwell et al., 2018). Several ML algorithms are a black box type, so it is not possible to look at an algorithm and decide theoretically which one will yield the best results for the dataset. This means that it is

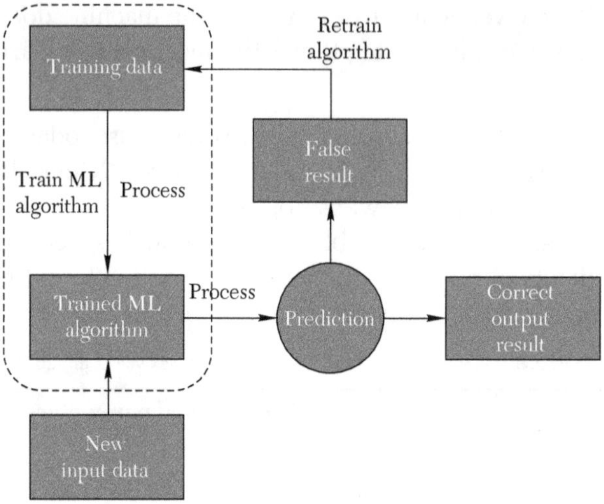

Figure 7.22. Steps used in a machine learning algorithm.

hard to assess how the algorithm is arriving at a specific result. Therefore, it is useful to first narrow down the choice of the algorithm based on the type of problem, and then apply those algorithms on a part of dataset and assess which one performs best. There are however two problems that always occur: (*a*) most of these ML algorithms are based on the given variables and lack the ability to extract information, and (*b*) the performance of ML algorithms is extremely case specific. Lawrence and Moran (2015) compared the performance of a wide range of ML classification algorithms using consistent procedures, and found that the RF had the highest average classification accuracy, 73.19%, significantly better than that of the SVM classifier. The RF is a kind of superior performance of ML algorithms, which uses a multiple tree classifier. Table 7.2 shows the advantages and disadvantages of ML-based classifiers. Table 7.9 presents the details of various traditional ML models with their potential applications.

Table 7.9. Various Traditional Machine Learning Algorithms with Applications

Processing stage	Machine learning methods	Application
Preprocessing	GMM, ANN	Denoising
	PCA, LLE (locally linear embedding), ISOMAP (isometric mapping)	Dimensionality reduction

Processing stage	Machine learning methods	Application
Processing/ analysis	Sparse coding	Sparse representation
	HOG (histogram of oriented gradients), SURF (speeded up robust features), SIFT (scale-invariant feature transform), decision trees, random forest, genetic algorithm, HMRF (hidden markov random field), SVM, MRF.	Feature selection
	Decision tree, multilayer perceptron, logistic regression, SVM, k-means, GP (Gaussian process), NN, ARIMA (autoregressive integrated moving average).	Segmentation and classification
	k-NN, SVM, SOM, GMM, k-means, fuzzy clustering, hierarchical clustering, hybrid clustering	Clustering
	Image transformations, correlation analysis	Change detection

Most of the ML algorithms are now widely employed in remote sensing fields by developing new algorithms. The problems in mapping land use and land cover, which seemed difficult and sometimes impossible, are being solved by new algorithms. It is not far-fetched to say that most analysis work done in the world today will be done by ML algorithms in the near future.

7.2.17 Deep Learning Classifiers

Machine learning research stems from the idea that a computer will have the ability to learn, as a human would do, without being explicitly programmed, while deep learning (DL) which is a subset of machine learning, refers to the application of a set of algorithms and their variants (Sisodiya, 2020). TDL has gained popularity over the last decade due to its ability to learn data representations in an unsupervised manner and generalize to unseen data samples using hierarchical representations. The DL algorithms, which learn the representative and discriminative features in a hierarchical manner from the data, have recently become in demand in the ML area (Maa et al., 2019).

The concept of DL originates from the ANNs. Prior to the development of DL, remote sensing work focused on NNs, SVMs, and RF for image classification and other tasks (e.g., change detection). The SVM received much attention due to its ability to handle high dimensionality data and perform well with limited training samples (Mountrakis et al., 2011), while RF gained popularity due to its ease of use (e.g., relatively insensitive to classification parameters) and generally high accuracy (Belgiu & Drăguţ,

2016). It has been introduced in remote sensing for big data analysis. The DL algorithms now have applications such as image preprocessing, pixel-based classification, image analysis tasks including land use and land cover classification, target recognition, high-level semantic feature extraction, and scene understanding. In addition, DL has been used in several other applications including fusion, segmentation, change detection, and registration.

Several commonly used DL models in remote sensing include the supervised CNN and recurrent neural network (RNN) models, and unsupervised autoencoders (AE) and deep belief networks (DBN) models, also including the recently popular generative adversarial network (GAN) model, which have been explained in Maa et al. (2019). Table 7.10 presents some DL methods, data requirements, their applications, and advantages. Despite its great potential, DL cannot be directly used in many remote sensing tasks, with one obstacle being the large numbers of bands, especially hyperspectral images.

Table 7.10. Some DL Methods and Their Characteristics (Sisodiya et al., 2020)

DL method	Data type	Application	Advantages
CNN	SAR, HIS, multispectral images, LiDAR data, RGB, pan-sharpening images	Image segmentation, urban area (building, road, etc.) classification and extraction, urban growth prediction, road extraction, oceanographic target classification, thematic classification, automatic target detection, super-resolution for sea surface temperature analysis, object detection and recognition, 3D object classification, human detection and activity classification, semantic labeling multisource earth observation data, anomaly detection, disease recognition in plants, denoising, poverty, multiscale classification	Automatic feature extraction for learning, from large labeled dataset provided as input. Easy to train. A pretrained model can be prepared using CNN and can be used for transfer learning. The CNN can be combined with other types of DL architectures for performance enhancement of the those architectures, for example, recurrent CNN, convolutional LSTM (long short-term memory), and so on.

DL method	Data type	Application	Advantages
RNN	Spatiotemporal and time-series data	Ocean and weather forecasting, tracking for multiple objects	Used for sequential data.
Recursive Neural Network	Spatiotemporal data	Precipitation nowcasting	The presence of memory unit helps it to work well with sequential data.
Deep Belief Network (DBN)	Polarimetric SAR, HSI, high-resolution image, time series data, radar data, spatiotemporal data	Classification, drought index prediction, object detection, prediction of traffic flow, urban LULC	Can be used with limited available labeled dataset.

DL can represent and organize multiple levels of information to express complex relationships between the data. In fact, DL techniques can map different levels of abstractions from the images and combine them from a low level to a high level, as shown in Figure 7.23. Considering the low-level features (e.g., spectral and texture) as the bottom level and the output feature representation from the top level of network, it can be directly fed into a subsequent classifier for pixel-based classification (Zhang et al., 2016). The DL algorithm in this figure includes three main components, the input data, the core deep networks, and the expected output data. In practice, the input–output data pairs are dependent on the particular application. For example, for target recognition and scene understanding, the inputs

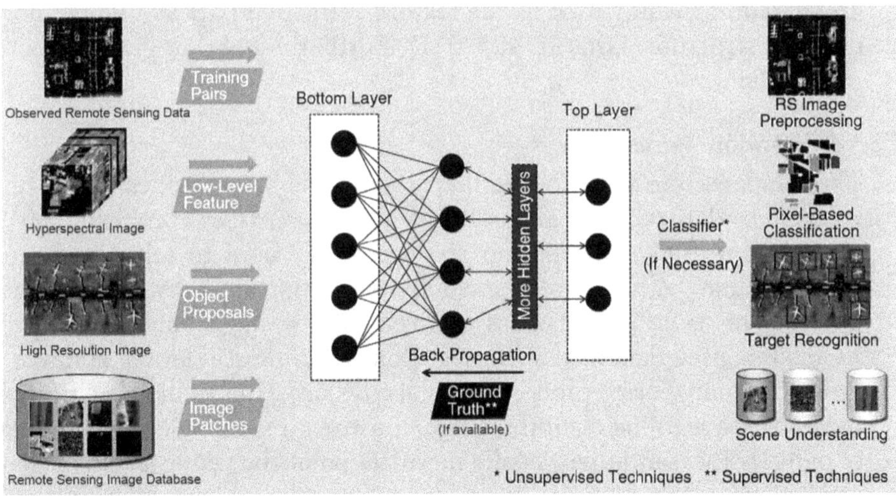

Figure 7.23. A general framework for DL in remote sensing data analysis (Zhang et al., 2016).

are the features extracted from the object proposals, as well as the raw pixel DN values of high-resolution images and remote sensing image databases, respectively. If sufficient training sample sets are available, such a deep network turns out to be a supervised approach. It can be further fine-tuned by the use of label information, and the top-layer output of the network is the label information rather than the abstract feature representation learned by an unsupervised deep network. When the core deep network has been well trained, it can be employed to predict the expected output data of a given test sample.

There are various tools and platforms available to experiment with DL (Bahrampour et al., 2016). The most popular are: Theano, TensorFlow, Keras (which is an application programming interface [API] on top of Theano and TensorFlow), Apache Spark, Caffe, PyTorch, TFLearn, Pylearn 2, and the Deep Learning MATLAB Toolbox. Some of these tools (i.e., Theano, Caffe) incorporate popular architectures, such as the AlexNet, VGG, and GoogleNet, either as libraries or classes. The two devoted frameworks,TensorFlow and Apache Spark, have been explained in this chapter under Section 7.3, Free and Open-Source Software.

A summary of well-known DL methods built in recent years is presented in Figure 7.24. These networks often intertwine, and many adaptations have been proposed for them. Although it may appear that most of the DL methods were developed during 2015–2017, it is important to note that some novel deep networks use most of the already developed methods as backbones, or accompanied from other types of architectures, mainly used as the feature extraction part of a much more complex structure (Osco et. al., 2021). Further details are given in Khan et. al. (2020).

7.2.18 Random Forest

The random forest (RF) classifier yields reliable classifications using predictions derived from an ensemble of decision trees (Breiman, 2001). Furthermore, this classifier can be successfully used to select and rank those variables with the greatest ability to discriminate between the target classes. This is an important asset given that the high dimensionality of remotely sensed data makes the selection of the most relevant variables a time-consuming, error prone, and subjective task. Ensemble classification methods are learning algorithms that construct a set of classifiers instead of one classifier, and then classify new data points by taking a vote of their predictions.

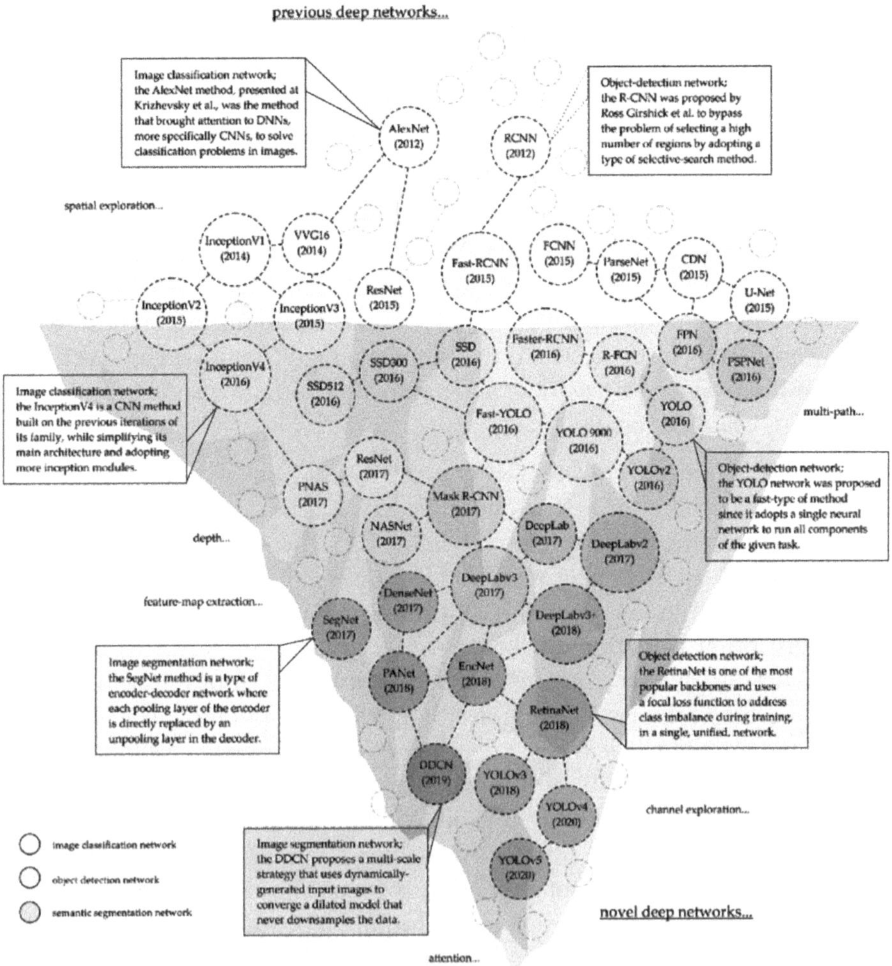

Figure 7.24. Time series of DL approaches with some popular architectures implemented in image classification (Osco et al., 2021).

Ensemble classifiers can be based on an individual supervised classifier or on a number of different supervised classifiers that are trained using Bagging or Boosting approaches, or variations of these approaches (Belgiu & Drăguţ, 2016). *Bagging* can be defined as a technique that is applied for prediction functions so as to reduce the variance of such functions. Breiman (1996) introduced the idea of bagging, which is short for "bootstrap aggregating." In the bagging algorithm, many bootstrap samples are drawn from a training dataset with replacement to learn a classifier, and a tree is constructed for each bootstrapped sample such that the successive trees are constructed independently from earlier trees, and a simple majority vote is taken for prediction (Gislason et al., 2006). In bagging, it has been proven that as the number of predictors increases, accuracy also increases until a certain point at which it drops off. The optimal number of predictors are generated that will yield the highest accuracy. For example, Pal and Mather (2003) were able to increase classification accuracy of remotely sensed data by bagging using multiple decision trees.

On the other hand, *boosting* uses iterative retraining, and the weights of incorrectly classified samples are increased as the iterations progress to make them more important in the next iterations. Boosting generally reduces both the variance and the bias of classification, and in most cases, it is considerably more accurate than the bagging classifiers. However, it has some disadvantages; it is slow and can overtrain, and also it is sensitive to noise (Gislason et al., 2006).

The optimization of an RF is mostly based on two parameters, mtry and ntrees, where *mtry* is the number of variables used in splitting a node and *ntrees* is the number of trees in an RF. The optimal value of mtry is determined by traversing all possible values. The generalization error of the RF converges as the number of trees increases, a characteristic absent in most other classifiers. In other words, the model performs better and better as the value of ntrees increases. Therefore, the optimization of ntrees is to balance the classification accuracy with computational effectiveness. The RF classifier is less sensitive than other streamline machine learning classifiers to the quality of training samples and to overfitting, due to the large number of decision trees produced by randomly selecting a subset of training samples and a subset of variables for splitting at each tree node. The most common recommendation is to set the ntree parameter to 500 and mtry to the square root of the number of input variables. The RF classification is robust to high data dimensionality (Belgiu & Drăguţ, 2016).

The RF is based on learning method which operates by constructing many decision trees. The tree construction in the RFs does not depend on the previous trees, but these are created independently by using bootstrap aggregation of the dataset. In general, a simple RF model is usually developed by using a low number of input variables so as to have them split randomly at each node (Breiman & Cutler, 2004). Sometimes, a slightly different splitting criteria, known as the Gini index, is used in place of information gain for trees in an RF. The development procedure of RF-based prediction models usually establishes a new set of values that are equal to the size of the originally spotted data. Each tree gives a classification, and that the tree "votes" for that class. An RF selects several samples randomly (with bootstrap) to build a decision tree (without pruning) every iteration and constructs the whole RF by building numerous decision trees. Then, all trees "vote" for the most popular class, the output. Figure 7.25 shows the general flow diagram to use the RF classifier.

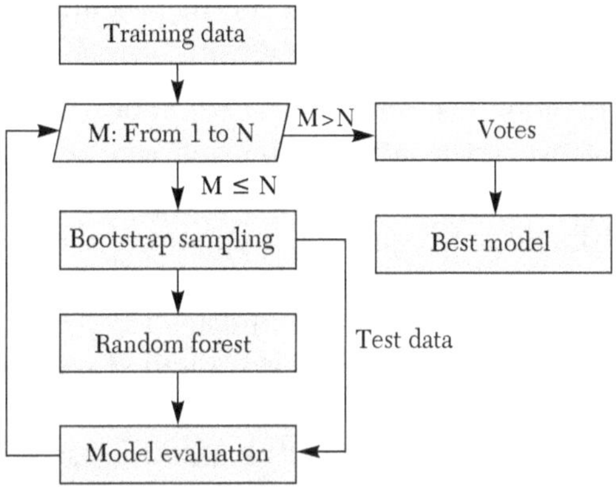

Figure 7.25. Method for using the RF classifier.

The RF uses a large number of decision trees which directly overcome the problem that any one tree many not be optimal, but by incorporating many trees, a global optimum may be obtained. Instead of splitting each node using the best split among all variables, RF splits each node using the best among a subset of predictors randomly chosen at that node, subject to a minimum error rate. Eventually, an average of the aggregating predictors is taken for regression prediction, while the majority vote is taken for prediction in classification (Liaw & Wiener, 2002). A new training dataset is created from the original data set with replacement. Then, a tree is grown using random feature selection. Grown trees are not pruned. The final decision is made based on the majority of the trees, and is chosen by the RF. The data not used in training are known as the out of bag (OOB) data, and can be used to provide an independent estimate of the overall accuracy of the RF classification. Furthermore, by systematically comparing the performance of the trees that use a specific band, and those that don't, the relative importance of each band can be evaluated (Maxwell et al., 2018).

RF-based feature selection (permutation) has been widely used in many domains, and is robust for variables involved with high dimension and high order correlation. In order to reduce the dimension of the data, the random forest cross-validation (RFCV) function for feature selection is widely used for the determination of the variable number. This function sequentially reduces the number of predictors (ranked by variable importance) via a nested cross-validation procedure (Belgiu & Drăguţ, 2016). In an ensemble classifier, the premise is that combining the ensemble classifiers is often more accurate than any one from the ensembles, thus avoiding the conflicts among the feature subsets. As a result, the RF classification is widely used in processing remotely sensed imagery. The RF algorithm has been used in many data mining applications, however, its potential is not fully explored for analyzing the remotely sensed images.

The RF technique can also handle big data with numerous variables. The RFs do not over fit as a predictor, and run fast and efficiently when handling large datasets, which gives it a superior predictive performance. Furthermore, RFs do not require assumptions of data distribution through the trees. Moreover, RFs can handle both continuous and discrete variables (Breiman, 2001). The RF presents estimates for variable importance, that is neural nets. They also offer a superior method for working with missing data, as missing values are substituted by the variable appearing the most in a particular node. They can automatically balance datasets when a class is more infrequent than other classes in the data. These methods also handle variables fast, making them more suitable for complicated tasks. Another advantage of RF is that because of the presence of multiple trees, the individual trees need not be pruned. Among many advantages of RF, the significant ones are: higher accuracy among current algorithms, efficient implementation on large datasets, and an easily saved structure for future use of pre-generated trees (Belgiu & Dragut, 2016). The RF model is generally less sensitive to parameter settings than some other predictors. The advantages and disadvantages of RF are given in Table 7.2. Figure 7.26 shows the results of various classifiers into four distinct classes: water, soil, forest, and agriculture. A comparison of the results can be found in Lowe and Kulkarni (2015).

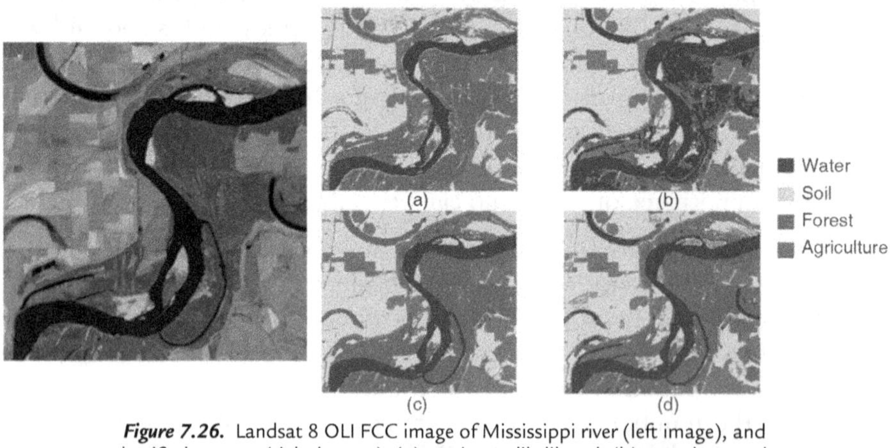

Figure 7.26. Landsat 8 OLI FCC image of Mississippi river (left image), and classified outputs (right images): (a) maximum likelihood, (b) neural network, (c) support vector machine, and (d) random forest (Lowe and Kulkarni et al., 2015) (please see color figures in companion files).

The main RF classification modeling tools are: the Waikato Environment for Knowledge Analysis (Weka), Fortran, R language, MATLAB, and EnMAPBox (Lowe & Kulkarni, 2015). The Weka is a free tool based in the environment of Java open-source machine learning and data mining software (see Section 7.3). The original RF algorithm is in Fortran language. The R language enables a statistical analysis of the language environment, and free source code with open software (explained in Section 7.3). The R language has two software packages that can run RFs, namely Random Forest and 4.6-7 party. The RF program based on the MATLAB platform is RF_MexStandalone-v0.02. The EnMAP-Box software is based on IDL. The package can run on ENVI and the software is embedded in the RF algorithm.

7.2.19 Support Vector Machine

The support vector machine (SVM) is a high-performing supervised machine learning technique based on statistical learning theory, in which the input space is mapped to a feature space (Vapnik, 1995). It is a nonparametric classifier which separates the classes using a linear boundary. This classifier assumes that there is no prior information on how to classify the data. The algorithm aims at finding a decision function, that minimizes the functionality. It permits the training, of nonlinear classifiers in high-dimensional spaces using a small training set. The SVM focuses exclusively on the training samples that are closest in the feature space to the optimal boundary between the classes (Pal & Foody, 2012). The subset of points that lie on the margin (called *support vectors*) are the only ones that define the hyperplane of maximum margin.

SVMs were originally designed to identify a linear class boundary (i.e., a hyperplane). This limitation was addressed through the projection of the feature space to a higher dimension, under the assumption that a linear boundary may exist in a higher dimensional feature space (Mountrakis et al., 2011). For classification, an SVM constructs a hyperplane or set of hyperplanes to differentiate classes (e.g., the presence and absence of class) in a high dimensional space. The SVM maximizes the margin by dividing the input space into two parts while minimizing the total classification errors. A typical SVM is divided into two-class and multiclass SVM (grouping of a chain of two-class SVMs); two-class SVM being the most commonly used model.

Figure 7.27 illustrates the scheme of the nSVM principle in which circles and squares denote two-class samples. The surface is often called

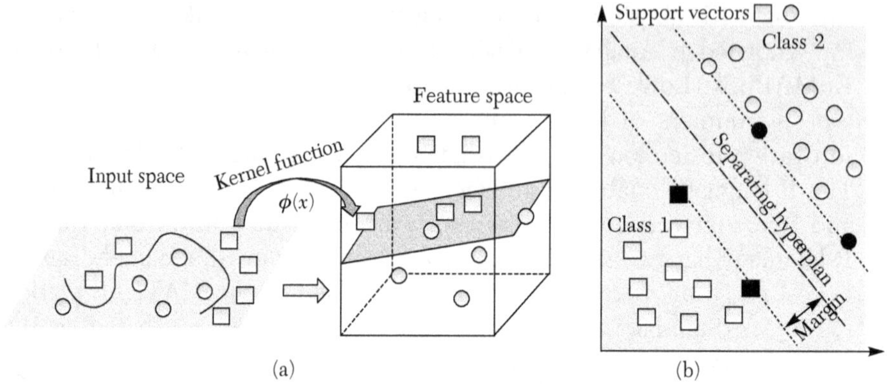

Figure 7.27. The process of SVM classifiers.

the *optimal hyperplane*, and the data points closest to the hyperplane are called *support vectors*. In order to classify linearly, the kernel function converts the input samples into a high-dimensional space. The separating hyperplane is one of the probable planes for separating the two classes; the space between the two dotted lines is the so-called *margin*.

SVMs are particularly useful in remote sensing due to their ability to successfully use small training datasets to produce higher classification accuracy than the traditional methods (Mantero et al., 2005). The data sample to be labeled in remote sensing classification is normally the individual pixel derived from the multispectral or hyperspectral image. Such a pixel is represented as a pattern vector, and for each image band, it consists of a set of numerical measurements. Elements of the feature vector may also include other discriminative variable measurements based on pixel spatial relationships, such as texture. Classes are not separated by statistical learning theory means as in the maximum likelihood classifier, but by geometric criteria areas called training sites on the image, which contain the predictor variables measured in each sampling unit (Mountrakis et al., 2011).

The SVM is popular due to its strong mathematical foundation based on statistical learning theory and structural risk minimization, its capacity for high dimensional datasets, its effective handling with nonlinear classification using kernel functions, and its accurate performance (Mantero et al., 2005). The SVM optimizes the use of training data, which is the biggest advantage of this classifier over other classifiers, like the MLC (Figure 7.26[a]). The SVM classifier is inherently binary, identifying a single boundary between two classes. However, this issue is resolved by repeatedly applying the classifier to each possible combination of classes, though this does imply that processing time increases exponentially as the number of classes increases

(Vapnik, 1995). The SVM classification output is the decision values of each pixel for each class, which are used for probability estimates. It performs classification by selecting the highest probability. An optional threshold allows reporting pixels with all probability values less than the threshold as unclassified. To identify the hyperplane, the central distance between the closest points of each of the two classes is measured (Pal & Mather, 2005).

Burges (2018) found that the SVM outperformed even the best neural networks. This is particularly advantageous in remote sensing applications since data acquired from remotely sensed imagery usually have unknown distributions, and methods such as maximum likelihood that assume a multivariate normal data model do not necessarily match that assumption. The other advantage is that the SVM-based classification is known to keep the right balance between accuracy attained on a given finite amount of training patterns and the ability to generalize to unseen data. The major limitation concerning the applicability of SVMs is the choice of kernels. Choosing a small value for the kernel width parameter (i.e., the kernel footprint in that multidimensional space) may lead to overfitting, while large kernel width values may lead to over-smoothening. Although many options are available, some of the kernel functions may not provide optimal SVM configuration for remote sensing applications. The choice of a kernel function often has a bearing on the results of analysis. A good explanation on SVM kernels and their functionality is presented by Kavzoglu and Colkesen, (2009). The SVM classifiers, characterized by self-adaptability, swift learning pace, and limited requirements on training size have proven a fairly reliable methodology in intelligent processing of data acquired through remote sensing (Mountrakis et al., 2011). The advantages and disadvantages of SVMs are given in Table 7.2.

The SVM deals with quadratic problems, hence it always gets to the global minimum. The advantage of SVM is that there is no need for repeating classifier training using different random initializations or architectures. Recently, the SVM classification algorithms have been used to classify satellite imagery (Figure 7.26[c]), face recognition in photos, and handwriting and object recognition. The SVMs have proven to be superior to most other image classification algorithms in terms of classification accuracy. Choice of the parameter value, which controls the trade-off between maximizing the margin and minimizing the training error, is also an important consideration in SVM application. No established heuristics exist for selection of these SVM parameters, which frequently lead to a trial-and-error approach (Pal & Foody, 2012).

7.2.20 Markov Random Field

The Markov random field (MRF), as a branch of probability theory, considers the spatial-contextual information for a variety of fields in image processing, such as image restoration, texture analysis, edge detection, image segmentation and classification, data fusion, object matching, and recognition (Yong et al., 2008). It is an effective method for semantic segmentation as it makes full use of spatial information constraints, which is suitable for spectral and texture processing of remote sensing imagery (Lu & Weng, 2007).

The MRF classifiers deal with the problem of intra-class spectral variations (Magnussen et al., 2004). These classifiers exploit the spatial information among neighboring pixels to improve the classification results. Contextual classifiers,such as MRF, spatial statistics, fuzzy logic, sgmentation, or neural networks may use smoothening techniques, as they incorporate contextual information as additional bands and then carry out classification using normal spectral classifiers. Thereafter, post-smoothening classification is done on classified images previously developed using spectral-based classifiers. It provides a methodological framework which allows the images from different sensors and map data to be merged in a consistent way. Spatial and temporal contextual information is also readily modeled within this framework for image classification problems. They are widely used for low-level applications, like filtering, segmentation, and classification (Lu & Weng, 2007). The flowchart of the proposed spectral– spatial dynamic classifier/ensemble selection (DEC/DES) method by Damodaran et al. (2015) is shown in Figure 7.28. This framework extracts the spectral information by using the DES method and the spatial information by applying the MRF model to classify the images.

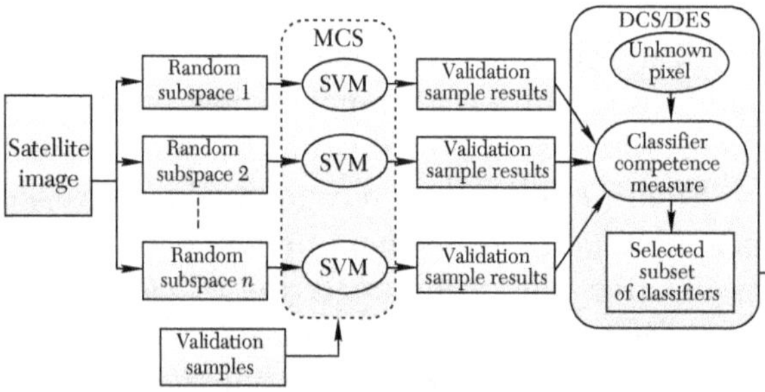

Figure 7.28. Flowchart of the proposed spectral–spatial classification method.

There are two sub-models in the MRF; one is the *feature field* that can effectively extract and model different features with a likelihood function.

It can also measure the probability of occurrence for the features of one pixel. The other sub-model is the *label field* that uses the potential function and the Markov property to reduce the heterogeneity of pixels within the same object by modeling the spatial neighborhood interactions between the class labels of pixels (Yong et. al., 2008). Both the information of the feature field and the label field are, respectively, treated as the observed information and the prior information, and then the posterior probability can be obtained by integrating them together, and the final segmentation can be provided according to the maximum a posteriori criterion (Yong et al., 2008). In the MRF model, the likelihood function of the feature field can consider the pixel-based features only, and the potential function of the label field can just model the spatial neighborhood interactions within a small spatial context, such as 8 pixels in the neighborhood of a 3 × 3 window. In the application of image semantic segmentation, the algorithm obtains the optimal result by solving the maximum posteriori estimation of the label field. The classical method of semantic segmentation based on the MRF model is pixel-based markov random field (PMRF; Lu & Weng, 2007). This method is utilized to measure the similarity between the pixels.

The main advantage of the MRF model is its regular context, which is convenient for spatial relationship description and model solution. However, due to the improvement of spatial resolution of remote sensing imagery, the model is not suitable for capturing complex visible features, and the computation is time-consuming (Zheng & Wang, 2015). Therefore, the multi-resolution MRF (MRMRF; Zheng et al., 2010) defined on the image pyramid structure extends the classic PMRF model. The method improves the computational efficiency and extends the descriptive spatial features to a certain extent, but it is still a pixel-level MRF. With the application and development of OBIA in remote sensing images, the object-based

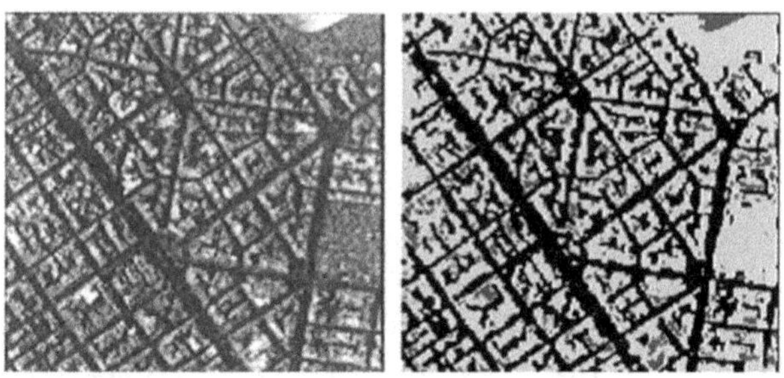

(a) (b)

Figure 7.29. (a) Image showing network of roads, and (b) results from MRF classifiers (Yong et al., 2008).

MRF (He et al., 2020) has been widely used. The object-based MRF model overcomes the limitation of the pixel-level model in the description of spatial features, and can better capture the visible features of the image. Figure 7.29 shows the results of the MRF classifier applied to an image. A road network is highlighted in the output image.

7.2.21 Spectral Angle Mapper

The spectral angle mapper (SAM) is a supervised classification method, widely utilized for remote sensing images. The algorithm determines the spectral similarity between two spectra by calculating the angle (called a *spectral angle*) between the spectra and treating them as vectors in a space with dimensionality equal to the number of bands (Langhans et al., 2007), as shown in Figure 7.30.

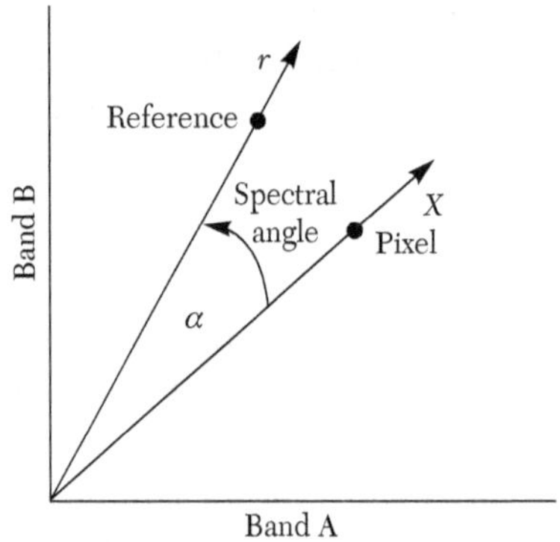

Figure 7.30. Representation of a spectral angle.

The SAM classification algorithm (Kruse et al., 1993) computes the spectral angle α to determine the spectral similarity between an image pixel spectrum t and a reference spectrum r in an n-dimensional feature space (n = number of spectral bands), using the following equation:

$$\alpha = \cos^{-1}\left(\frac{\sum_{i=1}^{n} t_i r_i}{\left(\sum_{i=1}^{n} t_i^2\right)^{1/2}\left(\sum_{i=1}^{n} r_i^2\right)^{1/2}}\right) \tag{7.1}$$

The SAM offers a user-definable threshold in the spectral angle to be used as a determinant for the assignment of a pixel to a spectral class. Smaller angles represent closer matches to the reference spectrum; a pixel is assigned to the class that exhibits the smallest spectral angle (Weyermann et al., 2009). The pixels further away than the specified maximum angle threshold are not classified. A pixel with minimum or zero spectral angles in comparison to the reference spectrum is assigned to the class defined by the reference vector. However, when the threshold for classification based on the spectral angle is modified, the probability of incorrect object detection may increase (Rashmi et al., 2014). A threshold needs to be defined carefully for each of the reference spectra so that it covers all pixels of the intended surface type or spectral class, neglecting differences caused by reflectance anisotropy, and at the same time excludes those of other spectral classes that exhibit similar spectra.

The SAM algorithm is based on an ideal assumption that a single pixel of a remote sensing image represents one certain ground cover material, and can be uniquely assigned to only one ground cover class. Each pixel of every band can be considered as a vector which has a certain length and direction. The SAM performs a band-wise comparison only of the vector's direction, so that the length of the vector does not influence the final spectral angle. A spectral class should contain all spectra that for all bands show a comparable reflectance factor, with the exception of differences caused by wavelength-specific reflectance anisotropy for this class. This ability makes the SAM a well-accepted method for classification. The advantages and disadvantages of SAM are listed in Table 7.2.

The SAM algorithm is simply based on the measurement of the spectral similarity between the two spectra. Object-induced reflectance anisotropy is wavelength-dependent for the majority of objects, especially for natural surfaces, and introduces a nonlinear relationship in the total illumination intensity between the objects of the same species composition but differing in illumination and/or viewing angles. This technique, when used on calibrated reflectance data, is relatively insensitive to illumination and albedo effects. The SAM algorithm therefore is sensitive to bidirectional reflectance distribution function (BRDF) effects (Langhans et al., 2007). Future research should focus on obtaining appropriate ways to compensate or correct variations in reflectance anisotropy as well as to improve image analysis tools accounting for illumination differences.

7.2.22 Spectral Mixture Analysis

Spectral mixture analysis (SMA) has long been recognized as an effective method for dealing with the mixed pixel problem. Sub-pixel classification SMA methods, which are based on collection of end-members from the satellite imagery, are considered very useful for detecting the mixed features in one pixel that complicate accurate classification. It makes several assumptions in estimating the information of each pixel of an image (Roberts et al., 1998):

1. Each pixel contains information about the proportion and spectral response of each component (i.e., mixtures of surface materials and shadows).
2. Brightness (i.e., DN, radiance, or reflectance at each wavelength) of an image pixel is a linear combination of the percentage of each end-member and the brightness of a pure sample of that end-member.
3. The spectral proportions of the end-members reflect proportions of the area covered by features on the ground.
4. Most of the pixels in the image contain some measurable amount of the end-members.

The end-members are regarded as the extremes of the triangles of an image scattergram, and are derived from the extremes of the image feature space, assuming they represent the purest pixels in the images (Mustard & Sunshine, 1999). The output of SMA is typically presented in the form of fraction images, with one image for each end-member spectrum representing the area proportions of the end-members within the pixel. The end-member selection is one of the most important aspects in SMA, and is helpful for improving classification accuracy (Lu et al., 2003). It is especially important for improving area estimation of land cover classes based on coarse spatial resolution data.

Different end-member selection methods are available (Mustard & Sunshine, 1999), however, for many remote sensing applications, the image-based end-member selection method is often used because it is easily obtained and represented by the spectra measured at the same scale as the image data (Roberts et al., 1998). Selection of suitable end-members and image bands are the two most important aspects to develop high-quality fraction images. Constrained or unconstrained solutions can be used to unmix the image into different fractions that represent the areal proportions of the end-members within a pixel (Lu & Weng, 2007). When selecting the end-members, caution needs to be taken to identify the outliers. End-members obtained from the actual image are generally preferred because no calibration is needed between the selected end-members and the measured spectra. Appropriate selection of image end-members is often an iterative process. Checking fraction images and rmse images are a feasible way to assess whether the selected end-members are appropriate or not.

After selection of end-members, an unconstrained least rmse solution can be used to unmix the uncorrelated images into end-member fraction images. However, the selection of end-members for natural systems can be exceedingly difficult because:

1. Potential end-members (surface features) sometimes do not occur in patches larger than the image resolution.
2. Inherent variability in nature, for example, rainfall, soil minerals, and growing cycle phase, makes it difficult to match image end-members with actual pixel composition on the ground.
3. End-members are not truly constant within an image and this creates a mismatch between the defined end-member and its actual form on the ground.
4. Shadow introduces nonlinearity. The shade end-member varies with terrain, vegetation type, and vegetation density. It should not be ignored in end-member selection because it is too common.

The SMA is regarded as a physically based image processing tool (Lu et al., 2003). It discriminates between spectrally similar materials, such as individual plant species, specific water types, or distinctive man-made materials. It supports repeatable and accurate extraction of quantitative sub-pixel information (Roberts et al., 1998). The SMA methods generally involve three steps (Langhans et al., 2007): (*a*) assessment of dimensionality or number of unique reflecting materials in a landscape to obtain the end-members, (*b*) identification of the physical nature of each end-members within a pixel, and (*c*) determination of the amount of each end-member in each pixel. Before implementing the SMA approach, it is necessary to reduce the correlation between the bands, if high correlations exist between bands. For example, the PCA can be performed for this purpose. The uncorrelated bands can be used in the SMA approach to convert the image signatures into physically based fraction images. The process of information extraction from satellite imagery using linear spectral unmixing methods may be as follows:

1. Determine observed radiance, surface reflectance, and the sensor that acquired the image (this is needed to calibrate the image and select end-members to define the spectral mixture of the image).
2. Apply an SMA model (linear unmixing code) to estimate end-member fractions.

3. Invert the model to break the image into fractions of end-members.
4. Compare models, and refine end-member selection (if necessary).
5. Classify the image using SMA output bands.

Table 7.2 lists the advantages and disadvantages of the SMA classifier. A detailed description of the SMA approach can be found in Mustard and Sunshine (1999).

7.2.23 Texture-Based Classifiers

Texture can be thought of as the spatial patterns of gray levels in an image. Texture in image analysis describes the placement and spatial arrangement of repetitions of tones, and is often employed to quantify the variability of pixels in a neighborhood (Li et al., 2014). It can be classified as smooth, fine, coarse, lumpy, stippled, mottled, and rippled in photogrammetric applications. Texture is one of the most important spatial features of an image. Compared to other important spatial features, such as shape and size, the texture is relatively simple to use because it does not require prior image segmentation. At the same time, it is a distinctive feature of selected land use and land cover classes, compared to other classes exhibiting significant spectral similarities.

Texture is a measure of surface property that characterizes the coarseness or smoothness of land features. A classic problem in pattern recognition is classification between the textures. The spatial arrangement of gray tones can provide information about the objects on the ground which can be used to identify and then to classify similar objects in an image. Texture extraction in remote sensing image can improve the classification accuracy through mitigating spectral confusion among spectrally similar classes. Texture-based methods are widely used in applications like face recognition, content-based image retrieval, pattern classification in medical imagery, and land cover classification of remotely sensed images (Suruliandi & Jenicka, 2015).

Pixel-based classification techniques classify a pixel of the image depending on the intensity of the current pixel, but the texture-based techniques classify a pixel based on its relationship with the neighborhood pixels. Texture measures therefore are capable of capturing micro and macro patterns by varying the size of the convolution filter used. Most of the texture-based methods are rotation, illumination, scaling, and color invariant, and are robust and susceptible to noise. It is observed that the texture measures along with a contrast measure characterizing the local neighborhood can yield accurate results, provided a sufficient number of precise samples for training, a suitable neighborhood for finding a pattern unit, and an optimal window size are selected.

The texture of an image can be defined with the primitives and their placement rules. Major texture extraction-based methods can be grouped into major categories: (*a*) structural, (*b*) statistical, (*c*) model based, and (*d*) transform. Structural methods attempt to examine the image textures through evaluating the predefined primitives and spatial arrangements of these primitives. Statistical methods include first-order statistics (i.e., mean, standard deviation) and second-order statistics, especially the gray-level co-occurrence matrix (GLCM) as proposed by Haralick (1979). The GLCM has proved to be a popular statistical method of extracting textural features from images. It is a matrix of second-order probabilities, and is used to identify periodicity and structure within object texture through a variety of texture statistics. A GLCM displays the frequency of transitions from pixels with gray tone i to pixels with gray tone j in a determined direction and distance. Step-by-step, a kernel of size ω is moved over the image, and in each step a GLCM is allocated to the respective central pixel. The algorithm for GLCM construction from a gray scale image is shown in Figure 7.31. The window size ω counts how many times pixels with gray tone i have the orientations as horizontal (0 degrees), vertical (90°) or diagonal (45° and 135°) nearest (distance = 1) neighbors to pixels with gray tone j. The numbers are entered in the appropriate cell $\#(i, j)$. Other statistics computed from co-occurrence matrices, such as entropy, angular second momentum, and inverse difference moment, can also be used. Such an adaptive window

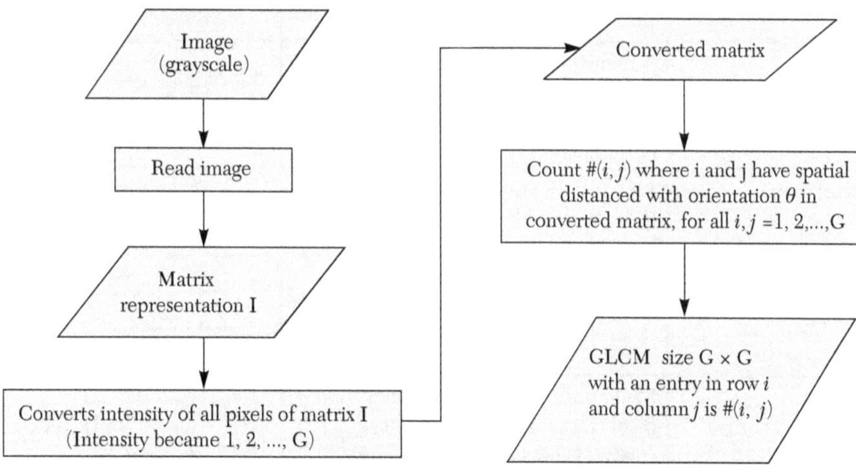

Figure 7.31. Construction of GLCM from gray scale image.

procedure results in an improved image for certain applications (Woodcock & Ryherd, 1989), preserving the boundaries between areas of low and high texture, while not enhancing or widening the edges, which is one of the drawbacks of traditional moving-window texture images (Suruliandi & Jenicka, 2015).

The third category of texture extraction is the model-based approaches, such as fractal models, auto-regressive models, and MRFs models. Finally, the transform methods include Fourier, wavelet transforms, and Gabor. When compared to Fourier and Gabor, the wavelet transforms perform better as they are based on multiple spatial resolutions, and a wide range of wavelet functions can be chosen to improve the classification accuracy (Suruliandi & Jenicka, 2015).

Haralick et al. (1973) proposed fourteen textural features that can be obtained from a GLCM. Table 7.11 presents the eight important second-order measures of image texture, where N corresponds to the range of gray levels in a quantized image; i and j represent the different gray tones in the window; $p(i, j)$ constitutes the normalized frequencies for the neighborhood of gray tone i and j in the respective direction and distance in the window; and μ is the mean. Those features are calculated for each window and its value is allocated to the central pixel.

Table 7.11. Eight Second-Order Measures of Image Texture Calculated from a GLCM

Second-order statistic	**Statistic description of behavior**	**Statistic formula**
Angular-second moment	High when the GLCM is locally homogenous. Similar to homogeneity.	$\sum_i \sum_j [p(i,j)]^2$
Contrast	A measure of the amount of local variation in pixel values among neighboring pixels. It is the opposite of homogeneity.	$\sum_{n=0}^{N-1} n^2 \left\{ \sum_{i=1}^{N} \sum_{j=1}^{N} p(i,j) \right\}$
Correlation	Linear dependency of pixel values on those of neighboring pixels.	$\frac{\sum_i \sum_j (ij) p(i,j) - \mu_x \mu_y}{\sigma_x \sigma_y}$
Dissimilarity	Similar to contrast and inversely related to homogeneity.	$\sum_{n=0}^{N-1} \left\{ \sum_{i=1}^{N} \sum_{j=1}^{N} p(i,j) \right\}$
Entropy	Shannon-diversity. High when the pixel values of the GLCM have varying values. Opposite of angular second moment.	$\sum_j p(i,j) \log[p$

Second-order statistic	Statistic description of behavior	Statistic formula
Homogeneity	A measure of homogenous pixel values across an image.	$-\sum_i \sum_j \frac{1}{1+(i-j)} p(i,j)$
Mean	Gray level average in the GLCM window.	$\sum_{ij=0}^{N-1} p(i,j)$
Variance	Gray level variance in the GLCM window.	$\sum_{ij=0}^{N-1} p_{i,j}(i-\mu_i)^2$

Per direction, distance, and window size, a new structural image arises. Figure 7.32 shows the results of variance, homogeneity, and entropy texture measures applied to satellite images.

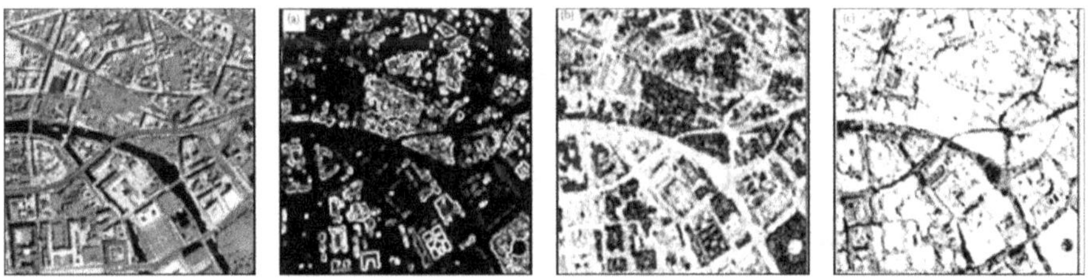

Figure 7.32. Original image, variance, homogeneity, and entropy, respectively (Chen & Cheng, 2016).

Many texture-based approaches have been developed to extract features, such as multivariate local binary pattern (MLBP), multivariate local texture pattern (MLTP), multivariate advanced local binary pattern (MALBP), wavelet and Gabor wavelet with k-NN classifier (Suruliandi & Jenicka, 2015). Among these algorithms, the MLBP, MLTP, and MALBP are computationally simple, and are scalable to capture the texture features from remotely sensed images, while, spectral methods like wavelet and Gabor wavelet extract high- and low-frequency components that describe patterns completely, and provide better cluster separability. In addition, wavelet and Gabor wavelet extract features in multiple scales and orientations to achieve the scaling and rotation invariance. Figure 7.33 presents an architecture of the MLBP technique for feature extraction and classification.

The attributes of texture, coarseness, contrast, directionality, line-likeness, regularity, and roughness, reproduce different textures through the use of a model (Cross & Jain, 1983). Texture information can be incorporated in the processes of image pre-classification (e.g., as an additional variable) and post-classification (e.g., image filtering). Several

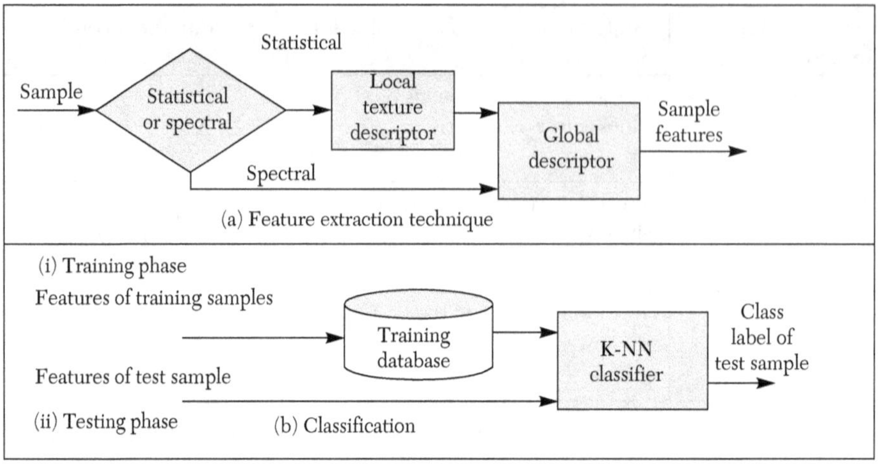

Figure 7.33. An overview of MLBP texture (a) feature extraction and (b) classification.

studies have established that the integration of textural information into remote sensing image classification can produce better classification accuracy. One limitation of texture extraction, though, is that unreliable classification results may exist, especially near the edges of different land covers.

7.2.24 Cellular Automata

An automaton is known as an abstract sequential machine with a finite state in the field of information science. The cellular automata (CA) is a class of automata defined on the simulation space divided into discrete areas called "cells." The theory of CA was first introduced by Von Neumann (1966). CAs are a discrete dynamic system in space, time, and state. The mapping is discrete in time evolution, and the value of state variables of CA is continuous, and therefore, the CAs have become popular in the field of discrete dynamical systems.

The procedure used in CA is to: (*a*) divide the simulation space into discrete areas; (*b*) define the state variables on each cell in digital quantity; (*c*) define the local neighbor rules and the transition rules, which expresses the local interaction among state variables on neighboring cells; and (*d*) make the transition state variables along discrete time steps according to the local neighbor rules or transition rules. The grid space which is composed of a series of cells can be 1D, 2D, or multidimensional (can be limited or infinite). A CA consists of grid cells normally in a matrix form that has the following basic features (Espinola et al., 2020):

States: Each cell can take an integer value that corresponds to its current state. There is a finite set of states. In a standard CA model, the state is usually used as the main attribute to describe the development of a cell. Any cell cannot take more than one state simultaneously, although the state can change from one to another in different periods. The essence of CA is that the states of the neighboring cells influence the state of the central cell (Couclelis, 1997). A simple model is to project the state of a central cell using a 3 × 3 window to count the distribution of states in its neighboring cells.

Neighborhood: A set of cells that interact with the current one.

Transition function f : Takes as input arguments the cell and neighborhood states, and returns the new state of the current cell.

Rules: The transition function *f* uses a set of rules that specify how the states of the cells change.

Iterations: The transition function *f* is applied to each cell of the grid across several iterations. The states of the cell evolve synchronously in discrete time steps according to a set of local rules, in which the cell state at time (t + 1) is a function of the cell state and the states of its neighboring cells at time (t). Transition rules are generally in the form of "if–then–else" statements, and are applied on input from a neighborhood template to evaluate the value of each cell (Torrens, 2000). Transition rules enable complex systems to be modeled using simple components that drive their dynamics (Batty, 2000). These rules are applied to each cell to determine what state it should change to during a time transition. The transition rules change each cell to the state for which it has the highest potential. The changes in cell states occur in a discrete time form. In each iteration, entire cells are checked, and rules are applied through the transition function *f* to each cell, taking into account the neighborhood to change its state. Therefore, CAs have an evolution process because the cells are always changing their states across different iterations. Due to this, in recent years CAs have become a powerful tool applied in remote sensing, especially to simulate the analysis from satellite images, image enhancement (noise-reduction filters), and edges detection (Al-Ageili, 2017).

The application of CA in satellite image classification processes is a relatively new field of investigation, which includes image acquisition, image pre-processing, image segmentation, feature extraction, and parameter measurement (Peng et al., 2019). Image pre-processing can suppress noise and enhance contrast, thus improving the quality of the source image. The object of interest is segmented from the background by image segmentation, and the parameters such as dimensions are important for feature extraction and parameter measurement. CA models can be used in applications that experiment with time evolution, like environmental simulations, complex social phenomena modelling, images treatment in artificial vision, cryptography of digital information, and AI in mathematical games. These models are also used as explanatory tools for real-world applications, such as modeling land use dynamics, simulating snow-cover dynamics, modelling vegetation dynamics, simulating forest fire spread, and modeling bio-complexity of the deforestation process (Batty, 2000).

The SLEUTH (slope, land use, exclusion, urban extent, transportation, hillshade) model is a typical calibrated CA model in which multiple land use datasets may be employed in the training period to define the patterns of land uses (Chaudhuri & Clarke, 2013). It is rapidly gaining popularity among urban planners and geographers for simulating urban areas and landscape (Li & Yeh, 1998). Figure 7.34 shows a general structure of the SLUTH model. Due to its scale independence, dynamic and future oriented structure, transportability, use under different conditions by modifying some initial conditions and changing input data layers, and application of all regions with different data sets, SLEUTH has become a popular tool in modeling urban spreading extent over time or forecasting growth into the future (Goldstein, 2004). However, its exhaustive search method makes the SLEUTH time-consuming, while the high number of parameters in this model increases the randomness of simulation results.

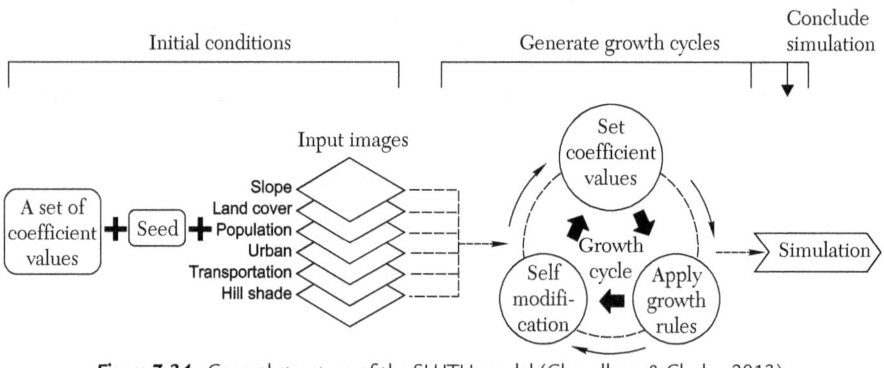

Figure 7.34. General structure of the SLUTH model (Choudhary & Clarke, 2013).

The rapid development of GIS helps to foster the application of CA in urban simulation. Current GISs are not designed for fast iterative computation, but CA can be used by creating batch files that contain iterative command sequences. By linking CA to GIS, some of the limitations of the current GIS can be overcome. In addition, the CA can benefit from the useful information provided by GIS in defining the transition rules. The data requirement of CA can be best satisfied with the help of GIS. The integration of GIS with CA serves as an analytical engine to provide a flexible framework for programming and running dynamic spatial models (Li & Yeh, 1998). Figure 7.35 shows a process by Markov chain analysis and CA in IDRISI

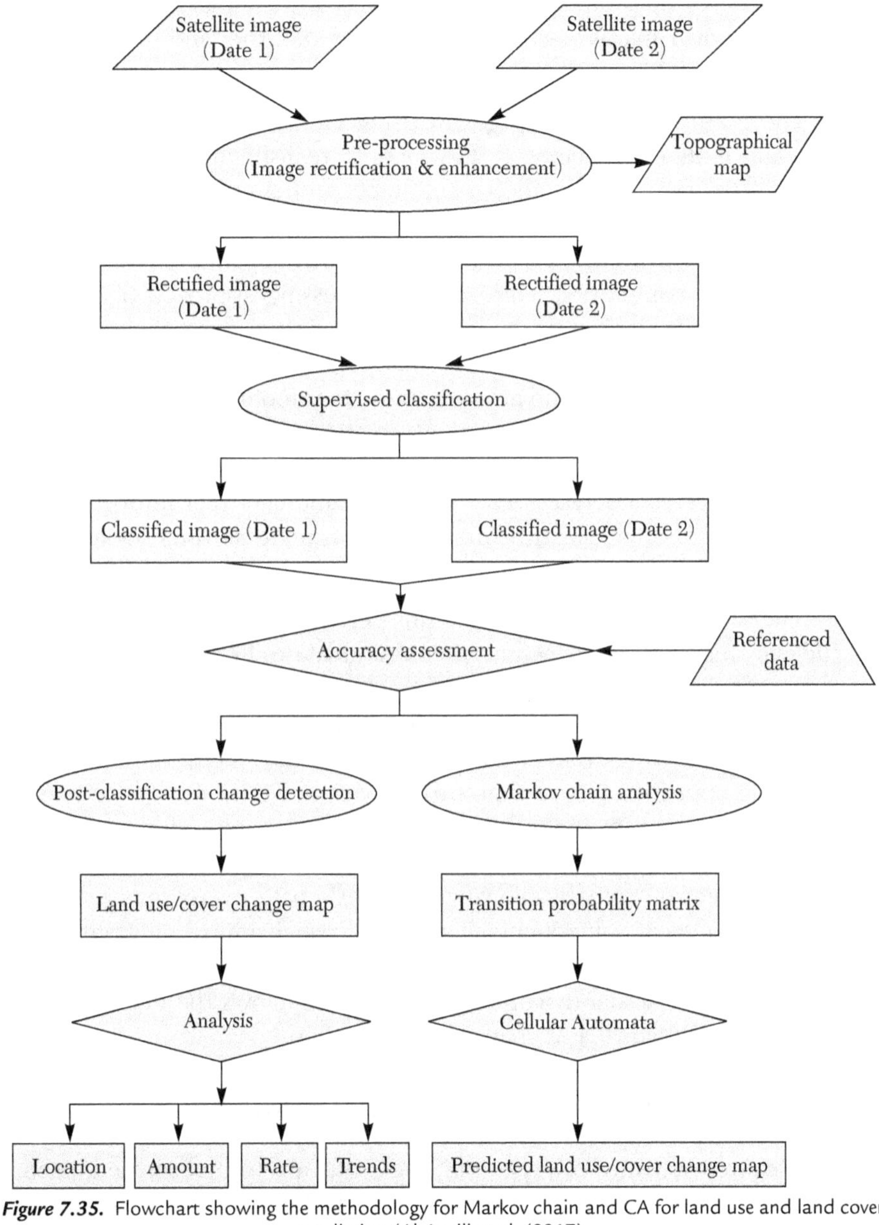

Figure 7.35. Flowchart showing the methodology for Markov chain and CA for land use and land cover prediction (Al-Ageili et al. (2017).

for the modeling and forecasting of land use/cover change in the future (Koranteng & Zawila-Niedzwiecki, 2015). For this study, the two land cover maps 1990 and 2000 were first employed to predict the land cover map of 2010. This predicted land cover map was compared with the actual land use/cover map of 2010. After the successful validation, the 1990 and 2000 land cover maps were again used to predict the land use/cover maps for 2020 and 2030.

There are several advantages of using CA (Torrens, 2000). The most important is their simplicity. In CA models, realistic and complex patterns emerge from simple rules and the interaction of simple components at the local level. In these models, complexity emerges from within the model, and is not a part of the model design. They are explicitly spatial, and make implicit use of spatial complexity. The parallel processing approach of CA makes them decentralized, and therefore treats the system dynamics as nonlinear interactions among its individual components. The CA models are more suitable with raster data, and are well suited to remotely sensed data and GIS work. Further, both CA and spatial information systems represent attributes in a layered fashion (e.g., class and spectral layers in remote sensing, themes in GIS, and cell state-spaces in CA), and manipulate that information with spatial operators (e.g., texture measures in remote sensing, neighborhood operators in GIS, and transition rules in CA; Al-Ageili et al., 2017).

The capability of CAs to handle fine-scale dynamics with computational efficiency makes them ideal for detailed simulation. For example, traditional urban models are relatively static and treat urban dynamics very poorly. The CA models represent a significant advance in the treatment of time, since they are interactively dynamic and move in short time steps. This makes them capable to approximate real-time urban dynamics. Also, CAs allow multiple time scales to be represented, facilitating their suitability for modeling the activities and events of cities that vary with time. The CA models can connect the subsystems of a city by allowing large-scale patterns to emerge from the interaction of local components. The highly visual simulation environments of CA models enhance the engagement and interaction of users with the model. Also, the visual dynamics of CA models allows the evolution of the system to be displayed as it changes through time.

7.3 FREE AND OPEN-SOURCE SOFTWARE (FOSS)

Open-source software is a type of "free" software to be accessed, used, or modified by their user groups and developers. One key feature to distinguish open-source software from other types (such as proprietary software and shareware) is their "free software licenses," which explicitly define the legal rights to users with freedom to run, study, change, redistribute, and access the source codes of the licensed software (Tsou & Smith, 2011).

GIS and remote sensing tasks are mutually interlinked, and division of remote sensing software with GIS software is often a difficult task (Halounova, 2007). Many GIS software are able to work with remote sensing data and to perform simple tasks with them. On the contrary, GIS functions are embedded in remote sensing software. This duality is a result of close relation between these, as several input to GIS will come from remote sensing data. Some of the commercially available remote sensing and GIS software along with their weblinks is presented in Table 7.12.

Although many private GIS vendors and software companies, such as ESRI, Microsoft, Google, and Intergraph, played an important role for GIS development in the past, open-source software has become a stronger player recently in the GIS industry. There has been a big shift over the years to build software in an open, collaborative way. Some open-source software are explained as follows in alphabetical order (Halounova, 2007; Tsou & Smith, 2011; *https://gisgeography.com/open-source-remote-sensing-software-packages/; https://monde-geospatial.com/top-14-open-source-remote-sensing-software/12/*).

Table 7.12. Commercial Remote Sensing and GIS Software

S No.	Commercial GIS Software	Websites
1.	Arc GIS (ESRI)	*http://www.arcgis.com/index.html*
2.	Bentley MAP	*https://www.bentley.com/en/products/*
3.	ERDAS Imagine (Intergraph)	*https://www.hexagongeospatial.com/products/powerportfolio/erdas-imagine*
4.	ENVI	*https://www.harrisgeospatial.com/*
5.	GEOMATICA 2018	*https://www.pcigeomatics.com/software/geomatica/*
6.	GEOMEDIA	*https://www.hexagongeospatial.com*
7.	Global Mapper	*https://www.bluemarblegeo.com/products*
8.	IDRISI SELVA (CLARK LABS)	*https://clarklabs.org/terrset/*
9.	Manifold GIS	*http://www.manifold.net/*
10.	Mapinfo Professional	*https://www.pitneybowes.com*
11.	Mapviewer AND SURFER	*https://www.goldensoftware.com/products/mapviewer*

(Contd.)

S No.	Commercial GIS Software	Websites
12.	MAPTITUDE	*https://www.caliper.com/maptovu.html*
13.	Small World	*https://www.smallworldfs.com/small-world/*
14.	Super GIS	*https://www.supergeotek.com/*

7.3.1 Apache Spark

Apache spark is an analytics engine for large-scale data processing that supports high-level programming APIs in various languages and tools for querying, graph processing, and machine learning. It is an open platform for processing large data sets (big data), and can perform 100 times faster than Hadoop MapReduce for big data processing (Proficz & Drypczewski, 2017). It delivers flexibility, scalability, and speed to meet the challenges of big data. It enables efficient analysis and processing of data from a variety of sources, including block and object file systems.

Apache Spark has the following features: (*a*) high processing speed (in comparison with Apache Hadoop MapReduce); (*b*) a parallel computing framework; (*c*) ease of use and direct support for Scala, Java, Python, and R languages; (*d*) can combine a variety of applications: SQL, machine learning, graphical modeling, data streaming, and processing; and (*e*) easy portability and integration with Hadoop YARN, Apache Mesos, Kubernetes, and the use of external data sources: HDFS, Swift, HBase, and Cassandra. It integrates two main libraries, SQL for querying large structured data and MLlib involving main learning algorithms with statistical methods (Gupta et al., 2017). The MLlib is Spark's open-source ML library which includes several efficient training functionalities. It also supports different languages and provides a high-level API that enriches Spark's ecosystem and facilitates the development of ML pipelines (Meng et al., 2016). One of the possible applications of Apache Spark in satellite data processing is selection of satellite products based on the information contained in the images and the metadata.

Big remote sensing data management and processing is currently one of the most important topics. Xu et. al. (2020) introduced ScienceEarth, a cluster-based data processing framework, with the aim to store, manage, and process large-scale remote sensing data in a cloud-based cluster-computing environment. The ScienceGeoSpark is an easy-to-use computing framework using Apache Spark as the analytics engine for big remote sensing data processing, as shown in Figure 7.36.

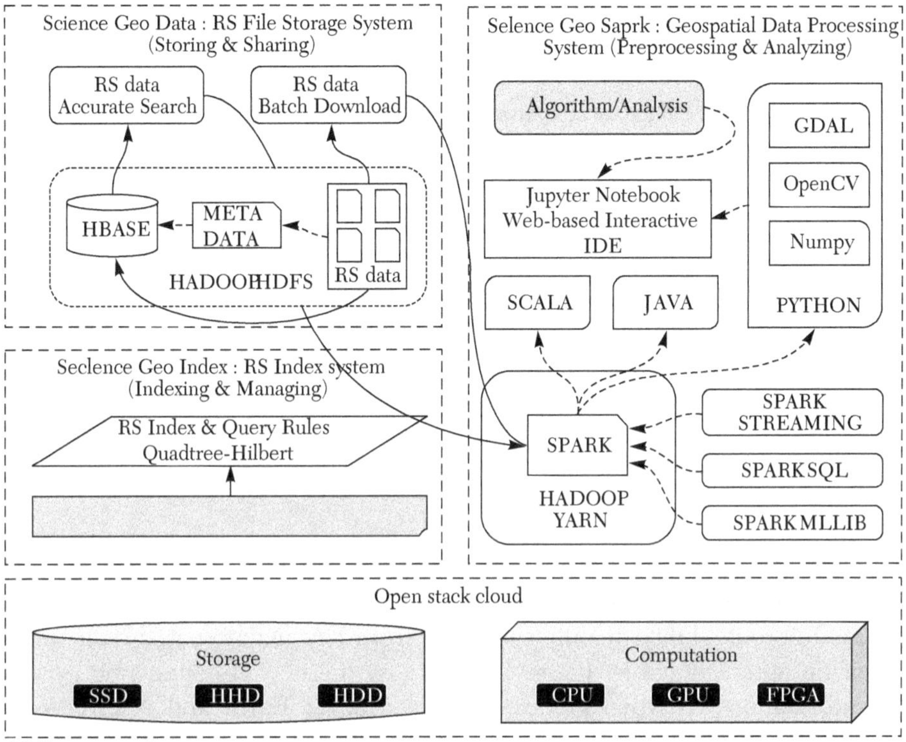

RS-Remote sensing

Figure 7.36. An architecture of ScienceEarth using Apache Spark (Xu et al., 2020).

7.3.2 CLAS Lite

It is designed specifically for crop monitoring from images for forest management (*https://claslite.org*). The software captures the images of dense tropical forest via satellite imaging; and converts the images into highly detailed maps that can be readily searched for deforestation, logging, and other forest disturbance events. The process of conversion goes through calibration, pre-processing, atmospheric correction, and so on.

7.3.3 E-foto

It is a digital photogrammetry software (*http://www.efoto.eng.uerj.br/en*). The core functionality is: photo-triangulation, stereoscopic modeling, and DEM extraction. This software provides a fully functional photogrammetry toolset to use with lots of tutorials and examples.

7.3.4 Geo Express

It is a tool built to compress a large number of images in MrSID (multi-resolution seamless image database) format (*https://www.extensis.com/geoexpress*). This format helps retain the quality of an image while compressing it in a compact size. It can be used to edit the image, create on orthophoto, and share it for further analysis.

7.3.5 Geographic Resources Analysis Support System

The Geographic Resources Analysis Support System (GRASS) is used for map design, analysis, and image and graphics design (*www.grass.osgeo.org*). It is also used for geospatial analysis and data management. It offers tools like image classification, PCA, edge detection, and radiometric corrections as well as processing and analysis of LiDAR points data to create contours and generate DEMs. It also has hi-tech network and satellite tools which could be used for a large range of applications including geography, landscape ecology, epidemiology, remote sensing, urban planning, biology, geophysics, hydrology, groundwater, flow modeling, vector network analysis, geo-statistics, and raster 3D volume (voxel).

7.3.6 GeoServer

GeoServer is an open-source server written in Java that allows users to share, process, and edit geospatial data (*http://geoserver.org*). Designed for interoperability, it publishes data from any major spatial data source using open standards. It provides an easy method of connecting existing information to virtual globes, such as Google Earth and NASA World Wind, as well as web-based maps such as OpenLayers, Google maps, and Bing maps.

7.3.7 GMT Mapping Tools

GMT is a free collection of ~60 UNIX tools that allow users to manipulate (x, y) and (x, y, z) datasets, including filtering, trend fitting, gridding, projecting, and so forth) (*http://gmt.soest.hawaii. edu/*). It can produce Encapsulated PostScript file (EPS) illustrations ranging from simple x-y plots through contour maps to artificially illuminated surfaces and 3D perspective views in B&W, gray tone, hachure patterns, and 24-bit color.

7.3.8 gvSIG

The gvSIG emerged in 2004 (*www.gvsig.org*), and has simple GUI and well-documented mobile application and 3D capabilities. It has powerful CAD tools with good visualization functionalities. The gvSIG is well suited for interactive data exploration and analysis. It can be used to perform supervised classification, band algebra, and decision trees, including creation of tassled cap and vegetation indices. The gvSIG's mobile app helps maintain track of fieldwork with its GPS tools.

7.3.9 Image Analyzer

It has both the most image enhancement features (*http://meesoft.logicnet.dk/Analyzer/*) found in conventional image editors as well as a number of advanced features not even available in professional photo suites. The functionality includes: automatic brightness, contrast, gamma, and saturation adjustment; Built-in conventional and adaptive filters for noise reduction, edge extraction, and so on, retouch tools; deconvolution for out-of-focus and motion blur compensation; easy red-eye removal; user specified filters in the spatial & frequency domain; resize, rotate, crop, and warping of images; morphological operations; color model conversion; RGB, CMY, HSI, Lab, YCbCr, YIQ, and PCA; and distance, Fourier, and discrete cosine transformation.

7.3.10 ImageJ

ImageJ allows users to open, display, edit, process, and analyze 8-bit, 16-bit, and 32-bit images in several formats: TIFF, GIF, JPEG, PNG, DICOM, BMP, PGM, and FITS (*http://rsbweb.nih.gov/ij/*). By downloading or writing additional plug-ins, additional image formats and numerous new functions can be processed. It includes comprehensive processing capabilities such as geometric transformations, image enhancement (edge detection, sharpening, smoothing, etc.), and color processing. The program can also stack multiple images in one display that are correlated spatially or temporally with the same size and bit depth, allowing the user to scroll through them interactively.

7.3.11 Integrated Land and Water Information System

The Integrated Land and Water Information System (ILWIS) is a user-friendly software with a combination of both a GIS and remote sensing package that can display, process, and analyze image (raster), vector, and thematic data (*https://52north.org/software/software-projects/ilwis/*). One of the key features of ILWIS is its stereoscopy, anaglyph, and photogrammetry tools. Features available for vector data include digitizing, display, interpolation, calculations, and more. For raster data, several functions include creation of digital elevation models, slope, aspect, distance calculation, and much more. With satellite images, the users can derive statistics, filters, mosaic, georeferencing, classifications, and histograms.

7.3.12 InterImage

It is a framework specialized in automatic image interpretation using knowledge-based classification of remote sensing images (*http://www.lvc.*

ele.puc-rio.br/projects/interimage/). The automatic image interpretation is OBIA which involves segmentation, exploring their spectral, geometric, and spatial properties, and then classification. The software also performs other functions like calculation of spectral, extraction of objects from an image and texture features analysis. It is interoperable, mainly through a command line interface.

7.3.13 Mapnik

It is a C++/Python library for rendering purpose (*https://mapnik.org/*). It is used by OpenStreetMap.

7.3.14 MapServer

MapServer is an open-source web-based platform designed especially for publishing spatial data and interactive mapping applications to the web (*https://mapserver.org/*). It runs on all major platforms (Windows, Linux, Mac OS X). The software builds upon many popular open-source or freeware systems, like Shapelib, FreeType, Proj.4, libTIFF, Perl, and others.

7.3.15 Maptitude

Maptitude is used for spatial analysis and geocode addresses, and creates highly interactive maps (*https://www.softwaresuggest.com/maptitude*). The tool is cloud-based and more of a location intelligence tool (e.g., geocode addresses) than just being a GIS tool. It can quickly and seamlessly analyze all types of spatial data. It is mainly developed for business users, wherever spatial analysis is needed. It is very easy to integrate it with Microsoft Office, but this tool requires knowledge of the programming language called Caliper Script.

7.3.16 MultiSpec

It is a multispectral image data analysis system. It results from an ongoing multiyear research effort which is intended to define robust and fundamentally based technology for analyzing multispectral and hyperspectral image data, and to transfer this technology to the users' community (*http://www.purdue.edu/*).

7.3.17 OpenEV

It is a program that displays and analyzes vector and raster data, and offers a library for developers to construct new applications (*http://openev.sourceforge.net/*). It was later procured by Microsoft and developed into the satellite imagery viewer. OpenEV can display large datasets from georeferenced images to elevation data in 2D and 3D mode. Some of the features include image enhancement, image comparison, overlay, and on-screen digitizing for image analysis.

7.3.18 OpenLayers

OpenLayers is an open-source JavaScript library for displaying map data in web browsers (*https://openlayers.org/*). It provides an API for building rich web-based geographic applications similar to Google maps and Bing maps. It supports GeoRSS, Keyhole Markup Language (KML), Geography Markup Language (GML), GeoJSON, and map data from any source using OGC-standards such as Web Map Service (WMS) or Web Feature Service (WFS).

7.3.19 Open-Source Software Image Map

The Open-Source Software Image Map (OSSIM) is a high performance remote sensing image processing, GIS, and photogrammetry software, compatible with over 100 raster and vector formats, and supporting a large number of sensors (*https://trac.osgeo.org/ossim/*). It supports rigorous sensor modeling, universal sensor modeling and a wide range of map projections. It is written in C++, employing the latest techniques in object oriented software design. A number of command line utilities, GUI tools and applications, and integrated systems have been implemented with the baseline. Some functions in the program include sensor modeling, native file access, ortho-rectification, elevation support, vector and shapelib support, histogram matching, and tonal balancing. ImageLinker is another tool built to work on top of the OSSIM software. Through their Visual Chain Editor (VCE), users can create, connect, and change "image chains" (multiple images). Users can also utilize and/or create plugins within ImageLinker to add new functionality to the program.

7.3.20 Optical and Radar Federated Earth Observation Toolbox

The Optical and Radar Federated Earth Observation (ORFEO) toolbox is an image processing software that has several algorithms for image filtering, image segmentation, and image classification with k-means and SVMs (*www.orfeo-toolbox.org*). It is a useful tool for 3D models in an urban environment using optical images: modeling, representation, and visualization; studying of movements: deformation, displacements, evolution (subsidence, glaciers, etc.); and high-resolution radar imaging in urban areas: radargrammetry, interferometry, and coupling with optical imaging. The ORFEO aims at high spatial resolution data with remote sensing tools, including change detection, radiometry, PCA, and pan-sharpening. It is mainly a framework which can be used through the command line. Also, there is an interface that allows for the interactive execution of applications. It provides integration with other software through a Python interface. The ORFEO has Large-Scale Mean-Shift segmentation (LSMC) object-based image analysis, which is a rare feature in other software.

7.3.21 Opticks

Opticks is a flexible remote sensing and image analysis software, and is a less complicated design to use (*www.opticks.org*). It supports many types of imagery and remote sensing data, such as motion imagery (videos), SAR, multispectral, and hyperspectral data. Opticks was originally created for hyperspectral analysis. It has extensive functions, such as displaying false color images, histogram production, adding annotations, creation of animations through a sequence of all available bands, linking frames, georeferencing, performing filters, automation, and creating customized algorithms.

7.3.22 Polarimetric SAR Data Processing (PolSARPro)

The Polarimetric SAR Data Processing (PolSARPro) software (*www.earth.eo.esa.int/polsarpro*) from the European Space Agency allows for easy accessibility and use of multi–SAR datasets. It is used to handle dual and full polarization SAR from satellites, like ENVISAT-ASAR, ALOS-PALSAR, RADARSAT-2, and TerraSAR-X. It consists of a wide range of tools, like radar decompositions, InSAR processing, and calibration. The PolSARPro software has a graph processing framework where users can automate workflow, a functionality similar to ArcGIS ModelBuilder.

7.3.23 Python

Python is a modern, powerful, high-level programming language that has been developed over the last 20 years into an extremely versatile tool. It is fairly easy to learn, and interfaces very easily with modules written in C, C++, Fortran, and so forth. The Python Data Analysis Library (Pandas) has incredible tools for data processing and working with statistical analysis and modeling, and is now being used in a huge range of applications. In combination, Python, NumPy/SciPy, and Pandas offer a powerful data processing ability. In Python, for instance, the usual Numpy, SciPy, Scikit-learn (and image), OpenCV, Keras, and Tensorflow libraries may be used to build the solution. In particular, a large library of Python modules has been written for scientific programming, making it ideally suited for data analyses, plotting, model fitting, and so on. It is a popular programming language for image processing and machine learning. The sequence of steps for information extraction from a large set of satellite images is given in Figure 7.37.

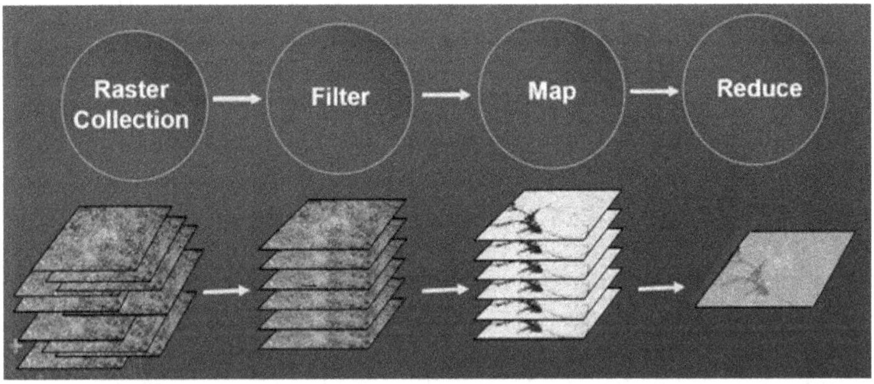

Figure 7.37. Workflow of processing a collection of satellite images using Python (Xu & Zhang, 2020).

The Arc GIS has two modules; ArcPy for image analysis and ArcGIS API for Python. A good place to start is with Google's free online python course: *https://developers.google.com/edu/python/*. There are a huge range of online examples, tutorials, mailing lists, and so forth to learn and understand Python (*www.diveintopython.com*).

7.3.24 Quantum GIS

Quantum GIS (QGIS) is a user-friendly open-source GIS software which runs on Linux, Unix, Mac OSX, windows, and Android, and supports numerous vector, raster, and database formats and functionalities (*www.qgis.org*). The program's core functionality can be extended through plug-ins (e.g., for spatial query, topology checking, zonal statistics, etc.). The semiautomatic classification plugin (SCP) of QGIS is one of the best plugins. It is especially useful to download satellite images (e.g., Sentinel, Landsat, ASTER, and MODIS) directly in the plug-in tools, as well as their pre-processing and post-processing. The QGIS gets high preference for its ability to automate map production, generate cartographic figures, and process geospatial data. Users can view, edit, and analyze their geospatial data in both vector and raster format.

The Python programming language is available in QGIS to automate the tasks. The QGIS can perform spatial data analysis on spatial databases and other OGR-supported formats. It currently offers vector analysis, sampling, geo-processing, geometry, and database management tools. It can be used for basic GIS tasks, like map production and geospatial analysis, as well as more specific tasks, like disaster risk reduction, terrain analysis, and environmental resource mapping. The QGIS also uses the GDAL (geospatial data abstraction library) to read and write raster data formats including ArcInfo Binary Grid, ArcInfo ASCII Grid, GeoTIFF, ERDAS Imagine, and many more. The interface is fairly similar to ArcGIS, which allows experienced GIS users to quickly customize their tools and plug-ins. It can also integrate with other open-source GIS systems to extend its functionality.

7.3.25 R

T R is a programming language and software environment for statistical computing and graphics supported by the R Foundation for statistical computing (*www.r-project.org or www.rstudio.com*). It compiles and runs on a wide variety of UNIX platforms and similar systems (including FreeBSD and Linux), Windows, and MacOS. R is an integrated suite of software with facilities for data manipulation, calculation, and graphical display. One of its strengths is the ease with which well-designed publication-quality plots can be produced, including mathematical symbols and formulae where needed. R and its libraries implement a wide variety of statistical and graphical techniques including linear and nonlinear modeling, classical statistical tests, time-series analysis, classification, clustering, and others. It includes an effective data handling and storage facility; a suite of operators for calculations on arrays in particular matrices; a large, coherent, integrated collection of intermediate tools for data analysis; graphical facilities for data analysis and display either on-screen or on hardcopy; and a well-developed, simple, and effective programming language which includes conditionals, loops, user-defined recursive functions, and input and output facilities. The R-ArcGIS community, for example, is a community-driven collection of free, open-source projects, making it easier and faster for R users to work with ArcGIS data, and ArcGIS users to leverage the analysis capabilities of R.

7.3.26 Sentinel Toolbox

The Sentinel Toolbox consists of three separate applications: Sentinel-1 Toolbox (SAR applications), Sentinel-2 Toolbox (high-resolution optical applications), and Sentinel-3 Toolbox (high-resolution optical applications; *https://sentinel.esa.int/web/sentinel/toolboxes*). Sentinel-2 has become popular for open satellite data. In order to process the vast amount of data collected by Sentinel-2A/2B, the Sentinel Toolbox can be used effectively.

The Sen2cor plug-in allows the users to correct for atmospheric effects and classify images. Sentinel-1 SAR data can be processed with the Sentinel-1 Toolbox, such as interferometry, speckle filtering, and co-registration.

7.3.27 SPRING

SPRING is a GIS and remote sensing image processing system with an object-oriented data model which provides for the integration of raster and vector data representations in a single environment. It works on the following platforms: Windows, Linux, UNIX, and Macintosh (*http://www.dpi.inpe.br/spring/english/*).

7.3.28 System for Automated Geoscientific Analyses

The System for Automated Geoscientific Analyses (SAGA) is a comprehensive stand-alone GIS with a modular plugin structure (*www.saga-gis.org*). It provides a large suite of functions for terrain analysis, such as hill-shading, watershed extraction, visibility analysis, hydrological modeling, and geo-statistics, which are also accessible from QGIS. It is ideal for most remote sensing needs because of its rich library grid, imagery, and terrain processing modules. As an interactive GUI, it enables users to visualize and manage geographic data with the help of maps, graphs, and histograms. SAGA has basic supervised classification, but it is not as intuitive and user-friendly as other remote sensing software. For analyses and manipulation, it flourishes with terrain tools like topographic position index (TPI), topographic wetness index (TWI), and soil classification. It also has some rudimentary tools for photogrammetry and SVM. Although it has a vast library of raster-based tools, the only drawback of SAGA GIS is the lack of documentation for some of them.

7.3.29 TensorFlow

TensorFlow is an open-source software library for numerical computation using data flow graphs (*https://www.tensorflow.org*). Its library is used for a number of services, such as speech recognition, photo search, and automatic responses for Gmail's Inbox. It was originally developed by the Google for internal use in machine learning and deep neural networks research, but the system is applicable in a wide variety of other domains as well.

TensorFlow is cross-platform which runs on GPUs and CPUs, including mobile and embedded platforms, and even tensor processing units (TPUs), which are specialized hardware to do tensor math (Goldsborough, 2016). It also offers users the capability to run each node on a different computational device, making it highly flexible. TensorFlow has automatic differentiation and parameter sharing capabilities which allow a wide range of architectures to be easily defined and executed.

TensorFlow is a C++ based deep learning framework along with python APIs developed for both deep neural networks research and complex mathematical computations, and it can even support reinforcement learning. Its uniqueness also lies in dataflow graphs, structures that consist of nodes (mathematical operations) and edges (numerical arrays or tensors). The Object Detection API is still a core machine learning challenge to create accurate machine learning models capable of localizing and identifying multiple objects in a single image. Open-source machine learning tools leverage solving problems by applying knowledge gained after working with a problem from a related or even distant domain. Depending on the task being performed, pretrained models and open datasets may not be accurate as compared to the customized ones, but they save a substantial amount of effort and time, and they don't require users to gather datasets.

TensorFlow has built-in support for DL and NN, so it makes it easy to assemble a network, assign parameters, and run the training process. It has a collection of samples of trainable mathematical functions that are useful for NN. Due to the large collection of flexible tools, TensorFlow is compatible with many variants of machine learning. In addition, it uses both CPUs and GPUs for computing, hence it makes the compilation much faster (Goldsborough, 2016).

7.3.30 Torch

Torch is an open-source software library which provides flexibility and speed in building scientific algorithms, making the process extremely simple (*http://torch.ch*). It consists of community-driven packages in machine learning, optimization, computer vision, signal processing, parallel processing, image, video, audio, and networking. Also, it has popular NN and optimization libraries which are simple to use, while having maximum flexibility in implementing complex NN topologies. Torch is often called the easiest deep learning tool for beginners. It is a scientific computational framework that also provides access to state-of-the-art speedups for convolutional operations. It has easy to use multi-GPU support and parallelizing packages that make it very powerful for training deep architectures. It has a simple scripting language, Lua, and a helpful community sharing an impressive array of tutorials and packages for almost any deep learning purpose. Despite using a less common language than

Python, Torch is widely adopted by Facebook, Google, and Twitter. Many companies also have in-house customized Torch for their deep learning platforms, which has contributed to its popularity in recent times.

7.3.31 Waikato Environment for Knowledge Analysis

The Waikato Environment for Knowledge Analysis (WEKA), an open-source software, can be accessed through a graphical user interface, standard terminal applications, or a Java API (*https://www.cs.waikato.ac.nz/ml/weka/*). It provides tools for data preprocessing, implementation of several ML algorithms, and visualization tools (as shown in Figure 7.38) to develop ML techniques and apply them to real-world data mining problems. There are many stages dealing with big data to make it suitable for ML. The raw data collected from the field may be cleaned using data pre-processing tools provided in WEKA. Depending on the kind of ML model to be developed, the Classify, Cluster, or Associate option may be selected. The Attributes Selection allows the automatic selection of features to create a reduced dataset. Under each category, WEKA provides the implementation of several algorithms to run them on the dataset. WEKA gives the statistical output of the model processed, and also provides a visualization tool to inspect the data.

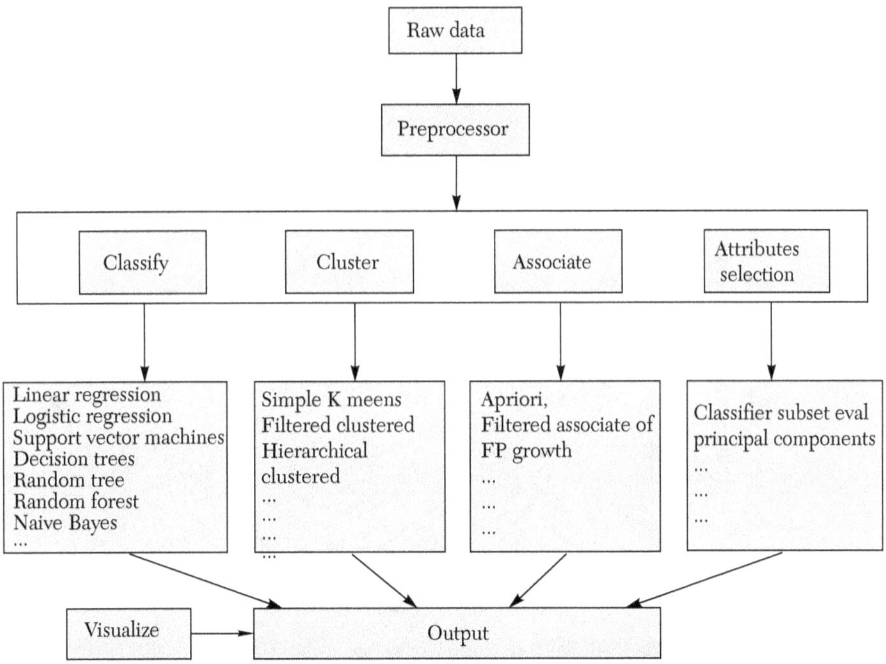

Figure 7.38. Tools of WEKA for various functionalities.

7.4 SELECTION OF TRAINING SAMPLES AND CLASSIFICATION ALGORITHMS

The selection of suitable sensor data is the first important step for a successful classification (Jensen & Cowen, 1999). It requires considering factors such as user's need, the scale and characteristics of the study area, the availability of various image data and their characteristics, cost and time constraints, and the analyst's experience in using the selected image (see Section 4.6). In general, a classification system is designed based on the user's need, spatial resolution of selected remotely sensed data, compatibility with previous work, image-processing and classification algorithms available, and time constraints. Such a system should be informative, exhaustive, and separable (Jensen, 2005).

A suitable classification system and a sufficient number of training samples are pre-requisites for many algorithms for their successful classification (Mather, 2004). Training samples are usually collected from fieldwork, or from high-resolution aerial photographs and satellite images. Different training data collection strategies, such as single pixel, seed, and polygon, may be used, as they may influence the classification results, especially when using high spatial resolution images (Lu & Weng, 2007). Singh et al. (2001) have established several curves indicating the relationship between optimum spatial resolution and the field size considerations for analysis, as shown in Figure 7.39. Large field size analysis requires medium-resolution images.

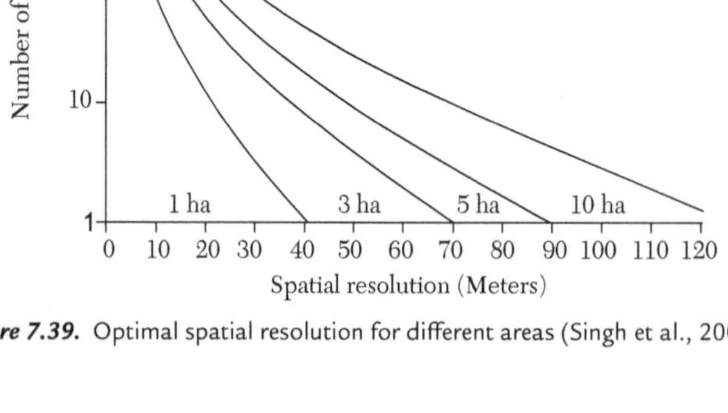

Figure 7.39. Optimal spatial resolution for different areas (Singh et al., 2001)

When the landscape of a study area is complex and heterogeneous, selecting sufficient training samples becomes difficult. This problem would be complicated if medium or coarse spatial resolution images are used for classification, because a large volume of mixed pixels may occur. In general, spatial resolution is the most important factor that affects classification details and influences the selection of a classification algorithm, as shown in Table 7.13. Therefore, selection of training samples must consider the spatial resolution of the remote sensing data being used, availability of ground reference data, and the complexity of landscapes in the area. Several factors, such as spatial resolution of data, different sources of data, a classification system, and availability of classification software, must be taken into account when selecting a classification method for an application. Since different classification methods have their own merits and demerits, the basic question remains which classification approach is suitable for a specific study. Different classification results of the same area may be obtained depending on the classifier(s) chosen.

Table 7.13. Relation between Spatial Resolution and Classification Approach (Prasad et al., 2015[a]).

High-resolution images	Objects are made up of several pixels.
	Object-based classification is superior to traditional pixel-based ones.
Medium/low-resolution images	Pixels and objects are similar in scale.
	Both pixel-based and object-based image classifications perform well.

Although many classification approaches have been developed, which approach is suitable for features of interest in a given study area is still not fully understood. Modern satellite image classification software provides a wide choice of algorithms for classification, so users are normally confused about the best among the algorithms. No algorithm is considered to be effective in all possible cases. Instead, each method has its own scope and limitations. One can only determine the best classification algorithm by trying out each of them and assessing the accuracy of the results, and then finally picking the method that gives the best results. However, it can be a tedious, lengthy, and time-consuming process, especially if images of large volumes are to be processed. In order to choose the right classification algorithm, it is important to understand the mathematical basis of algorithms which would help deciding when to use or not to use any classification algorithm. Ukrainski (2017) presented a diagram in Figure 7.40 which could help in selecting some of the most common tools for supervised classification.

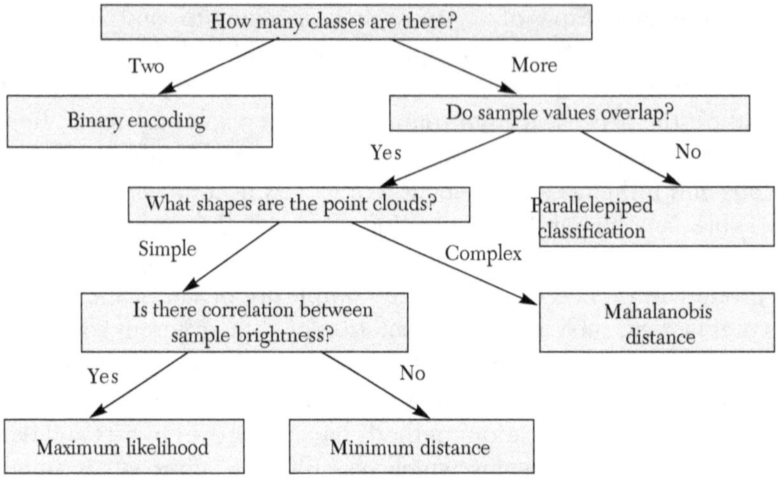

Figure 7.40. Approach to select a supervised classification algorithm.

Classification algorithms can be per-pixel, sub-pixel, and per-field. Per-pixel classification is still most commonly used in practice, however, the accuracy may not meet the requirements because of the impact of the mixed pixel problem. Sub-pixel algorithms have the potential to deal with the mixed pixel problem, and may achieve higher accuracy for medium and coarse spatial resolution images. For fine spatial resolution data, although mixed pixels are reduced, the spectral variation within land classes may decrease the classification accuracy. Per-field classification approaches are most suitable for high spatial resolution images. When using multisource data, such as a combination of spectral signatures, texture and context information, and ancillary data, advanced nonparametric classifiers such as NN, DT, and knowledge-based classifiers may be more suited to process these complex data, and thus are increasingly being used in various remote sensing applications in recent years.

Selection of a suitable classifier also requires consideration of factors such as classification accuracy, algorithm performance, availability of ancillary data, the analyst's experience, and available computational resources (DeFries & Chan, 2000). In many cases, contextual-based classifiers, per-field approaches, and machine-learning approaches provide a better classification result than the MLC, although some trade-offs exist in classification accuracy, time consumption, and computing resources. The selection of a proper algorithm is a challenging issue to get satisfactory results. In some cases, with multiple dimensions (features) but limited training data, the SVMs might work better, but, if there is a lot of training data but fewer dimensions, the NNs might yield a better prediction/classification accuracy. In addition, it is necessary to fine-tune various parameters for different algorithms (i.e., variable importance for RF, number of hidden layers and neurons for NNs, and "decision function shape" for SVMs etc.; Ukrainski, 2017).

For a variety of application requirements, a single classifier classification accuracy is not high and easily leads to fitting problems. However, if a multiple classifier combination is more accurate than a single classifier up to the collective decision, it is able to resist more noise. Sometimes, a better accuracy may be achieved by combining multiple algorithms together, which is called ensemble algorithms. It is also possible to combine SVM and NNs or SVM and RF or any other combination to improve prediction accuracy. Prediction accuracy might change based on the particular feature which is being used for classification/prediction purposes. Another factor which affects the choice of the algorithm is whether the available data is linearly separable or not. For instance, linear classification algorithms (e.g., SVM, logistic regression) expect that the data can be divided by a straight line in linear space. Assuming that the data is linearly separable might work for most of the cases but will bring down the prediction/classification accuracy for some scenarios. Hence, it must be ensured that the algorithm used is able to handle the available data.

CHAPTER 8

Applications of Remote Sensing

8.1 INTRODUCTION

Exploration of Earth's natural resources is vital for the socioeconomic development of any country. State-of-the-art remote sensing technology offers timely and accurate information on various natural resources, such as land, water, forests, mineral resources, and so on for their mapping, monitoring, management, and sustainable planning (Garg, 2019). Remote sensing data have inherent advantages of synoptic viewing, repetitive imaging, capability to study the inaccessible areas, at relatively low cost and near real time (Gibson, 2000). Remote sensing image classification for an application is a complex process, and requires consideration of many factors. The major steps of image classification may include: (*a*) selection of a suitable classification system, (*b*) image preprocessing methods, (*c*) selection of training samples, (*d*) selection of suitable classification approaches and post-classification processing, and (*e*) accuracy assessment. Additionally, the user's need, scale of the study area, economic condition, and analyst's skills are important factors influencing the selection of remotely sensed data, the design of the classification procedure, and the quality of classification results (Lu & Weng, 2007).

Remote sensing applications primarily involve interpretation and analysis of EMR interacting with Earth and its atmosphere from data collected by sensors onboard aerial or satellite platforms. Remote sensing data is now employed in a wide variety of disciplines, such as natural resource management, urban planning and development, climate and environmental assessment, weather forecast, agricultural drought modeling, rainfall prediction, flood mapping, crop growth monitoring, forest fire mapping and urban applications, traffic monitoring, vehicle navigation, and urban development projects (Garg, 2021). These data have also been extensively used in a number of Earth science applications for societal benefits. Some of the applications are explained in detail in this book through Chapters 8–15.

Some of the frequently used coarse resolution data for land cover monitoring are MODIS sensors and SPOT Vegetation data. These data are usually free, well archived, and available soon after their acquisition. The details of other remote sensing data and geospatial data available free of cost are given in Chapter 1, but, they may not be optimal for monitoring the extent and rates of change of many land use and land cover classes, such as urban areas. However, the advantages of coarse resolution images are that these can be processed quickly, and provide a valuable complement to finer resolution data. Very high spatial resolution data, generally provided by commercial satellites, are usually very expensive. However, they are perhaps the only option available, other than repeated aerial/drone or field surveys, to use for monitoring small-size areas such as water bodies, minor rivers, habitats, dam sites, agricultural land, and so forth (Ilie & Vasile, 2011). However, it would be very expensive and its to conduct nationwide monitoring of these features from commercial satellites. In order to use remote sensing data for an application, the user has to have information about the data characteristics, region of interest, and the type of results expected. Based on this knowledge, the user can select the most suitable image and its processing technique, and develop a methodology to solve the application problem. Table 8.1 presents the difference in characteristics of various remote sensing data collected from LiDAR, optical/mechanical and SAR.

Table 8.1. Characteristics of LiDAR, Optical/Multispectral, and SAR Data

Parameters	LiDAR	Optical/Multispectral	SAR
Platform	Airborne	Airborne/spaceborne	Airborne/spaceborne
Radiation	Own radiation	Reflected sunlight	Own radiation
Spectrum	Infrared	Visible/infrared	Microwaves
Frequency	Single frequency	Multifrequency	Multifrequency
Polarimetry	Not available	Not available	Polarimetric phase
Interferometry	Not available	Not available	Interferometric phase
Acquisition time	Day/night	Day only	Day/night
Weather	Blocked by clouds	Blocked by clouds	Penetrate through clouds

8.2 SOME USEFUL APPLICATIONS

Satellite remote sensing has found applications in a wide variety of fields of interest, both for civil and military purposes. With the availability of higher resolution images taken by various satellites, it has become possible to prepare maps at a scale of 1:5,000. Potential applications of aerial photography and satellite imagery are numerous. Some important applications of remote sensing systems are shown in Figure 8.1.

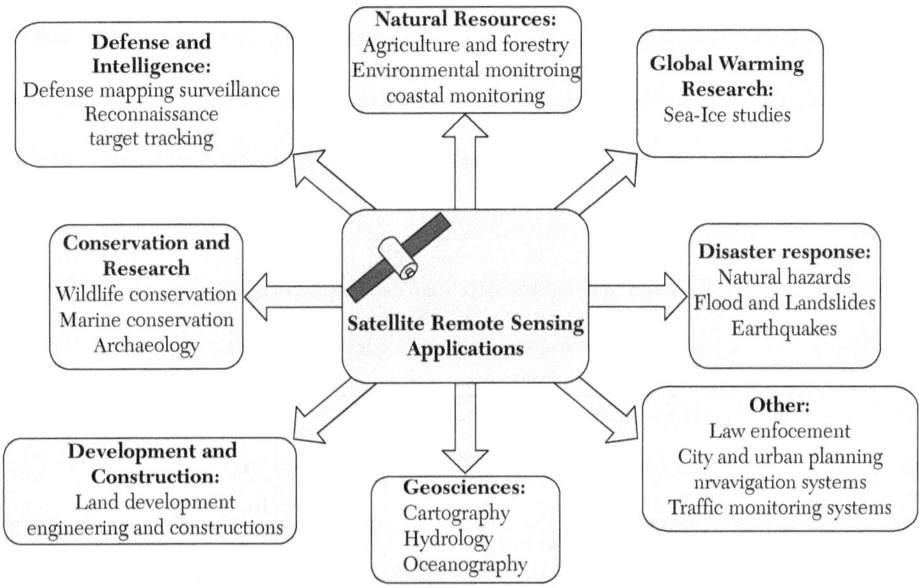

Figure 8.1. Various applications of remote sensing (Ilie & Vasile, 2011).

Remote sensing plays a significant role in GIS also, as thematic maps or DEMs generated from these images are used directly as the input in many applications. As the maps prepared from satellite images are location-based, they are specifically useful for planning developmental activities in GIS. The information derived from the satellite imagery is also useful for government decision-making, civil defense operations, police, and health sectors. These days, the use of satellite imagery has also become almost mandatory in all government plans and works. Therefore, there is a further requirement for development of new sensors for various land and ocean applications, some of which are given in Table 8.2.

Spatial and temporal resolution of data requirements may vary widely for monitoring terrestrial, oceanic, and atmospheric features and processes (Kadhim et al., 2016). Each application of remote sensing sensors has

Table 8.2. Future Requirements of Earth Observation Sensors for Land and Sea Applications

Area	Sensor requirement	Specifications	Possible applications
Land	High spatial and temporal resolution Multispectral	4 bands in VNIR-SWIR; resolution: 20–30 m, 5 days repetivity, >700 km swath	Multiple crop forecasting
	High temporal and moderate spatial resolution	VNIR, SWIR, MIR and TIR bands; resolution: 50–100 m (VNIR), 250–500 m (TIR), daily sensor repetivity, wide swath ~1000 km	Regional vegetation monitoring; parameter retrieval
	Very high spatial resolution pan	Panchromatic resolution: <40 cm, -8 km swath, spot imaging stereo pairs	1:1000 scale mapping, urban and local area planning, facility management, city 3D modeling
	Hyper-spectral sensors	Large number of narrow bands (0.4–2.5 (μm –10 nm bandwidth, 20–30 m resolution, 20–25 days repetivity	Applications in forestry, agriculture, coastal zone and inland waters, soil, geology
	Atmospheric corrector	Coarse resolution (500 m); 4 bands (0.485, 0.94, 1.625, 2.1 μm)	Need for simultaneous measurement along with other multispectral sensors for atmospheric correction
	Synthetic aperture radar (SAR)	Dual frequency (L and C band) polarimetric system	Crop monitoring, soil moisture, tree canopy
Ocean	Scatterometer	Spatial resolution: 25 km (open), 1 km (coast); temporal resolution: 6 hourly	Wind speed and direction, sea ice
	Altimeter	Spatial: 3 km; 50–100 km (coast); temporal: daily	Sea level, wave height
	Thermal sensor	Bands at 11 and 12 μm; spatial resolution ~1 km and NEDT-0.05 K	Sea surface temperature
	SAR	Microwave sensor	Bathymetry, wave-spectra, wave height
	Salinity sensor	Frequency: 1 GHz; spatial resolution: 50 km (open)	Ocean salinity
	Ocean color monitor	12 bands between 400 and 1100 nm; 300 m spatial and daily temporal resolution	Retrieval of ocean bio-geo-chemical parameters; atmospheric correction

its own unique resolution requirements and, thus, there are trade-offs between spatial resolution and coverage, spectral bands, and so forth as shown in Figure 8.2. For example, a high temporal resolution is essential for emergency situations, such as floods, as emergency situations require frequent observations on the day to monitor the rapid changes. On the other hand, urban infrastructure planning applications may require spatial data over a longer period, for which annual observations are often sufficient. However, in both cases, sometimes high spatial resolution images may be required to observe detailed processes in small areas. High temporal resolution is required for applications such as weather that changes rapidly. Operational weather forecast, therefore, requires satellite observations with high temporal resolution, often with poor spatial resolution.

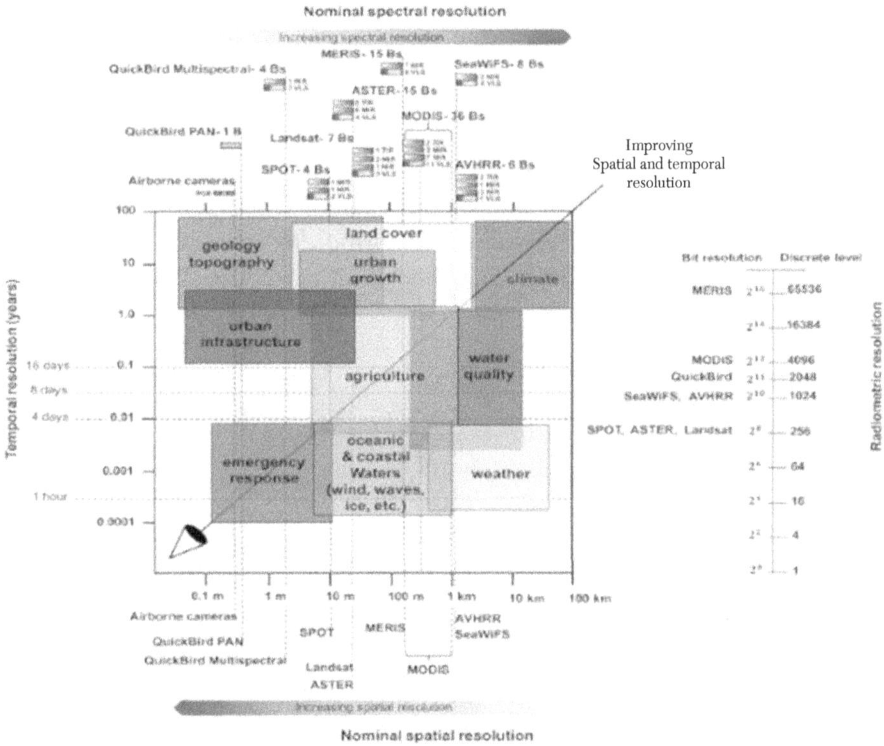

Figure 8.2. An overview of spectral, spatial, temporal, and radiometric resolution of different optical satellite systems, along with their possible application areas (Kadhim et.al. 2016). B is the number of spectral bands, VLS is the visible light spectrum (please see color figures in companion files).

Some important applications are discussed as follows:

8.2.1 Agriculture Development

There has been increased emphasis on the potential utility of using remote sensing platforms to obtain real-time assessments of agricultural crops. With the increasing population across the world and the need for increased agricultural production, there is an urgent need for proper management of the world's agricultural resources. It is therefore necessary to obtain reliable data on the types, quality, quantity, and location of these resources (Jung et al., 2021). Agriculture is the one area where remote sensing has been used successfully to: (*a*) map the crop types, (*b*) assess the crop damage, and (*c*) estimate production, and so on. Preharvest acreage and production estimation of major food grains, oilseeds, and crops can be made by using temporal satellite images. Figure 8.3 shows the collection of time-series agricultural crop data available from various sources, ground, GPS, airplane, and satellite, to be used for crop management.

Precision agriculture is a production system that promotes variable management practices within a field, according to the site conditions. This system is based on new tools and sources of information provided by modern technologies of GPS, GIS, yield monitoring devices, soil, plant and pest sensors, and remote sensing data (Seelan et al., 2003). Various insurance companies require accurate data on crop growth and yield, as well as crop

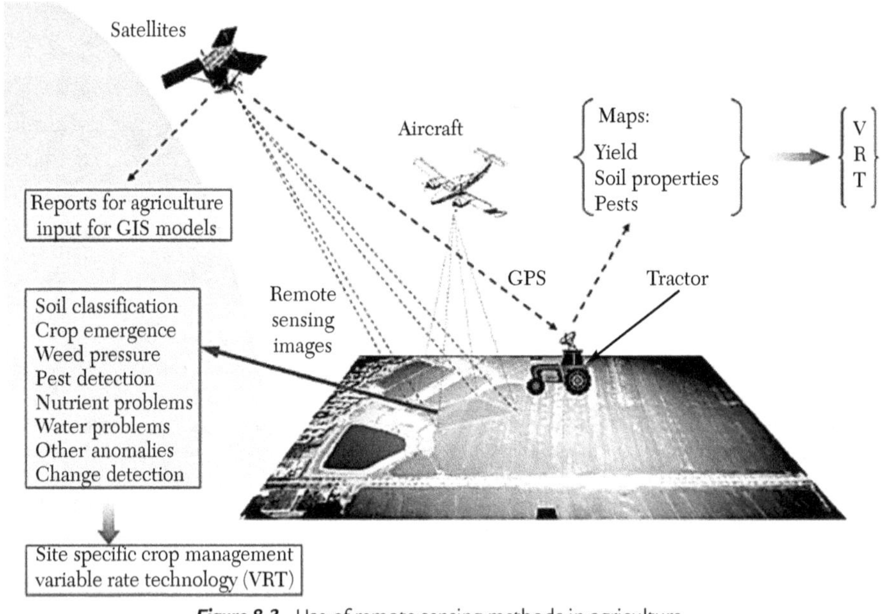

Figure 8.3. Use of remote sensing methods in agriculture.

damage in a particular area. This information can be used to verify seeded crops and as verification against crop insurance fraud. Multi-crop inventory and detection of crop diseases and pests require short visit measurements at high spatial resolution with hyperspectral technology. Vegetation stress, disease, and pest detection using hyperspectral techniques such as red edge shift and mineral targeting are some of the important applications, that would require a hyperspectral spectrometer onboard satellites on an operational basis. Satellite imagery and NDVI technologies are used in order to monitor crop growth (Jackson et al., 1980). Health of crops can be measured using remote sensing in order to save almost 10 percent of fertilizer as well as money and time invested on that fertilizer.

Different approaches to retrieve soil moisture have used a wide range of satellite data. The basic difference among these approaches are the wavelength band used, source of electromagnetic energy, response measured by the sensor, and model to relate the signal response to soil moisture content. Different wavelength bands from the optical, thermal, and microwave regions are used for crop and soil moisture study. For example, MODIS imagery has been used frequently to map soil types of large areas for agricultural predictions and planning. Active and passive sensors are used in order to determine soil moisture content. The Radarsat-2 is another example of an active sensors which illuminates the targeted area in order to measure backscatters. It has a comparatively high resolution but with reduced accuracy. The Soil Moisture and Ocean Salinity (SMOS) is an example of a passive sensor which is highly accurate but comparatively holds poor resolution. The capabilities and salient features of various sensors for soil moisture estimation are presented in Table 8.3 (Shukla & Garg, 2014).

Table 8.3. Summary of Remote Sensing Techniques for Near-Surface Soil Moisture Estimation

Sensors	Spectral range	Properties observed	Suitability	Advantages	Limitations
Optical	0.4–0.9 μm	Soil reflection.	Vegetated area.	Fine spatial resolution. Broad coverage.	Limited surface penetration. Cloud contamination. Smoke, fog, rain effect. Weather dependency.

(Contd.)

Sensors	Spectral range	Properties observed	Suitability	Advantages	Limitations
Thermal	3–5 μm and 8–14 μm	Radiant temperature.	Bare land.	Fine spatial resolution. Broad coverage.	Limited surface penetration. Cloud contamination. Rain effect. Perturbed by meteorological conditions and vegetation.
Active microwave	1.0–100 cm	Backscattering coefficient. Polarization. Dielectric constant.	Vegetated area and bare land.	Weather free operation. Moderate to high surface penetration (1 cm–1 km). High spatial resolution.	Varying resolution from near to far range. Sensitive to roughness and moisture in vegetation. Speckles.
Passive microwave	0.15–30 cm	Brightness temperature. Dielectric constant.	Vegetated area and bare land.	Weather free operation. Moderate surface penetration.	Coarse resolution. Sensitive to roughness and moisture in vegetation.

8.2.2 Base/Thematic Mapping

Remote sensing data has been used extensively for the preparation of base maps showing all basic details, such as roads, rails, rivers, canals, ponds, urbanization, forest, open land, agriculture, and so on for assisting planners and engineers. These data have also been used for updating the existing topographic maps. The amount of detail that an orthoimagery produces using high resolution satellite imagery is of immense value. Moreover, the ortho-images can be used straightaway for accurate measurements, like on a topographic map (Garg, 2019). Cadastral mapping for rural planning is an important thrust area. Very high-resolution images (<1 m) are very useful for preparation of rural development plans and creation of land information system at a 1:1,000–1:4,000 scale.

8.2.3 Digital Terrain Mapping

The terrain information can be acquired from remote sensing data by generating a digital terrain model (DTM). A DTM is a 3D representation of a terrain surface in digital form. It also includes information about the location of objects/features on the ground. The term DTM has been used interchangeably with the digital surface model (DSM) in civil engineering, yet they are different (Burrough & McDonnell, 1998). The term DTM refers to the bare ground elevation points without any non-ground points (e.g., buildings and vegetation); however, the DSM includes all objects on the Earth's surface. There are three general sources of data for DTM generation: (*a*) ground survey data, such as terrestrial laser scanning, total stations, and GPS; (*b*) available topographic maps; and (*c*) remote sensing data, both airborne and satellite photogrammetric methods, airborne laser systems (LiDAR), and those collected from other flying platforms. Specifications of very high resolution optical satellites with the capability of producing stereo-images for generation of DEMs are summarized in Table 8.4.

The DTMs can be generated from a stereo-pair of images with an overlapping region but taken from different viewpoints. Figure 8.4(a) shows various steps used for the creation of DTMs/DSMs from remote sending images, while Figure 8.4(b) illustrates the procedure to create DTMs/DSMs from airborne LiDAR data (Karan et al., 2014). Both the approaches require preprocessing the data, creation of DTMs/DSMs, and post-processing of data. Many image processing and GIS packages can be used to generate the DTMs from remotely sensed/geospatial data (Garg, 2021). On the other hand, DEMs that are freely available or at very low cost from USGS and Google Earth websites can also be used. However, such data are not updated frequently and are available at relatively low spatial resolutions. In terms of visualization and sharing real-time information through the Internet, Google Earth has been ranked as the most popular source of satellite images for research and applications (Chen et al., 2009).

Assessment of terrain for specific developmental activities can be made through terrain evaluation. The DTMs can facilitate the identification of alternative alignments of canals, roads, pipelines, transmission lines, and so on. For example, the construction of a dam requires the knowledge of terrain for proper planning, location, construction, and maintenance of engineering facilities. DTMs have been used successfully in various areas of civil engineering and construction, including volumetric calculations in cut-and-fill problems, insite planning, 3D landscape modeling, and visualization for construction engineering tasks (El-Sheimy et al., 2005). High spatial resolution satellite data along with DTMs can be analyzed for delineation of various landforms, mapping of soil classes, engineering construction, soil relationships, and grouping of landforms with various physiographic setting or terrain associations required for watershed analysis.

Table 8.4. Specifications of Very High Resolution Optical Satellites Producing Stereo-Images (Deilami & Hashim, 2011).

Satellite/Sensor	Date of launch	Resolution (stereo-image) Min/Max (m)	Swath width Min/Max (km)	Stereo	B/H ratio
IKONOS	Sep 24, 1999	Pan (Nadir) 0.8, MS (N) 3.2 Pan-sharpened 0.8–1.0	11 × 11	Along-track	0.54–0.83
EROS -A1	Dec 5, 2000	Pan (Nadir) 1.9	14 × 14	Along-track, Across-track	Variable
QuickBird	Oct 18, 2001	Pan 0.61, MS 2.4	16.5 × 16.5	Along-track	0.6–2.0; most collections between 0.9 and 1.2
SPOT-5	May 4, 2002	Pan 2.5–5, MS 10	60 × 60	Along-track, Across-track	Variable
Cartosat -1	May 5, 2005	Pan 2.5	26 × 26	Along-track	0.62
ALOS (PRISM)	Jan 24, 2006	Pan 2.5	35 × 35	Along-track triplet of images	1
EROS–B1	Apr 25, 2006	Pan 0.7	7–21	Along-track, Across-track	Variable
KOMPOSAT-2	July 28, 2006	Pan 1, MS 4	15	Across -track	Variable
WorldView-1	Sep 18, 2007	Pan (Nadir) 0.5 (20° off-Nadir) 0.55	17.6 × 17.6	Along-track	Variable
GeoEye-1	Sep 6, 2008	Pan (Nadir) 0.5 MS (Nadir) 2 Pan-sharpened	15.2 Contiguous area 224 × 28	Along-track	Variable
WorldView-2	Oct 8, 2009	Pan 0.46 (20° off-Nadir) 0.52	48 × 110	Along-track	Variable

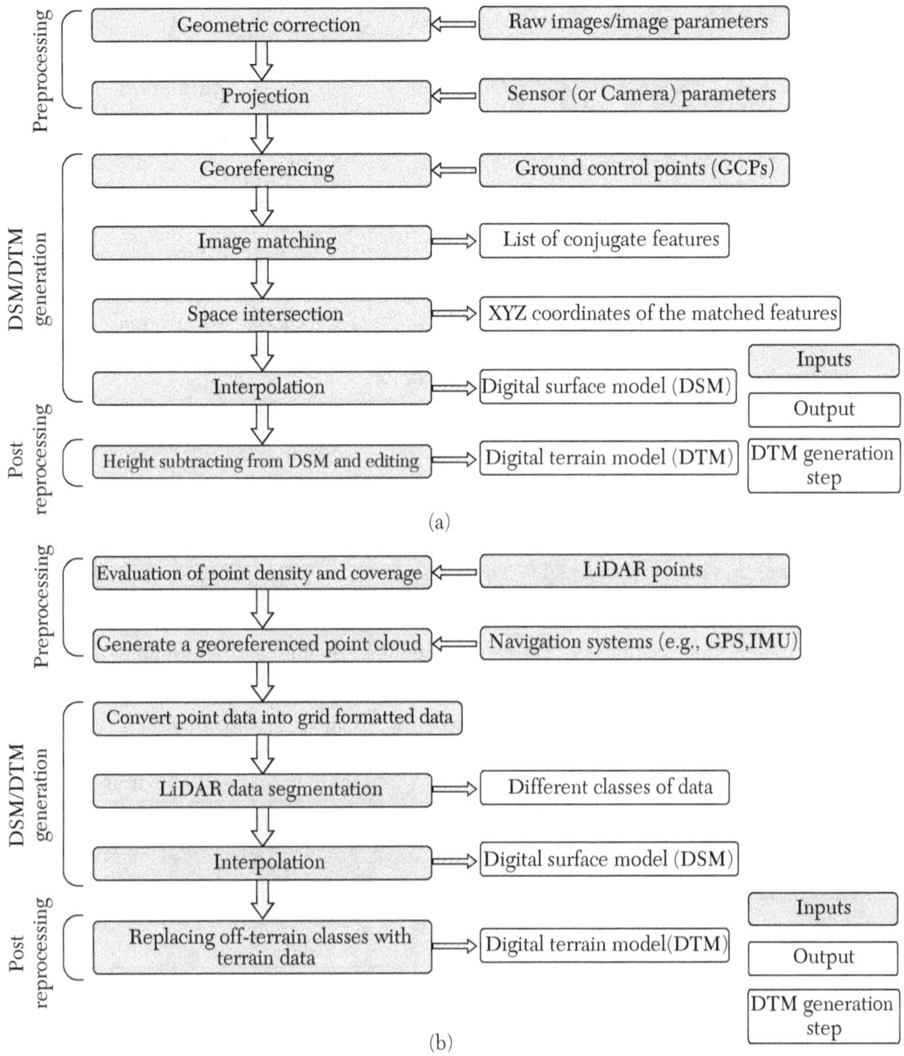

Figure 8.4. Procedure to create DTMs/DSMs from (a) airborne satellite images, (b) airborne LiDAR data (Karan et al., 2014).

8.2.4 Disaster Mitigation Planning

Natural calamities can be devastating and at times difficult to assess. Remote sensing can be used to study the damages caused by various disasters such as earthquakes, volcanoes, landslides, floods, and so forth. Remote sensing along with other devices have also been used to predict the occurrence of some natural hazards, such as landslides. Table 8.5 summarizes the remotely sensed data types and image processing techniques which could

Table 8.5. Remotely Sensed Data Types and Image Processing Techniques for Various Disaster Related Applications (Gähler, 2016).

Data type	Sensor examples	Technique	Application
Optical (multispectral)	Worldview, Pleiades, QuickBird, IKONOS, Rapid Eye, SPOT, ASTER, Landsat, ALOS	Manual interpretation	Infrastructure and property damage due to flooding, earthquakes, landslides, etc.
		Spectral classification	Location and extent of flooding, landslides, volcanic debris, fire scars
		Semi-variogram analysis and other textural classifiers	Damage due to earthquakes; location of landslides
		Image thresholding and band ratios	Location and extent of flooding, landslides, volcanic debris, fire scars
		Image differencing	Location and extent of flooding, landslides, volcanic debris, fire scars
		Post-classification change detection	Location and extent of flooding, landslides, volcanic debris, fire scars
		DEM generation	DEM is used as supplementary information in variety of studies
Thermal	ASTER, MODIS, AVHRR	Split window	Crater lake temperatures, lava flow, precursor to earthquake activity, temperature and size of fire hotspots
		Dual band	Crater lake temperatures, lava flow, precursor to earthquake activity, temperature and size of fire hotspots
Synthetic	TerraSAR-X, TANDEM-X,	Coherence	Coherence change detection due to landslide, flooding, fire, etc.
Aperture Radar (SAR)	Cosmo-SkyMed, Radarsat-1/2, JERS-1, ERS-1/2, ENVISAT, ALOS	Backscatter intensity	Coherence change detection due to landslide, flooding, fire, etc.
		Interferometry/ DEM generation	Change detection due to landslide, flooding, fire, etc.
		Differential interferometry	Surface deformation due to volcanic or tectonic activity; velocity and extent of slow moving landslides
		Polarimetry	Landcover classification and change detection

be used for information extraction about natural disasters (Gähler, 2016). In general, the availability of appropriate data with respect to acquisition time, image extent, and spatial, temporal, and spectral resolution is an important consideration for most applications of disaster events. Particularly, there are numerous examples requiring the availability of remote sensing data at shorter intervals, like damage assessment mapping for earthquakes, landslides, or flooding. However, for monitoring the spread of an oil spill or the extent of wildfire, the temporal resolution of images is very relevant.

Around the globe, landslides cause noticeable death and wealth loss, in addition to the damage to environment. Remote sensing technology is heavily implemented in order to assess the possible damage from hazards in order to develop a proper response system. Geologists can identify the rock types, geomorphology, and changes from natural events such as a flood or landslide. The potential landslides are studied using stereo- and optical images with slope/DTM information. Remote sensing techniques can be used to develop a system to provide early warning regarding potential landslides (Garg, 2019).

Assessment of the affected area from an earthquake is essential for planning rescue measures in a quick and accurate manner. This information has to be prepared and executed quickly and with great accuracy. Object-based image classification using change detection (pre- and post-event) is a quick way to acquire damage assessment data.

Periodic droughts are very common in many countries due to failure or irregular onset of monsoon, as the agriculture in such countries is mostly dependent on rainfall. Early famine signs are important for governments and planners so that they can plan to deliver food in areas affected by the shortage. Remote sensing images are capable of understanding the drought phenomenon. These data have been used for the assessment and monitoring of agricultural drought using a satellite data-based vegetation index to provide timely information on drought severity.

Meteorological drought depends on precipitation deficit and duration of period with precipitation deficit. The study by Choudhary et al. (2013) used meteorological-based drought indices such as normalized deviation, De Martonne's index, pluvothermic quotient, negative moisture index, and standard precipitation index (SPI) values to get the spatial pattern of meteorological-based drought in Jodhpur district, India. The methodology used is presented in Figure 8.5. From crop yield and production trend, an equivalent NDVI threshold is identified to get the agricultural drought risk where the occurrence is likely to be high. The Landsat-7 ETM+ and

Landsat-5 TM satellite sensor data for a period of 21 years (1991–2011) are used for calculating the brightness temperature (BT) and land surface temperature (LST). The BT values are converted to the vegetation condition index (VCI) and the temperature condition index (TCI), which are useful indices for the estimation of vegetation health and monitoring the drought. From the SPI analysis, it is found that in 2002 the entire area was affected with greater intensity of meteorological drought.

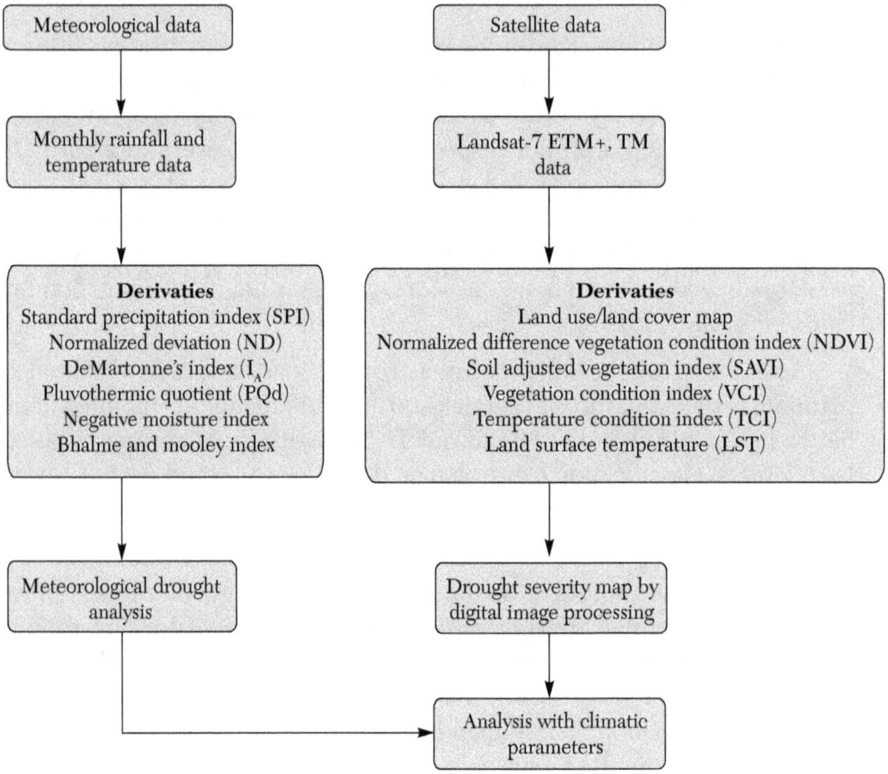

Figure 8.5. Flow diagram of methodology used for drought assessment (Choudhary et al., 2013).

Remote sensing provides comprehensive, reliable, and timely information on flood inundated and drainage congested areas, as well as extent of damage to crops, structures, and so on. River configuration, silt deposits, and vulnerable areas of bank erosion and flood risk zones can be studied using remote sensing data. However, comprehensive flood monitoring would require integration of ground measurements and remote sensing data and flood-plain characteristics, including the topography/DTM. Flood risk zone maps can be updated with high spatial resolution data and DEMs. Flood damage vulnerability analysis requires integration of information on satellite-derived physical damage and socioeconomic data. However, presence of cloud cover during the flood-monitoring period is a limitation for optical remote sensing data. In such cases, microwave data, such as from ERS1/2 and Radarsat satellites, are effectively used during flood events (Yang et al., 2015). The WiFS sensor with 5 days repetivity on board in the IRS-1 C/D satellite provides a unique opportunity to map and monitor flood events frequently over large river basins.

8.2.5 Geology and Minerals

Remote sensing data is helpful for updating existing geological maps, rapid preparation of lineament and tectonic maps, identifying sites for quarrying minerals, and locating fossil fuel deposits. Some of the remote sensing applications in geology include: bedrock, lithological, and structural mapping. Multispectral remote sensing data can provide information on rock composition based on their spectral reflectance pattern (Gupta, 2013). Radar data provides an expression of surface topography and roughness, and thus is extremely useful, especially when integrated with another data source to provide detailed relief.

Remote sensing data have been used widely to study bedrock, and structural geological features like faults/folds/fractures, geomorphology, and lithology, and are in turn used in mineral/oil exploration, ground water exploration and development, engineering geology, environmental geology neotectonics, seismo-tectonics, and so forth (Sabins, 1996). During the pre-feasibility and feasibility stages of mineral exploration, it is important to know about the mineral potentiality of the area to be considered for mineral extraction. In such scenarios, satellite remote sensing-based mapping and its integration in a GIS platform help mapping the mineral potential zones.

Hyperspectral remote sensing technology helps to build potential maps of these minerals, which have unique chemical compositions and spectral reflectance. With the help of spectral analysis of satellite images, it is possible to quickly identify and map mineral availability through indirect indicators. Radar and multispectral spectral reflectance can be used to derive valuable information in the field. It enables exploration to narrow down geophysical, geochemical, and test drilling activities to high potential zones (Birk & McCord, 1994).

8.2.6 Healthcare

The application of satellite data and techniques in healthcare supports more accurate and timely decision-making (Garg, 2021a). Temporal satellite data are able to visualize changes in patterns, environmental impacts, identification, and changes within high-risk disease areas. These images have been used to analyze the rates of spread of infectious diseases and other medical conditions across the globe, for example in the recent COVID-19 pandemic. In the challenging times of the COVID-19 outbreak, satellite monitoring has been playing a pivotal role in minimizing the spread of infection, by identifying the risk zones and facilitating the quick response. Several investigators in the health community have explored remotely sensed-based environmental factors that might be associated with disease-vector habitats and human transmission risk. Remote sensing has been found to be helpful to provide early warning of outbreaks, and to analyze the accessibility of medical services across regions. UAVs/drones, used for disinfection of areas by spraying chemicals in the areas during the COVID-19 outbreak, selected the roads and areas directly from Google maps to perform the task efficiently.

Cloud computing platforms for remote sensing data combined with analysis-ready datasets (Proficz & Drypczewski, 2017) and high-level data products have made satellite remote sensing more accessible for vector-borne diseases. For example, long-term remote sensing observations facilitate the analysis of climate–malaria relationships, and high-resolution images can be used to assess the effects of agriculture, urbanization, deforestation, and water management on malaria. Figure 8.6 shows a pathway through which satellite data provide information to predict the geographic patterns

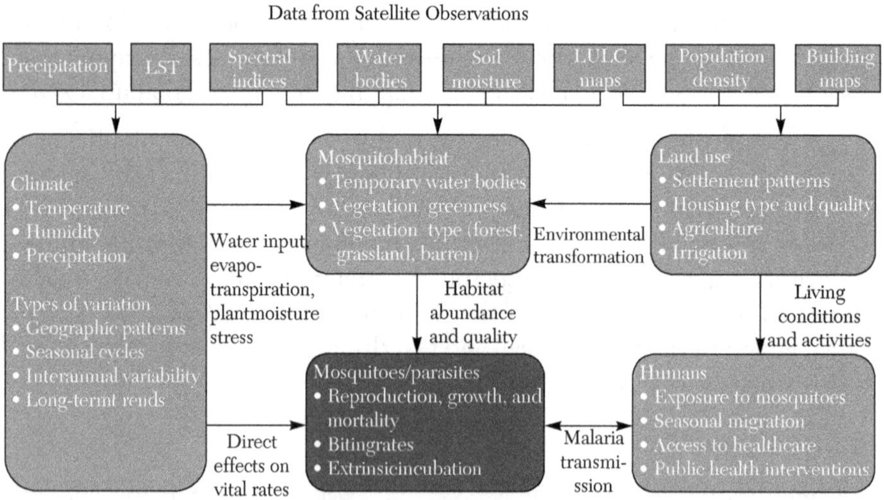

Figure 8.6. Information from satellite images about malaria (LST-land surface temperature, LULC–land use and land cover; Wimberly et al., 2021).

and changes over time in climate factors, mosquito habitats, and human land use for malaria (Wimberly et al., 2021). These environmental variables influence malaria transmission through their effects on mosquitoes, parasites, and humans. Very high resolution satellite imagery and SAR data can increase the precision and frequency of observations. Geospatial data products including remote sensing images can be used to support global efforts toward malaria control, elimination, and eradication.

8.2.7 Infrastructure Development and Planning

Up-to-date information regarding road networks is important for emergency services, urban planning, and commuting. Remote sensing data along with GIS and GPS have been extensively used to find out suitable corridors/alignment of roads, railways, pipelines, canals, tunnels, and so forth. High-resolution satellite imagery is used to derive information about the existing roads as well as damaged roads so that the engineers know the exact locations where immediate road repair is required. Assessment of road conditions with high accuracy is now possible using various geospatial and remote sensing data in GIS. Its proper assessment and maintenance management saves a lot time and funds of organizations involved with repair and maintenance (Kumar, 2020). Multispectral images and the OBIA algorithm can be used to develop automated road networks that serve useful information for planning in large areas.

Remote sensing technology allows the preparation of maps (2D or 3D) of buildings. Google and Bing maps have already implemented such functionality. Earth observation satellites are useful for monitoring and planning urbanization as well as identifying encroachment on government land. With the availability of very high resolution data from QuickBird, IKONOS, and other satellites, it is possible to prepare maps at a 1:4,000/2,000 scale which helps in a big way for all infrastructure projects (Garg, 2021b). City guide maps at a 1:10,000 scale have also been prepared using high-resolution satellite images. Base map updating, urban land use change and growth monitoring, master plan development, facilities management, taxation applications, and urban information systems are other major studies where remote sensing data have made tremendous contributions (Fauvel, 2007).

8.2.8 Land Use and Land Cover Mapping

Remote sensing data and technologies have been used mostly for land use and land cover (LULC) mapping, such as agriculture, open land, forest, snow cover, urban area, rivers, ponds, and so on. The outcomes of LULC mapping are used for global, regional, local mapping, change detection, landscape planning, and driving landscape metrics (Bunruamkaew & Murayama, 2012). LULC mapping is one of the most important applications of remote sensing data, as these maps provide basic information for most of the developmental projects on the Earth's surface. Land cover corresponds to the physical condition of the ground surface, for example, forest, grassland, concrete pavement and so on, while land use reflects human activities, such as the use of land, for industrial zones, residential zones, agricultural fields, and so forth. Temporal images are used to determine the changes in various LULC classes and the nature of the change as well as the utilization of land. Land cover change detection is also required for the management and sustainable planning of land resources (Garg, 2021b).

Satellite remote sensing, in conjunction with the GIS, has been recognized as a powerful and effective tool in detecting changes in LULC. It provides cost-effective multispectral and multi-temporal data which are used to derive the information valuable for understanding and monitoring land development patterns. Satellite imagery has been used to monitor discrete land cover types by spectral classification or to estimate biophysical characteristics of land surfaces via linear relationships with the spectral reflectances or indices (Steininger, 1996). Information on LULC change patterns is directly useful for determining and implementing a developmental policy, for example, municipal authorities for tax assessment. Governments also require this information for the general protection of national resources, parks, wetlands, wildlife sanctuaries, and so on. Urban planners use LULC change data to model an accurate measure of population growth (Chaudhuri & Clarke, 2013). A general supervised classification approach to classify satellite images into LULC maps is shown in Figure 8.7.

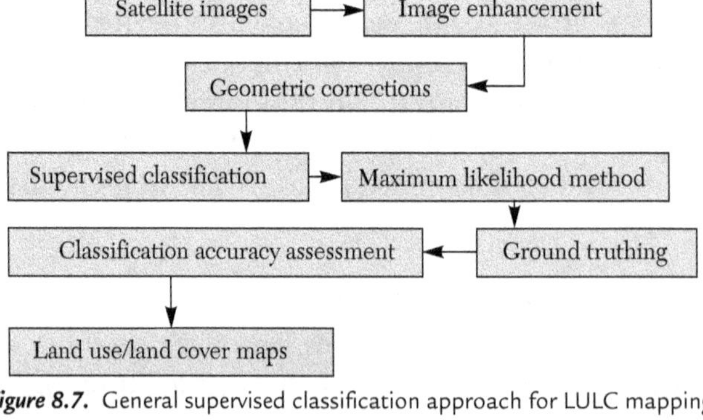

Figure 8.7. General supervised classification approach for LULC mapping.

8.2.9 Location Based Studies

Location based studies are important for many applications, like hiring a cab, tracking a criminal or emergency services, or food delivery. GPS is very helpful in providing accurate 3D locations of objects. The locational data superimposed on satellite images can help create digital versions of the physical world via various technologies for reality capture. These data can be visualized to understand the complex world, make evidence-based decisions, improve businesses, implement plans, and measure outcomes (Garg, 2019). Tracking of vehicles can be monitored on high-resolution satellite images with the help of integrated GPS and GIS technologies.

8.2.10 Ocean/Coastal Studies

This is one of the large scale applications of remote sensing technology. Remote sensing data have been used for developing capabilities for retrieval of oceanic parameters and monitoring of coastal zones. The SEASAT was the first Earth-orbiting satellite (launched on June 28, 1978) designed for remote sensing of the Earth's oceans using active microwave wavelength instruments. It collected data on sea-surface winds, sea-surface temperatures, wave heights, internal waves, atmospheric water vapor, sea ice features, and ocean topography (Cracknell, 2018). The CZCS (Coastal Zone Color Scanner) flown on the NIMBUS-7 (launched on October 24, 1978) was a six channel scanning radiometer with a resolution of 800 m. The CZCS was used to monitor algal blooms with the benefit of the temporal frequency of NOAA satellites but with enhanced spectral resolution. The development of CZCS enabled sea surface data to be obtained with a spatial, temporal, and spectral resolution on marine chlorophyll and suspended sediment concentration for the development of marine algal blooms over time. As ocean color data is related to the presence of phytoplankton and particulates, these data can be used to calculate the concentrations of material in surface waters and the level of biological activity (Gansvind, 2019).

The SMMR (Scanning Multichannel Microwave Radiometer) with a resolution of 60 km was flown on SEASAT and NIMBUS-7. The objective of SMMR was to obtain ocean circulation parameters such as sea surface temperatures (SST), low altitude winds, water vapor and cloud liquid water content, sea ice extent, sea ice concentration, snow cover, snow moisture, rainfall rates, and differentiation of ice types (Mumby et al., 1999). In India, "Potential Fishing Zone" advisories are generated for the entire Indian coast and disseminated to the fishermen based on the integration of SST derived from NOVA-AVHRR data and ocean color from the OCM (Ocean Color Monitor) of IRS P4.

Mapping of coastal zone features such as LULC, wetland/land form, mangroves, and shoreline changes is essential. Methodologies for retrieval of various ocean parameters such as surface winds, wave forecast model, internal waves, and wave spectrum have been developed using remote sensing images (Al-Tahir et.al. 2006). Remote sensing data are used to measure sea levels with accuracy. The OSCAR (Ocean Surface Current Analysis Real-Time), a remote sensing technology, is used to monitor them with good accuracy (*https://podaac.jpl.nasa.gov/dataset/OSCAR_L4_OC_third-deg*). Remote sensing satellites such as MODIS, CERES, AMSRE, TRMM, and MOPITT have made it possible to observe climate changes and global pollution. Carbon monoxide is considered as the major pollutant in global pollution. It cannot be measured by IR radiation using a spectrometer used by MOPITT satellite. Some diseases are strongly related with climate, land use, and air. Remote sensing technologies have been used to assess climatic changes in areas where these diseases are more likely to happen so that preventive measures could be planned.

8.2.11 Online Mapping Services

A large number of users have now used images from Google Earth, Bing Maps, or Open Street Maps for various purposes. These maps are derived from high-resolution remote sensing images. For example, these maps can provide a quite efficient way to check out locations along with the directions for navigation purposes. GPS is used in these systems for providing user interface on the screen. Maps are great for bringing teams together, remotely. They can be used to deliver services that work better for the users.

8.2.12 Site Investigations and Planning

Site investigations, in general, require topographical and geological information, which can easily be obtained through remote sensing data. Remote sensing is extensively used in site investigations for dams, bridges, roads, canals, pipelines, townships townships, and so on (Eastman, 1999). High-resolution images can be used to locate construction materials like sand, gravel, boulders, quarry rock, and sand–clay mixtures for projects, like dams, bridges, and so forth, across rivers. By analyzing multi-date remotely sensed data of pre- and post-dam construction, the forest and other land at different water levels can be assessed. This would also help in preliminary investigation of locating suitable areas for human resettlement. In selecting suitable sites for bridges and dams, an important consideration is the stability of slopes, upstream and downstream of a river. Past data of river erosion and sedimentation would give clues needed for locating the sites where scour is likely to occur (Yang et al., 2015). High spatial resolution satellite data can also be used for the planning of large projects like airports, harbors, industrial towns, and recreational sites.

The flowchart in Figure 8.8 determines the area best suited for ecotourism development by using five decision criteria and nine factors (Bunruamkaew & Murayama, 2012). The nine GIS-based layers are landscape/naturalness (visibility, land use/cover), wildlife (reservation/protection, species diversity), topography (elevation, slope), accessibility (proximity to cultural sites, distance from roads), and community characteristics (settlement size). The data layers are integrated in GIS to get various suitable sites in the area for ecotourism development.

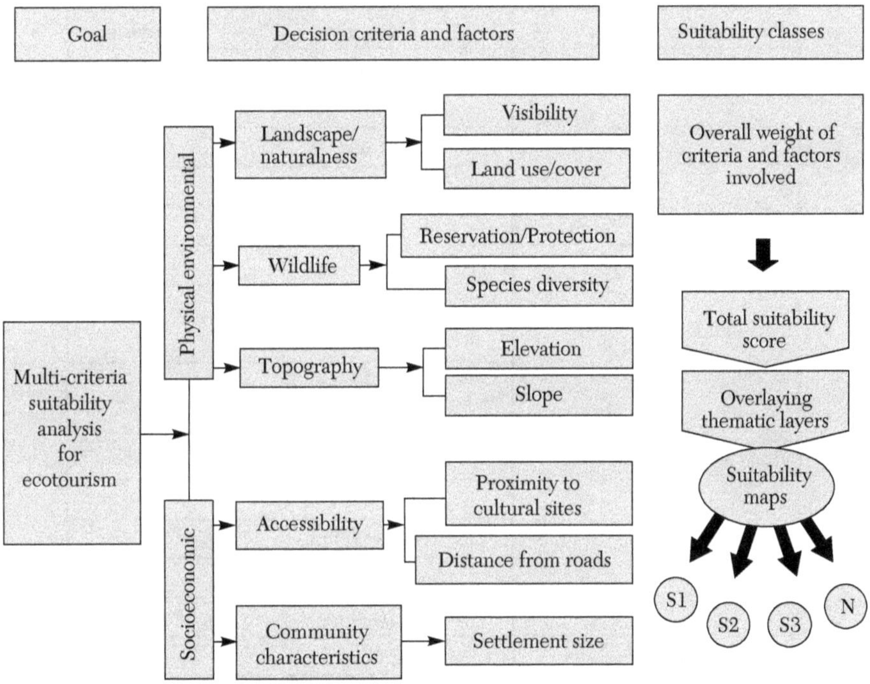

Figure 8.8. Land suitability approach for ecotourism using remote sensing and other data (Bunruamkaew & Murayama, 2012).

The selection of an industrial site involves a complex array of critical factors involving economic, social, technical, and environmental issues. Industrialization is a dynamic phenomenon that requires a lot of data to support the decision and to satisfy human needs. Thus, it is important to plan and monitor the industrial process in a systematic manner and carry out suitability analysis for industrial sites. For proper planning, accurate and timely data are required. The GIS and remote sensing provide a broad range of tools for industrial area mapping, monitoring, and management.

A study has been carried out for a part of Banda and its surrounding areas in UP state, located between 24° 53'–25° 55' north latitudes and 80° 07'–81° 34' east longitudes, for identifying suitable sites for industrial development (Johar et al., 2013). The Survey of India toposheets, Cartosat data, and IRS LISS III data are used primarily for mapping rivers, urban areas, road networks, railway lines, and land use and land cover. A DEM of the area is also generated. The GIS-based approach has been developed to find out suitable sites for industries. An example of the parameters used in GIS along with their scoring is presented in Table 8.6.

Table 8.6. GIS-Based Parameters and Their Scoring for Industrial Siting

Parameters	Category	Scoring
Land use	Wasteland	5
	Agriculture land	4
	Forest	0
	River	0
	Ponds	0
Railway accessibility	0–2,000 m	5
	2,000–5,000 m	4
	5,000–10,000 m	3
	10,000–20,000 m	2
	> 20,000	1
Road accessibility	0–200 m	5
	200–500 m	4
	500–1000 m	3
	1,000–2,000 m	2
	>2000 m	1
Distance from river	>2,000 m	5
	1,000–2,000 m	4
	500–1,000 m	3
	250–500 m	2
	0–250 m	1

Parameters	Category	Scoring
Proximity to urban areas	>2,000 m	5
	1,000–2,000 m	4
	500–1,000 m	3
	250–500 m	2
	0–250 m	1

8.2.13 Snow and Glaciers

Remote sensing and GIS-based techniques along with the field data have been developed to map and monitor snow and glaciers. Glaciers act as a means to determine the change in the climatic conditions. The change in the position of the snout over a specific time period represents the effect on the glacier by the change in atmospheric conditions. For comparing the physical change with the atmospheric change, constant monitoring of the glacier is required. Presently many supervised and unsupervised techniques, such as SMA, Landsat NDSI, MODIS, NDSI, and linear snow cover reflectance models are available in order to derive the snow cover area (SCA; Manjul & Kulkarni, 2004). Higher resolution satellite images are very useful in the detailed study of snow and glaciers.

The GRACE (Gravity Recovery and Climate Experiment) satellite of NASA is responsible for monitoring the melting level of glaciers located in the Alaskan and Polar regions, since rapid melting ice is scarily causing increased sea levels around the globe. Snow cover in parts of the Himalayas has been monitored using high repetivity WiFS data at regular 5-day intervals. Several of the Himalayan glaciers in India are retreating, mostly due to the increase in temperature from the global warming phenomenon and also due to changes in climatic conditions (Kulkarni et al., 2002).

A study by Nijhawan et al. (2016) monitors the glaciers of the Alaknanda river basin of the Himalayan region, Uttarakhand state, using Landsat TM and ETM+ satellite data between for the years 1998, 2002, and 2003. Three classifiers, the sub-pixel classification algorithm, indices-based supervised classification, and the object-based algorithm are used to compare the classification accuracies of glacier change detection. The flow diagram of the methodology used is presented in Figure 8.9. It is observed that the shadow effect is not removed by sub-pixel based classification, whereas it is removed by the indices-based supervised classification method. The sub-pixel based approach gives less accuracy mainly due to the shadow effect. The object-based method produces the best results in mountainous and shadowed covered regions.

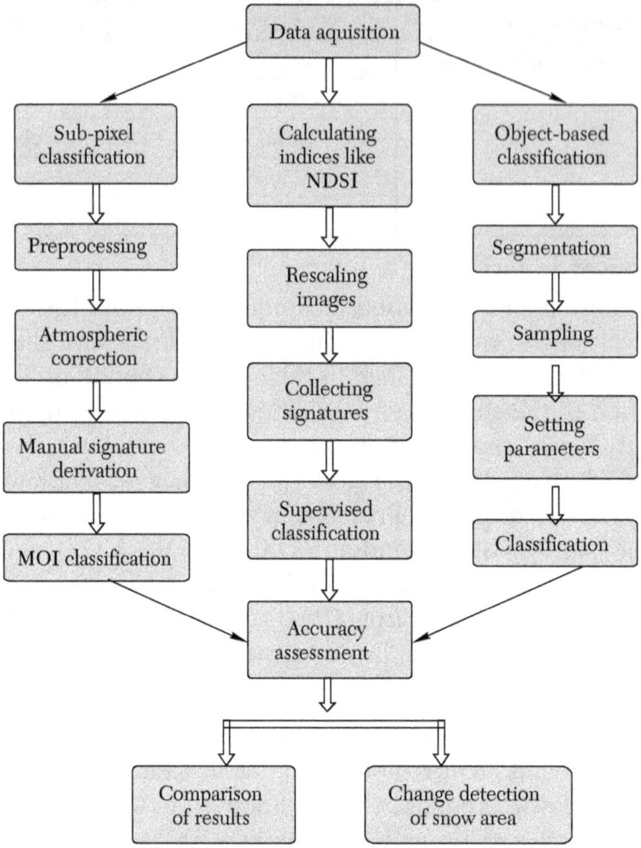

Figure 8.9. Flow diagram of methods used.

8.2.14 Transportation Network Mapping

The transport system is efficient if the design and schedule of the transit network is efficient. Urban planners, emergency services, and navigation systems require up-to-date information on road networks. In hilly areas, conventional survey for rail/road alignment is extremely time-consuming and tedious work which can be easily accomplished now with the help of high-resolution satellite images. High-resolution spatial data can be used for mapping existing road and rail networks, and planning new ones. This also facilitates locating the optimal routing for the transport of materials and people. Object-based classification using multispectral images has automated the tedious process of generating linear features such as road networks (Yong et al., 2008).

A study by Johar et al. (2014) presentes a review of the application of the genetic algorithm (GA) in transit network design and scheduling.

It is found that the problems related to design and scheduling of transit networks are highly complex and nonlinear in terms of decision variables and are difficult to achieve using classical programming. The GA as an optimization technique is computationally more efficient to solve problems requiring a large number of resources and service related constraints, such as design and scheduling of a transit network. It is observed that GAs have an advantage over traditional optimization techniques, as they work with clusters of points rather than a single point. Due to simultaneous processing of more than one string, they increase the possibility of a global optimum solution. However, limitations still exist; the solution for a complex problem is efficient if the evaluation of the fitness function is good, but it is difficult to evaluate a good fitness function.

8.2.15 Urban Development

Temporal remote sensing images are necessary for systematic mapping and periodic monitoring of urban land use required for proper planning and management. Master plans for development and management of urban settlements and other necessary facilities/amenities can be prepared efficiently using high-resolution images (Al-Ageili et al., 2017). Sustainable development of urban agglomeration, optimal urban land use plans, and resource development models need to be generated by integrating the information on natural resources mapped through satellite data along with the other environmental and socioeconomic data in a GIS (Jenson & Cowen, 1999).

In urban planning and management, three levels of observation are commonly used, as shown in Figure 8.10 (Tran, 2011): (*a*) the level of urban spot or urban area (agglomeration) to distinguish mineral from natural spaces; (*b*) the level of the urban district characterized by several urban fabric patterns; and (*c*) the level of urban objects (buildings, trees, roads etc.). For each level of observation, remotely sensed images are quite useful. At the level of the urban area (urban spot), the concept of the urban morphological area is commonly used, and the urban area is classified as a built-up area distinct from the rural zone, with medium spatial resolution (such as Landsat MSS). At the level of the urban district, different urban structures are observed according to their spatial distribution between the built-up area and vegetation. They can be identified based on the texture (spatial organization of gray levels) according to homogeneity criteria. Using the contrast and the spatial frequencies of urban LULC classes, the spatial resolution requirements for detailed urban area analysis for different structures may be identified. For example, a spatial resolution of 30 m

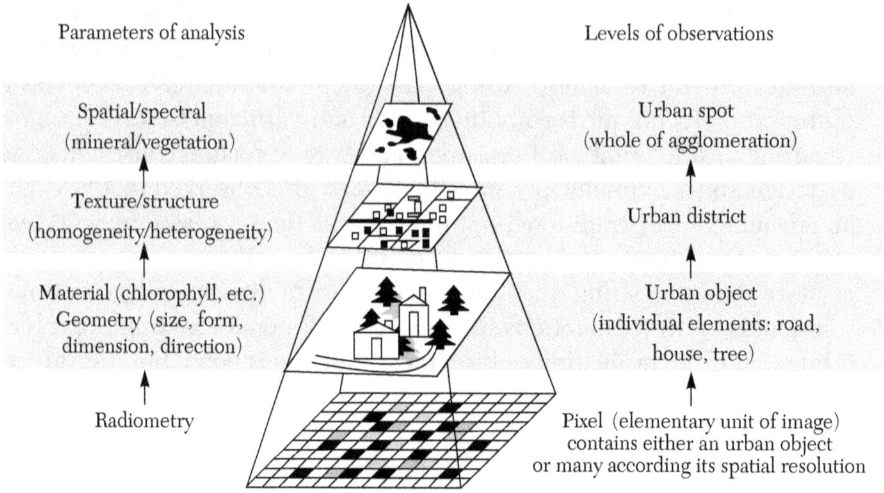

Figure 8.10. Levels of observations in urban studies and parameters of analysis from remotely sensed imageries (Tran, 2011).

(Landsat TM) is adapted to identify industrial and service sector patterns; a spatial resolution of 20 m is suitable to extract collective and residential urban fabric; a spatial resolution of 10 m allows identifying a low or high density of built-up urban fabric; and a spatial resolution of < 5 m is adapted to extract the spontaneous urban fabric.

At the level of urban objects, all constitutive elements of an urban block are then defined by their dimensions and shapes (Tran, 2011). The extraction of individual elements (buildings, trees, roads etc.) is based on the identification of both material (spectral response) and geometry (spatial response) characteristics of objects. The optimum spatial resolution to identify and analyze an urban individual object remains to be defined, which may not be identical for all the objects at this level. The relevant spatial resolution between 5–30 m generally corresponds to intra-urban analyses in developed countries because of the coarse size of urban objects. The identification of urban districts in developing countries often requires a higher spatial resolution because of their small parcels, compact structures, and narrow street patterns. Moreover, the optimum spatial resolution seems to depend on urban context, geometric form of urban objects, and spatial organization of urban areas.

Automatic built-up area extraction from high-resolution satellite imagery remains an open research area in the field of remote sensing, computer vision, and machine learning. A study by Kumar et al. (2018) proposes a system for increasing the degree of automation in extraction of building features from high-resolution multispectral satellite images. A novel image segmentation method is proposed in Figure 8.11, which involves the computation of textural parameters from high-resolution multispectral imagery that are combined with spectral bands for extracting spectral-structural characteristics. Hence, in addition to the spectral information, tone, texture, and shape information is evaluated for an object-oriented analysis. The SVM classification rules are applied for large scale building extraction.

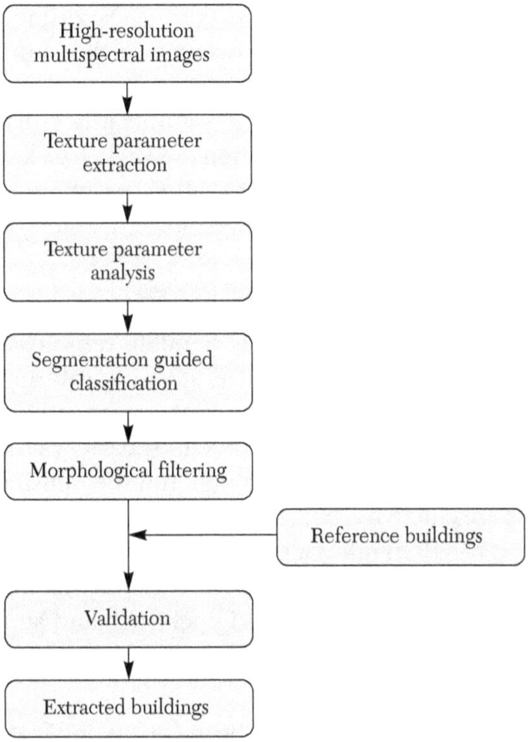

Figure 8.11. Method proposed for building extraction.

8.2.16 Water Resources

Water as a resource is essential to support human existence. The availability of fresh water for human use is important for future planning as the demand of a growing population is increasing. In this context, there is an urgent need to monitor and obtain a better understanding of its use for the development of effective water management strategies and infrastructure. One of the earliest applications of remote sensing in the water resources area is surface water inventory. Presently, remote sensing data are being used for a large number of applications, namely, irrigation water management, command area surveys, monitoring and management, dam/reservoir site assessment, canal alignment, snow-melt runoff forecasting, snow/glacier studies, surface water bodies mapping/monitoring, river/water pollution assessment, reservoir sedimentation/siltation assessment and capacity evaluation, water quality mapping and monitoring and so on (Perotti et al., 2015).

Satellite remote sensing has been used for base line inventory of irrigated areas, cropping patterns, crop condition and productivity assessment in irrigation systems, monitoring irrigation status throughout the season, optimum design of crop cutting experiments, and so forth. For example, the 5 days repetitive coverage of the WiFS sensor provides necessary monitoring capability across the command area throughout the season (Singh et al., 2012). It may be significantly enhanced with higher spatial resolution data and the additional SWIR band. The WiFS sensor helps in concurrent monitoring during the irrigation season, generating near real-time information on the sowing progress and crop conditions for appropriate irrigation water delivery.

Many irrigation projects worldwide suffer from adverse impacts of excessive irrigation and poor drainage. Satellite remote sensing can help in mapping areas affected by salinity/alkalinity and waterlogging. Monitoring of crop condition is essential to ensure that the irrigation requirements are adequately met by the canal releases. The severity of such soil limitation can be either directly detected through remote sensing or seen by the impact on crop productivity. The stereo coverage from the PAN sensor can provide contours at 10 m intervals or better, helping in preliminary analysis of canal alignment and land/infrastructure development. The 1 m and 61 cm resolution data from IIKONOS, QuickBird, and so on could be used for delineating the details of canal networks and field level crop inventory (Inglada et al., 2015).

Hydrological models are generally used for watershed management, but they often suffer problems of data scarcity or lack of quality input data. Remote sensing technology is a promising tool to integrate with these models for providing temporal input data in data scarce regions. The DEMs prepared by using remote sensing technology are used to represent the flow of water bodies and plan proper drainage systems. Remote sensing data helps in delineating the potential ground water occurrence zone in a more cost-effective manner than with conventional methods. Remote sensing technology has been used for preparation of hydro-geomorphological maps to delineate groundwater prospective zones in order to provide potable drinking water to problematic villages/areas.

Assessment of water quality is an important activity for multiple uses, such as irrigation, water supply, and so forth, and also for the protection of the aquatic and shoreline environment. Water bodies are polluted through sedimentation, industrial effluents, municipal sewage, agricultural runoff, eutrophication, and oil pollution. Remote sensing can complement ground efforts in mapping and monitoring point and nonpoint pollution sources, the influx and dispersal of pollutants in the aquatic environment, and consequent impact, such as algal blooms and weed growth. Growth of aquatic weeds and algal blooms indicating the eutrophication status of water bodies can be mapped from satellite data. High-resolution data from IKONOS, and QuickBird and hyperspectral data from MODIS are found to be useful in assessing water quality. Oil spill monitoring (in the ocean), turbidity assessment, and so on have also been carried out successfully using remote sensing.

The sustainable development and management of groundwater resources requires precise quantitative assessment based on scientific principles and modern techniques. The study by Agarwal et al. (2013) demonstrated the use of remote sensing and GIS for groundwater potential mapping in the Unnao district, UP, India. The methodology used is presented in Figure 8.12. To identify the groundwater potential zone in the study area, thematic layers on geomorphology, geology, slope, rainfall, land use, soil, groundwater depth, lineament, and drainage density are generated using topographic maps, thematic maps, field data, and satellite images in a GIS environment. The Landsat ETM+ image of October 11, 2002, at 15 m spatial resolution; ASTER DEM of October 17, 2011, Geological Survey of India (GSI) map and soil map; and so on are integrated using multi-criteria decision making (MCDM) techniques and the analytical network process (ANP). The ANP and the analytical hierarchy process (AHP) are used to determine the weights of various themes which are applied in a linear combination to obtain five different groundwater potential zones. It is observed that about as 153.39 km^2, area has very good groundwater potential, which is only 3.37% of the total study area. However, the area having very poor groundwater potential is about 850 km^2 which is about 19.63% of the study area. A water quality index (WQI) has been calculated to find the suitability of water for drinking purposes. The overall analysis of the WQI in the study area revealed that most of the area with a > 50 value exhibited poor, very poor, and unsuitable water quality (Agarwal et al., 2012).

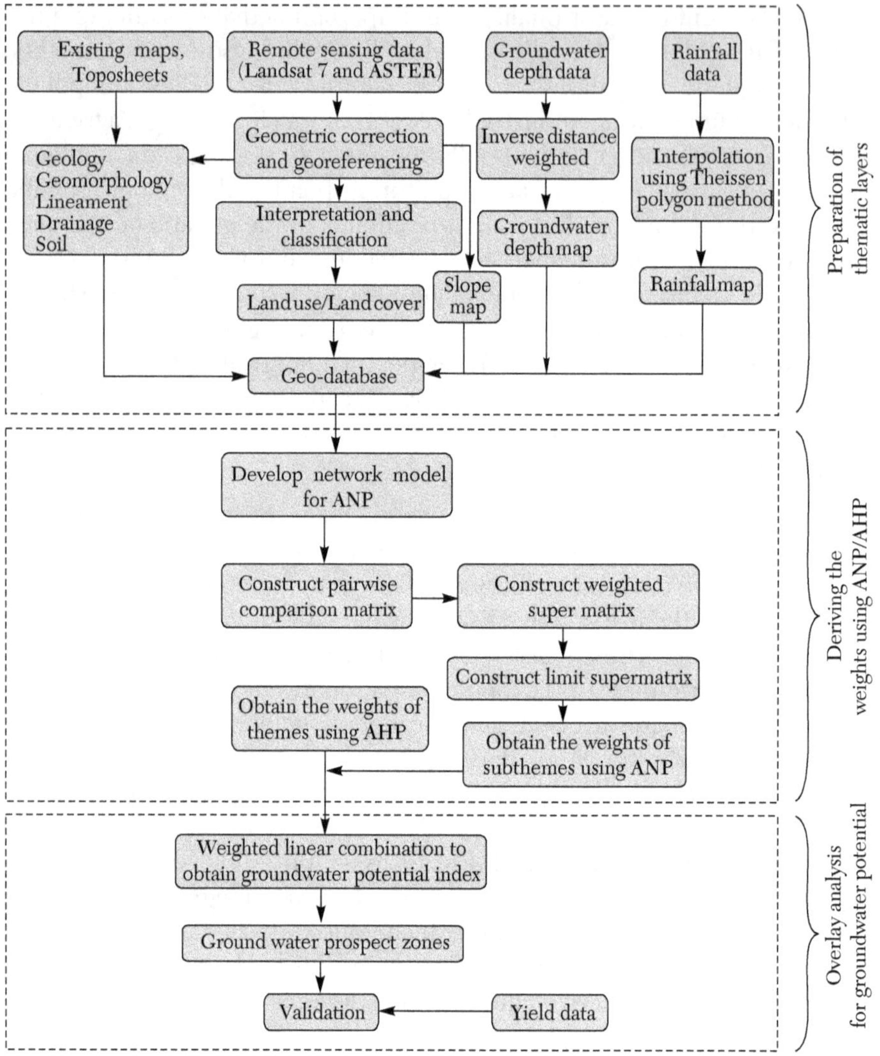

Figure 8.12. Flowchart of methodology used for deriving groundwater potential.

8.2.17 Watershed Planning and Management

Judicious management and conservation of soil and water resources in a watershed is a prerequisite for sustaining productivity. Characterization and prioritization of watersheds are essential steps toward integrated management. Watershed characterization involves measurement of geological, hydrogeological, geomorphological, hydrological, soil, and land cover/land use parameters. Remote sensing can effectively be used for watershed characterization and assessing the watershed priority; evaluating problems, potentials, and management requirements; and periodic monitoring (Garg, 2019).

Remote sensing data greatly facilitates the mapping of land use/cover, geology, soils, and other features of the watershed, which would assist in the study of land use patterns, water potential, degradation, and so on. This along with ground-based information can be used for assessing the land capability classes, irrigation suitability classes, potential land uses, suitable water harvesting measures, monitoring the effects of watershed conservation/development measures, correlation of runoff and sediment yields from different watersheds, and monitoring land use changes and land degradation (Perotti et al., 2015). Remote sensing techniques also provide significant information inputs toward calibration, validation, and use of various hydrologic models like the soil conservation service (SCS) model for rainfall runoff prediction, the sediment yield index (SYI) and the universal soil loss equation (USLE) for prioritization of watersheds,.

Watershed health can be assessed through some indirect metrics such as change in rainfall-runoff response, depletion of groundwater, groundwater pollution, and degradation of geomorphological characteristics (Agarwal et al., 2010). In a study by Jat et al. (2008), remote sensing and GIS technologies have been used for the assessment of the health of two urbanized sub-watersheds, that is Anasagar and Khanpura in the Ajmer urban area, Rajasthan state, India, over a period of 29 years (1977–2005). Investigations include estimation of urbanization and resulting changes in watershed characteristics such as change in surface runoff response, groundwater level, groundwater quality, and morphological characteristics. Eight remote sensing images of Landsat MSS, TM, ETM+, IRS-1B LISS II, IRS-1C, and-1D LISS III for various years have been used for extraction of LULC and urban growth. Change in surface runoff characteristics has been estimated using a physically based distributed storm water management model (SWMM). The rate of land development in Ajmer from satellite image analysis is observed to be outstripping the rate of population growth. From 1977–2005, the population of Ajmer has increased by 59%, while urbanization grew by 200%. In the last 29 years, the urban area in the Anasagar sub-watershed has increased from 1.26% in 1977 to 12.36% in 2005; however, peak runoff has increased by more than 42% corresponding to urban area (built-up) of 2005. Such situations create many hydrological problems, like overflow of drains, flooding, erosion, and interruption of water at many locations, resulting unhygienic conditions and degradation of watershed health.

8.2.18 The 3D City Models

The 3D city models are digital models of urban areas that represent terrain surfaces, sites, buildings, vegetation, infrastructure, and landscape elements as well as related objects belonging to urban areas. Their components are described and represented by corresponding 2D and 3D spatial data. The 3D city models support presentation, exploration, analysis, and management of tasks in a large number of different application domains. Various urban/rural planning departments require 3D GIS data for areas, like drainage, sewerage, water supply, canals, designing BIM, and many more (Garg, 2021). Major companies such as Google and Microsoft use high-resolution satellite images and GPS and GIS technologies on the Internet for applications from finding a house to 3D city modeling (Dollner, 2005). Nowadays, the current trend to create 3D terrain maps is to use a laser scanning technique, but this technique also has some limitations, like higher cost of equipment, point cloud data size, editing, storing, and management of data.

Remote sensing is a prominent data source during the acquisition of 3D city models. The majority of the currently available 3D city models have major contributions of remote sensing products, such as aerial photographs, high-resolution satellite images, or LiDAR point clouds data (Li & Yeh, 1998). During the data acquisition phase, data fusion approaches can be used where highly detailed 3D urban geometry would be integrated with data from, for example, hyperspectral, thermal, or microwave remote sensing images (Pohl & Van Genderen, 1998). In the processing phase, remote sensing techniques such as interferometric SAR, or 3D extensions of parameter retrieval, NDVI and image classification, might be applied for modeling or simulating 3D city scenarios. Applications of integrating remote sensing with 3D city modeling include all those analyses where detailed 3D geometry from city models with features extracted from remote sensing data leads to new insights. They might include urban water management, combining 3D topography with soil permeability, urban heat island modeling, and building and vegetation objects in a 3D model.

8.3 CONCLUSION

Several applications require high temporal measurements at high spatial resolution with hyperspectral technology. Multi-crop inventory and detection of crop disease, vegetation stress, and pest detection using hyperspectral techniques such as red edge shift and mineral targeting are some of the important applications which would require a hyperspectral spectrometer on board satellites on an operational basis. Cartographic and urban/rural planning from high-resolution satellite images is another important thrust area. There is further need for high-resolution mapping for preparation of rural development plans and creation of land information system, with land parcel mapping at 1: 1,000–1: 4,000 scale.

Future applications in remote sensing will combine the available resources from space/aerial/UAV platforms with ground-based data. The prerequisites of such resource integration are (Zho et al., 2017): (*a*) the spatial resolution of the satellite data is high enough to match ground-based data; (e.g., both spatial data and ground data are of the same order of accuracy) and WorldView-3 has achieved a 30-cm spatial resolution, which is comparable with ground-based sub-centimeter data accuracy (e.g., 2 cm in a mobile laser point cloud); and (*b*) the cloud-based calculation supports big datasets from crowd-sourced remote sensing resources. The current situation shows promising support for the integration of multiple sources of remote sensing data, and is expected to generate new applications in the coming years.

CHAPTER 9

Land Use and Land Cover Mapping and Modeling

9.1 INTRODUCTION

The terms *land use* and *land cover* are often used interchangeably, but each term has its own unique meaning. Land use refers to the purpose the land serves, for example, recreation, wildlife habitat, or agriculture, while land cover refers to the surface cover on the ground, like vegetation, urban infrastructure, water, bare soil, and so on (Singh, 2015). Land use is the description of how the land is being utilized for socioeconomic and other activities. Land cover provides the baseline information for activities like thematic mapping and change detection analysis. It is the physical material at the surface of the Earth.

When used together the term *Land use* and *Land cover* (LULC it generally refers to the categorization or classification of human activities and natural elements on the landscape within a specific time frame based on established scientific and statistical methods of analysis (Parker et al., 2002). Drawing relations between the two is not straightforward, as multiple relations exist in the pathways of change within and across LULC categories, as shown in Figure 9.1. For example, changes in land use can occur with or without a conversion of the broad land cover class. Land cover changes may not necessarily result from direct human activities and land use alone, but also from natural processes.

Anthropogenic LULC change is a major cause of global environmental change, as human-induced activities are continuously converting natural

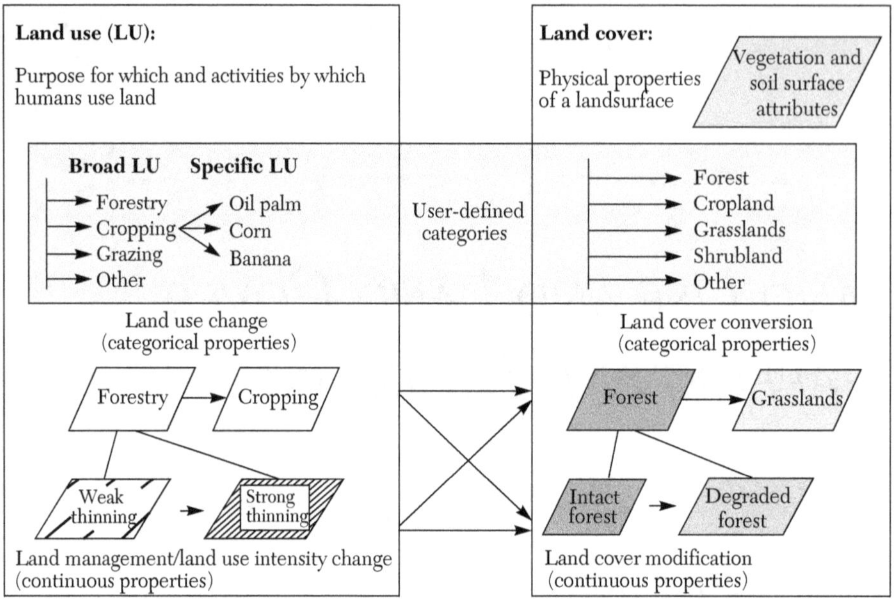

Figure 9.1. Relations between land use and land cover (Joshi et al., 2016).

lands into various human uses (Behera et al., 2012). The transition of forests and grasslands to croplands and pastures, and subsequently conversion of cropland into habitation is the most prevalent of these changes. Quantifying the LULC remains a challenge, partly since the dynamics of change are complex and fast-evolving, and partly since robust methods for analyses are still in development for many LULC processes.

Considerable progress has been made toward assessing the LULC using air- or paceborne remote sensing data due to their ability to provide spatial and temporal data across large areas, as compared to field-based assessments (Alshari & Gawali, 2021). Remote sensing sensors operating in different parts of the EMS provide a variety of information on land characteristics. With the advancements in remote sensing, monitoring networks, and GIS, the availability of spatial data is rapidly increasing. These geospatial data include not only LULC maps, but also multiple attributes of data, such as socioeconomic data from the census, land value, and crop statistics. The LULC information is extracted based on integration with ground knowledge or user interpretation. Improvements in the use and accessibility of multi-temporal, satellite-derived thematic maps have contributed significantly to LULC modeling. Some advantages of remote sensing in LULC mapping are (Houet & Hubert-Moy, 2006): (*a*) provides reliable, accurate baseline information, and generalized LULC classification for large areas; (*b*) repetitive coverage of the same area provides details on the structure and dynamics of LULC; (*c*) monitoring of long-term LULC for optimal use is possible; (*d*) multispectral imagery enhances the LULC information; and (*e*) LULC maps can be prepared speedily, accurately, and economically.

9.2 NEED FOR LULC MAPS

The LULC maps play a significant and prime role in planning, management, and monitoring the programs at local, regional, national, and international levels. This type of information not only provides a better understanding of land utilization aspects, but also it plays an important role in the formation of policies and programs required for development planning (Singh, 2015). For ensuring sustainable development, it is necessary to monitor the impact of ongoing processes on LULC over a period of time. The LULC maps also help study the changes that are taking place in the ecosystem and environment.

Accurate and up-to-date land cover change information is required to understand both human and environmental consequences with respect to change (Garg, 2019). It contributes significantly to Earth-atmosphere interactions and biodiversity loss, and is a major factor in sustainable development. Inventory and monitoring of LULC changes are needed to understand the change mechanism and model the impact of the environment and associated ecosystems at different scales. Accurate and up-to-date LULC change information is necessary in understanding and assessing the environmental consequences of such changes.

The study of LULC is very important to have proper planning and utilization of natural resources and their management. Some significant uses of LULC maps are (Houet & Hubert-Moy, 2006): (*a*) frame and mplementing land policies regarding existing and future land use; (*b*) planning, management, and monitoring of natural resources; (*c*) planning infrastructural development; (*d*) damage assessment from various disasters; (*e*) tax and revenue collection and boundary delineation; (*f*) achieving sustainable urban development and checking the haphazard development of towns and cities; and (*g*) providing input parameters in many fields, such as agriculture, geology, hydrology, demography, the environment, and so forth.

9.3 ROLE OF REMOTE SENSING

The LULC classification is a fundamental task in remote sensing. Aerial photography served as a primary source of information on LULC before the availability of satellite imagery, and remains even now an important source of land cover information with the availability of high-resolution photography from UAV/Drones. The ability to identify the LULC classes by analyzing aerial imagery saves significant cost and time to carry out these studies as compared to conventional field-based approaches (Carranza-García et al., 2019).

Remote sensing data have a long history of being used for creating land cover maps after the launch of the first Landsat satellite in 1972. Today, land cover information from a large number of satellites at multiple spatial, thematic, and temporal resolutions provides a direct relevance to develop models for forecasting the future LULC. Unlike land cover, land use typically can be derived using a combination of remote sensing, regional and local knowledge (including field observations), and other ancillary information that links a given land cover in a region with a given land use. In addition, active sensors such as LiDAR or RADAR offer the ability to obtain measurements of land surfaces at any time or season, and are also often used for monitoring landscape conditions. A wide variety of available datasets (e.g., SAR, multispectral, and hyperspectral imagery, as given in Table 9.1) has opened up the possibilities to generate new techniques for LULC pattern classification.

Table 9.1. Summary of Multisource Datasets and Their Application Scenarios with Limitations (You et. al. 2020)

Data type	Image attributes	Application scenarios	Application restrictions
AR images	• Microwave penetrability and multi-polarization scattering characteristics represent different geographic texture features • Contain corresponding amplitude and phase information of only one band microwave	All-day and all-weather scenes, even under adverse weather conditions	• Relatively low signal-to-noise ratio • Amplitude information is deeply affected by noise • Detection results are susceptible to be interfered with by geometric distortion and electromagnetic interference. Labeling information requires expertise

Data type	Image attributes	Application scenarios	Application restrictions
Multispectral images	• Consist of bands with magnitude of 10 • Otherness exists in the spectral characteristics of different targets • Possible to distinguish features according to different morphological attributes	Most accessible and intuitive remote sensing images	• Only images without severe atmospheric interference, such as clouds and fog, and reflection effect, namely under conditions of snow or ice, are applicable. • Surrounding shadows and distractors around the concerned targets affect the change results. • Shooting times influence overall spectral characteristics of multi-temporal images
Hyperspectral images	• Consist of dozens, hundreds, or even thousands of continuous bands • Contain triple information of space, radiation and spectral • Higher dimensional spectral detail is conducive to fine classification and recognition of change features	Without strong requirements to recognize category and detail change information	• Difficult to perform image preprocessing, annotation, and end-element decomposition • High spectral resolution and low spatial resolution • Contain a lot of redundant information
Heterogeneous images	• Consist of images captured by different sensors • Different temporal images possess inconsistent statistical characteristics and features	Multi-temporal images of higher temporal frequency	• Need to perform image preprocessing and registration operation on the multi-temporal images according to their respective characteristics • Different information is difficult to compare directly

Multi-temporal remote sensing data can be used to reduce the effects of climate change on vegetation, as well as examine the effects of government policy on land use, allowing LULC planners to develop qualitative and quantitative relationships between a policy driver and impacts on LULC change. Satellite remote sensing techniques have been widely used in detecting and monitoring the LULC change at various scales, which requires utilization of effective and automated change detection techniques.

The creation of more accurate maps is always a necessity. Classification methods for extracting accurate LULC data from remote sensing images are generally based on the process of assigning land cover classes to pixels to categorize them into, for instance, water, forest, buildings, cropland, grasslands, wasteland, and so on. Accurate classification often proves to be a challenging task due to the heterogeneous nature and high intraclass variance of classes, the large number of spectral bands, and the limited training samples available for the analysis of images. The Anderson classification scheme (Table 9.2) presents a standard system for classifying remote sensing data into various levels of land use and land cover information (Anderson et al., 1976). It provides guidelines to classify images into various useful classes at a particular scale. This scheme originally includes nine main categories and four different levels, but as an example only two levels of classifications are given in Table.9.2. Level I is suitable for 1:250,000–1:150,000 scale images, like using MODIS and Envisat MERIS data. Level II is useful for higher spatial resolution satellite sensor images with a scale of 1:80,000, such as Landsat MSS. Level III is suitable for 1:20,000 to 1:80,000 scale images such as Landsats 4-7 TM/ETM+, SPOTs, and IRS. Level IV is the most useful for images at scales larger than 1:20,000 (IKONOS, KOMPSAT, RapidEye, GeoEye, WorldView, and aerial photos). The Level IV categories are designed to be adaptable to local needs. The Anderson classification scheme was utilized within a large number of models in the context of land physical dynamics and natural risk assessment.

Table 9.2. Anderson Classification System Used in USGS LULC Datasets

Level I	Level II
1 Urban or Built-up land	11 Residential
	12 Commercial and Services
	13 Industrial
	14 Transportation, Communications, and Utilities
	15 Industrial and Commercial Complexes
	16 Mixed Urban or Built-up Land
	17 Other Urban or Built-up Land

Level I	Level II
2 Agricultural land	21 Cropland and Pasture
	22 Orchards, Groves, Vineyards, Nurseries, and Ornamental Horticultural Areas
	23 Confined Feeding Operations
	24 Other Agricultural Land
3 Rangeland	31 Herbaceous Rangeland
	32 Shrub and Brush Rangeland
	33 Mixed Rangeland
4 Forest land	41 Deciduous Forest Land
	42 Evergreen Forest Land
	43 Mixed Forest Land
5 Water	51 Streams and Canals
	52 Lakes
	53 Reservoirs
	54 Bays and Estuaries
6 Wetland	61 Forested Wetland
	62 Non-forested Wetland
7 Barren land	71 Dry Salt Flats
	72 Beaches
	73 Sandy Areas other than Beaches
	74 Bare Exposed Rock
	75 Strip Mines, Quarries, and Gravel Pits
	76 Transitional Areas
	77 Mixed Barren Land
8 Tundra	81 Shrub and Brush Tundra
	82 Herbaceous Tundra
	83 Bare Ground
	84 Wet Tundra
	85 Mixed Tundra
9 Perennial Snow or Ice	91 Perennial Snowfields
	92 Glaciers

Since the early 1970s, two broad types of classification approaches utilizing satellite images, supervised and unsupervised classification techniques, have evolved and being used mostly even today, despite of the developments of many other advanced classification techniques (Alshari et al., 2021). In recent years, numerous other approaches of these two basic classification methods have been developed, such as decision trees, neural networks, fuzzy classification, and mixture modeling under supervised classification; and progressive generalization, classification through enhancement, and post-processing adjustments under unsupervised techniques. Figure 9.2 presents various methods/techniques which could be used for LULC classification from remote sensing images.

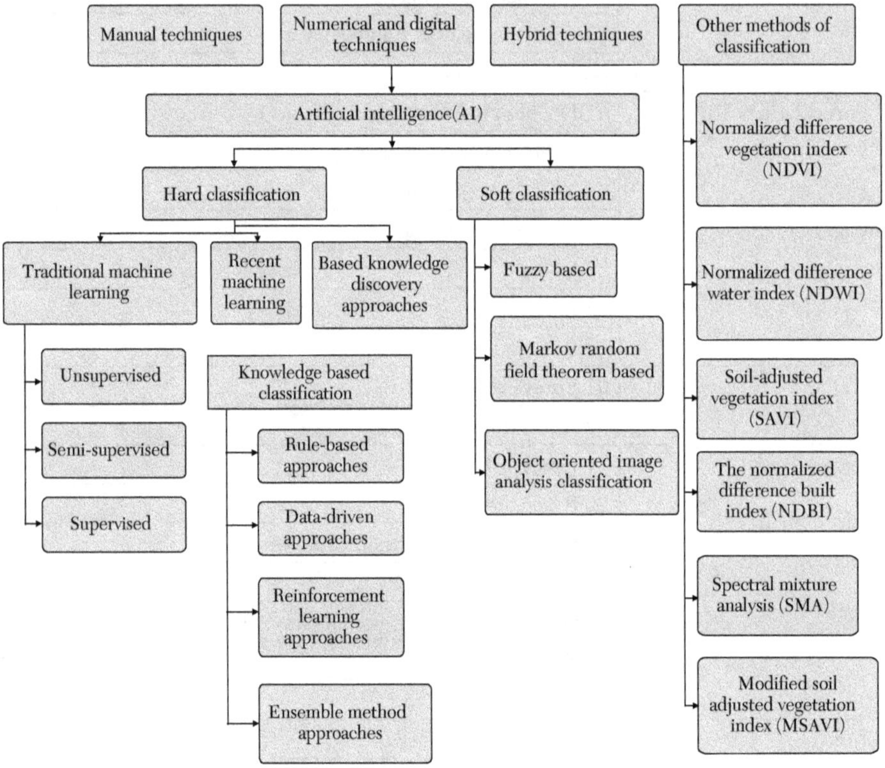

Figure 9.2. Different techniques of LULC classification in remote sensing (Alshari & Gawali, 2021).

The automated extraction of LULC information currently plays an essential role in many remote sensing applications such as change detection, environmental monitoring, urban expansion control, infrastructure planning, and biodiversity analysis. Automating LULC classification may reduce errors, resources, and time requirement manifolds while ensuring consistency in methodology (Kerins et al., 2021). At the same time, automation can dramatically expand the scope of the mapping, particularly for large areas. An automated system could regularly generate LULC maps based on new images, providing up-to-date monitoring of resources.

Commercial satellite imagery offers superior spatial resolution, but its high cost often introduces major challenges. Ground truth information about the precise location and nature of features on the ground is normally also limited. The volume of imagery and ground truth data needed to use an automated approach, along with the associated computing demands, has posed another technical hurdle in automated classification of images.

But this situation is now slowly changing. Landsat images, and Sentinel-2 images, which are available freely, provide regular global coverage with improved spatial and spectral resolution. New machine learning techniques can help in automation across a host of domains, from classification to predictive Internet searches to autonomous vehicles. The open-sources of available ground truth provide crucial geospatial LULC data for automation in classification (Carranza-Garcia et al., 2019). Many machine learning systems "learn" to recognize patterns by "looking at" training samples (ground truth). These systems iterate using such examples and continually adjust the underlying mathematical equations, that is the model, in order to link inputs to expected outputs ever more closely. Commercial cloud computing could be employed, which has provided huge amounts of processing power and storage with a modest cost.

9.4 GLOBAL LULC DATASETS

Currently, there are a few global land cover datasets that contribute to the distribution of ecosystems, produced by various remote sensing data (Table 9.3). However, none of the global datasets are directly comparable with each other, as they all are derived from different satellite systems and/ or use different classification systems (Strand et al., 2007). The SPOT and MODIS datasets could be used to detect changes in smaller areas.

Table 9.3. Global Datasets of LULC Information (Strand et al., 2007)

Dataset	Website	Spatial resolution	Sensor used	Year
Global Land Cover-2000	*http://www-gvm.jrc.it/glc2000/*	1 km	SPOT	2000
MODIS Land Cover	*http://lpdaac.usgs.gov/modis/mod12q1v4.asp*	1 km	MODIS	Starting from 2001
FAO Global Forest Resources Assessment	*www.fao.org/forestry/site/global-fra/en*	Statistics only, no maps	Varied, compiled from country reports	2005

(Contd.)

Dataset	Website	Spatial resolution	Sensor used	Year
AVHRR Global Land Cover	*http://glcf.umiacs.umd.edu/data/landcover/*	1 km, 8 km, 1 degree	AVHRR	1993
IGBP DISCover	*http://edcsns17.cr.usgs.gov/glcc/*	1 km	AVHRR	1993

Some organizations have recently begun commercialization of information related to urban LULC. For example, DigitalGlobe, formerly a privately owned satellite imagery company, now part of Maxar, offers categorized building footprints, the location, outline, and type of individual structures, on a cost basis. Orbital Insight, a geospatial intelligence company, offers a land use classification service (Kerins et al., 2021). Both LULC data sources rely on high-resolution imagery that these companies are selling. There have also been academic research efforts outside of the commercial sector. The CORINE Land Cover inventory of the Copernicus Land Monitoring Service (CLMS) utilizes an extensive typology, and, though not global, is publicly available (*https://land.copernicus.eu/pan-european/corine-land-cover*). The Global Land Survey from the National Aeronautics and Space Administration (NASA), U.S. Geological Survey (USGS), and the University of Maryland focus on tree cover (*https://www. usgs.gov/land-resources/nli/landsat/global-land-surveygls?qt-science_support_page_related_ con=0#qt-science_support_page_related_con*). The Climate Change Initiative Land Cover Project by the European Space Agency (ESA) deploys a rich typology with dozens of categories (*https://climate.esa.int/en/projects/land-cover/*). OpenStreetMap is one of the very few projects that incorporates LULC data, but quality and especially coverage are highly variable in different areas, making it unsuitable for systematic use (*https://www.openstreetmap.org/#map=5/21.843/82.795*).

The accuracy of global land cover datasets and global land cover change datasets needs to improve. The collection and distribution of land cover and land cover change validation (ground truthing) data is necessary to guide the classification process and assess the accuracy of LULC maps and the change maps. Unfortunately, there remains a lack of international coordination and standardization in the ground truthing component (Wulder et al., 2006). Globally, there are several independent efforts for collecting the ground truth data, but they are typically not shared. The reasons for not sharing could be that the validation data are typically collected for a specific project, and are not available in a format that is suitable for easy query and used by others in any region. The other reason could be that remote sensing satellite systems for land cover assessment are operated independently by several countries. By the time these raw images are processed and analyzed, indicators of biodiversity change. Also, some projects don't want to share their data, in order to maintain a competitive edge over others. Currently, there are no widely accepted standards for the collection, archiving, and distribution of validation data. Continuous long-term satellite data from a single platform are needed to achieve the highest levels of accuracy for detecting LULC change globally.

9.5 LULC CHANGE ASSESSMENT

Land use change is locally pervasive but globally significant. The LULC change is a major driver of global change, including changing of ecosystem processes, biological cycles, and biodiversity. Both anthropogenic and environmental forces largely affect the behavior of changes in LULC (Liu et al., 2009). These changes have been accelerating as a result of socioeconomic and biophysical drivers (Lambin et al., 2003), and are closely linked with sustainability. Figure 9.3 shows a LULC map of India, prepared for 2005 at 100 m resolution from Landsat 4 and 5 TM, ETM+, and MSS data, as well as IRS Resourcesat LISS-1 and LISS-III data (Roy et al., 2016). This study also created LULC maps of the years 1985 and 1995 to assess the changes in LULC from 1985–2005.

Changes in land use can be categorized by the complex interaction of structural and behavioral factors associated with technological capacity, demand, and social relations that affect both environmental capacity and the demand, along with the nature of the environment (Lambin et al., 2003). These changes in land use have important implications for changes in the Earth's climate in the future. The LULC change analysis is essential for better understanding of interactions and relationships between human activities and natural phenomena. This understanding is also necessary for improved resource management and improved decision-making. The LULC change is the mean quantitative changes in spatial extent (increase or decrease) for a specified period of time for a specific type of LULC. The change detection study may provide the changes in area and change rate, spatial distribution of changed types, change trajectories of land cover types, and accuracy assessment of change detection results.

According to Mas (1999), change detection procedures can be grouped under three broad headings characterized by the data transformation procedures and the analysis techniques used to delimit areas of significant

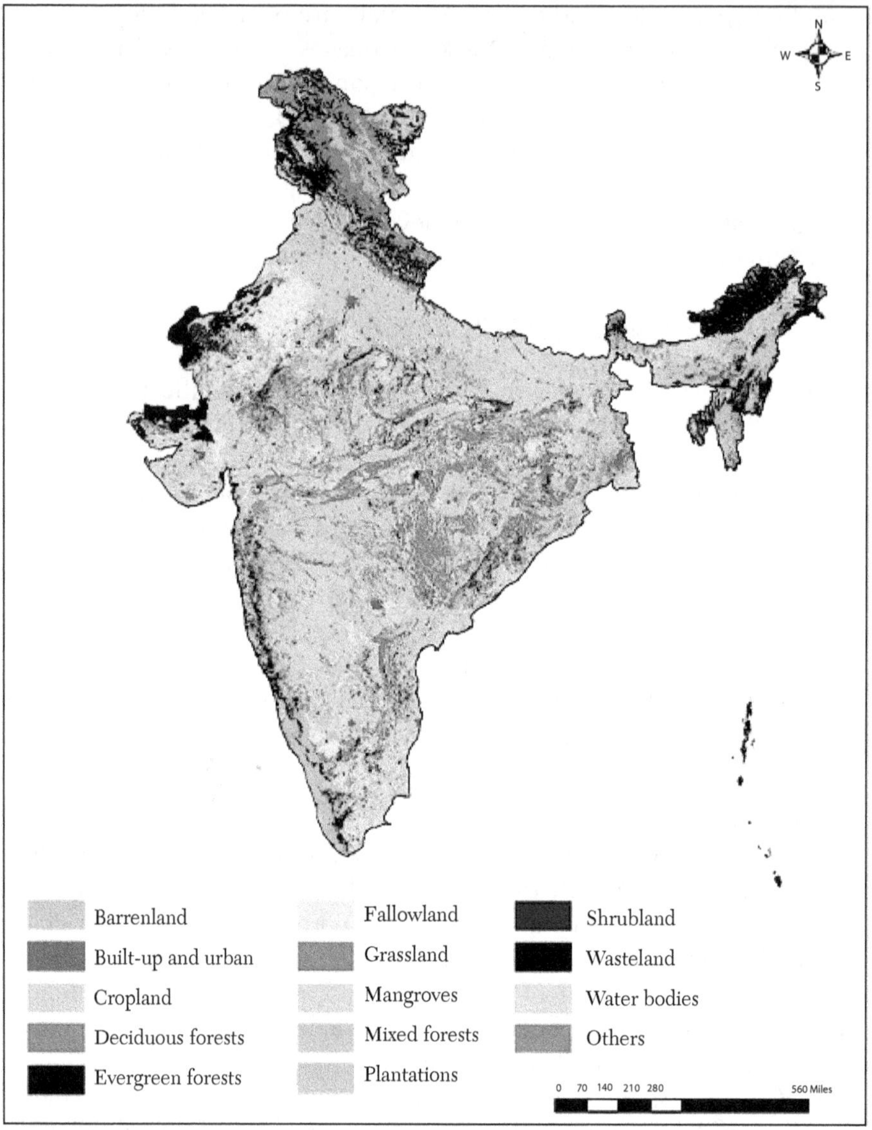

Figure 9.3. LULC map of 2005 (Roy et al., 2016) (please see color figures in companion files).

changes: (*a*) image enhancement, (*b*) multi-date data classification, and (*c*) comparison of two independent land cover classifications. Lu et al. (2004) suggested that implementation of LULC change detection analysis requires precise registration of multi-temporal images, precise radiometric and atmospheric calibration or normalization between multi-temporal images, similar phenological states between multi-temporal images, and selection of the same spatial and spectral resolution images, if possible.

Land use change models are defined as tools to support the analysis of causes and consequences of land use change. There has been a growing trend in the development of change detection techniques using remote sensing data to assess the LULC changes. Various methods have been developed to compare multi-temporal signatures for change assessment. Mas (1999) used Landsat MSS images to apply six change detection techniques as follows: image differencing, vegetative index differencing, selective principal components analysis (SPCA), direct multi-date unsupervised classification, post-classification change differencing, and a combination of image enhancement and post-classification comparison. Figure 9.4 shows the general procedure to assess the changes in LULC.

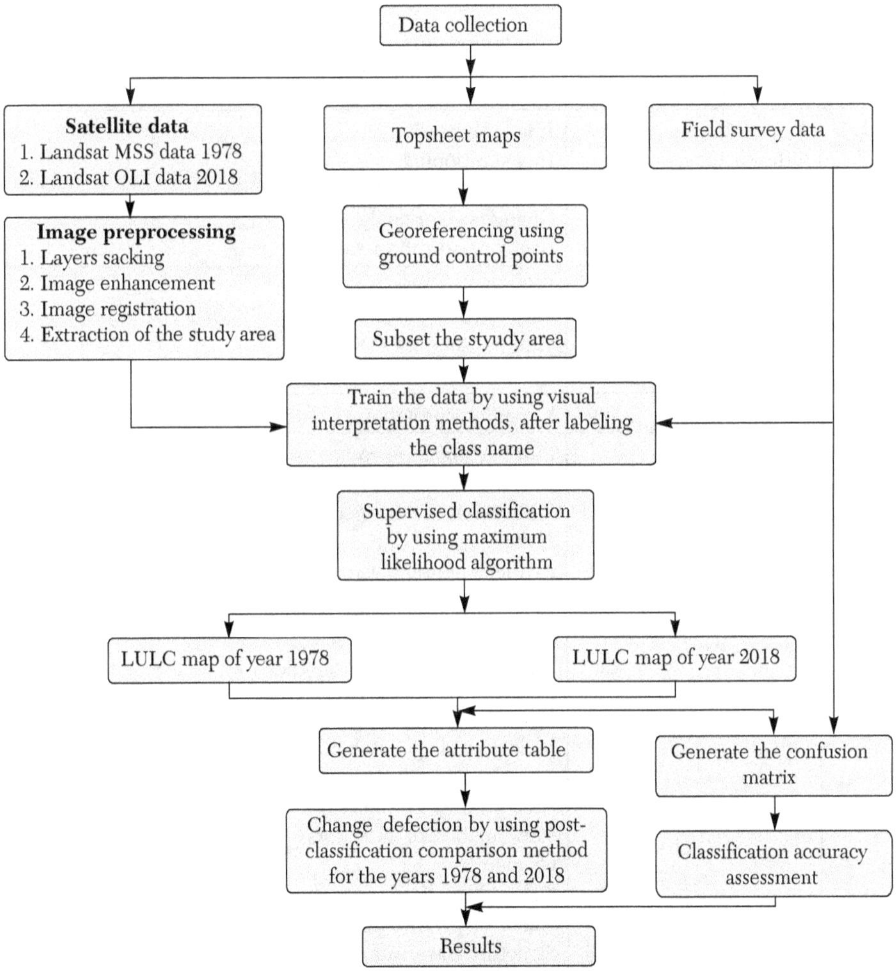

Figure 9.4. General procedure for LULC change detection.

Lu et al. (2004) grouped change detection techniques into seven categories: (*a*) algebra, (*b*) transformation, (*c*) visual analysis, (*d*) classification, (*e*) advanced models, (*f*) GIS approaches, and (*g*) other approaches. The details of these techniques are summarized in Table 9.4. They also provided, for the first six change detection categories, the main characteristics, advantages and disadvantages, key factors affecting change detection results, and some application examples. They concluded that image differencing, PCA, and post-classification comparison are the most common methods used for change detection. In recent years, some new techniques for change detection applications, such as spectral mixture analysis, ANN and integration of remote sensing, and GIS have become

Table 9.4. The Change Detection Techniques (Lu et al., 2004)

S. No.	Categories	Techniques
1.	Algebra image differencing	Image regression Image rationing Vegetation index differencing Change vector analysis Background subtraction
2.	Transformation	Principal component analysis Tasseled cap transformations Gramm-Schmidt (GS) Chi-square
3.	Visual analysis	Visual interpretation
4.	Classification	Post classification comparison Spectral-temporal combined analysis EM detection Unsupervised change detection Hybrid change detection Artificial neural networks (ANN)
5.	Advanced models	Li-Strahler reflectance model Spectral mixture model Biophysical parameter method
6.	GIS-based approach	Integrated remote sensing and GIS methods
7.	Other approaches	Measures of spatial dependence Knowledge-based vision system Area production method Combination of indicators Change curves Generalized linear models Curve theorem-based approach Structure-based approach Spatial statistics-based method

very popular. It is observed that no single approach is optimal which is directly applicable to all the cases.

Land use dynamics can be accurately determined by fusing the datasets acquired from various remote sensors. Particularly with the multiple datasets of freely available images (e.g., optical and radar images from the Sentinel satellites), data fusion brings the benefits of higher spectral resolution, compensating the limitations of using single data products of higher spatial resolution (Joshi et al., 2016). The complementary nature of these two data types is able to provide enhanced information on LULC types. The energy reflected by vegetation in visible regions is dependent on leaf structure, pigmentation, and moisture, while active microwave energy scattered by vegetation is dependent on the size, density, orientation, and dielectric properties of elements comparable to the size of the radar wavelength. Optical data may provide more robust and interpretable images for delineating broad LULC classes, which, added to the information provided by radar images on surface roughness and moisture, can allow more detailed characterization of land. Furthermore, techniques such as interferometric SAR (InSAR) make use of differential phases of reflected signals to detect land surface changes, and can be used for mapping various LULC properties. The development of adequate data fusion techniques is an important field of ongoing research.

The analysis and modeling of LULC dynamics in a watershed is important due to changes in the performance characteristics of the watershed (water balance components), including water infiltration rate, runoff, and base flow (Behera et al., 2012). Changes in land use in a watershed can affect the water quality and quantity. Land use pattern change due to watershed development frequently results in increased surface runoff, increased erosion, reduced ground water recharge, and transfer of pollutants. Thus, the assessment of land use patterns and their changes is crucial to planning and management of water resources of the watershed. Behera et al. (2018) stated that a reliable LULC change model is required if one needs to know the future state and the spatial distribution of the LULC in the watershed by using the knowledge gained from the previous years of data (historical LULC change).

The representative technologies, image attributes, and disadvantages of the various methods which could be used for LULC change analysis are listed in Table 9.5. Practically, since there is no one-size-fits-all solution, it is possible to complement the existing methods. For example, the output of feature transformation methods are available to act as the input of neural

Table 9.5. Algorithms Used for Change Analysis Using Different Remote Sensing Data (You et al., 2020)

Method		Representative Technologies	Attributes	Disadvantages	Performance
Mathematical analysis	Algebraic analysis	Algebraic difference, Change vector analysis	• Concern about the differences of pixels • Low implemented difficulty and convenient operation	• Easy to produce high-dimensional CVs from multispectral images • Unstable for disturbances • Change threshold of change must be set artificially	Low accuracy, low efficiency
	Statistical analysis	IR-MAD, PDF, NDWI	• Obtain regional difference and features according to statistical results • Robust to noise	• Ignore the existence of variance between neighbor pixels • Difficult to recognize small-scale changes • Potential relationship hidden in the training samples is not taken into account	Medium accuracy, low efficiency
Feature Space transformation		PCA, SVD, nonlinear filters, wavelet transform, self-expression layer	• Extract the hidden relationship in the samples and reduce the feature dimension of the original input data • Sparse noise and assist in denoising • Emphasize change features	• Must make artificial adjustment based on the reference parameters • May eliminate real change information • Difficult to deal with high-resolution images	Medium accuracy, high efficiency
Change classification		Post-classification comparison	• The existing object detection and classification methods can be utilized • Suitable for multi-category change detection • Intuition in revealing the object categories of multi-temporal images	• Low efficiency • Excessive reliance on the results of previous object detection	High accuracy, medium efficiency

Method		Representative Technologies	Attributes	Disadvantages	Performance
		Support vector machine	• Nonlinear mapping of high-dimensional space for the low-resolution remote sensing images • Implement small sample learning Excellent generalization performance	• Not sensitive to outliers • Limited to suitable kernel functions • Requirement of data storage space • Long training duration of Big Data	High accuracy, medium efficiency
		Decision trees	• Suitable for wide-area change detection datasets • Fast training speed, with the ability to solve nonlinear data • Stable detection results	• Easy to overfit • Not robust to noise • Require comparatively bigger storage space	High accuracy, medium efficiency
Feature clustering		K-Means, FCM, FLICM, DBSCAN, Hierarchical clustering	• Suitable for unsupervised training of the datasets without annotation • Reduce the effect of imprecise image registration • Robust to scattered noise	• Difficult to determine the cluster centers and the number of clusters • Sensitive to initial parameters • Clustering results are stochastic, and prone to be trapped in the local optimum	Medium accuracy, medium efficiency
Deep Neural Network	Naive DNN	Stacked autocoder, Deep belief network	• Flexible construction with multi-functional neurons • Strong learning ability and nonlinear fitting ability	• Powerless for relatively complex situations • Optimization function is prone to fall into the local optimal solution • As the number of network layers increase, "gradient disappears" phenomenon may happen	High accuracy, high efficiency
	Spatio-temporal DNN	CNN, RNN, Convolution long short-term memory	• Consider the spatial and temporal relevance of multi-temporal images • Available to copy with high-dimensional data • Fault-tolerant ability and strong anti-interference	• Cannot work with insufficient data • Difficult to explain the inference procedure and reason network • Long term dependency cannot be resolved with RNN	High accuracy, high efficiency

(Contd.)

Method		Representative Technologies	Attributes	Disadvantages	Performance
	Features generation DNN	Generative adversarial networks (GAN), Conditional GAN	• Confrontation training with discriminator and generator • A wide range of application scenarios for unsupervised and semi-supervised learning • As long as there is a changing standard, it can be applied to the discriminator for adversarial learning	• Unstable training, prone to gradient disappearance and mode collapse • The distribution of the generative model is difficult to express mathematically	High accuracy, high efficiency

Note: IR-MAD–Iteratively regularized multivariate change detection, PDF-Probability density function, SVD–Singular value decomposition, FLICM–Fuzzy local information c-means, DBSCAN–Density based spatial clustering of applications with noise, DNN–Deep neural networks.

networks, and the results of the mathematical analysis can be applied for later clustering and classification. It is observed that the combination of the shallow mathematical feature extraction method (e.g., PCA, IR-MAD) with the deep image semantic feature extraction method (e.g., SVM, DT, NN) can improve the performance of the results to some extent (You et al., 2020).

Satir and Berberoglu (2012) carried out the classification with various classifiers (MLC, MD, LDA, ANN, and SVM) using Landsat TM and ground truth data, and the results are shown in Figure 9.5. They found that the MLC produced the most overall accuracy when the strong training dataset is used. However, LDA with a weak training dataset performed accurately because of its distance separation algorithm. On the other hand, the unsupervised k-means classifier was the least accurate due to the fact that no training pixels were used. The SVM has a reasonable performance compared to other data dependent classifiers using a weak training

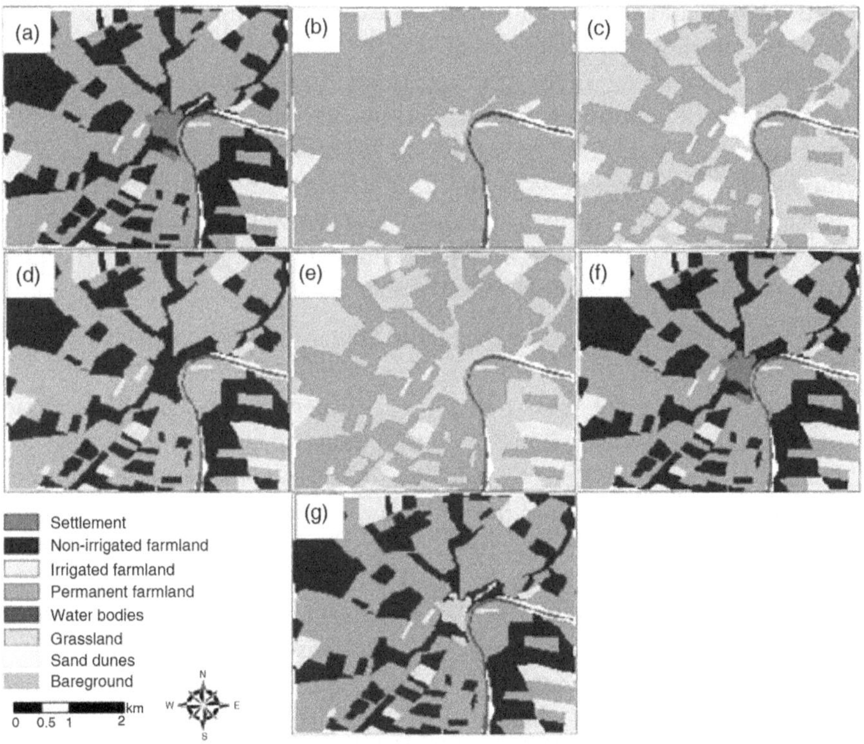

Figure 9.5. (a) Ground truth data, classification from Landsat TM image using (b) the naximum Likelihood classifier, (c) minimum distance, (d) linear discriminant analysis, (e) an artificial neural network, (f) a decision tree, and (g) support vector machine classifiers (Satir & Berberoglu, 2012) (please see color figures in companion files).

dataset. However, the largest accuracy resulted from the DT classifier using a strong dataset. The DT, LDA, and SVM showed a reasonably good performance with both weak and strong training datasets. In general, data dependent classifiers performed well with weak training datasets. SVM was especially, successful in vegetative area separation. It is clear that if a more detailed classification scheme is required (e.g., forest tree species) using a weak training dataset, SVM might be first option in terms of classification accuracy. On the other hand, application of SVM is time-consuming when using standard PCs and laptops.

9.6 LULC PREDICTION MODELING

First and foremost in LULC change modeling is the generation of scenarios (Behera et al., 2012). This is because the relationship of the people with the land use is very strong, as the human beings have modified their surroundings frequently to meet their needs. Remote sensing data of both history and the present time are extremely important for evaluating and monitoring the changes in LULC parameters, which are quite helpful in modeling LULC through scenario development, driving-force analysis, model parameterization, and model validation. Sohl et al. (2010) discussed the need to address several "foundational elements" of LULC modeling, which are: (*a*) geographic context, (*b*) regional land-use history, (*c*) representation of drivers of change, and (*d*) representation of local land-use patterns. Some of these elements can be obtained directly through the use and analysis of remotely sensed data.

Monitoring and characterizing the spatial patterns of LULC are vital for understanding and predicting the change (Figure 9.6). Future land use scenarios are required for various studies, such as land use impacts on greenhouse gas emissions, climate, biodiversity, urban planning, water resources, and hydrologic change. The LULC scenarios require two things: knowledge of present conditions and an understanding of how drivers of change interact to create historical landscapes. In both, remote sensing data play a crucial role for developing the LULC changes and predicting future scenarios (Sohl & Sleeter, 2012). The LULC changes over a long time period will help understanding of the relationships between driving forces such as population growth, economic development, and technological innovation and LULC change. The identification of the driving forces responsible for these changes is needed for projecting future land cover trajectories. The rate and type of historical LULC changes occurring within socioeconomic and biophysical settings will provide basic understanding to

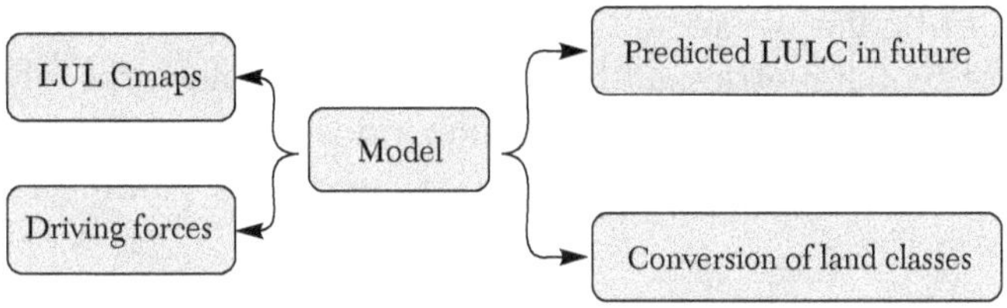

Figure 9.6. Input and output to LULC models.

develop alternative scenarios of the future. While being relatively simple in design, these types of scenarios are still dependent on LULC history, which is based primarily on satellite observations.

The LULC models that link remote sensing observations with socioeconomic data can greatly improve our understanding of the determinants of LULC change. One of the great challenges in LULC modeling is linking socioeconomic analyses with remotely sensed data (Sohl et al., 2010). The ultimate goal of any LULC change studies using remote sensing data is identifying the primary driving forces responsible for that change. One of the common uses of remote sensing data for analyzing the driving forces includes regression-based approaches employing empirical/statistical-based modeling. Remote-sensing data are commonly used for both dependent variables (LULC type) and independent variables (topographic variables, climate variables, landscape structure information derived from LULC data, etc.) in regression analyses to produce probability-of-occurrence or suitability maps for LULC and LULC change.

Many modeling frameworks depend directly on consistent, historical remote sensing data for model parameterization. For example, the SLEUTH-CA (slope, land cover, exclusion, urban growth, transport, and hill-shade cellular automata) model is typically parameterized and calibrated with patterns of historical LULC change, with the historically derived parameters driving the modeling of future urban growth patterns (Sohl & Sleeter, 2012). The SLEUTH has become one the most popular simulation models of urban growth and land use change. The model is open-source and runs under Unix, Linux, and Cygwin, a Windows-based Unix emulator (Clarke, 2008). The model uses two land use maps with a consistent classification scheme along with at least four urban maps to represent the unique historical pattern of growth during the model's calibration and application. The two land use layers are used in calculating the class-to-class transition matrix among the different land use classes (Clarke, 2008). The exclusion layer is included to control urban growth in areas where urbanization is restricted according to local land use policies (e.g., water bodies, protected areas, etc). Model calibration is also built on a series of spatial metrics, and many modeling frameworks rely on remote-sensing data for establishing transition rules and transition probabilities, with mapped historical LULC often driving model parameterization such as ANN.

The LULC models need both historical and current land cover maps coupled with data representing the driving forces of change. Availability of spatially and temporally consistent data representing those driving forces is a primary challenge for LULC modeling (Parker et al., 2002). Remote sensing data with synoptic coverage and with consistent observation at a relatively low cost make them ideal for modeling the changes. There are several methods for predicting future land cover changes, which are classified as mathematical equation based, system dynamic, statistical, expert system, evolutionary, cellular, and hybrid models (Katana et al., 2013). However, the most common are the cellular and agent-based models or a hybrid of the two. For example, the Markov chain and cellular automata analysis (CA-Markov) model is a hybrid model which is suitable for land cover change detection and simulations (Hyandye & Martiz, 2017).

The CA-Markov model has high efficiency, is simple to calibrate and has high ability to simulate multiple land covers and complex patterns. This model takes into account the spatial and temporal components of land cover dynamics (Houet & Hubert-Moy, 2006). It enables a more comprehensive simulation as compared to other LULC change models, such as GEOMOD and conversion of land use and its effects (CLUE) at a small regional extent (Mas et al., 2014). The limitations of the CA-Markov model in LULC change predictions include its inability to combine the social, human, and economic dynamics in the simulation, which can be realized in agent-based modeling systems (Hyandye & Martiz, 2017). This model fails to recognize new developments taking place in the area; in addition, the socioeconomic and urban planning concerns for urban growth are not quantitatively incorporated into the model. Failure to incorporate the new development plans leads to the decrease of prediction of CA-Markov when it is used to predict for longer periods of time due to the fact that a uniform transition rule is used by the model throughout the simulation period. Table 9.6 presents some LULC models used for either change analysis or future prediction.

Table 9.6. Some Popular LULC Models

Category	Details
Markov Chain	Probabilistic state-transition models, with LULC at time $(t + 1)$ strictly a function of LULC at time t • Transition rules for a given LULC type are often dependent on historical transition probabilities • Transition probabilities typically independent from status or dynamics of adjacent cells
Geo-statistical-Empirical (e.g., GEOMOD, CLUE)	Development of suitability or probability maps for modeled LULC types to guide placement and location of LULC change • Regression-based analyses often used for development of probability surfaces • Artificial neural networks (ANNs) one subcategory
Cellular Automata (e.g., SLEUTH)	Spatial-temporal extension of Markov transition models • State-transition model with neighborhood component • Transition rules defined by current state of a cell, but also by status of neighboring cells
Agent-based (e.g., PALM, SAMBA)	Recognizes and attempts to model the role of human decision-making in LULC change • Models behavior and interaction of "agents" (individuals, businesses, governmental bodies, or other entities with power to influence change) • Agents influence LULC change at a given location • LULC patterns emerge from interactions between human and natural processes
Integrated (e.g., IMAGE)	Integration of multiple modeling approaches and/or frameworks • May include tightly coupled models with significant feedback or loose model coupling focusing on passing data between models • Advantage of potentially incorporating both spatial and aspatial modeling approaches • Typically include econometric or other economic modeling framework
LUCAS (Land Use Change Analysis System)	Transition probability matrix (TMP) of change in land cover Model shows process, output, and impact. Uses open-source GIS software

The CA-Markov model is one of the most accepted methods for modeling LULC change using the current trends. It is able to simulate changes in multiple land-use types (Houet & Hubert-Moy, 2006), hence giving possibilities of simulating the transition from one category of LULC to another (Behera et al., 2018). The GIS and remotely sensed data provide the means to define CA-Markov initial conditions, parameterization of the model, computations of the transition probabilities, and determination of the neighborhood rules (Wang & Zhang, 2001). It uses evolution from “t_1} to “t_2” to project probabilities of LULC changes for the future date “t_3.” The model is based on probability that a given land will change from one mutually exclusive state to another (Houet & Hubert-Moy, 2006). These probabilities are generated from past changes and then applied to predict the future changes (Behera et al., 2012). The steps are shown in Figure 9.7.

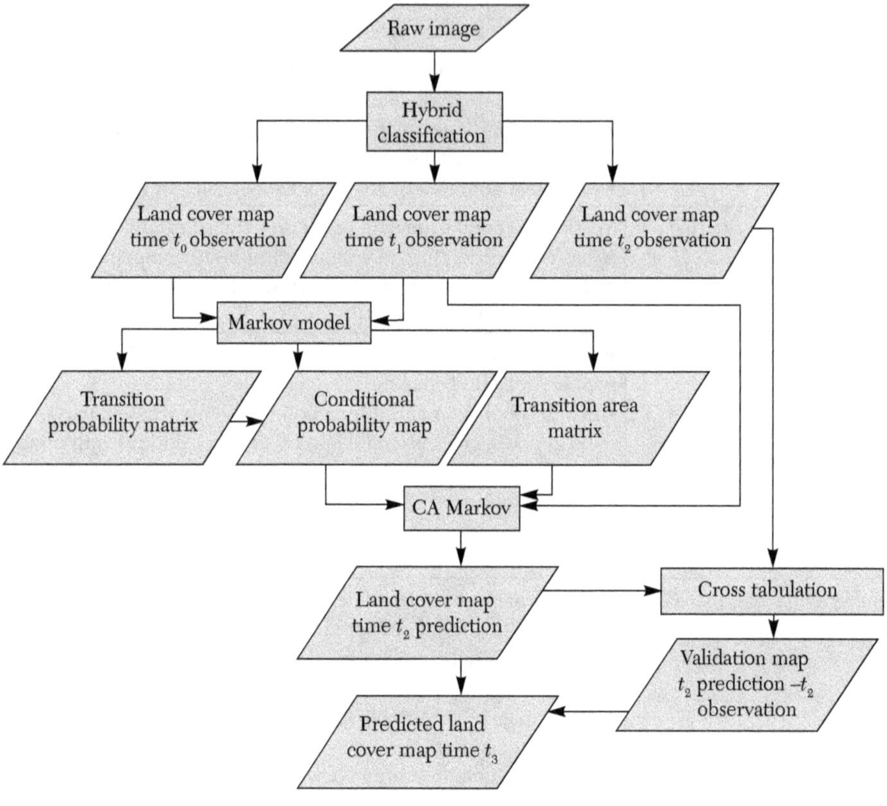

Figure 9.7. Steps leading to the prediction of land-cover classes (modified from Adhikari & Southworth, 2012).

Model validation is an essential component of LULC modeling. The problem for validation of predicting the LULC by the model lies more often with the availability of data. Validation data obviously are not available for future predictions, so LULC models typically depend on modeling a historical period to perform the model validation. These models often rely on the validation of landscape patterns where the model fit can be determined by the proportion of pixels correctly predicted in a local neighborhood or by comparison of generated landscape metrics between reference and modeled LULC (Mridha et al., 2021). Several methods for validation of future LULC change prediction models are available (Hyandye & Martiz, 2017), such as (*a*) chi-square f-tests of two images for variance, (*b*) kappa coefficient (κ) and Cramér's V, and (*c*) quantity disagreement and allocation disagreement approach.

9.7 CASE STUDIES

Two case studies are presented as follows to predict the future LULC using remote sensing and other data.

9.7.1 LULC Prediction

Mondol et al. (2013) carried out a study in a part of the Brahmaputra River basin, Kamrup Metropolitan district in Assam state, India, spreading over an area about 413.94 km^2. The elevation of the area ranges from 49.5 to 638 m above the mean sea level, and is located on Survey of India Topographical maps 72 N/12 and 72 N/16 at 1:50,000 scale. Digital satellite data of Landsat-5 TM image of December 1987, IRS-1C LISS III image of March 1997, and IRS-P6 LISS III image of December 2007 have been used for this study. In addition to field data, a master plan prepared by Guwahati Metropolitan Development Authority (GMDA) also have been used for this study.

The methodology adopted to prepare the LULC maps from satellite images in this study involves following phases: preprocessing of satellite images, development of a classification scheme, formation of a training dataset, spectral separability analysis, satellite image classification, and accuracy assessment. The supervised maximum likelihood classification technique has been used after selecting the training areas of homogeneous spectral response (10–12 training set for per class). In the classification, the signature separability functions are used to examine the quality of the training sites and class signature, before performing the classification. The overall accuracy of the LULC maps of 1987, 1997, and 2007 containing 14 LULC classes are found to be 84.77%, 85.55% and 87.50%, respectively, at a confidence level of 95%, and with overall Kappa statistics for 1987, 1997, and 2007 as 0.8011, 0.8111, and 0.8363, respectively.

The CA-Markov model is calibrated using the LULC raster image/map series generated from the classification of various images. The Markov chain transition probability matrix has been calculated for the time period of 1987–1997 and for the prediction LULC of year 2007. The transition probability areas matrix records the number of pixels that are expected to change over the specified time. The predicted LULC map of year 2007 using a 5 × 5 contiguity filter and 10 iterations is shown in Figure 9.8.

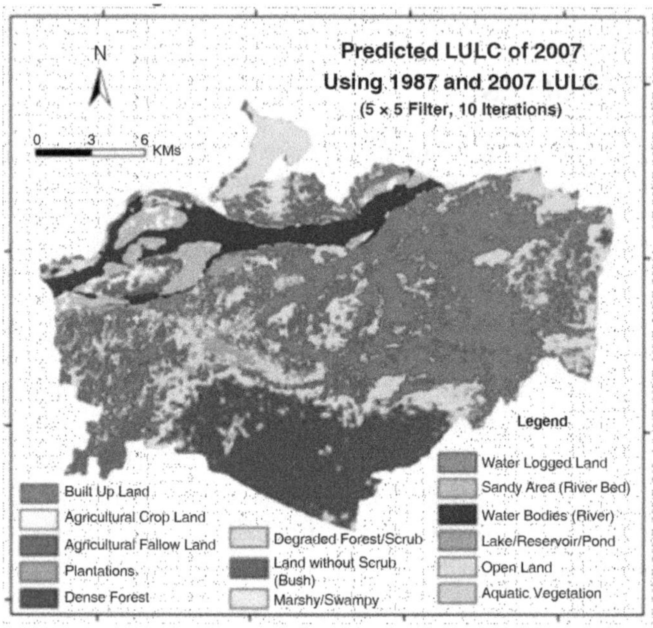

Figure 9.8. Predicted LULC of 2007 using CA Markov model (please see color figures in companion files).

According to the underlying LULC change dynamics between years 1987 and 1997, a series of suitability maps (evidence likelihood maps) consisting of various LULC are standardized between 0 and 255. Analysis from the quantitative figures of 14 simulated scenarios for 2007 is slightly different from LULC derived from the LISS III image of 2007. Relative difference in predicted LULC of 2007 and LULC derived from LISS III image of 2007 ranges between only +15.88 km^2 and −16.26 km^2. This difference is small, that is ±1.34 km^2/year in a study area of 413.98 km^2. The LULC classes such as built-up land, agricultural crop land, agricultural fallow land, dense forest, waterlogged areas, and rivers/lakes/reservoirs/ponds are relatively under-estimated, and on the other hand some classes, such as plantation, sandy areas, open land, aquatic vegetation, degraded forest land with or without scrub, and marshy/swampy land have the tendency to be overestimated. Correlations between two predicted two LULC classes are strong, where r = 0.983 and R^2 = 0.967 (Figure 9.9).

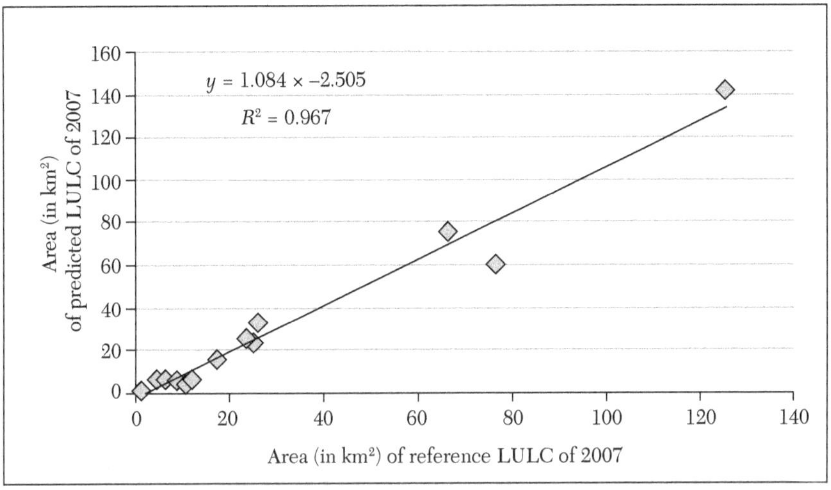

Figure 9.9. Relationship between predicted LULC of 2007 and derived LULC from LISS III image of 2007.

The kappa for Llcation ($K_{location}$) statistic measures the goodness-of-fit between two images based on the grid cell-level location of categories, given that the category quantities are specified. The validation methods give the traditional kappa index of agreement (KIA), which is also denoted by $K_{standard}$. In addition to calculating the standard KIA, three more statistics are used to validate: (*a*) kappa for no information (denoted K_{no}), (*b*) kappa for grid-cell level location (denoted as $K_{location}$), and (*c*) kappa for stratum-level location (denoted as $K_{location\ strata}$). All of these statistics are linear functions of points in the validated output. The simulated map of 2007 is compared to the reference map of 2007. The statistics for the location shows K_{no} as 0.8347, $K_{location}$ as 0.8591, $K_{location}$ strata as 0.8591 and $K_{standard}$ as 0.7928. The results indicate that the CA-Markov models' ability to specify grid cell-level location of future change is nearly perfect (here $K_{location}$ value is 0.859, where $K_{location}$ value of 1 is perfect).

9.7.2 Urban Growth Prediction

Understanding dynamic phenomena such as urban sprawl/growth requires land use change analyses, and urban sprawl pattern identification. Computation of landscape metrics is very important. In this study, Jat et al. (2008) have analyzed satellite images over a period of 25 years (1977–2002) to extract information about areas of impervious surfaces and their spatial and temporal variability, which are related to the urban sprawl of Ajmer city, Rajasthan state, India. The standard image processing techniques, such as image extraction, rectification, restoration, and classification, had been used for the analysis of four satellite images (1977, 1989, 2000, and 2002) from Landsat MSS, TM, ETM+, and IRS LISS-III. Built-up areas obtained from the classification had some errors due to mixed land use class pixels. In this study, supervised classification has been used, which does not deal with sub-pixel classification. Therefore, results are further refined using a knowledge-based approach by reducing the problem of mixed pixels.

Urban sprawl/growth over a period of 25 years is obtained from the classified images and results are compared with the settlement maps prepared by Ajmer Town Planning Department. To understand the urban sprawl pattern, different landscape metrics (Shannon's entropy, patchiness, and map density) are calculated using the demographical and built-up area statistics. Decadal population growth trends are obtained by plotting the data and fitting different types of distributions like linear, logarithmic, exponential, power, and polynomial. The quadratic model, given as follows, has been found to be the best fit for the population growth of Ajmer city, as it has the highest correlation coefficient (0.97) as compared to the linear, exponential, logarithmic, and power distribution.

$$P = 1.7556X^2 + 55.087X + 168.220 \tag{9.1}$$

where P is the population in thousands and X is years in decades (1961 onward).

Urban sprawl and its spatial and temporal characteristics have been derived from the classified satellite images. The Shannon's entropy and landscape metrics (patchiness and map density) have been computed in terms of spatial phenomena, in order to quantify the urban form (impervious area). Further, multivariate statistical techniques have been used to establish the relationship between urban sprawl and its causative factors. Results reveal that land development in Ajmer is more than three times (160.8%) the population growth (50.1%). Shannon's entropy and landscape metrics have revealed the spatial distribution of urban sprawl over a period of the last 25 years.

Defining the dynamic urban sprawl phenomenon and its future prediction is a greater challenge than its quantification. Urban sprawl dynamics have been further analyzed by considering some of the basic causative factors, like population (P), population density (PD), population density for the built-up (a density, αD), and population growth rate (PGR). In this study, population and related densities are used as independent variables for modeling the future urban sprawl, and the developed relationships are given in Table 9.7.

Table 9.7. Coefficient of Casual Factors and Percentage Built-up by Multivariate Linear Regression Analyses

Dependent variable	Independent variable	Equation	Standard error of estimate	Correlation
Percentage built-up (PB)	P and PD	PB = – 3.1 × 10-5P + 0.708PD – 11.5	0.71	0.99
PB	PD and αD	PB = 0.3564PD – 0.00688αD – 3.504	0.48	0.99
PB	P and αD	PB = 4.19 × 10 – 5P – 0.00688αD – 3.504	0.48	0.98
PB	P, PD and αD	PB = – 3.1 × 10 – 5P + 0.616PD – 0.00688αD – 3.504	0.60	0.98
PB	PD, αD and PGR	PB = 0.00285PD – 0.0447αD + 0.000567PGR + 27.797	0.00	0.98
PB	P, PD and PGR	PB = – 3.1 × 10 – 5P + 0.6854PD – 8.4 × 10 – 5PGR – 9.610	0.00	0.99

It is observed that although the correlation coefficient is almost the same for all the relationships, the relationship of PB with PD and αD (second one) is found to be most suitable as its significance F is smallest (2.62×10^{-5}). To project the impervious area (built-up) from years 2011–2051 (decadal growth) of Ajmer city within the notified municipal area, the corresponding population has been computed using Eqn. (9.1). Figure 9.10 presents the predicted population as well as the built-up area. It is estimated that the built-up area in year 2051 would be 33.9%, as compared to 17.9% in 2011 (Figure 9.10). This implies that by year 2051, built-up area in the municipal limits would rise to 2889 ha, which may be nearly 129.3% more than the built-up area (1259 ha) in year 2002. Thus, the pressure on land would further grow and the vegetal areas, open grounds, and region around the highways would be likely to become prime targets for urban sprawl.

The relationship between the urban sprawl and its causative factors may be useful for local development authorities and municipalities to determine the spatial distribution of urban sprawl. Also, such relationships can be used to predict and quantify the urban sprawl, which are useful for optimal planning of land and natural resources, zonal and regional planning, and designing of urban drainage infrastructure. Remote sensing is an indispensable tool for dealing with dynamic phenomena, like land use and urban sprawl as well as predicting the future scenario.

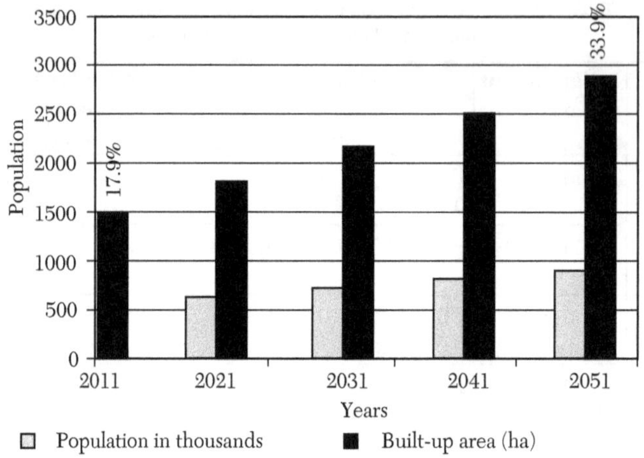

Figure 9.10. Prediction of population and built-up area (impervious area) for 2021–2051.

9.8 CONCLUSION

Timely and precise information about LULC change detection of the Earth's surface is extremely important for understanding the relationships and interactions between humans and natural phenomena for better management of resources and taking appropriate actions (Lu et al., 2004). Determining the effects of land use change on biodiversity depends on understanding the past land use practices, current land use patterns, and projections of future land use, as affected by population size and distribution, economic development, technology, and other factors. Meaningful use of remote sensing technology for LULC change detection largely depends on understanding the characteristics of the area, satellite data types, and the algorithms used for information extraction for change detection as well as prediction of future LULC patterns. The LULC change detection analysis has become a major application of remote sensing data worldwide, because of repetitive coverage of satellite data at short intervals and consistent image quality (Mas et al., 2014). With the availability of historical remote sensing data and higher resolutions from various satellites, remote sensing technology will make even greater impact on monitoring the LULC changes as well as developing various prediction models at different scales (Kerins et al., 2021).

There is a need to recognize the best classifiers and to study the best classification for LULC characterization. Land classification using spectral-spatial information together may result in more accurate and reliable classification results. In order to achieve future scenarios of LULC, an expert understanding of the interactions and the influence of factors with LULC is very important (Hyandye & Martiz, 2017). Further studies are required to develop the criteria for suitability maps of LULC categories, which might improve the accuracy in simulating their future dynamics.

Significant progress in the quantification and understanding of LULC changes has been achieved over the last decade. Much remains to be learned to fully assess and project the future role of LULC change in the functioning of the Earth's system and identify conditions for sustainable land use. Improved understanding of the complex dynamic processes underlying land use change will allow more reliable projections and more realistic scenarios of future changes (Lambin et al., 2003). Global land use/cover change transforms the land which influences climate change and reduces biotic diversity; hence, there is a need to develop interest in deforestation, desertification, and other changes in natural vegetation. The research methods applied in LULC change research were initially largely influenced by advances in remote sensing. This technology has led to an emphasis on short-time scales, because the Earth observation data have been available only for a few decades. Recently, a wide range of other methods have been used to reconstruct long-term changes in landscapes. This change in temporal frame has led to a greater consideration of the long-term processes of ecological restoration and land use transition.

CHAPTER 10

Remote Sensing Platforms for Agricultural Applications

10.1 INTRODUCTION

Farming is the most important economic activity in many developing countries. Food is the basic and compulsory human requirement. The increasing population has put immense pressure on food, water, and other resources. By 2050, the global population is expected to reach 9.8 billion (United Nations, 2015). Food security for the growing population is of prime importance, which requires dramatic changes in the agriculture sector, particularly for food production and management using modern technologies. The Food and Agriculture Organization (FAO) projections indicated a growth rate of world consumption of agricultural products of 1.1% per year for the period from 2005–2050 (Alexandratos & Bruinsma, 2012). In order to meet this projected global demand, agricultural production has to be increased by 60% from 2005–2050. However, limited land and water resources, climate change, and an increase in extreme events are likely to pose a significant threat for achieving the sustainable agriculture goal. Given these challenges, food security is included in the United Nations' Sustainable Development Goals (SDGs; United Nations, 2015).

At current growth rates, most developing countries' agricultural sectors are likely to fail to meet these targets due to challenges posed by, among others, climate change, limited land availability, competition for uses of arable land, and an aging population of farmers (CTA, 2018). To meet the world's food demand in a productive and sustainable way also presents enormous opportunities for innovations using ICTs for agriculture. This information is critical for determining crop health and productivity, and serves as an essential measure for agriculture planning and management.

Despite expansion in agricultural areas, food security may continue to be a major problem, particularly in the developing nations, mainly due to improper management of resources and policies related to the pricing of food and irrigation water use. The effects of anthropogenic climate change may further restrict the crop yield, thereby limiting the management of food and water systems in the near future (Alexandratos & Bruinsma, 2012). Extreme events, such as severe floods and droughts may also influence the availability of food, access to safe food, food prices, and sustainable food utilization, which regulate global food security. The changing climate conditions can induce prolonged droughts in the future, which can increase crop dependency on groundwater resources for irrigation, thereby affecting their sustainability (Shanmugapriya et al., 2019). Climate change can also significantly impact crop yield production from rain-fed crops. The deficit in irrigation may result in wilting of plants, which ultimately reduces the crop yield, while excess irrigation can also affect crop growth in the form of increase in salinity due to evaporation of standing water. In addition, pests and diseases may significantly reduce crop productivity, further adding to global food insecurity.

Agricultural practices, such as multiple cropping and agro-forestry, also increase soil organic matter and soil carbon. Thus, multiple cropping and agro-forestry practices reduce soil erosion through cover crops, deep-rooted crops, and varieties. Poor crop management practices are directly related to inducing water erosion. This situation can be improved by implementing crop management strategies such as planting cover crops, minimum tillage, and adding organic matter to enhance water infiltration by improving the availability of soil moisture. In addition, these strategies may also help to mitigate the impacts of severe rainfall and drought events or from water erosion (Nellis et al., 2009).

Remote sensing has long been used in monitoring and analyzing agricultural activities. During the early stages of satellite remote sensing, most of the studies in agricultural applications are focused on the use of data for classification of crop types. Remote sensing of agricultural canopies has provided valuable insights into various agronomic parameters. Remote sensing technology has the potential for detection and characterization of agricultural productivity based on biophysical attributes of crops and/or soils. Data recorded by remote sensing satellites can be used for yield estimation, crop phenological information, and detection of stress situations and disturbances. In recent years, the work in agricultural remote sensing has focused more on characterization of plant biophysical properties. Remote sensing-based repeated information can also be used for precision agricultural applications.

10.2 POTENTIALS OF REMOTE SENSING

Satellite-based remote sensing imagery has been used in agriculture since the early 1970s with the launch of Landsat-1 in 1972. Remote sensing applications in agriculture are based on the interaction of EME with soil or plant material. The amount of radiation reflected from plants is inversely related to radiation absorbed by plant pigments, and varies with the wavelength of incident radiation (Mulla, 2013). Plant pigments such as chlorophyll absorb radiation strongly in the visible spectrum from 0.40–0.70 μm (Colwell, 1983), particularly at wavelengths such as 0.43 (blue) and 0.66 (red) μm for chlorophyll-*a* and 0.45 and 0.65 μm for chlorophyll-*b*. In contrast, plant reflectance is high in the NIR region (0.70–1.30 μm) as a result of leaf density and canopy structure effects. This sharp contrast in reflectance behavior between the red and NIR portions of the spectrum gave rise to the development of spectral indices that are based on ratios of reflectance values in the visible and NIR regions (Bannari et al., 1995). Thermal remote sensing for water stress in crops is based on emission of radiation in response to the temperature of the leaf and canopy, which varies with air temperature and the rate of evapotranspiration. These spectral indices are often used to assess various attributes of plant canopies, such as leaf area index (LAI), biomass, chlorophyll content, or nitrogen content.

Remote sensing has been used to inform agricultural policy decisions; to monitor drought situations; for crop damage assessment and crop planning purposes; to tailor agronomic practices; and for demand-based irrigation scheduling. Monitoring of several indicators, such as crop growth and crop health, irrigation, soil condition, and the spread of diseases, requires direct or indirect measurement in space and time (Karthikeyan et al., 2020). The remotely sensed data is often used to retrieve variations in the vegetation by extracting information on photosynthesis, phenology, disturbances, recovery, and human interventions. Satellite remote sensing is being increasingly used to provide information on these indicators at several spatial and temporal scales. Figure 10.1 shows the relationship between the spatial resolution and the effective revisit time of the most relevant sensors for crop monitoring. The color of the bubbles indicate the sensor type on board the satellites and the size of the bubble is proportional to the number of years that the sensor(s) have been operating (Guerschman et al., 2016).

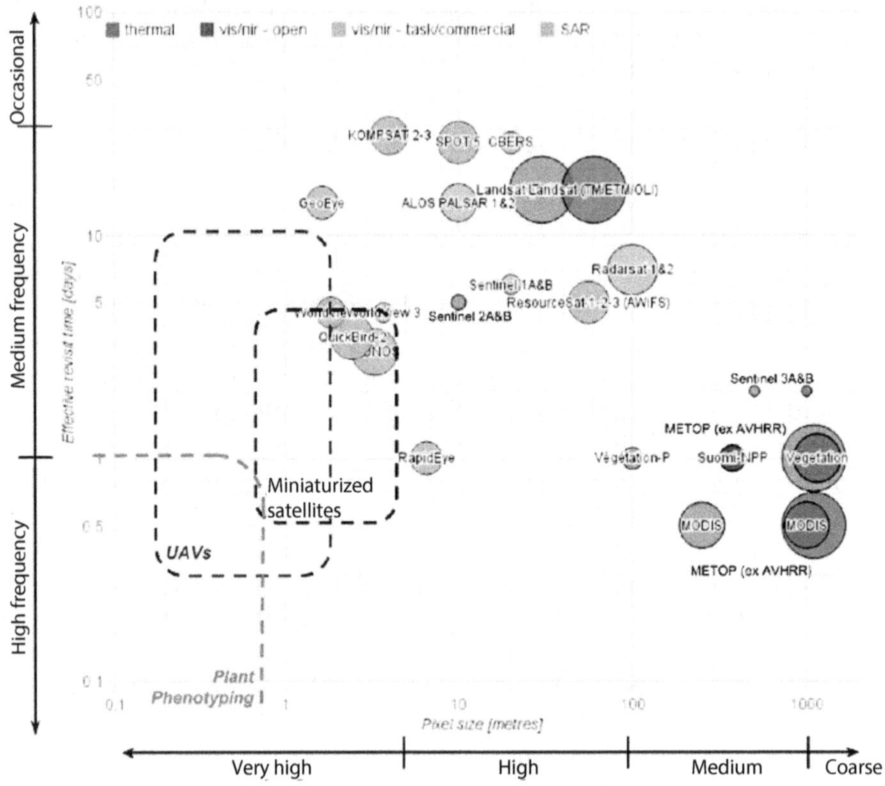

Figure 10.1. Relationship between spatial resolution and effective revisit time of the most relevant sensors/satellites for agriculture monitoring. The area potentially covered by miniaturized satellites and UAVs is shown by dotted lines (Guerschman et al., 2016) (please see color figures in companion files).

Remote sensing data in agriculture have been used for a wide range of activities, such as crop yield and biomass, crop nutrient and water stress infestations of weeds, insects and plant diseases; and soil properties, such as organic matter, moisture and clay content, or salinity. Plant responses to water stress are numerous and complex. Figure 10.2 shows the most important relationships between primary plant stresses, the induced plant responses, and multi-/hyperspectral remote sensing techniques for the detection of stresses (Gerhards et al., 2019). Hyperspectral remote sensing has also helped to enhance more detailed analysis of crop classification. Remote sensing along with GIS is highly beneficial for creating spatio-temporal layers which can be successfully applied to hydrological modeling, land use changes, crop growth monitoring, and stress detection.

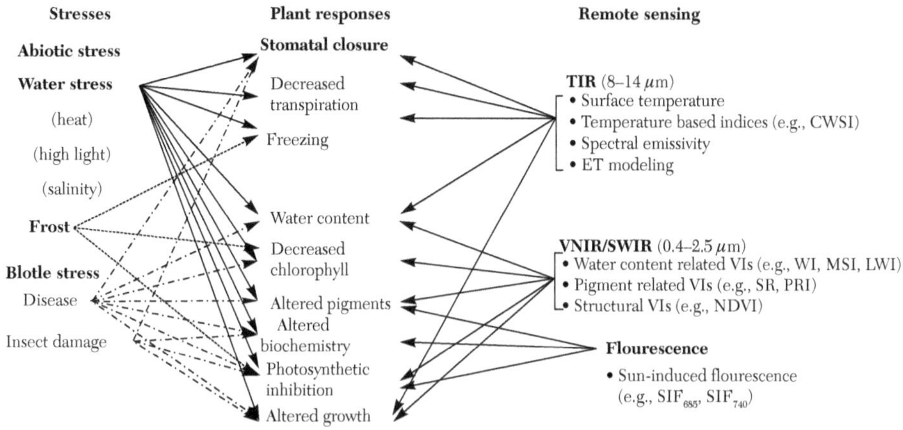

Figure 10.2. Relationships between primary plant stresses, the induced plant responses, and multi-/hyper-spectral remote sensing techniques for the detection of stresses (Gerhards et al., 2019).

Remote sensing applications in agriculture can be classified according to the type of platforms and sensors, including ground, aerial, and satellite based platforms (Mulla, 2003). These platforms and sensors have their own utility based on the altitude of the platform, spatial resolution, and spectral and temporal resolutions of the images (some are given in Table 10.1). Poor spatial resolution makes it difficult to identify various agricultural crops. Temporal resolution is important to monitor plant characteristics for its complete cycle. Several trends are seen from Table 10.1. Firstly, the spatial resolution of imaging systems has improved from 80 m with Landsat to

Table 10.1. Examples of Some Satellite Remote Sensing Platforms and Their Suitability for Agriculture (Mulla, 2013)

Satellite (year)	Spectral bands	Spatial resolution (m)	Temporal resolution (days)	Suitability for agriculture
Landsat 1 (1972)	G, R, two IR	80	18	Low
AVHRR (1978)	R, NIR, two TIR	1,090	1	Low
Landsat 5 TM (1984)	B, G, R, two NIR, MIR, TIR	30	16	Medium
SPOT-1 (1986)	G, R, NIR	20	2–6	Medium
IRS-1A (1988)	B, G, R, NIR	72	22	Medium
ERS-1 (1991)	Ku band altimeter, IR	20	35	Low
JERS-1 (1992)	L band radar	18	44	Low
LiDAR (1995)	VIS	0.10	NA	High

(Contd.)

Satellite (year)	Spectral bands	Spatial resolution (m)	Temporal resolution (days)	Suitability for agriculture
Radarsat (1995)	C-band radar	30	1–6	Medium
IKONOS (1999)	PAN, B, G, R, NIR	1–4	3	High
SRTM (2000)	X-band radar	30	NA	Medium
TERRAASTER (2000)	G, R, NIR and 6 MIR, 5 TIR bands	15–90	16	Medium
EO-1 Hyperion (2000)	400–2,500 nm, 10 nm bandwidth	30	16	High
QuickBird (2001)	PAN, B, G, R, NIR	0.61–2.4	1–4	High
MODIS (2002)	36 bands in VIS-IR	250–1,000	1–2	Low
RapidEye (2008)	B, G, R, red edge, NIR	6.5	5.5	High
GeoEye-1 (2008)	PAN, B, G, R, NIR1, NIR2	1.6	2–8	High
WorldView-2 (2009)	P, B, G, Y, R, red edge, NIR	0.5	1.1	High

Note: P refers to purple, B to blue, G to green, R to red, IR to infrared, NIR to near infrared, MIR to mid-infrared, TIR to thermal infrared.

0.5m with GeoEye and WorldView. Secondly, the temporal resolution has improved from 18 days with Landsat to 1 day with WorldView. Thirdly, the number of spectral bands available for analysis has increased from four bands of Landsat to more than 100 bands with hyperspectral imaging systems such as Hyperion. Finally, as the spatial and spectral resolutions of satellite imagery have improved, the suitability of using these images for agricultural applications has increased. The most appropriate spatial and spectral resolution for agricultural applications depends on factors such as crop management objectives, capacity of farm equipment to vary farm inputs, and farm unit area.

10.3 VARIOUS VEGETATION INDICES

Biophysical features of plants can be characterized spectrally by vegetation indices defined as unitless radiometric measures. They are calculated as ratios or differences of two or more bands in the VIS, NIR, and SWIR wavelengths (Garg, 2019). The usefulness of a vegetation index is determined by its high correlation with the biophysical parameters of plants and low sensitivity to factors restricting remote sensing data interpretation, for example, soil background, relief, non-photo synthesizing elements of plants, atmosphere, and viewing and illumination geometry (Wójtowicz et al., 2016). Vegetation indices, computed from combinations of visible red and NIR spectral measurements, have been developed primarily for vegetation study. In addition to identifying the occurrence of significant phenological events, spectral-temporal profiles of vegetation index values over time can be used to classify different vegetation types and to map their spatial distributions. The advantages of using vegetation indices include: minimizing soil and other background effects, reducing data dimensionality, providing a degree of standardization for comparison, and enhancing the vegetation signal (Chavez et al., 1994).

Different indices have been developed to estimate various crop parameters (e.g., crop density, biomass, chlorophyll, nitrogen, water content, etc.). These include simple ratio (SR), normalized difference vegetation index (NDVI), visible atmospheric resistant index (VARI), Enhanced Vegetation Index (EVI), soil adjusted vegetation index (SAVI), moisture stress index (MSI), leaf water content index (LWCI), Water Index (WI), global vegetation moisture index (GVMI), shortwave infrared water stress index (SIWSI) etc. The NDVI is the most frequently used index to determine the condition, developmental stages, and biomass of cultivated plants and to forecast their yields. The phenologic behavior of different broad vegetation types can be observed, analyzed, and mapped using multi-temporal NDVI profiles (Chavez et al., 1994). Many efforts have been made aiming to develop further indices that can reduce the impact of soil background and atmosphere on the results of spectral measurements. Some available vegetation indices used in agricultural studies are summarized in Table 10.2 (Wójtowicz et al., 2016).

The vegetation indices such as crop water stress index (CWSI), surface temperature (ST), water deficit index (WDI), and stress index (SI) describe the relationship existing between water stress and thermal characteristics of plants. The leaf area index (LAI), which is defined as the ratio between the one-sided green leaf area and ground surface area, is used to assess leaf characteristics and crop biomass (Mulla, 2013). It influences the canopy reflectance, and generally ranges between 0 (bare ground) to more than 10 (evergreen coniferous forests; Karthikeyan et al., 2020). The NDVI values tend to saturate when the LAI values are very high (greater than 8), so hyperspectral vegetation indices have exhibited an ability to reduce the saturation effect. The hyperspectral sensors measure the reflectance from the red-edge vegetation spectrum, which is the sharp slope between the low reflectance red spectrum and high reflectance NIR spectrum, between 0.35 and 1.05 μm wavelengths.

Table 10.2. Some Available Vegetation Indices Used in Agricultural Studies (Wójtowicz et al., 2016)

Index	Formula	Spectral bands or wavelengths (nm)	Level/Sensor	Application
Advanced Normalized Vegetation Index	$ANVI = \frac{NIR - BLUE}{NIR + BLUE}$	BLUE: 400–500 NIR: 700–900	Airborne (RMK TOP 15 camera)	Mapping Fidolfia segetum patches in sunflower crop
Aphid Index	$AI = \frac{NIR1 - NIR2}{RED1 + RED2}$	RED 1: 712 RED 2: 710 NIR 1: 761 NIR 2: 908	Ground-based (ASD Field Spec3 spectrometer)	Identification of aphid infestation in mustard
Chlorophyll Index	$CI = \frac{NIR}{GREEN} - 1$	GREEN: 520–600 NIR: 700–900	Ground-based (Exotech radiometr) Satellite (Quick Bird)	Plant nitrogen status estimates
Continuum Removed (CR) Spectral Index	$A_{CR(1200)} \int_{\lambda_{1116}}^{\lambda_{1284}} \left(1 - \frac{NIR}{NIR_{CR}}\right)$ $A_{CR(1200)} = 1 - \frac{(NIR1 - NIR2)\left(1 - \frac{NIR3}{NIR3_{CR}}\right)}{2}$	NIR: 1110–1284 NIR 1: 1267 NIR 2: 1156 NIR 3: 1210	Laboratory (ASD Field Spec Spectroradiometer) Airborne (MIVIS)	Estimation of water content at leaf and landscape level
Damage Sensitive Spectral Index	$DSSI = \frac{RED - NIR - BLUE - GREEN}{(RED - NIR) + (BLUE - GREEN)}$	BLUE: 509 GREEN: 537 RED: 719 NIR: 873	Ground-based (ASD FieldSpec Handheld Spectroradiometer)	Determine the sunn pest damage on wheat
Effective Leaf Area Index	$ELAI = -0.441 + 0.285\frac{NIR}{RED}$	RED: 610–680 NIR: 720–890	Ground-based (CIMEL 313 radiometer)	Winter oilseed rape yield prediction

Index	Formula	Spectral bands or wavelengths (nm)	Level/Sensor	Application
Green Normalized Difference Vegetation Index	$GNDVI = \frac{NIR - GREEN}{NIR + GREEN}$	GREEN: 557–552 NIR: 720–920 and/or GREEN: 520–600 NIR: 700–900	Airborne (Multispectral Digital Camera) Satellite (IKONOS)	Corn yield predictions
Green Red Vegetation Index	$GRVI = \frac{GREEN - RED}{GREEN + RED}$	GREEN: 520–590 RED: 620–680	Ground-based (GER 1500 spectroradiometer)	Estimation of damage caused by thrips
Healthy-Index	$NI = \frac{GREEN - RED1}{GREEN + RED1} 0.5RED2$	GREEN: 534 RED1: 698 RED2: 704	Airborne (MCA-6 and Micro-Hyperspec Tetracam)	Early detection of verticillium wilt in olive
Leaf Rust Disease Severity Index 1 Leaf Rust Disease Severity Index 2	$LRDSI\_1 = 6.9\frac{RED1}{BLUE} - 1.2$ $LRDSI\_2 = 4.2\frac{RED2}{BLUE} - 0.38$	BLUE: 455 RED: 605 RED: 695	Ground-based (ASD FieldSpec spectrometer)	Detection of wheat leaf rust
Modified Soil-Adjusted Vegetation Index	$MSAVI\ 2 = 0.5\ [2NIR + 1 - \sqrt{(2NIR + 1)^2 - 8(NIR - RED)}]$	RED: 630–690 NIR: 760–860	Satellite (Terra ASTER)	Prediction of corn canopy nitrogen content
Normalized Difference Infrared Index	$NDH = \frac{NIR - SWIR}{NIR + SWIR}$	NIR 1: 845–885 NIR 2: 1650–1700	Airborne (MASTER)	Detection of diurnal orchard canopy water content variation
Normalized Difference Water Index	$NDWI = \frac{NIR1 - NIR2}{NIR1 + NIR2}$	NIR 1: 841–876 NIR 2: 1230–1250	Satellite (MODIS)	Estimation of plant water content

(Contd.)

Index	Formula	Spectral bands or wavelengths (nm)	Level/Sensor	Application
Normalized Pigment Chlorophyll Ratio Index	$NPCI = \frac{RED1 - BLUE1}{RED2 + BLUE2}$	BLUE: 460 RED: 600	Ground-based (Exotech and CropScan radiometers)	Estimation of leaf chlorophyll content
Optimized Soil-Adjusted Vegetation Index	$OSAVI = \frac{NIR - RED}{NIR + RED + 0.16}$	RED: 640–720 NIR: 770–880	Satellite (IKONOS)	Nitrogen status estimation of winter wheat
Ratio Vegetation Index	$RVI = \frac{NIR}{RED}$	RED: 630–690 NIR: 760–900	Ground-based (ASD) FieldSpec Handheld Spectroradio-meter)	Estimating nitrogen status of winter wheat
Relative Reflectance Index	$RRI = \frac{NIR_a/VIS_a}{NIR_r/VIS_r}$	VIS: 400–700 NIR: 740–820	Ground-based (quantum sensor LI-190s and LI-220s)	Indication of drought of field grown oilseed rape
Shortwave Infrared Water Stress Index	$SIWSI(6,2) = \frac{SWIR - NIR1}{SWIR + NIR1}$ $SIWSI(6,2) = \frac{SWIR - NIR2}{SWIR + NIR2}$	NIR 1: 841–876 NIR 2: 1230–1250 SWIR: 1628–1652	Satellite (MODIS)	Indication of canopy water content
Simple Ratio	$SR = \frac{RED}{NIR}$	RED: 648 NIR: 747	Airborne (Hyperspectral camera)	Detection of pest infestation at regional scale
Structure Insensitive Pigment Index	$SIPI = \frac{NIR - BLUE}{NIR - RED}$	BLUE: 445 RED: 680 NIR: 800	Ground-based (ASD FieldSpec Handheld Spectroratdiometer)	Determine the sunn pest damage on wheat

Index	Formula	Spectral bands or wavelengths (nm)	Level/Sensor	Application
Transformed Soil-Adjusted Vegetation Index	$\text{TSAVI} = \dfrac{a(\text{NIR} - a\text{RED} - \text{b})}{a\text{NIR} + \text{RED} - ab + X(1 + a^2)}$	RED: 610–720 NIR: 760–950	Airborne (Fushbroom camera), Satellite (SPOT HRV)	Assimilation of remote sensing data into sugar beet yield prediction model
Triangular Greenness Index	TGI = 0.5[(RED–BLUE) (RED – GREEN)–(RED–GREEN) (RED–BLUE)]	BLUE: 450–520 GREEN: 520–600 RED: 630–690	Ground-based (ASD) FieldSpec spectrometer Airborne (AVIRIS), Satellite (Landsat TM)	Crop nitrogen requirements detection

10.4 MODERN TRENDS

Remote sensing, by means of satellites, airplanes and unmanned aerial vehicles (UAVs)/drones provides large-scale images of the agricultural environment (Kamilaris & Prenafeta-Boldú, 2018). Multispectral and hyperspectral imaging, synthetic aperture radar (SAR), and thermal and near-infrared cameras have been used in various applications related with agriculture. In recent years, UAVs have been used for agricultural crop mapping, monitoring, crop yield estimations, damage assessment, and so forth (Garg, 2021a). The UAVs are typically low cost, lightweight aircraft that are well suited for agricultural applications (Wójtowicz et al., 2016). Several advantages include the following: they can be deployed quickly and repeatedly, they are flexible in terms of flying height and timing of missions, and they can obtain very high resolution imagery. Providing a swath width of 50–500 m and a spatial resolution of 1–20 cm, UAV platforms are able to provide the high-resolution input necessary for observation of individual plants, patches, gaps, and patterns over the landscape as well as site-specific crop management. Figure 10.3 shows that Drones have a cost advantage in imaging up to a field size of 20 hectares, but in larger fields high-resolution satellite images are economical.

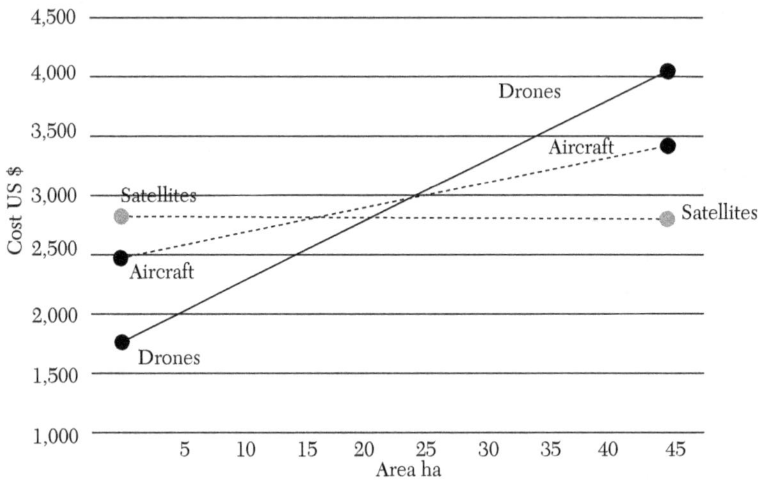

Figure 10.3. Cost comparison for satellite, aircraft, and agriculture drone imaging (Reger et al., 2018).

Artificial intelligence (AI) refers to the ability of a computer or a computer-enabled robotic system to process information and produce outcomes in a manner similar to the thought process of humans in learning, decision-making, and solving problems (Garg, 2021a). The goal of AI systems is to develop systems capable of tackling complex problems in ways similar to human logic and reasoning. For agriculture applications in developed countries, AI and robotics have been used mainly for monitoring crop conditions, plant populations, and soil moisture content (Shehzadi, 2017). Additionally, they are used for automated irrigation, crop health monitoring, crop spraying using drones, facial recognition of cows, crop harvesting, and early warning systems. Agriculture-based AI tools operate in a very unpredictable environment—for instance, there are frequent weather changes; soil quality can vary in a single field; and pests and diseases affect different sections of a farm, and the variations are even greater when the scope changes from a locality to a country or continent (Cai et al., 2019). However, the potential for AI application in agriculture has been recognized but, its application has not succeeded in practice due to a lack of adoption by small field holders in most of the developing countries. The most common techniques used for image analysis acquired from UAVs or satellites include machine learning (K-means, SVM, and ANN, among others), wavelet-based filtering, vegetation indices such as the NDVI, and regression analysis (Shanmugapriya et al., 2019).

There are a growing number of ICT-based initiatives to improve farming and food production. Emerging ICT technologies relevant for understanding agricultural ecosystems and improving food production include remote sensing data, field based measurement through sensors, UAV images, Internet of Things (IoT), cloud computing, and big data analysis. In many countries, required infrastructure is being developed to speed up the use of ICTs to improve agriculture and livelihoods (Sharma & Garg, 2020). Universal access to ICTs to a greater population in a country (World Bank, 2011) has not yet been fully achieved, especially in rural areas of developing countries where the majority of agricultural production is carried out. In addition, the cost of Internet services remains high in developing and least developed countries with prices reportedly three times the global average (CTA, 2018).

Several remote sensing systems can be used to continuously collect the data of the Earth's surface about a phenomenon, leading to generation of huge point cloud data, also known as big data. Big data is described as "high-volume, high-velocity and/or high-variety information assets" (Garg, 2021), and has in recent years helped in improved agricultural decision-making and subsequent production. The open agricultural data is poised to have a significant impact on farming activities globally. For instance, large-scale farmers and other stakeholders in developed countries are deriving the benefits of big open data for prediction and analysis in the practice of smart farming activities. These users are better informed and equipped to make use of data collection and analysis tools, like the Open Data Kit and Hadoop—a software library that allows for the distributed processing of large data. There are a lot of open data available for agriculture from a great number of sources such as The World Bank (World Bank, 2016) and the Global Open Data for Agriculture and Nutrition (GODAN; Global Open Data, 2016). However, small-holder farmers don't have access to big open data, mainly due to the lack of suitable tools for the collection, processing, and retrieval of such information.

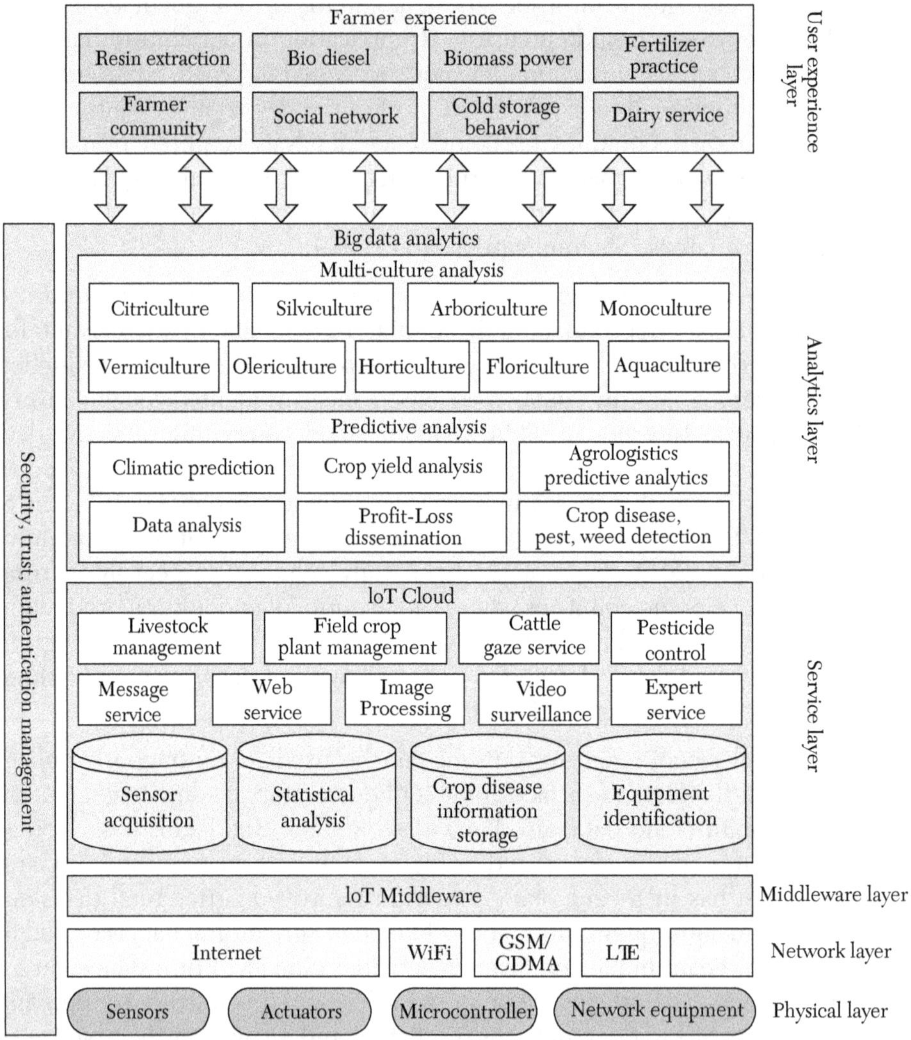

Figure 10.4. IoT based agricultural framework.

The IoT refers to the implementable machine-to-machine (M2M) communication which is a crucial component of recent growth in the digital market (Sharma & Garg, 2020). A detailed framework to cater to full-fledged agricultural solutions using IoT is given in Figure 10.4. The framework is a six-layered concept which includes hardware facilities, Internet and allied communication technologies, IoT middleware, IoT enabled cloud services, big data analytics, and farmer experience. The service layer plays an important role in providing cloud storage and services to agricultural problems. Crop management, pesticide control, and livestock management can be effectively done from agriculture data. Farmers can obtain information through Web services, message services, and expert services. The analytics layer performs big data processing to do predictive analysis and multicultural analytics. Prediction involves estimation of the climatic condition of the area, including soil moisture, temperature, rainfall, and so on, and crop yield productivity in advance, as well as probability of occurrence of various crop diseases based on the past data in order to save the crop. The multicultural analytics involves formulating, processing, and efficiently managing a few forms of farming. For example, Big Data can enhance the forest growing process by utilizing environmental data analysis, and can also be used to know how these plants grow and respond to their environment through statistical modeling. The inclusion of IoT is envisioned to be useful for advancing the agricultural and farming industries by introducing new dimensions.

10.5 APPLICATIONS IN AGRICULTURE

There are various applications of different types of remote sensing data, as presented by Guerschman et al. (2016) in Figure 10.5. However, the estimation of crop yield and forecasts of crop yields are one of the major applications of remote sensing in agriculture. It is shown in the figure that the modeling of the remote sensing signals into surface or canopy parameters can meet the needs of various applications. The remote sensing modeling of crop yield can be done with or without a physiological model. Forecasts based on remote sensing are usually statistical models that link past and current season indices, whereas the use of remote sensing indicators as input variables into crop simulation models simply enrich the potential of such models by providing gridded information.

A more complex approach may be using remote sensing indicators such as LAI, soil moisture, or the crop development stage for model initialization and calibration. The forecast solution based on remote sensing simply

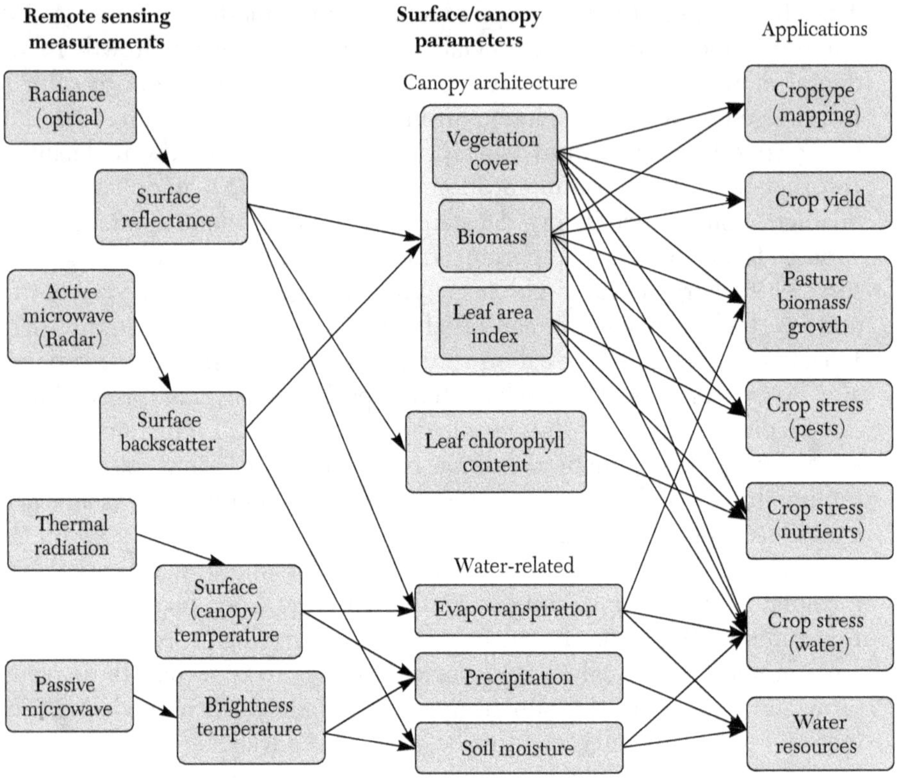

Figure 10.5. Applications of remote sensing in agriculture.

relates a time series of crop yields to vegetation indices (such as the NDVI and the enhanced vegetation index, or EVI) through empirical regression analysis (i.e., statistical models). An improved solution is to complement the exploratory variables of the regression with other information, such as satellite-derived surface temperature, rainfall, solar radiation, or soil moisture.

Some of the applications are discussed as follows:

10.5.1 Crop Condition Assessment

Remote sensing can play an important role in agriculture by providing timely spectral information which can be used for assessing the biophysical indicators of plant health. The physiological changes that occur in a plant due to stress may change the spectral reflectance characteristics, resulting in the detection of stress amenable to remote sensing techniques (Shanmugapriya et al., 2019). Remote sensing data can be used to measure the variability in the reflectance spectra of plants resulting from the occurrence and severity of pests and disease. Large-scale crop condition monitoring can provide decision-making information for working out agricultural policy and commercial trade. Spectral characteristics of healthy and infested plants are significantly different than the diseased plants (Gehards et al., 2019). Ground-based spectral reflectance (e.g., use of spectro-radiometer) proves to be very helpful in detection of pest damage in crops.

Crop growth monitoring at regular intervals is necessary to know the probable loss of production due to any stress factor, and to take appropriate measures. The crop growth stages and its development are influenced by a variety of factors, such as available soil moisture, date of planting, air temperature, day length, and soil condition, which are responsible for the plant conditions and their productivity. For example, corn crop yields can be negatively impacted if temperatures are too high at the time of pollination, so knowing the temperature at the time of corn pollination could help predict better corn yields (Nellis et al., 2009). The occurrence of drought also makes the land unproductive. Agricultural drought monitoring through satellite-based information has been carried out globally with varying ecological conditions (Shanmugapriya et al., 2019).

Crop growth and its condition are often characterized through the use of various vegetation indices such as simple ratio, NDVI, vegetation condition index (VCI), perpendicular vegetation index (PVI), transformed vegetation index (TVI), greenness vegetation index (GVI), land surface water index (LSWI), temperature-vegetation dryness index (TVDI), soil adjusted vegetation index (SAVI), water deficit index (WDI), and so forth. The LSWI employs the SWIR and NIR bands, and SWIR provides powerful luminosity assimilation by liquid water, and therefore it is well recognized to be susceptible to the entire quantity of liquid water in vegetation and its soil. The TVDI is acquired from spatial land surface temperature (LST)-NDVI and used to mark soil humidity and vegetation water pressure. The SAVI takes into consideration the visual soil characteristics on the plant cover reflectance. The WDI represents the relative rate of hidden heat change, so it illustrates a rate of “zero” for a totally wet surface and a value of “one” concerning dry surfaces where there is no hidden heat change. Other indices are given in Table 10.2.

Meteorological satellites can provide the information of the terrestrial process daily, which makes it possible to monitor crops continuously and dynamically. The NDVI calculates vegetation density through evaluating the variation between NIR and red wavelengths. The temporal NDVI profiles of crops can reflect the change of crop NDVI values from planting, seedling, tassel, to maturation and harvesting (Quarmby et al., 1993). So the crop condition and its growing trends can be obtained by analyzing the characteristics of its time series NDVI images, as shown in Figure 10.6.

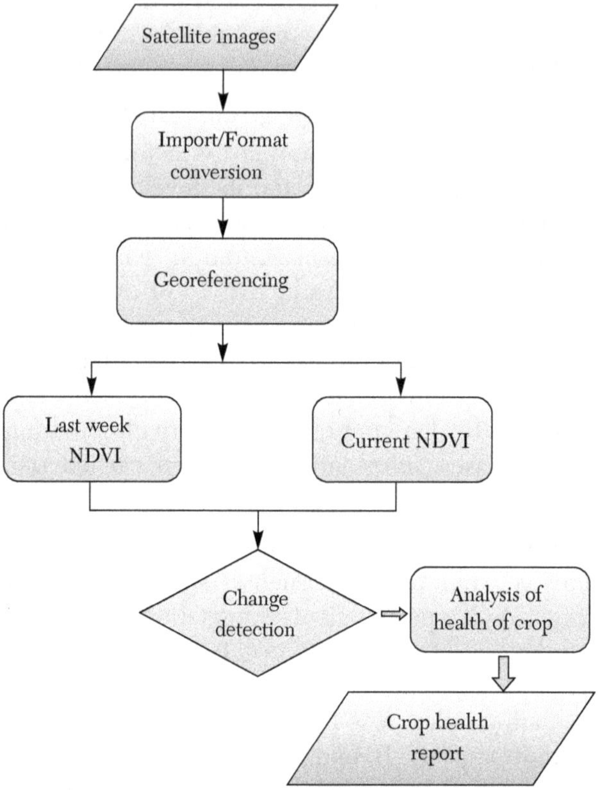

Figure 10.6. Monitoring crop health with NDVI images.

Crop growing models realize the dynamic monitoring of the crop growing process by simulating the crop growth process. These models can actually monitor the crop condition accurately. All crop growing models use the interception of solar radiation for the vegetation canopy and the photosynthesis that produces dry biomass (Ji-hua & Bing-fang, 2008). Remote sensing data can be applied to crop growing models in large areas, but the application of models needs a large number of agro-parameters. In addition, the model must be calibrated by local field data to be used in different regions. These requirements sometimes put a limitation on the application of these models.

10.5.2 Crop Yield and Production Forecasting

Information on expected yield is very important for national food policy planning, government agencies, commodity traders, and producers in planning harvest, storage, transportation, and marketing activities. *Crop yield* is generally used to represent the outcome of an agriculture crop, and is defined as the weight of crop output at certain soil moisture content per unit harvestedarea of the crop (Karthikeyan et al., 2020). Crop yield is dependent on meteorological conditions, seed quality, soil types, water and nutrient availability, the amount of absorbed photosynthetically active radiation, crop variety, influence by weeds, pest and disease infestation, weather parameters, and so forth. Remote sensing-based measurements have been used to forecast crop yields, primarily based upon statistical–empirical relationships between yield and vegetation indices (Ji-hua & Bing-fang, 2008).

Quarmby et al. (1993) used AVHRR-NDVI over the growing season for estimation of the crop yield of wheat, cotton, rice, and maize crops through linear regression. Some studies also used a nonlinear regression approach to estimate the crop yield using a vegetation index (Mulla, 2013). The NDVI, RVI, TVI, or meteorological variables have also been used to build yield estimation regression models. Most vegetation indices require radiometric calibration or surface reflectance retrieval of remote sensing images to decrease the influence of atmospheric effects. This requirement limits the usage of these vegetation indices for yield estimation. The accuracy of a vegetation index is found to be influenced primarily by the crop type, vegetation saturation, soil, and atmospheric effects, among others. Attempts have also been made to use meteorological variables along with a vegetation index as predictors in the statistical regression model to estimate the crop yield (Cai et al., 2019). Figure 10.7 shows the systematic steps to compute yield from rice crops using remote sensing and ground data.

Sophisticated machine learning techniques are also used to model crop yield and LAI using satellite sensor-based vegetation indices. Some of them include ANNs, SVMs, and random forests (Karthikeyan et al., 2020). While more than 250 crop simulation models are currently available, the inclusion of biotic factors in the model is still restricted to limited models, the purpose of which is to evaluate yield reduction due to weeds, pests, and diseases or yield increases induced by intercropping. Some of the major crop models being used are APSIM (the Agricultural Production System Simulator), CROPSYST (the cropping systems simulation model, DSSAT (decision support system for agrotechnology transfer), EPIC

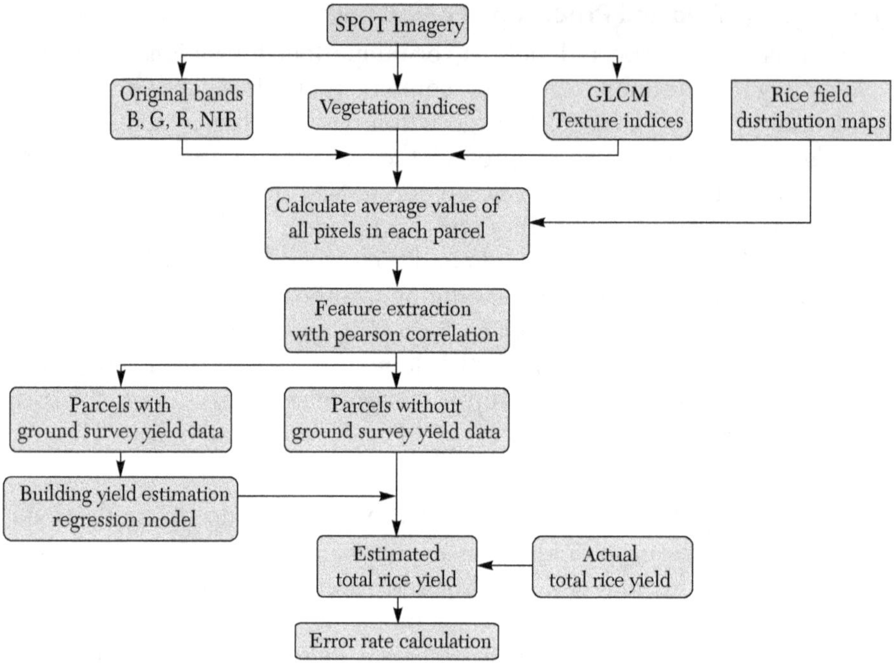

Figure 10.7. Flowchart of rice yield estimation.

(environmental policy integrated climate), ORYZA (from the Latin word for rice), STICS (simulateur multi-discplinaire pour les cultures standard), and WOFOST (world food studies model). Di Paola et al. (2016) have reviewed approximately seventy models, classifying them in terms of model type, sub-models, scale and time paths, crops addressed, place of application, and IT system used. It is important to realize that crop yield models are applied at very different scales, ranging from the field to the continent.

10.5.3 Precision Agriculture

Precision agriculture (PA) is one of the top revolutions in agriculture (Sharma & Garg, 2020), although it has only been practiced commercially since the 1990s. PA generally involves better management of farm inputs, such as fertilizers, herbicides, seed, and fuel (used during tillage, planting, spraying, etc.) by undertaking the right management practice at the right location and time. Globally, there is little documented information about the rates of adoption of these technologies for PA in the developing world. With PA, fields can receive customized management inputs based on their soil types, climatic conditions, landscape position, and crop history in order to improve crop productivity and farm profitability through improved management of farm inputs (Guerschman et al., 2016).

The practices of PA also reduce the risk of over or under applying production inputs, such as agrochemicals and fertilizers. In commercial farming sectors, the PA practices boost the labor productivity because of the increased output per unit input. Therefore, the precise use of pesticides, herbicides, and fertilizers reduces losses and consequently lowers input costs. With PA, input resources such as labor and fertilizers are optimized, increasing the productivity of the farm significantly. PA results in up to 60 percent savings in agrochemicals, such as pesticides and close to 30 percent savings in fertilizers (Rider et al., 2006).

Pollution of ground and surface water resources by agrochemicals is of major concern globally. PA practices play an important role in reducing the misuse of chemicals that are harmful to the environment. Soil degradation through erosion is also minimized on and around farms where PA practices are implemented.

PA is also known as "smart farming" or "precision farming," and combines remote sensing, global navigation satellite system (GNSS), GIS, robotics, data analytics, AI, and other new technologies into an integrated crop production system. PA involves both data collection/analysis and information management, as well as technological advances in computer processing, field positioning, yield monitoring, remote sensing, and sensor design (Mulla, 2003) to optimize production by accounting for variability and uncertainties within agricultural systems (as shown in Figure 10.8). PA requires intensive information in space and time collected through various remote sensing approaches and processes the data, leading to improved crop production (Mulla, 2013). A detailed review on sensors related to agricultural crops was reported by Ruiz-Altisent et al. (2010) as: (*a*) electromagnetic sensors, spectroscopic sensors, and computer vision; (*b*) mechanical contact and acoustic sensors; (*c*) biosensors; and (*d*) wireless sensor networks.

Remote sensing technology is a key component of precision farming and is being used by an increasing number of scientists, engineers, and large-scale crop growers. The greatest challenges in remote sensing and PA resources emanate from their lack of availability, and often large costs are involved with acquiring and using them (Wójtowicz et al., 2016). The alternative low-cost technologies involving small UAVs and low altitude remote sensing platforms could be used, which provide high spatial and temporal resolution, and flexibility in image acquisition. Remote sensing and GIS can be used for PA activities to monitor crop growth, the maturity period, and crop stresses such as nutrient and water stress, disease, and pest

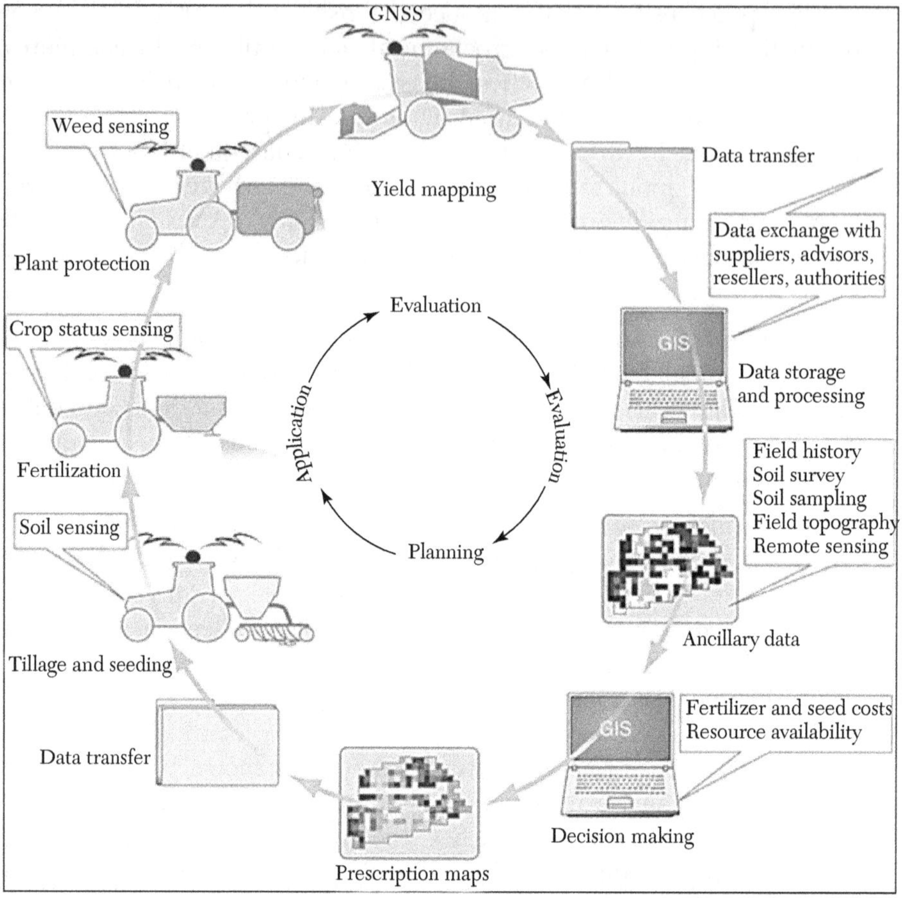

Figure 10.8. Precision agriculture information flow in crop production (Gebbers & Viacheslav, 2010).

and weed infestation (Shanmugapriya et al., 2019). Information gathered using different sensors and referenced using GPS can be integrated to create field management strategies for nutrient, chemical, and water application; cultivation; and harvest. Other uses by farmers include more precise hybrid selection, rental agreements that are better aligned with actual soil productivity, better matching of fertilizer applications to crop yield potential, and lower consumption of chemicals and fuels.

10.5.4 Crop Insurance

Agricultural insurance is a financial tool to transfer the production risk associated with farming to a third party via payment of a premium that reflects the true long-term cost of the insurer assuming those risks. Relying on spatial information technology, the smart agricultural insurance system realizes the disaster assessment, survey, damage determination, and risk prediction of agricultural insurance, combining remote sensing images, big data, IoT technology, and cloud computing technology, as shown in Figure 10.9. It helps accurate quantification of claims, scientific risk prediction, timely monitoring of disasters, and scientific business management, with a strong basis and high accuracy (Garg, 2021). In other words, it provides support for insurance companies to improve efficiency at reduced costs.

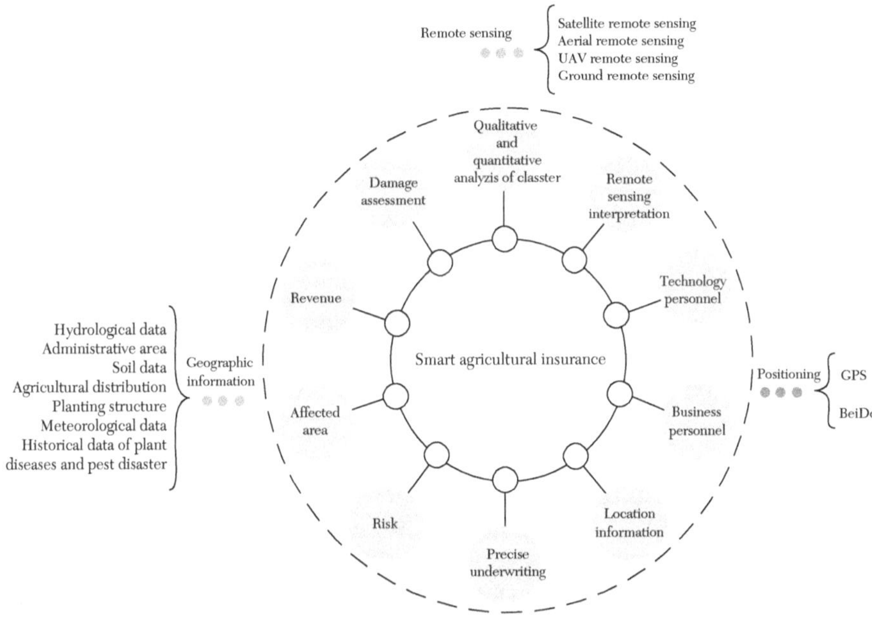

Figure 10.9. Using geospatial technology for smart agriculture insurance.

Spatial information technology is composed of remote sensing technology, spatial positioning technology, and GIS technology. It combines these technologies with agricultural insurance business to achieve disaster assessment, survey and damage assessment, and risk prediction of agricultural insurance. Remote sensing data such as image, color infrared, hyperspectral, point cloud, satellite remote sensing, UAV, and so forth, are used to carry out qualitative and quantitative analysis of the disaster situation, damage determination, and so on. Large-scale disaster can be quickly assessed through the analysis of remote sensing images in terms of degree of disaster, the extent of the disaster, the number of affected agricultural fields, and the prediction of the amount of claims to provide data support for insurance companies for making rapid decisions (Karthikeyan et al., 2020). GIS data such as administrative division data, plot data, farmers distribution, planting structure, meteorological data, and so on, can be used for spatial analysis to achieve accurate evidence-based support for crop insurance claims.

Pradhan Mantri Fasal Bima Yojana (PMFBY) of the government of India, launched on February 18, 2016, has emphatically stressed the use of remote sensing images and UAV/drone technology to capture the details and status of agriculture crops (*https://pmfby.gov.in*). These technologies are found to be very useful to settle the claims of the farmers quickly. In a recent locust attack in the country and damage to agriculture crops in year 2020, UAVs/drones have been used to spray pesticides to kill locusts.

10.6 CASE STUDIES

Two case studies are presented as follows.

10.6.1 Crop Yield Modeling

The advancement of crop yield simulation models began after the accomplishment of a model created by Ritchie (1972) to estimate evapotranspiration from crops. The model was more empirical as it required seasonal variation in LAI as an input. Several simulation models such as DSSAT, WOFOST, CROPSYST, and INFOCROP have been used in recent years to anticipate the regional yield forecast for various crops (Di Paola et al., 2016). There is a need to assess the efficiency capability of the sugarcane crop in various agro-climatic zones of India. Various sugarcane models such as APSIM, CANEGRO, and QCANE have been used in several studies worldwide (Verma et al., 2021), but very few studies have been reported using the simulation crop growth model in India due to a lack of understanding of their capabilities and lack of data requirements for calibration. Sugarcane in India is a high priority crop for the government, given the fact that India is the second largest producer of sugar in the world.

A study was undertaken by Verma et al. (2021) to predict the sugarcane yield by the DSSATCANEGRO model for early, mid-, and late planting, and to determine the best time of planting sugarcane in Chhapar village of the Muzaffarnagar district (Uttar Pradesh) of India. The study area of 11.43 km^2 has an average elevation of 242 m above mean sea level, and is dominated with the commercial sugarcane crop (Figure 10.10). The village revenue maps and crop yield data were collected from Muzaffarnagar district, toposheets at scale 1 : 50,000 (covering parts of 53G and 53H), collected from Survey of India, Dehradun; ground data for LAI were collected from the field in September 2013 and January/February 2014; and Resourcsat-2 LISS IV satellite images at 5.8 m spatial resolution for 2013, 2014, and 2015 were collected from the National Remote Sensing Centre (NRSC) Hyderabad. Various generated files and the flow of data in the DSSAT model are shown in Figure 10.10.

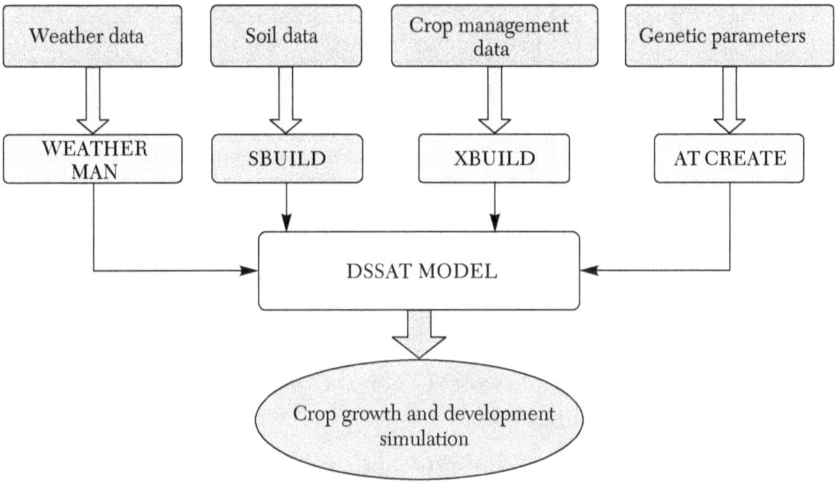

Figure 10.10. Data used and files generated in the DSSAT model.

This study established an empirical relationship between the IRS LISS-IV satellite derived LAI and farm scale sugarcane yield (Verma et al., 2018). A strong exponential relation $y = 0.1202e^{5.538x}$ ($R^2 = 0.861$) is observed between the ground measured LAI (y) and the LISS-IV- based NDVI (x). The sugarcane yield model is developed using regression analysis between plot-wise sugarcane yield data (y) with LISS-IV LAI data (x). This empirical yield model has been found to give a reasonably fair indication of $y = 110.05x + 339.85$ ($R^2 = 0.714$) of expected yield of sugarcane.

Table 10.3. Results from the DSSAT-CANEGRO Model

Year	Observed sugarcane yield (q/ha)	Model	CANEGRO predicted sugarcane yield (q/ha)	Error (%)
2013 (Calibration)	729.44	Early planting model	693.05	4.98
		Mid-planting model	723.17	0.85
		Late planting model	579.83	20.51

(Contd.)

Year	Observed sugarcane yield (q/ha)	Model	CANEGRO predicted sugarcane yield (q/ha)	Error (%)
2014 (Validation)	718.76	Early planting model	656.76	8.62
		Mid-planting model	698.63	2.80
		Late planting model	574.05	20.13
2015 (Validation)	759.40	Early planting model	686.43	9.60
		Mid-planting model	719.91	5.20
		Late planting model	519.06	31.64

Further, the calibration of the DSSAT-CANEGRO model was performed for twelve different locations of the study area during spring season 2013–2014, under three different dates of planting (early planting, mid-planting, and late planting). Data on growth, development, and yield were recorded for 180 treatment combinations using five sugarcane varieties (CoS-767, CoSe-95422, CoS-8436, CoSe-92423, and CoSe-98231). The results of the models are given in Table 10.3.

These results of the model are validated for 2014 and 2015 data. The predicted sugarcane yield by the CANEGRO sugarcane model shows the percent error from +4.98 to +9.60 for early planting, from +20.13 to +31.64 for late planting, and from +0.85 to +5.20 for mid-planting. The sugarcane yield simulation mid-planting model presented the highest R as 0.81 and 0.83, and the Willmott index of agreement (D) as 0.88 and 0.91, and the lowest *rmse* as 8.37 q/ha and 10.70 q/ha, and the mean absolute percent error (MAPE) as 1.10% and 1.30% for the years 2014–15 and 2015–16, respectively, as compared to the models of the other two dates. The predicted sugarcane yield by the CANEGRO model shows a percent error varying from +2.80 to +5.20 for mid-planting seasons. The results of the mid-planting thus proved superior over early and late planting. Sugarcane requires about 25–32°C for good germination, and therefore, early and late planting may be the cause of yield loss due to low and high temperatures, respectively. It is also observed that the variety CoSe-98231 produces a higher yield, whereas variety CoS-767 gives the lowest yield for all three planting dates. Thus, remote sensing images could be successfully used for the modeling of sugarcane crop yield predication with fairly good accuracy.

10.6.2 Crop Classification

For planning purposes, current and accurate crop type maps are essential. Different advanced classification techniques such as machine learning and

ANN algorithms have been adopted to discriminate crops. A study was undertaken by Shukla et al. (2018) to carry out the comparative performance analysis of popular classical ensemble models—bagging/ARCing, RF, gradient boosting, and importance sampled learning ensemble (ISLE) with a traditional single model (decision tree) for large area crop classification. The study area comprises fertile land lying along the river Yamuna covering the entire Mathura district, India, the major parts of districts (>70% of total district area): in Aligarh, Bharatpur, Hathras, and some parts of districts: in Agra, Bulandshaher, Etah, Faridabad, Firozabad, and Gurgaon. It lies between 27°02'42" N to 28°13'01" N latitude and 76°58'33" E to 78°22'20"E longitude, covering about a 12,421 km^2 area out of which 11,551.53 km^2 (93%) is the net irrigated area and the remaining 869.47 km^2 land is under multiple uses.

The data used for the study included Landsat-8 data of March 9, 2015 at 30 m resolution to generate vegetation, soil, and parent material variables for *rabi* crops (October–March); MODIS NDVI product MOD13Q1 of an entire year (January 2015–December 2015) retrieved from USGS Earth Explorer; ALOS Global Digital Surface Model at 30 m obtained from the website to generate topographic, soil, and parent materials; two different polarization modes; linear and circular SAR data to incorporate backscattering responses of feature classes; C-band SAR data of Indian satellite RISAT-1 in RH (Right circular transmit and Horizontal receive), and RV (Right circular transmit and Vertical receive) and Sentinel-1A of European Space Agency) in VH (Vertical transmit and Horizontal receive); VV (Vertical transmit and Vertical receive), and climatic data (precipitation, temperature). The area is classified into several *rabi* crops namely wheat, mustard, gram, red lentils (*masoor*), potato, and fallow land. To incorporate dependency of a large area crop in different variables, namely parent material and soil, phenology, texture, topography, soil moisture, vegetation, climate, and so on, thirty-five digital layers were prepared using different satellite data. The field survey was carried out to prepare the training data.

To evaluate the performance of all the classifiers employed (RF, RF-OOB, CART Decision Tree, CART-Ensembles bagger, CART-Ensembles ARCing, ISLE, and TreeNet Gradient Boosting Machine), the quality assessment measures used include: (*a*) overall success rates (OSR), (*b*) marginal rates, (*c*) F-measure and Jaccard's coefficient of community (Jaccard, 1912), (*d*) classification success index (CSI), and (*e*) agreement coefficients (AC). The score is calculated to rank the various algorithms used, and these values are shown in Table 10.4. The RF is found to be the best performer, followed by gradient boosting for crop classification. Other ensemble methods ARCing, bagging, and ISLE are in decreasing order of performance. The traditional non-ensemble method decision tree scored higher than the ISLE classifier.

Table 10.4. Performance Measures of Various Classifiers Used

Classifiers	Fallow land	Gram	Masoor	Mustard	Potato	Wheat	Non-crop	Score
RF-Test								
F-measure	0.9787	0.9811	0.9841	0.9365	0.9655	0.9444	0.9679	35
JCC	0.9583	0.9630	0.9688	0.8807	0.9333	0.8942	0.9378	33
ICSI	0.9574	0.9630	0.9687	0.8732	0.9333	0.8883	0.9360	33
Score	35	35	35	33	35	33	35	
RF-OOB								
F-measure	0.9950	0.8496	0.7867	0.9152	0.9063	0.9618	0.9611	32
JCC	0.99	0.7385	0.6484	0.8436	0.8286	0.9264	0.9252	28
ICSI	0.99	0.7219	0.6389	0.8318	0.8198	0.9242	0.9248	28
Score	35	28	24	32	32	35	35	
CART Decision tree								
F-measure	0.9787	0.9286	0.9841	0.90625	0.8754	0.92865	0.96265	34
JCC	0.9583	0.8667	0.9688	0.82864	0.77783	0.86674	0.92795	30
ICSI	0.9574	0.8595	0.96885	0.81324	0.75693	0.85734	0.92525	30
Score	35	32	35	32	29	33	35	
CART-Ensembles Bagger								
F-measure	0.9787	0.9454	0.9841	0.9036	0.9333	0.9293	0.9679	35
JCC	0.9583	0.8966	0.9695	0.8241	0.8754	0.8679	0.9378	31
ICSI	0.9574	0.8915	0.9688	0.8082	0.8667	0.8585	0.9360	31
Score	35	33	35	33	33	33	35	
CART-Ensembles ARCing								
F-measure	0.9787	0.9811	0.9841	0.92	0.9333	0.94	0.9704	35
JCC	0.9583	0.9630	0.9688	0.8519	0.8754	0.8868	0.9426	32
ICSI	0.9574	0.9630	0.9688	0.8417	0.8667	0.8802	0.9413	32
Score	35	35	35	32	33	33	35	
ISLE								
F-measure	0.9677	0.7222	0.9841	0.9032	0.7179	0.945	0.9399	31
JCC	0.9375	0.5652	0.9687	0.8235	0.561	0.8867	0.8867	24

Classifiers	Fallow land	Gram	Masoor	Mustard	Potato	Wheat	Non-crop	Score
ICSI	0.9357	0.5407	0.9687	0.8117	0.5167	0.8802	0.8836	24
Score	35	21	35	32	21	33	33	
TreeNet Gradient Boosting Machine								
F-measure	0.9892	0.9454	0.9841	0.9253	0.9655	0.935	0.9677	35
JCC	0.9787	0.8966	0.9688	0.8611	0.9333	0.8692	0.9375	32
ICSI	0.9787	0.8915	0.9688	0.8519	0.9333	0.8602	0.9358	32
Score	35	33	35	33	35	33	35	

10.7 CONCLUSION

Remote sensing is highly useful in assessing various stresses in different crops and is also very useful in detecting and management of various crop issues (Wójtowicz et al., 2016). The ability to use remote sensing data to assess fertilization needs of plants based on the nutrient content of crops and soils helps to increase yields and improves crop profitability. Governments and other stakeholders can make use of remote sensing and other IoT based data in order to make important decisions about policies to tackle national issues regarding agriculture.

Remote sensing is used to assess the water needs of plants, making it easier to manage crop production under conditions of water stress. Hyperspectral imagery has revolutionized the ability to distinguish multiple crop characteristics, including nutrients, water, pests, diseases, weeds, biomass, and canopy structure. There is a need to develop state or district level information systems based on various crops derived from remote sensing and GIS in order to effectively utilize the information on crops for improvement of the economy. Combining remotely sensed data with existing crop simulation models will improve the reliability of decision support systems and will contribute to modernized agricultural production management.

Precision farming has progressed through many stages. There is a significant potential in PA by combining historical remote sensing data with the DEMs, soil series map data, crop yield maps, and real-time soil moisture data for improved agricultural management. More recently, there has been an increasing emphasis on real-time on-the-go monitoring with ground based sensors. Implanting nano-chips in plants can be helpful in near-real time to monitor them. The challenge for the future is to develop precision farming approaches that can provide customized management of farm inputs for individual plants (Mulla, 2013).

A crop may be subjected to variety of stresses which change over its growth cycle. It is challenging to detect pests and diseases in the early stages, since most of them originate from the base of the plant. The crop would have already been destroyed by the time leaves get affected by disease, which could be detected with temporal remote sensing/NDVI. Multi-look angle sensors may aid in early detection of pests and droughts. Efforts have to be made to differentiate the stresses using satellite sensors so as to precisely strategize crop protection.

The ICT is playing a key role in transforming agricultural activities. However, inadequate infrastructure and ICT based services, gadgets, and management and maintenance challenges affect the use and investment in ICT initiatives in agriculture, particularly in developing countries. While designing an IoT based system for agriculture, the variations in temperature, soil, and water properties may be incorporated to increase crop production. The IoT uses advanced sensor technology to measure various parameters in the field, while cloud computing is used for collection, storage, preprocessing and modeling of a huge amount of data collected from various heterogeneous sources. The four technologies (i.e., remote sensing, IoT, cloud computing, and big data analysis) could create novel applications and services in the future that could improve agricultural productivity and increase food security, for instance by better understanding climatic conditions and changes. The ICT will continue to contribute in the future to agricultural transformation as their potential is further harnessed.

Spectral indices should continue to be developed that simultaneously allow assessment of multiple crop characteristics (e.g., LAI, biomass, etc.) and stresses (e.g., water, weeds, and insects, etc.). There is a need to develop robust algorithms that determine the crop yield in heterogeneous agricultural systems. Advanced techniques such as SVM, KNN, K-means clustering, Wavelet-based filtering, Fourier transform, and CNN, may be explored in identification of crop phenology, soil and leaf nitrogen content, irrigation, plant water stress detection, pest detection and herbicide use, identification of contaminants, diseases in crops, and so forth. (Kamilaris & Prenafeta-Boldú, 2018). More advanced and complex models such as recurrent neural networks (RNN) or long short-term memory (LSTM) architectures may also be attempted to exhibit the dynamic temporal behavior of agricultural activities. The use of UAV-based aerial images may help in the future to monitor the effectiveness of the seeding process, increase the quality of production by harvesting crops at the right time for the best maturity levels, monitor animals and their movements, identify possible diseases, and many other scenarios.

CHAPTER 11

Disaster Monitoring and Management Using Remote Sensing Technology

11.1 INTRODUCTION

Environmental, financial, and human losses caused by sudden and unexpected events is called natural disaster. Various disasters like earthquakes, landslides, floods, fires, tsunamis, volcanic eruptions, and cyclones are natural hazards that kill a large number of people and destroy property and infrastructure every year, worldwide (ARDC, 2016). The rapid increase in the population and its increased concentration, often in hazardous environments, has escalated both the frequency and severity of natural disasters. The world population is estimated to be between seven and ten billion by the year 2050 (World Bank, 2016). Due to increased awareness of the international community on disaster management, the decade 1990–2000 was designated as the "International Decade for Natural Disaster Reduction (IDNDR)" by the General Assembly of the United Nations.

Disasters are not inevitable all over the world (ADRC, 2016). Successful mitigation of natural disasters requires adequate knowledge about the expected frequency, character, and magnitude of hazardous events. *Disaster management* refers to the comprehensive strategy in all phases of disaster for effectively reducing the impact of the disaster (Bhanumurthy & Behera, 2008). Disaster management consists of four phases: two phases that take place before a disaster are disaster mitigation (prevention) and disaster preparedness, and the remaining two phases that happen after the occurrence of a disaster are disaster response and disaster recovery (Figure 11.1; Westen, 2000). Disaster management is represented here as a cycle, since the occurrence of a disaster event will eventually influence the way society is preparing for the next one.

The impact of natural disasters can be reduced through proper disaster management, including disaster prevention (hazard and risk assessment, land use planning and legislation, enforcement of building bylaws and codes), disaster preparedness (forecasts, warning, prediction), and rapid and adequate disaster relief (Nirupama & Simonovic, 2002). It will, however, require several activities: (*a*) identifying the areas that are likely to be affected, monitoring the environment, and mapping the areas, including flood plains; (*b*) planning mitigation measures which include hazard mapping; (*c*) adoption and enforcement of land use and zoning practices, and implementing and enforcing building codes; and (*d*) organizing disaster mitigation public awareness programs. In countries where warning systems exist and building codes are followed, remote sensing data have been successfully used to predict the occurrence of disastrous phenomena and to warn people on time.

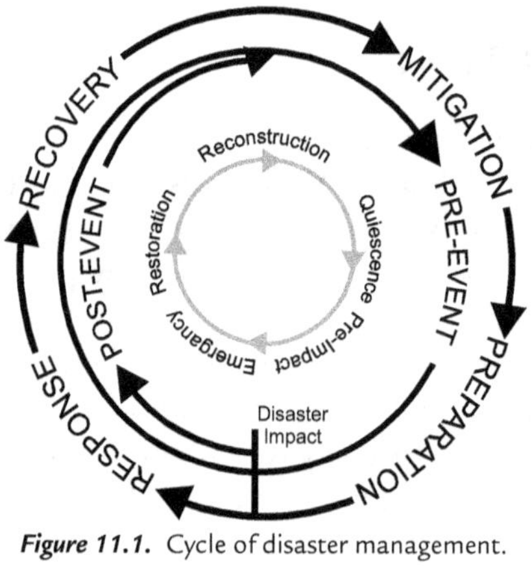

Figure 11.1. Cycle of disaster management.

11.2 TYPES OF DISASTERS

Disasters can be classified into two broad categories as follows and shown in Figure 11.2 (ASOSAI, 2015):

1. *Natural disasters*–which are caused by purely natural phenomena, damaging human societies, such as floods, earthquakes, landslides, volcanic eruptions, hurricanes, and so on.

2. *Man-made disasters*–which are caused by human activities such as explosions, fire, structure failure, atmospheric pollution, industrial chemical accidents, oil spills, major armed conflicts, nuclear accidents, oil spills, and so forth.

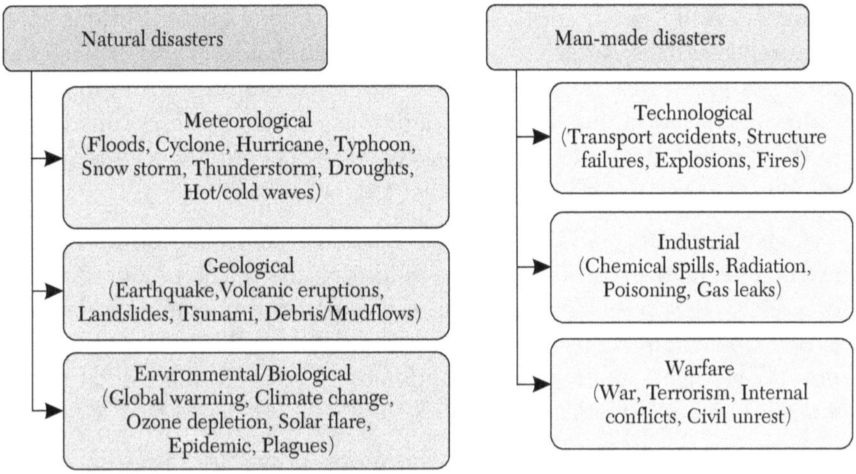

Figure 11.2. Types of disasters (ASOSAI, 2015).

Before analyzing disasters, it is important to understand the difference between the terms "disaster" and "hazard." As shown in Table 11.1, a *hazard* is a dangerous situation or event that carries a threat to humans, whereas a disaster is an event that actually harms humans and disrupts the operations of society. A potentially damaging disaster (hazard) such as an earthquake may not be considered a disaster if it occurs in uninhabited areas, whereas it is called a disaster if it occurs in an habited area, creating damages and losses (Westen, 2000). A hazard is less severe and has fewer critical consequences. It is caused by negligence, and takes shape after a series of events that are about to take place. A hazard leads to a disaster and possesses a risk.

Table 11.1. Difference between a Hazard and a Disaster (Wither, 2017)

Hazard	**Disaster**
A dangerous situation needing to be heeded as it can lead to a disaster.	A dangerous situation that becomes out of control is called a disaster.
A threat that can be managed by observing warning signs and judiciously managing the environment.	An international danger and threat to humanity that needs intervention to bring the situation under control.
Hazards are known to have specific warnings, usually man-made, to prevent disastrous events.	Disasters are the outcomes of hazards when warning signs were ignored.
Hazards can lead to disasters.	A disaster is the result of a hazard but at the same time is also a hazardous event.
Hazards are not used to describe everyday mishaps. They are specific occurrences with appropriate warning signs.	Disasters, in literal terms, are more severe than hazards.

Disasters often occur in a short period of time, creating serious effects. Natural disasters such as earthquakes, landslides, avalanches, flood, drought, forest fire, cyclones, and so on are caused as a result of different behaviors of nature due to many conditions. These disasters constitute one of the greatest threats to development and socioeconomic well-being of the people, as they are difficult to predict (ARDC, 2016).

The frequency and magnitude of natural and human-induced disasters are increasing every year, and thus hindering the development and growth of society. It is possible to prevent a hazard from becoming a disaster by proper management of the environment (Wither, 2017). An awareness of potential dangers and taking precautionary steps is often needed to prevent a hazard becoming a disaster.

Geospatial data has had an enormous utility for disaster mapping, monitoring, and management. The application of satellite remote sensing to disaster management has been supported under the IDNDR framework, when satellite-based Earth observations for rapidly assessing disaster situations were started globally. According to ARDC (2016), Asia has been seriously damaged by natural disasters over the last thirty years (1987–2016), as shown in Figure 11.3. Disasters occurring in Asia comprise 39% of the total disasters in the globe. This is further aggravated by its large population, which is more than half of the world's population. The Asia region suffers 61% of global fatalities and has 88% of the total victims associated with such disasters. About 95% of the natural disaster-related deaths occur in the developing world. Economic losses due to natural hazards in developing countries may represent as much as 80% of their gross national product.

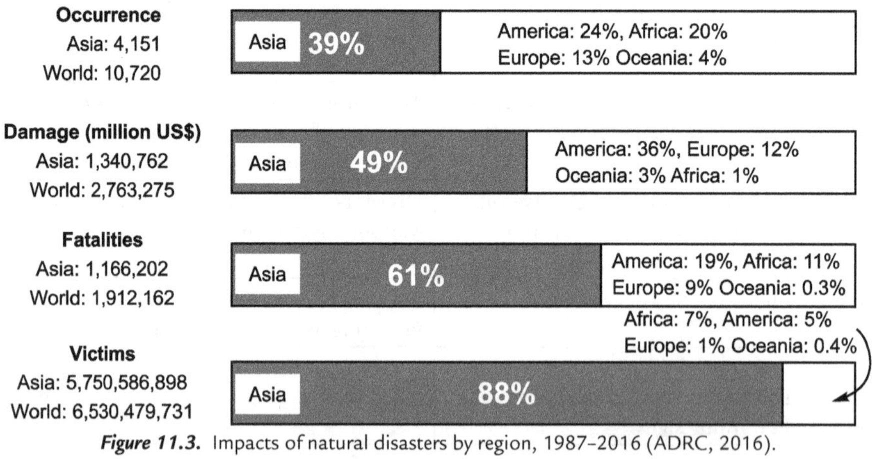

Figure 11.3. Impacts of natural disasters by region, 1987–2016 (ADRC, 2016).

11.3 GEOSPATIAL DATA FOR DISASTERS

In general, various types of geospatial data are required for disaster management (Nirupama, & Simonovic, 2002), such as regarding: (*a*) disastrous phenomena (e.g., landslides, floods, earthquakes, their location, frequency, magnitude, etc.), (*b*) environment in which the disastrous events took place (e.g., topography, geology, geomorphology, soils, hydrology, land use, vegetation, etc.), (*c*) elements that are destroyed/affected (i.e., infrastructure, settlements, population, socio-economic data, etc.), and (*d*) emergency relief resources (e.g., hospitals, fire brigades, police stations, warehouses, etc.). Remotely sensed data collected from various sources can be used very effectively for quickly assessing the severity and impact of damage due to earthquakes, landslides, flooding, forest fires, cyclones, and other disasters.

According to Rathje et al. (2008), there are several opportunities for remote sensing to make an impact in an earthquake study, such as pre-event activities, rapid post-earthquake activities, and long-term post-event activities (as shown in Figure 11.4). Pre-event activities generally comprise pre-earthquake loss estimation modeling for different urban areas, which

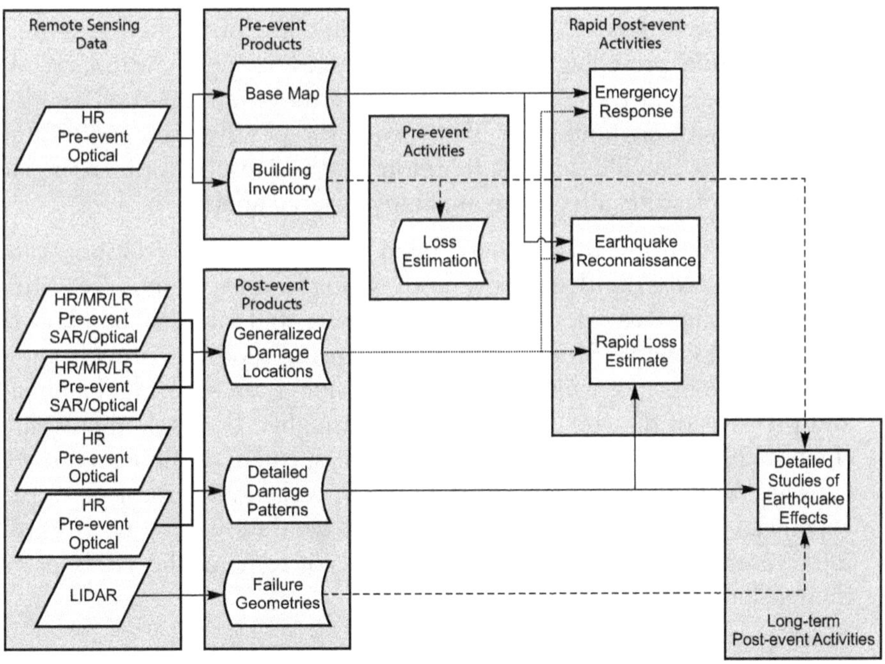

Figure 11.4. Flowchart of remote sensing processing and analyses for rapid and long-term studies (Rathje et al., 2008).

in turn are often used in emergency response planning. Rapid post-event activities include emergency response, earthquake reconnaissance, and rapid loss estimation, while long-term post-earthquake activities represent detailed studies of earthquake effects. This figure shows various sources of remote sensing data, the data products that can be derived from them, and the use of these data products in pre- and post-earthquake activities. The pre-event remote sensing data products are generally derived from high-resolution (HR) optical data and include base maps of affected regions and building inventories, while post-event remote sensing data products are generally derived from high-resolution, medium-resolution (MR), and low-resolution (LR).

11.3.1 Various Satellites and Sensor Images

Since the launch of first Earth resource satellite, Landsat-1, in 1972, several remote sensing satellites have been launched and used for Earth observations, but the area covered and their temporal resolution vary from satellite to satellite. Each satellite carries more than one sensor onboard that take measurements in different wavelength regions. Polar-orbiting satellites, flying at low altitude, provide relatively high spatial resolution, but they offer the advantage of collecting repetitive data of the same point after every few days. Geostationary satellites are positioned at a much higher altitude, providing low-resolution images frequently with large area coverage (Garg, 2019). The advantage is that they view a part of the globe continuously throughout their life, providing repetitive data every fifteen minutes or so. Satellite images, therefore, are very useful for mapping and monitoring disasters affecting comparatively large areas.

Remote sensing images have been used to map vegetation, water, soil, urban areas, and geology, both spatially and temporally. Multi-temporal data have demonstrated the potential of bringing out the damage assessment. Moreover, remote sensing methods provide multi-date satellite images, which in turn greatly help in monitoring the change and progress of disaster events, such as droughts, floods, and landslides (Lewis, 2009). In addition, many disasters, such as floods, drought, cyclones, volcanic eruptions, and so forth, will have certain precursors which can be detected very effectively and accurately using post-disaster satellite images. Table 11.2 provides details of some of the satellites and their salient features.

Table 11.2. Detailed Description of Satellites and Sensors (Nirupama & Simonovic, 2002)

Satellite	Country	Repeat Cycle	Sensor	Frequency/Band (mm)
RADARSAT	Canada	24 days	SAR	5.3 GHz (C-Band)
LANDSAT	USA	16 days	TM/ETM MSS MTB PAN	0.45–0.52, 0.52–0.60, 0.63–0.69, 0.76–0.90, 1.55–1.75, 2.08–2.35 0.5–0.6, 0.6–0.7, 0.7–0.8, 0.8–1.1, 10.4–12.6, 10.4–12.5, 0.5–0.9
SPOT	France	26 days (5 days with pointing capability)	HR VIR (1)	0.50–0.59, 0.61–0.68, 0.79–0.89, 1.58–1.75
			HR VIR (2)	0.61–0.68
			VMI	0.43–0.47, 0.50–0.59, 0.61–0.68, 0.79–0.89, 1.58–1.75
IRS	India	24 days	LISS-III VNIR SWIR PAN WiFS	0.52–0.59 0.62–0.68, 0.77–0.86, 1.55–1.70 0.50–0.75, 0.62–0.68, 0.77–0.86
RADARSAT	Canada	24 days	SAR	5.3 GHz (C–Band) Standard reso– lution, Fine resolution
ERS	Europe	35 days	AMI Active Microwave Instrumentation	
			SAR Image	5.3 GHz (C–Band)
			SAR Wave	5.3 GHz (C–Band)
			Scatterometer	5.3 GHz (C–Band)
			Radar Altimeter	13.5 GHz (KU–Band)
			ATSR-M Along Track Scanning Radiometer with Microwave Sounder	
			Infrared Radiometer	1.6, 3.7, 11, 12
			Microwave Sounder	23.5 GHz
			GOME	(1) 0.24–0.295 (512 Bands) (2) 0.29–0.405 (1024 Bands) (3) 0.40–0.605 (1024 Bands) (4) 0.59–0.79 (512 Bands)
			AATSR	0.65, 0.85, 1.27, 1.6, 3.7, 11.0, 12.0

(Contd.)

Satellite	Country	Repeat Cycle	Sensor	Frequency/Band (mm)
NOAA	USA	12 hrs	AMSU-A	23.8 (GHz), 31.4, 50.3, 52.8, 53.33, 54.4, 54.94, 55.5, 57.29 (6 Channels), 89.0
			AMSU-B	89.0, 166.0, 183.31 (3 Channels)
			AVHRR/3	(1) 0.58–0.68, (2) 0.72–1.00, (3A) 1.58–1.64 (day), (3B) 3.55–3.93 (night), (4) 10.3–11.3, (5) 11.5–12.5
			HIRS-3	
			20 IR Bands	Wavelength at Band center (1) 14.95, (2) 14.71, (3) 14.49, (4) 14.22, (5) 13.97, (6) 13.64, (7) 13.35, (8) 11.11, (9) 9.71, (10) 12.47, (11) 7.33, (12) 6.52, (13) 4.57, (14) 4.52, (15) 4.47, (16) 4.45, (17) 4.13, (18) 4.00, (19) 3.76, (20) 0.69
MOS	Japan	17 days	MESSR	0.51–0.59, 0.61–0.69, 0.72–0.80, 0.80–1.1
			VTIR (1)	0.5–0.7, 6.0–7.0
			VTIR (2)	10.5–11.5, 11.5–12.5
			MSR (1)	23.8 GHz
			MSR (2)	31.1 GHz
JERS	Japan	44 days	SAR OPS VNIR SWIR	1275 MHz (L–Band) (1) 0.52– 0.60), (2) 0.63–0.69, (3) 0.76– 0.86+, (4) 1.60–1.71++, (5) 2.01–2.12, (6) 2.13–2.25 , (7) 2.27–2.40
IKONOS	USA	2.9 days 1.5 days	PAN VNIR	0.45–0.90 0.45–0.52, 0.52–0.60, 0.63–0.69, 0.76–0.90
QuickBird	USA	3.5 days	PAN Multispectral	0.45–0.9 0.45–0.52, 0.52–0.6, 0.63–0.69, 0.76–0.9

Note: AATSR—Advanced Along Track Scanning Radiometer, AMI—Active Microwave Instrumentation, AMSU—Advanced Microwave Sounding Unit, GOME+– Global Ozone Monitoring Experiment Band 3 and 4 make stereo pair, HIRS-3—High Resolution Infrared Radiation Sounder, HRVIR IR—High Resolution Visible Infrared Infrared, MESSR MTB—Multispectral Electronic Self- Scanning Radiometer Multispectral Thermal Band, MSR++ – Microwave Scanning Radiometer Band 4 is for forward viewing (15.33 degrees), OPS—Optical Sensors, VIS VMI—Visible Vegetation Monitoring Instrument, VTIR—Visible and Thermal-Infrared Radiometer.

Remote sensing has been used extensively to map the extent of the impact caused by earthquakes, landslides, floods, tsunamis, hurricanes, and forest fires (Westen, 2000). High-resolution images (e.g., IKONOS, QuickBird), SAR, and LiDAR have demonstrated their capabilities in quantifying post-disaster damage and monitoring the progress of infrastructure during rehabilitation (Ramiya et al., 2017). Although none of the existing satellites and their sensors has been designed solely for the purpose of observing natural disasters, the variety of images in visible, NIR, IR, SWIR, TIR, and microwave provide adequate spectral, spatial, and temporal coverage to be used for disaster mapping, monitoring, and management. In general, remote sensing data can be used very effectively for (Nirupama & Simonovic, 2002): (*a*) quickly assessing severity and impact of damage due to flood, earthquakes, landslides, forest fires, and other disasters; (*b*) planning escape routes; (*c*) finding the quickest routes for ambulances to reach victims; (*d*) locating safe places for shelter; (*e*) assessing the population affected in disaster-prone areas; (*f*) developing early warning of potential disasters; (*g*) planning for timely evacuation and recovery operations during a crisis; (*h*) monitoring construction or rehabilitation activity after a major disaster, and (*i*) developing accurate base maps.

Different satellites and sensors can provide unique information about properties of the surface of the Earth (Garg, 2019). For example, measurements of reflected solar radiation give information on albedo, thermal sensors measure surface temperature, and microwave sensors measure dielectric properties and hence provide better information on moisture content of surface soil or snow. In many of the weather-related disasters, obtaining cloud free images for damage assessment is often a severe problem, so microwave satellite images are preferred, as these can penetrate into clouds and fogs (Lewis, 2009). For disasters such as floods, debris flows, or oil spills, SAR images are promising, whereas for disasters like landslides, earthquakes, and wildfires, optical images could be used. Table 11.3 presents the utility of wavelength bands useful for disaster mapping and monitoring, such as thermal sensors for spotting active fires, infrared sensors for floods, and microwave sensors for Earth deformations before and during earthquakes or volcanic eruptions. The nature of the natural disasters determines the suitability of sensor types, spectral bands, active or passive radar data, and their spectral, temporal, and spatial resolutions.

Disaster mapping is useful for assessing, storing, and conveying information on the geographical location and spread of the disasters

(Sharma et al., 2010). Disaster maps show the affected areas due to loss of life, property, and infrastructures, which is useful for planning relief measures. These maps generally show risk zones as well as disaster impact zones that would be affected increasingly with the increase in the magnitude of the disaster. These could include landslide hazard maps, flood zone maps, seismic zone maps, forest fire risk maps, industrial risk zone maps, and so on.

Table 11.3. Applications of different wavebands for disaster management (Lewis, 2009)

Wavelength	Waveband	Useful for	Example sensors
Visible	0.4–0.7 μm	Vegetation mapping	SPOT; Landsat TM
		Building stock assessment	AVHRR; MODIS; IKONOS
		Population density	IKONOS; MODIS
		Digital elevation model	ASTER; PRISM
Near infrared	0.7–1.0 μm	Vegetation mapping	SPOT; Landsat TM; AVHRR; MODIS
		Flood mapping	MODIS
Shortwave infrared	0.7–3.0 μm	Water vapor	AIRS
Thermal infrared	3.0–14 μm	Active fire detection	MODIS
		Burn scar mapping	MODIS
		Hotspots	MODIS; AVHRR
		Volcanic activity	Hyperion
Microwave (radar)	0.1–100 cm	Earth deformation and ground movement	Radarsat SAR; PALSAR
		Rainfall	Meteosat; Microwave Imager (aboard TRMM)
		River discharge and volume	AMSR-E
		Flood mapping and forecasting	AMSR-E
		Surface winds	QuikScat radar
		3D storm structure	Precipitation radar (aboard TRMM)

Note: Acronyms: Satellite Pour l'Observation de la Terre (SPOT); Thematic Mapper (TM); Advanced Very High Resolution Radiometer (AVHRR); Moderate Resolution Imaging Spectroradiometer (MODIS); Advanced Spaceborne Thermal Emission and Reflection Radiometer (ASTER); Panchromatic Remote-sensing Instrument for Stereo Mapping (PRISM); Synthetic Aperture Radar (SAR); Phased Array type L-band SAR (PALSAR); Tropical Rainfall Measuring Mission (TRMM); Global Precipitation Measurement (GPM); Advanced Microwave Scanning Radiometer (AMSRE); Atmospheric Infrared Sounder (AIRS).

Remote sensing and geographic information system (GIS) techniques are effectively used to map the extent of a disaster, and facilitate its management through computer-generated visualization models to monitor the effect of the disaster. In addition, GIS-based models are used to mitigate the impact of a disaster, find safer places, deploy rescue teams, and undertake rehabilitation activities. In recent years, development in the areas of wireless sensors coupled with GIS and remote sensing has facilitated the advancement of susceptibility mapping, risk assessment, and risk mitigation.

11.3.2 Unmanned Aerial Vehicle Images

UAVs or drones are a futuristic technology and a potential solution to many humanitarian problems (Aljehani & Inoue, 2019). Disaster monitoring can be facilitated by the use of UAVs/drones, which are becoming more and more popular due to various advantages, such as small size, low cost of operation, access to dangerous environments, and high probability of mission success without the risk of the life of the aircrew (Garg, 2021a). UAVs have demonstrated the ability to compete with aerial and satellite data due to their operational flexibility and high spatial resolution data. They can be used to enhance the ability of network-assisted disaster prediction, assessment, and response (Erdelj & Natalizio, 2016), such as network with IoT-based sensors.

UAV/Drone applications supporting disaster management can be used in three major activities: pre-event, during the event, and post-event. In pre-disaster activity, drone applications can support the prevention or early detection of activity. For example, an escalated forest fire can be avoided by undertaking periodic UAV/drone flight patrols and timely measures. During disaster, drones can support with real-time monitoring of the area, providing quick and relevant information regarding intervention or mitigation. They also act as first responders during emergency to save time, money and, most importantly, lives. Thus, the effect of the disaster can be mitigated more effectively, and UAVs/drones can support making better decisions. In post-disaster activity, UAVs/drones can support quick damage assessment and help recovery (Garg, 2021a). They are utilized for acquiring the data of the area that is unsafe or hard-to-reach disaster stricken areas, as shown in Figure 11.5.

When a disaster occurs, the most important issue that needs to be attended to save human lives. In this context, the first seventy-two hours after a disaster hits are the most critical, which means that search and rescue operations must be conducted quickly and efficiently (Erdelj & Natalizio,

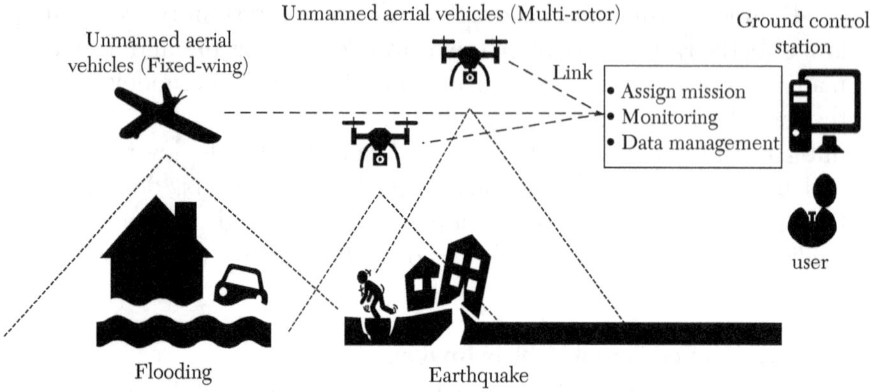

Figure 11.5. Scenario of a post-disaster mapping mission undertaken by drone/UAV (Aljehani & Inoue, 2019).

2016). UAVs/drones provide a more efficient means of data collection than the traditional ground-based and manned aerial methods. Effective disaster relief operations start with a sound situational assessment. Officials taking measures to save lives and minimize damages require accurate geographic information to coordinate rescue operations. UAVs/drones can be quickly deployed over the disaster-stricken zones to produce 3D maps and assess the damaged infrastructure, which are essential in emergency disaster response. They can support other technologies such as the IoT, M2M communication systems, networking, and satellite systems. Furthermore, UAVs/drones can be used to (Garg, 2020): (*a*) provide rapid situational awareness with mapping technology, (*b*) identify hot spots and assess property and infrastructural damage, (*c*) capture high-resolution images for planning, (*d*) conduct search and rescue operations for survivors, and (*e*) create pre- and post-disaster maps of the impacted area.

11.3.3 Point Cloud Data

Traditional stereo-photogrammetry based 3D data capturing techniques are being replaced by rapidly growing LiDAR techniques for fast capturing of dense point cloud data of land surface features and their geometry (Garg, 2021a). The LiDAR data can be collected with ground-based, mobile-based (vehicle), and airborne systems. During a disaster, airborne LiDAR data is mostly used, as the other two means of data collection may not be feasible during a disaster event. UAV-assisted LiDAR technologies are a relatively cost-effective means to capture and represent the point cloud data by recording their 3D coordinates as x, y, and z (Vetrivel et al., 2016). The processing of LiDAR point clouds is significantly different from the existing image processing software. For example, UAV-assisted LiDAR data captured over a forest area consists of return signals from both the trees and bare ground below, and the extraction of tress from the bare ground would require specialized processing of data. With the dynamic growth of an urban landscape, 3D city models generated by point cloud data are gaining prominence in various applications, such as smart cities, facility management, utility management, disaster management, noise modeling, and transport planning (Ramiya et. al., 2017). Therefore, there is an increasing demand for street-level 3D views of the urban landscape due to 3D geospatial data based products, such as those employed by Google and Apple.

Structural damage assessment is an important process to be carried out immediately after the earthquake disaster event for effective planning and execution of response and recovery actions. Detailed information of an affected area can be provided in a short time using a variety of sensors such as optical, SAR, and LiDAR (Ramiya et al., 2017). Geometrical deformations, such as partial/complete collapse, pancake collapse, inclination, and broken and dislocation of elements can be easily derived by 3D geometric information, while damages, such as cracks, can be derived from the pre-event and post-event point cloud data directly, as shown in Figure 11.6. These damaged elements, called missing structural elements, can be identified by

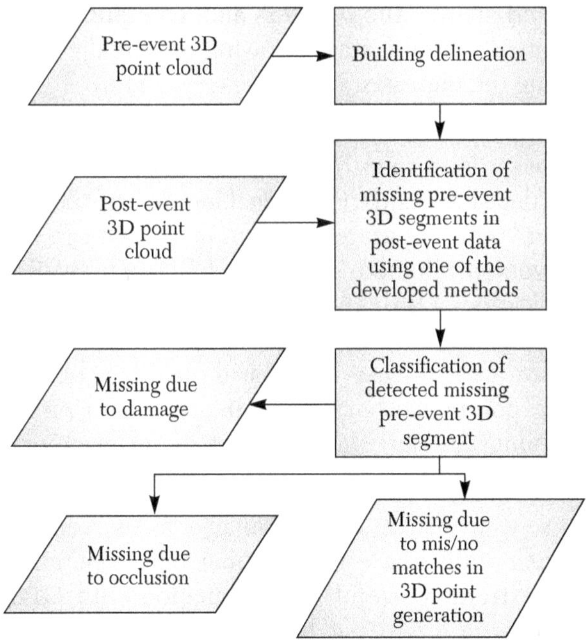

Figure 11.6. Building damage assessment methodology using point cloud data (Vetrivel et al., 2016).

comparing the pre- and post-event 3D point clouds (Vetrivel et al., 2016). The missing elements due to damage are detected using the pre- and post-event images and 3D point clouds derived from them by three processes. Firstly, the individual buildings are delineated from the pre-event 3D point cloud. Secondly, the delineated buildings in the pre-event data are compared to the post-event 3D point cloud to detect changes. Lastly, the changes are classified to isolate the changes caused by damage.

11.4 ATA INTEGRATION IN GIS

Disasters, such as earthquakes, landslides, tsunamis, volcanoes, cyclones, storms, and floods endanger the population and their surrounding environment. Mitigation of natural disasters requires detailed knowledge about the expected frequency, character, and magnitude of hazardous events in an area as well as the geology and topography of the area. Information needed in natural disaster management can be derived from various thematic maps, aerial photographs, satellite images, GPS data, LiDAR data, rainfall data, meteorological data, and so on. (Westen, 2000). Remote sensing is emerging as a popular means of thematic map preparation, while GIS can be used for storage, analysis, and retrieval of information for change detection (Sharma et al., 2010). Remote sensing and GIS can be used to map and analyze the disasters and, combined with other ancillary information, create hazard maps showing potentially dangerous areas as well as assessing the damages.

The GIS is an effective tool for modeling, analyzing, and identification of the potential hazardous zones from various disasters as well as the planning required for decision- and policy making (Garg, 2019). Planners and decision-makers require hazard zonation maps that can form the basis for disaster management. Remote sensing and GIS in mitigating disasters have potential applications (Nirupama & Simonovic, 2002): (*a*) establishment of susceptibility of the land and vulnerability of the society; (*b*) mapping potential hazard areas for use in physical planning (hazard zoning maps); (*c*) monitoring potentially hazardous situations and processes, providing advanced warning; and (*d*) improvement in management of emergency situations following a disaster.

Remote sensing can assist in damage assessment and monitoring, providing a quantitative base for planning relief operations. For disaster relief, GIS is extremely useful in combination with GPS in search and rescue operations in areas that have been devastated or damaged. The main advantage of the use of GIS is the creation of several hazard and risk scenarios for decision-making about the future development of an area, and planning its protection from future natural disasters (Westen, 2000). Integration of remote sensing with GIS and web technology makes it an extremely powerful tool to identify indicators of potential disasters. Information sharing through the Internet reduces data acquisition time, thus providing an efficient way to carry out real-time disaster predictions (floods, forest fires, tsunamis, and hurricanes etc.).

11.5 DISASTER MANAGEMENT USING REMOTE SENSING AND GIS

Remote sensing and GIS have become well developed and successful tools in disaster management. Although none of the new satellites was specifically designed to be used in disaster mitigation, most of them also have demonstrated their usefulness in disaster prevention, preparedness, and relief. GIS is used in managing the huge amount of data required for vulnerability and hazard assessment. Recurring occurrences of earthquakes, floods, landslides, and forest fires can be studied using advanced geospatial technology to design effective preventive measures. In disaster preparedness, remote sensing, GIS, and GPS tools are used for planning evacuation routes, planning emergency operations, planning search and rescue operations, and the design of disaster warning systems. In disaster rehabilitation, GIS is used to organize the damage/loss information and post-disaster census information, as well as identify the best sites for reconstruction (Sharma, 2010).

Disaster mitigation involves preparation of master plans, hazard zonation and vulnerability analysis, location specific analysis of the disaster problem, methods available for disaster analysis including research, and so on (Bhanumurthy & Behera, 2008). There is a need to make a strong GIS-based mitigation plan for disaster-prone areas, which involves those activities which are carried out during real emergencies. Data from remote sensing and GPS can be used to locate damaged infrastructures, identify the type and amount of damage, and establish priorities for action and major restoration work. In case of an earthquake or a flood, these data can be useful to determine accessible roads for movement, locate damaged houses, identify the suitable ground for landing a helicopter, and so forth.

Remote sensing and GIS tools can be used for developing robust emergency route planning during evacuation (Nirupama & Simonovic, 2002). An important aspect in satellite-based monitoring involves the

Table 11.4. Role of Remote Sensing and GIS in Various Phases of Disaster Management (Lewis, 2009)

Disaster	Mitigation	Preparedness	Response	Recovery
Cyclone	Risk modeling; vulnerability analysis.	Early warning; long-range climate modeling.	Identifying escape routes; crisis mapping; impact assessment; cyclone monitoring; storm surge predictions.	Damage assessment; spatial planning.
Drought	Risk modeling; vulnerability analysis; land and water management planning.	Weather forecasting; vegetation monitoring; crop water requirement mapping; early warning.	Monitoring vegetation; damage assessment.	Informing drought mitigation.
Earthquake	Building stock assessment; hazard mapping.	Measuring strain accumulation.	Planning routes for search and rescue; damage assessment; evacuation planning; deformation mapping.	Damage assessment; identifying sites for rehabilitation
Fire	Mapping fire-prone areas; monitoring fuel load; risk modeling.	Fire detection; predicting spread/direction of fire; early warning.	Coordinating fire-fighting efforts.	Damage assessment.
Flood	Mapping flood-prone areas; delineating flood-plains; land-use mapping.	Flood detection; early warning; rainfall mapping.	Flood mapping; evacuation planning; damage assessment.	Damage assessment; spatial planning.
Landslide	Risk modeling; hazard mapping; digital elevation models.	Monitoring rainfall and slope stability.	Mapping affected areas.	Damage assessment; spatial planning; suggesting management practices.
Volcano	Risk modeling; hazard mapping; digital elevation models.	Emissions monitoring; thermal alerts.	Mapping lava flows; evacuation planning.	Damage assessment; spatial planning.

assessment of damage incurred during the disaster. The locational data about the extent of damage is urgently required for decision-makers to provide fast rescue service to trapped people or to provide immediate medical support to injured people. For example, the GIS can help to estimate the number of paramedical staff and their postings where they are to be located. GIS and remote sensing data can provide details about the damages to land, infrastructure, and habitation, which is important information required for the government to plan and release economic relief measures (Sharma et al., 2010).

For mapping and management of natural disasters, multi-temporal remote sensing data is the ideal tool, as it offers information over large areas at frequent intervals (Garg, 2019). The large amount of remote sensing data is required to be properly analyzed and results combined with other data in GIS to devise disaster management strategy (Kaku, 2019). Although remote sensing has great potential to be utilized in the various phases of disaster management (Table 11.4), in reality it has mostly been used for mapping and monitoring disasters. During the last decades, remote sensing has become an operational tool in disaster preparedness and issuing warnings. Requirements for applying satellite remote sensing to disaster management support can be derived from a holistic viewpoint, including human factors, as shown in Figure 11.7.

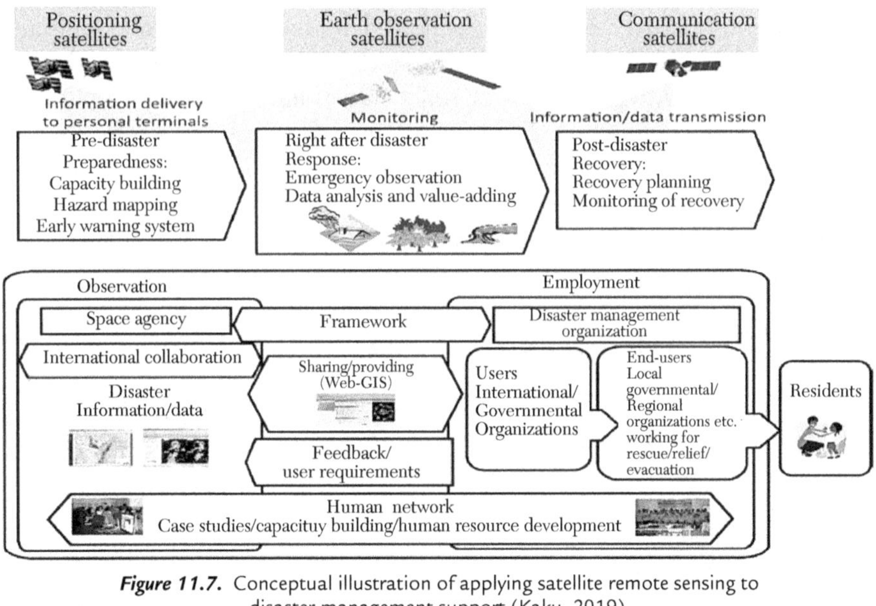

Figure 11.7. Conceptual illustration of applying satellite remote sensing to disaster management support (Kaku, 2019).

High-resolution satellite images, topographic maps, and DEMs are very critical in reconstruction and rehabilitation processes after the disaster has taken place. These processes include rebuilding infrastructures, healthcare, and rehabilitation, as well as developing policies and practices to avoid similar situations in the future. Today, due to major scientific advancements, it has become easier to carry out these operations efficiently. Advanced and highly sophisticated remote sensing and GIS techniques, coupled with IoT and big data, help in several ways: zoning areas according to disaster occurrences and risk magnitudes, simulating disaster damage scenarios, disaster mitigation action plans, and so on. These techniques cannot completely eliminate the risk of disasters, but their effect can be minimized by taking appropriate scientific measures. Considerable losses of life and property could be minimized through the latest information about the risk and onset of disasters, improved risk assessment, planning, and disaster monitoring.

In many cases, the disasters act as catalysts in the adoption of new and emerging technologies. Driven by the need to rapidly collect vital information for disaster management, technology innovations have often helped the government to assess the impact of large disasters more efficiently and rapidly, as well as track and monitor the progress in critical response and recovery operations. The impact of natural disasters can be reduced through proper disaster management, including disaster prevention (hazard and risk assessment, land use planning and legislation, building codes), disaster preparedness (forecasts, warnings, predictions), and rapid and adequate disaster relief. Geospatial data plays a big role in disaster management, as disasters' affected features can be located with their geographic addresses. Remote sensing and GPS are vital to provide locational information, while satellites and UAVs/drones are capable of providing the latest images of the disaster event as well as its happening in real time, which are helpful in decision-making (Garg, 2019).

11.6 APPLICATIONS IN VARIOUS DISASTERS

Remote sensing and GIS technology have the great potential to be used in almost all types of disasters, man-made or natural. Some selected applications have been discussed here.

11.6.1 Cyclones

Meteorologists have used satellite images to monitor storms for decades. For example, the Tropical Cyclone Program of the World Meteorological

Organization uses satellite observations, together with meteorological measurements and modeling to produce cyclone warnings (Lewis, 2009). This program can estimate the storm's position, direction, and speed, maximum wind speeds, areas likely to be affected, and likely storm surges. In India, the Bay of Bengal is particularly vulnerable to cyclones. The sensors of the INSAT-3A satellite collect meteorological data in visible, near infrared, and short-wave infrared, and provide data on cloud motion, sea surface temperature, and rainfall. A network of cyclone warning centers analyze the data and then issue timely warnings of impending cyclones. The warnings provide information on the cyclone itself as well as likely damage and suggested actions.

11.6.2 Drought

Drought is a major disaster in the developing world. Unlike many natural disasters, drought sets in slowly and can often be predicted months in advance. Long-term climate forecasts, derived from satellite observations, can help develop various scenarios before or during the early crop-growing season (Kaku, 2019). Throughout the season, satellite data can help monitor growing conditions and predict soil moisture. At the end of the season, satellite-observed vegetation indices (e.g., NDVI) can be used to monitor vegetation vigor and density and assess the likely crop production and yield. It helps in relating agricultural drought with food shortage.

For large area monitoring, daily observed coarse resolution (1.1 km) NOAA-AVHRR data may be used for a detailed assessment of drought (Bhanumurthy & Behera, 2008). The assessment of the agricultural drought situation takes into consideration the following: (*a*) seasonal NDVI, that is transformation of NDVI images from the beginning of the season over agricultural areas; (*b*) comparison of the agricultural area NDVI profile with previous normal years; (*c*) weekly rainfall status compared to normal; and (*d*) weekly progression of the sown area compared to normal. The relative deviation of NDVI from that of normal and the rate of progression of NDVI during the season give an indication about the agricultural drought situation, which is then verified by the ground situation as evident from rainfall and the sown area. The agricultural drought information thus derived may be shared with stakeholders.

11.6.3 Earthquakes

Earthquakes are hard to predict, and there is no generally accepted operational method for predicting earthquakes. Although there is some mention of observable precursors for earthquakes in literature, such as variations in the electric field or thermal anomalies, these are heavily disputed and not used in remote sensing. Remote sensing, however, could be used to improve forecasts using, for example, interferometric synthetic aperture radar (InSAR) data (Lewis, 2009). This technique combines two or more sequential radar images to measure the ground motion between them very accurately at scale of a few centimeters (or even millimeters). It allows for a better understanding of fault mechanisms and strain. InSAR instruments such as PALSAR (phased array type l-band synthetic aperture radar) are already being used after earthquakes to assess the damage and extent of ground movement and deformation. SLR (satellite laser ranging) and VLBI (very long base baseline interferometry) have also been used for the monitoring the crustal movement near active faults. In the measurement of fault displacements, even GPS technology has become very important (Garg, 2019).

The areas affected by earthquakes are generally very large, but they are restricted to well-known regions. One of the basic elements in assessing seismic hazards is to recognize seismic sources that could affect the particular location at which the hazard is being evaluated (Bhanumurthy & Behera, 2008). The most important data for seismic hazard zonation is derived from seismic networks. The parameters, such as distance from active faults, geological structure, soil types, depth of the water table, topography, construction types of buildings, fault rupture, damage due to ground shaking, liquefaction, landslides, fires, and floods, also play an important role (Westen, 2000). In seismic micro-zonation, the data may be derived from accelerometers, geotechnical mapping, groundwater, and topography, at large scales. In addition, high-resolution satellite data can map some geomorphic changes, if any, at the micro-level after the earthquake has occurred.

Remote sensing data is used for mapping lineaments and faults, study of the tectonic setting of an area, and neotectonic studies (Gupta, 2013). It can provide the basic inputs on the structural characteristics of the terrain. Remote sensing plays a major role in facilitating emergency relief and assessing damage after an earthquake. Very high-resolution satellite data can be used for post-disaster assessment in dense urban clusters. Such data can be used to create very large-scale base information of the terrain for carrying out disaster assessment and relief measures (Bhanumurthy & Behera, 2008). High-resolution images can help search and rescue teams navigate around the cities as well as estimate economic losses. The near real-time capability for the assessment of damage and the location of possible victims have become possible with the availability of very high resolution images, such as from IKONOS or QuickBird. These images can also help hazard mapping to develop building codes and disaster preparedness strategies. Apart from earthquake data, geological factors, structural design, soil data, and so forth can be used to develop building codes for the design

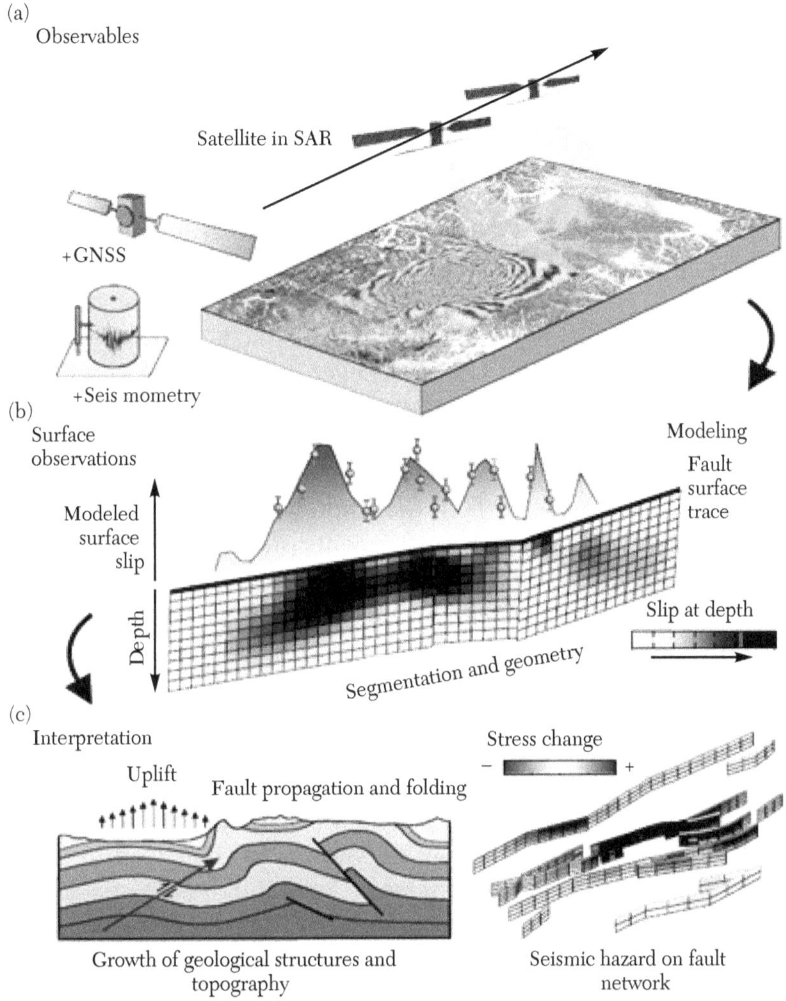

Figure 11.8. Workflow of observing and interpreting earthquake faulting and deformation using satellite data: (a) use Earth observations to model and interpret earthquake ruptures and the geometries of faults and their slip distributions; (b) infer the slip across the fault at depth, as well as constrain the geometry and segmentation of faulting; and (c) establish the relationship between faulting and the growth of geological structures such as folds and topography, as well as explore the potential control of lithology on both coseismic and postseismic slip (Elliott et al., 2016).

of earthquake-resistant structures in the region (Sharma et al., 2010). The earthquake risk or possible damage may be due to a combination of seismic hazards, vulnerability of the built surroundings, and exposure.

Elliott et al. (2006) studied earthquakes using satellites, mapping surface ruptures and estimating the distribution of slip on faults at depth for most earthquakes, as shown in Figure 11.8. This study directly links the earthquakes to their causative faults and suggests how resulting changes in crustal stress can influence future seismic hazard. Space-based observation is driving advances in models that can explain time-dependent surface deformation and long-term evolution of fault zones and tectonic landscapes. InSAR measurements are supporting seismological observations: while they do not contain information on the slip history of a rupture, they provide much higher resolution spatial constraints on the fault slip patterns for shallow earthquakes. High-rate GNSS offers the potential to span this temporal divide by providing unsaturated dynamic offsets from passing surface waves as well as the permanent deformation, although there are still only a few regions on Earth that have a GNSS network that is sufficiently widespread and dense enough for this technique. The longer repeat period of InSAR can be reduced by an increasing number of radar satellite constellations. As InSAR measurements image the displacement of the ground surface (relative to some assumed far-field undisturbed region), they are useful in determining the particular fault that ruptured at depth. The sensitivity to discontinuities in the phase measurements present in InSAR data enables mapping of small displacements across other fault displays away from the main fault rupture, allowing the identification of previously unmapped faults triggered by an earthquake.

11.6.4 Forest Fire

Recurrent fires potentially harm vegetation dynamics, and may disturb ecosystems. Forest fires can generate greenhouse gases and aerosols, and may contribute to climate change. The fire alert systems in many countries mainly depend on fire watchtowers, information collected by guards and communicated through wireless sets. Satellite remote sensing with its synoptic and temporal coverage can augment the ground operations in terms of fire detection, damage assessment, and planning mitigation measures, timely and economically (Bhanumurthy & Behera, 2008). Satellite systems are useful in forest fire detection, active fire progression monitoring, near real-time damage assessment, and mitigation planning. However, there are some gaps in the existing Earth observation capabilities in capturing certain disaster events due to temporal and spatial domains.

Coarse resolution sensors (e.g., INSAT VHRR/CCD, METEOSAT, NOAA AVHRR) having higher repetivity can capture a fire event to some extent, but spatial resolution is their main limitation. The MODIS Rapid Response System provides daily satellite images in near real time, and identifies hotspots and triggers requests to other satellites to collect additional information on active fires. The MODIS produces global fire maps that show active fires over the past ten days. The MODIS data-based fire products can be generated every day within four to five hours of the satellite ground pass. This active fire mapping system is being used by a wide array of fire monitoring programs (Lewis, 2009).

11.6.5 River Floods

Floods are the most common and widespread of all natural disasters. Globally, flood disasters not only cause substantial economic damage but also are responsible for taking many human lives. Through a combination of satellite images and ground sensors as well as other observation data, the water levels of rivers can be monitored in real time, and accurate evaluation of flood risk becomes possible. For the evaluation of flood hazards, the parameters to be taken into account may include: depth of water during flood, flood duration, flow velocity, rate of rise and fall, as well as the frequency of flood occurrence (Westen, 2000).

Geosynchronous satellites can be used in the planning phase of disaster prevention by mapping geomorphologic elements, historical events, and inundation areas, including frequency, duration, and depth of inundation. These satellites are also used extensively in the phases of preparedness/warning and response/monitoring (Lewis, 2009). The use of such data for flood mapping has serious limitation due to extensive cloud cover during a flood event. The SAR and Radarsat data have proven to be very useful for mapping flood inundation areas, due to their all-weather working capability. Satellites such as the TRMM (Tropical Rainfall Monitoring Mission) or MODIS can measure and map the rainfall, helping to forecast heavy rains and floods. Data from GOES satellites can be used for meteorological evaluation, interpretation, validation, and assimilation into numerical weather prediction models to assess the hydrological risks.

In flood disaster mitigation, the information requirements include existing structural and nonstructural measures, severity of the flood problem, suitable flood control measure, and so on. This information is important as it would be needed to reduce the impact of the disaster, saving loss of life and property (Bhanumurthy & Behera, 2008). High-resolution visible images and SAR images are good for extracting topographic and land use features of an area. The DEM and land use map are essential inputs to flood simulation models. The GIS can be used for the generation of detailed topographic information using high precision DEMs derived from GPS, stereo-aerial photography, SPOT, LiDAR, or SAR for the prediction of floods in river channels and flood plains, as well as assessing the damage. Remotely sensed data, hydrologic models, and GIS techniques can be combined to simulate potential flooding (Nirupama & Simonovic, 2002), and develop flood hazard controls in highly urbanized areas. The flood-related problems could be solved to a certain extent through detailed mapping of flood plains and its proper planning through remote sensing and GIS, supported with field data. Flood risk assessment includes both the chance of a disaster event taking place and its potential impact. Proper land use planning using remote sensing data, supported by flood-plain management plans generated using GIS, can be helpful to reduce flood risk. Flood risk management measures could include construction of check dams or levees, which can change the behavior of river flood water, or enforcement of building regulations in the area, which can protect buildings as well as humans against harm caused by floods.

In underdeveloped or developing countries, the floods cause heavy recurring losses, largely due to nonavailability of any flood warning systems. Flood warnings are generally based on well calibrated rainfall-runoff models, which require hydrological data over a watershed (Figure 11.9). However, acquisition of appropriate hydrological data is often insufficient

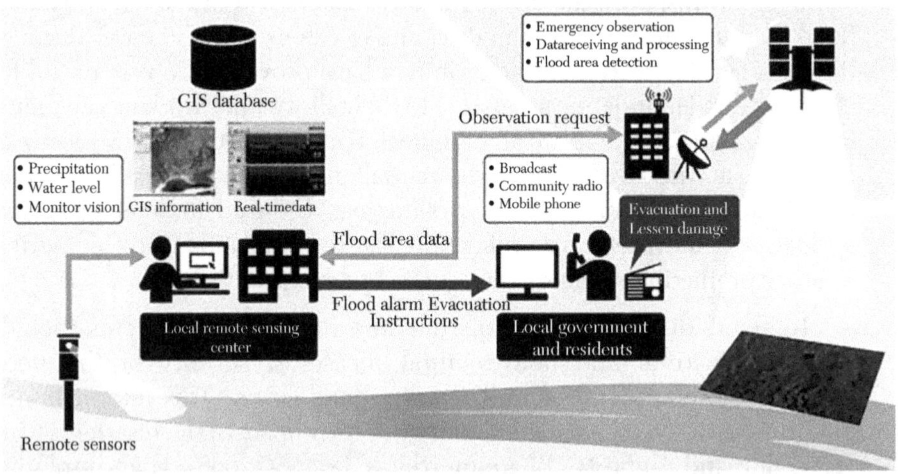

Figure 11.9. Flood monitoring and warning system (NEC, 2020).

in developing countries, which makes operation of flood warnings difficult. As an alternative, satellite remote sensing can detect and monitor spatio-temporal evolution of floods. The remotely sensed data has also been used worldwide for hydrologic prediction over sparsely or ungauged regions.

11.6.6 Landslides

Landslides are very frequent in certain mountain regions. Landslides occur in various forms, depending on the type of movement (slide, topple, flow, fall, spread), the speed of movement (slow to fast), the material involved (rock, debris, soil), and the triggering mechanism (earthquake, rainfall, human interaction; Westen, 2000). Landslide mapping includes: an inventory of landslides, seismic records, large-scale geological mapping, extensive geotechnical data on rock properties, high resolution DEM, and high-resolution remote sensing data. The hazard risk maps can be prepared from remote sensing and other field data, which can assist in emergency preparedness planning and in making rational decisions regarding development and construction in areas susceptible to slope failure (Bhanumurthy & Behera, 2008).

Landslide hazard zonation mapping involves a detailed assessment and analysis of the past occurrences of landslides with respect to their locations, sizes, and incidences associated with various environmental factors that have caused landslides and mass movements. A landslide hazard zonation map is of a probabilistic nature, which may also require information regarding slope steepness, land use/land cover, geology or lithology, density of drainages, and rainfall. So, preparation of a landslide hazard zonation map needs a huge quantity of data on these variables distributed over a large area. The use of high-resolution satellite images, LiDAR, and GPS help in the collection of these data. After the level of risk of landsliding is assessed, zonation maps are prepared, which have different uses, such as (Sharma et al., 2010): (*a*) preparation of development plans for cities, dams, roads, and other development works; (*b*) preparation of master plans and land use plans; (*c*) identifying new areas for development free from hazard; and (*d*) quick decision-making in rescue and relief operations.

Satellite imagery supported by field data have been used for landslide inventory and mapping of factors related to the occurrence of landslides, such as lithology, geomorphological setting, faults, land use, vegetation, and slope. These images provide information on the relevant parameters involved (soils, geology, slope, geomorphology, land use, hydrology, rainfall, faults, etc.). Stereo imagery is used in geomorphological mapping, or terrain classification (Cuartero et al., 2004). The airborne and satellite InSAR techniques can be used to generate DEMs, and produce detailed slope information, which would allow a more accurate interpretation of slope morphology and regional fracture systems with topographic expressions (Bhanumurthy & Behera, 2008). A warning system for landslides can be developed using satellite images and other supporting data, but at present these are operational only at a few places in the world. It requires a high density of information (landslide data as well as daily rainfall in order to establish rainfall thresholds; Fausto et al., 2020). However, the use of Meteosat and NOAA combined with rain-gauge data for predicting these thresholds is being investigated.

11.7 CASE STUDY: FLOOD HAZARD MAPPING

Kuldeep et al. (2021) have carried out flood hazard mapping of a part of the Yamuna river (Haryana district, India) using a 10 m resolution DEM. The DEM used in this study has been generated from Cartosat-1 satellite stereo-pairs. The Yamuna river, which flows from a north-east to south-west direction, has a total length of 32 km within the study area. The river geometry has been delineated from the high-resolution imagery and DEM. The river cross-sections have been extracted from the 10 m DEM. The Manning's n value has been obtained from the land cover map of the flood plain.

After generation of all geometry layers in the HECGeo-RAS model, the same have been imported to the HEC-RAS environment. The geometry file has been created by incorporating all the geometry layers. The discharge data has been used to create the flow file. The downstream boundary condition has been estimated considering the normal depth, which is computed from the slope of two consecutive points at the downstream location. The simulation for a 100 year flood has been carried out assuming the steady flow state. The simulation results are imported back to the HECGeo-RAS within GIS to obtain the flood depth and flood extent corresponding to the peak discharge profile. The workflow, as shown in Figure 11.10, summarizes the complete process of generating flood layers.

Flood modeling has been carried out using a 1D hydrodynamic model in HEC-RAS and HEC-GeoRAS within the GIS environment. Flood modeling has generated two outputs, namely, flood extent and flood depth. The flood depth layer has further been utilized for generating the flood risk map of the area. To validate the flood extent, RISAT-1 satellite data of the flood date is used.

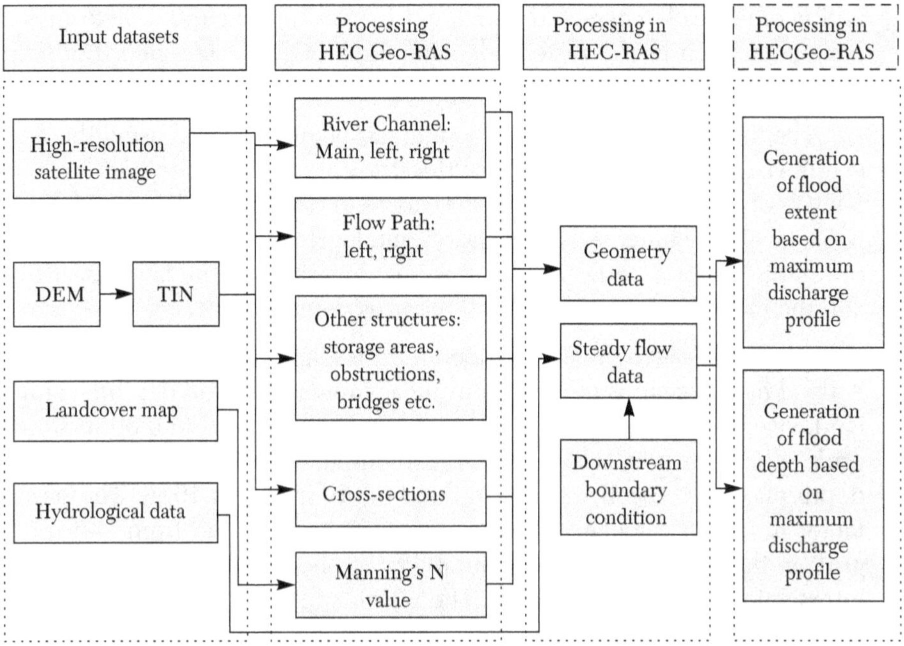

Figure 11.10. Process of flood inundation mapping.

The workflow for obtaining the flood inundation extent from the SAR data is summarized in Figure 11.11. Before using the SAR data for analysis, it was preprocessed to eliminate the noise introduced during the image acquisition process. The Lee-Sigma filter was applied for removing the SAR data speckle by eliminating high-frequency noise and retaining the edges and sharp features in the SAR data. Thereafter, the image was geometrically corrected. The HH band grid data contains the incident angle at each grid location, which has been imported into ArcGIS as a point shapefile. To obtain the incident angle image, the point shape has been interpolated using the inverse distance weighted (IDW) method. Speckle suppressed

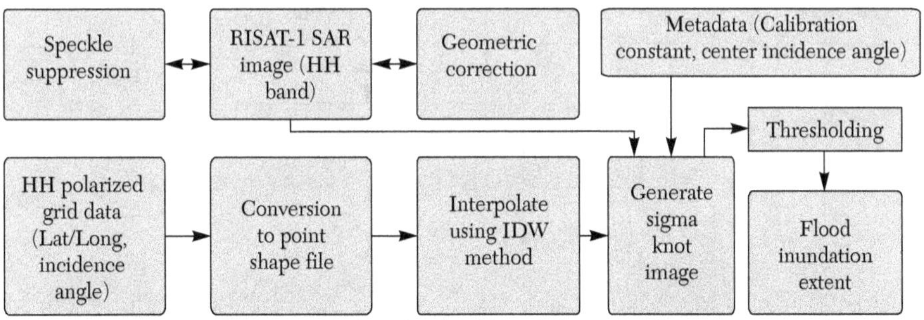

Figure 11.11. Methodology for obtaining flood inundation extent from RISAT-1 SAR data.

and geometrically corrected HH band images and incident angle images have been used in the raster calculator in ArcGIS to generate the Sigma knot ($\sigma 0$) image.

The damage caused by the flooding has been estimated using the flood extent layer and land cover map of the study area. The land cover layer has 3 levels of classification. There are 8 classes present in level-1 classification, while level 2 is composed of 13 classes and level-3 classification contains 30 different classes. The flood layer has been overlaid on the land cover map and the inundated area of the land features determined.

The workflow to obtain the flood hazard map is shown in Figure 11.12. A flood hazard map is generated using as inputs the flood depth layer and land cover map. The entire study area has been divided into 50 m square grids. For each grid, a depth value is computed with the available flood depth map, as the cell size of the flood depth map was 10 m. The hazard factor has been calculated by assigning an integer value from 1–6 for all grids in the study area. The vulnerability has been obtained from the grid intersection with the land cover. The weights for land cover classes within each grid have been calculated using the analytic hierarchy process (AHP) method, and the vulnerability factor corresponding to each grid is obtained based on the area coverage by different land cover types in a grid. The risk factor has been obtained by multiplying the hazard factor with the vulnerability factor.

Safe islands refer to the islands which remain uninundated with the maximum discharge during the highest flood event. To identify such islands

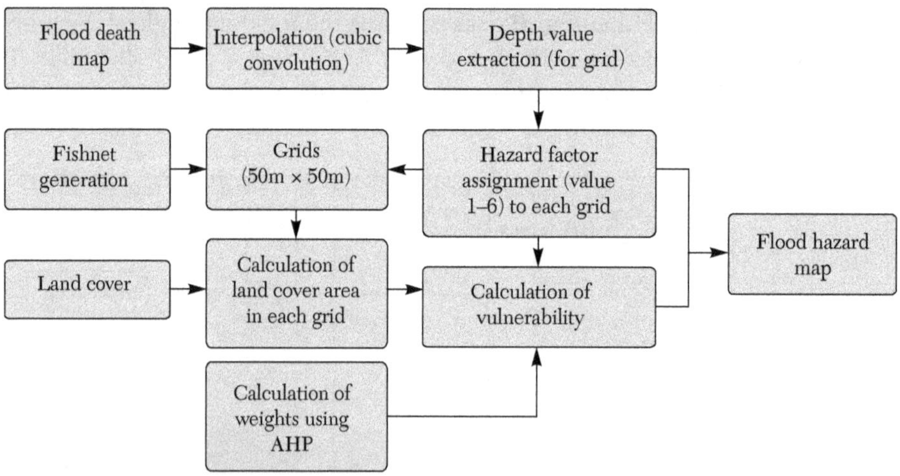

Figure 11.12. Methodology for flood hazard mapping.

in the study, flood inundation mapping of the study area is required. The flood extent map is overlaid on the classified map to identify the inundated islands. The depth of inundation of islands is estimated by overlaying the flood depth layer onto the classification map.

11.8 CONCLUSION

Natural disasters cause damage to life and property all over the world in various forms. The pressure on the Earth's resources caused by the increased population has resulted in increased vulnerability of humans and their infrastructure to natural hazards (Westen, 2000). Various natural disasters like earthquakes, landslides, floods, fires, tsunamis, volcanic eruptions, and cyclones are natural hazards that destroy property and infrastructure, and cause human death every year. For several types of disasters, the use of satellite remote sensing techniques has become operational in the monitoring phase and in issuing warning within a relatively short period of time. Present Earth observation satellites have been designed with general purpose instruments, but are not specifically designed to suit disaster prediction, monitoring, and mitigation related requirements.

The GIS combined with high-resolution images allows rapid assessment of land, infrastructure, and buildings affected. The events like drought, land degradation, and so on, are easy to capture by the existing Earth observation systems, while the disasters like earthquakes, cyclones and floods are relatively difficult to capture in real time. Observations of terrain features, ecological fragility, and socioeconomic status do provide the valuable information, especially on vulnerability and risk. These observations can support to a certain extent developing the scientific understanding and knowledge about various aspects of natural disasters, which help in disaster prediction, early warning, and mitigation (Bhanumurthy & Behera, 2008). Many tropical regions experience frequent cloud cover that is dependent upon the time of day. For effective and timely monitoring of any region on the Earth's surface, observation at least once each day from space is an urgent requirement in order to be able to act quickly to mitigate the effects of disasters.

Space technology can help the disaster mitigation process through better future scenario predictions, detection of disaster prone areas, location of protection measures and safe alternate routes, and so forth. Post-disaster satellite data acquisition and analysis help in disaster recovery, the damage claim process, and fast compensation settlement. Natural hazard information should be included routinely in developmental planning and large projects. They should include cost/benefit analysis of investing in hazard mitigation measures and weigh them against the losses that may occur if these measures are not taken. The people, government, and private insurance industry may benefit from derived maps showing the areas at high risk and at relatively lower risk due to disasters. Effective use of space technology will bring about a reduction in disaster damages, better prediction;,curate and timely damage estimation; and improved decision-making in planning stages. The future is promising with the new generation of very high-resolution satellites, like IKONOS and Quickbird and many more coming in the future to map, monitor, and manage natural and human-made disasters.

Disaster prevention is a long-term phenomenon, which can be best managed with the help of analysis of temporal satellite images. Disaster preparedness focuses on warnings and forecasts of impending disasters and often includes processes which are quite dynamic. The space technology and disaster mitigation communities should work together in developing effective and accurate methods for prevention, preparedness, and relief measures. Further, efficient communication plays an important role in collecting the required ground data in real time in order to disseminate forecast and warning information. In several advanced countries where warning systems and building codes are more robust, remote sensing has been found successful to predict the occurrence of disastrous phenomena and to warn the people well before time (Nirupama & Simonovic, 2002).

Various concepts are evolving over time to meet the demands of disaster management authorities. For example, the countries having developed satellite-based technologies are focusing more on developing intelligent and autonomous missions. These missions are configured with (*a*) various types of intelligent and smart sensors and detectors, (*b*) high data rate transmission and high-speed network communications, (*c*) the most powerful onboard data processing capabilities, and (*d*) autonomous operations and control of satellite and UAV systems. Intelligent and autonomous missions are aimed essentially at enabling simultaneous, global measurements and real-time analysis of images for all stakeholders. In the future, it is expected that when autonomous spacecraft (e.g., UAVs) detect an event, for example, a forest fire, floods, volcanoes, and so on, the sensing satellite will rotate its sensing system into position and change its coverage area to continuously capture the event. With more advances in space technology, sophisticated sensors, and more capabilities, it would be possible to undertake better management of natural disasters in the future.

CHAPTER 12

Remote Sensing of Snow Cover

12.1 INTRODUCTION

Snow is a type of precipitation in the form of crystalline ice. It is a granular material composed of small ice particles. The density of snow when it is fresh is 30–50 kg/m^3, but when it is compacted, the density becomes about 400–830 kg/m^3. Snow becomes glacier ice when density is 830–910 kg/m^3 (SAC, 2016). Snow when it remains there for many years, becomes glacier ice. Density increases due to remelting and recrystallization and reduction in air spaces within the ice crystals. Snow is a form of precipitation, but in the study of hydrology it is different because of the lag between when it produces snow, when it falls, groundwater recharge, and other issues involved in hydrologic processes.

Snow is highly variable both in time and space. This variability is especially enhanced in mountainous areas, as there it is governed by both the variability in the accumulation processes, that is amount of solid precipitation, wind speed and direction, and air temperature and humidity. In mountainous terrain, complex topography and high spatial variability in land cover (e.g., forest, rocks, vegetation) also contribute for a high spatial and temporal variability in both accumulation and ablation processes in snow cover properties. The required atmospheric conditions for snowfall are met at higher latitudes and altitudes of the Earth. There are three major types of snow cover, that is, temporary, seasonal, and permanent. Temporary and seasonal snow cover occur in winters while permanent snow cover is retained for many years. Permanent snow cover occurs mainly in

Antarctica, Greenland, and above the permanent snow line in mountainous areas. For example, 40–50% of the northern hemisphere is covered with snow during midwinter (Pepe et al., 2005).

Precipitation falling as snow in cold regions is temporarily stored as snowpack or icepack until the beginning of melt season. Monitoring accumulation and ablation of seasonal snow cover is an important requirement for various applications. Snow cover is important for local water availability, river runoff, and groundwater recharge, especially in middle and high latitudes (Jain et al., 2008). It is considered as an important parameter for numerous climatological and hydrological applications. Snow cover exhibits the greatest influence on the Earth's radiation balance in spring (April to May) when incoming solar radiation is greatest over snow cover areas (Tsai et al., 2019). Snow cover reflects incoming solar radiation, as fresh snow normally has an albedo between 0.8 and 0.9 while most land surfaces have an albedo ranging between 0.1 and 0.3; therefore, snow influences the regional and global energy balance. A decrease in snow cover extent and duration leads to a reduced albedo of the land surface, which increases the warming process and further accelerates the snowmelt process. Snowmelt is the source of freshwater required for drinking, domestic, agricultural, and industrial sectors, especially in middle and high latitudes (Jain et al., 2008). The snow precipitation also plays an important role in feeding the glaciers of the world.

Snow cover is one of the most easily identified features of water resources from satellite imagery. Remote sensing is a valuable tool for modeling and predicting snowmelt runoff. Mapping and monitoring of seasonal snow cover can be best done by remote sensing due to large area coverage and high temporal frequency. Snow is easily identified by remote sensing because its spectral signature is different from other surrounding features. The reflectance of snow is high when it is a small grain size snow or dry snow. Some examples of opportunities in using remote sensing data are (Meier, 1983): (*a*) forecasts of snowmelt runoff may be improved, and thus the reservoirs are operated more efficiently; (*b*) snowmelt flood potential may be anticipated more accurately and subsequent damage alleviated; (*c*) freezing and breakup of rivers and lakes may be monitored over large areas, permitting better control of river navigation and reservoir operation to minimize losses; (*d*) sea ice formation, movement, and breakup may be monitored to optimize vessel traffic schedules and to carry out improved marine geophysical surveys for oil and minerals; (*e*) changes in climate may be detected over regional or global areas, and these data used for many scientific and economic purposes; and (*f*) data on seasonal snow and sea ice cover may be used in general circulation and other climatic models to predict climatic variations.

12.2 SPECTRAL CHARACTERISTICS OF SNOW

Spectral reflectivity and scattering characteristics of snow depend on several factors, such as snow grain size and shape, liquid water content, snow depth (SD), impurity of snow, temperature, ice content, snow metamorphism from low to moderate density, and consistency of the surface beneath the snow cover. The spectral response of snow also depends upon the orientation and elevation of the sun, topographic position of snow in terms of slope, orientation, health of snow, and the atmosphere. The spectral reflectance curve is also affected by factors, such as soil nutrient status, snow grain size, spectral albedo, and color of the soil (Dietz et al., 2012).

Snow reflects a high proportion of the radiation in visible wavelengths. Depending on the impurity, grain size, and age of the snow, this proportion can reach 80% to 90% for freshly fallen, pure snow. The impurity of the snow cover increases with time, leading to decreased reflectance. The fallout of atmospheric aerosols is the main contributor to the impurity of the snow surface. Impurities in the snow cover, such as carbon soot, volcanic ash, and continental dust, have a strong effect on reflectance, mostly in the visible region (0.4–0.7 μm; Hall & Martinec, 1985). The presence of dark particles and impurities in the snow (like wind-transported soil particles) may be the main reason for the reduction in reflectance in the visual part of the wavelength.

This phenomenon can be observed in the season of late snowmelt with a mixture of bare ground and snow cover, and will generally increase with time until the snow has totally melted.

The melting and refreezing processes within the snow contribute to an increased grain size, which leads to reduced reflectance. Liquid water content in the snowpack has the effect of increasing the effective grain size, thus lowering the albedo. The spectral albedo of a surface is the upflux divided by the downflux at a particular wavelength. Freshly fallen snow almost immediately begins to compact and metamorphose, and this changes the snowpack characteristics. The albedo of fresh snow in the visible region of the spectrum remains high but decreases slowly with its age; however in the NIR region, the albedo of aging snow decreases considerably as compared to fresh snow (Singh et al., 2011). The snow albedo may decrease by >25% within just a few days as grain size grows (Hall et al., 2004). The albedo of a snow cover is also influenced by the albedo of the land cover that it overlies, especially when the snowpack is thin. Additionally, cloud cover normally causes an increase in spectrally integrated snow albedo due to multiple reflections caused by the clouds.

The grain size may be estimated using remotely sensed data. With the onset of snow surface melting and associated grain size increase, the NIR reflectance decreases dramatically. The NIR albedo of snow is very sensitive to snow grain size, while visible albedo is less sensitive to grain size, but is affected by snow impurities. Effective snow grain radii typically range in size from ~50 μm for new snow, to 1 mm for wet snow, consisting of clusters of ice grains (Hall et al., 2004). Different grain sizes lead to high variability in the reflection properties of snow, especially in regions around 1 μm and 1.2–1.3 μm, whereas the characteristics remain similar in the region below 0.8 and 1.5 μm. Grain size affects the reflectance mostly in the near and middle IR (0.7–3.0 μm), leading to lower reflectance for larger grain sizes. This region is therefore important for grain size determination (Dietz et al., 2012). For longer wavelengths, the reflectance of snow declines significantly, reaching near-zero values in the NIR.

In the optical region, snow reflectance is higher as compared to other land features, such as grass, rock, and water. However, in the SWIR region, snow reflectance is lower than rock and vegetation. Therefore, snow on satellite images appears white in the visible and black in the SWIR region (SAC, 2016). Figure 12.1 presents the spectral behavior of different snow and ice surfaces in the range 0.4–1.2 μm. Fresh snow reflects up to 100%, and reduces at longer wavelengths. It has high reflectance in the red part of the visible spectrum but decreases in the SWIR to 1.7 μm with a minor peak at approximately 1.09–1.10 μm, 1.83 μm, and 2.24 μm with a strong depression around reflectance 1.95 μm and 2.05 μm (Negi et al., 2009). The snowpack can change its spectral characteristics considerably within even a few hours. Maximum reflectance in a snow cover region varies between 0.45–0.55 μm. For ice surfaces, this amount is further reduced, reaching a minimum for dirty glacier ice with only 15–20% reflection. A snow spectrum has typically high reflectance values in the visible part with the highest values around 0.50 μm where the reflectance ranges between 90–100% (Dozier, 1989). It decreases in the NIR, with the steepest between 1.2–1.5 μm. Wavelength ranges from visible to 1.35 μm show almost higher reflectance in the 0.7–1.0 μm range. There are several peaks around 1.80 and 2.25 μm; however, the reflectance in SWIR varies greatly with the grain size. The grain size of 50 μm has a reflectance of about 40% with a peak at 1.8 μm, whereas it is decreased to 3% for 1,000 μm grain size. So, reduction of wavelength in the visible spectrum is observed for reduced grain size. There is only 5% reflectance for 0.50 μm (Meier, 1980).

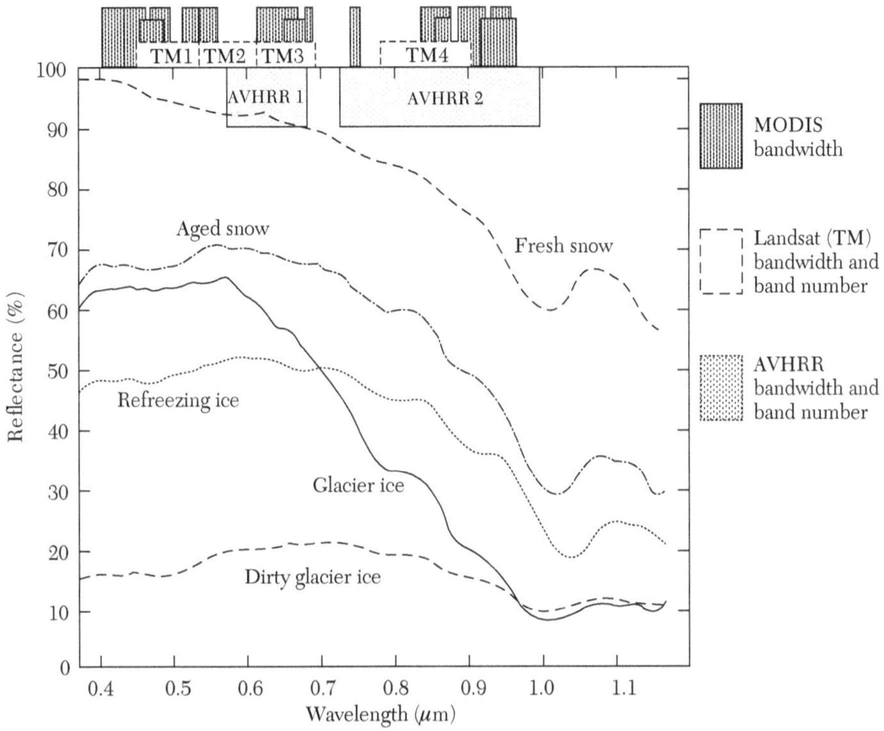

Figure 12.1. Reflectance of different types of snow cover (Dietz et al., 2012).

12.3 SATELLITES AND SENSORS FOR SNOW STUDIES

To monitor the evolution of snow cover for large areas, remote sensing observations, either ground-based, airborne, or spaceborne, are used with different wavelengths, employing either active or passive remote sensing sensors (Largeron et al., 2020). Satellite remote sensing technology has virtually revolutionized the study of snow cover. The high albedo of snow presents a good contrast with most other natural surfaces (except clouds), and therefore it is easily detected by many satellite sensors. A wide range of satellite sensors are available with various spatial resolutions and wavelengths, however the choice of remote sensing data for snow cover studies varies with the area and desired temporal and spatial resolutions.

Optical remote sensing is useful to determine the extent and albedo of snow cover, and some inference of snow depth based on the existing vegetation of known height. Infrared sensors can provide information on snow surface temperature, which may be a useful parameter for hydrological modeling. Microwave measurements have the ability to respond to the bulk properties of a snowpack as well as variation in other surface and subsurface features. As microwaves can penetrate the snowpack, they thus provide information on snow depth and snow water equivalent (SWE) when the snowpack is dry (Hall et al., 2004). Both passive and active microwave data have been found to be useful for mapping snow and determining snow wetness and SWE. The SWE is of critical importance for water resources and hydrologic and general circulation models. Microwave absorption within dry snow is low, resulting in volume scattering of the snowpack (Dietz et al., 2012). Vertically polarized data are more sensitive to the snow volume, and are therefore capable of mapping shallow snow cover. However, because there could be confusion between snow and underlying dry soils, horizontally polarized data are usually used to map the snow cover. Examples of spaceborne instruments or satellites used in snow cover studies with their general properties classified by type of satellites sensors are given in Table 12.1.

Table 12.1. Satellite/Sensor Data Used in Snow Cover Studies (Largeron et al., 2020)

	Mono optical	**Stereo optical**	**Active optical**	**Passive microwave**	**Active microwave**
Examples of space mission	MODIS, Landsat-8, Sentinel-2/3, Meteosat 8-11, GOES-16–17 VIIRS	SPOT-6/7, Pleiades 1A/1B, Worldview 1–4	ICESat, IC-ESat-2	SSM/I, SMMR, AMSR-E, AMSU-A/B	ALOS 2, Sentinel-1, RADARSAT-2, TerraSAR-X
Spatial Resolution	High (10–100 m)	Very High (1 m)	High (10–100 m)	Coarse (km)	Very High (1–20 m)
Temporal Resolution	High (daily to 5 days)	On request	Very low	High (daily)	High (daily)
Limits	Day/no cloud	Day/no cloud	Revisit time	Coarse resolution	Geometrical distortion
Retrieve data	SCF/SCA/ Albedo, No direct SWE	Snow depth, No direct SWE	Snow depth	Snow depth/ SWE, SWE with Tb, Only dry snow	Snow depth/ Snow wetness, No direct SWE

	Mono optical	Stereo optical	Active optical	Passive microwave	Active microwave
Accuracy	10% for SCA/ SCF, 5–15% for albedo	0.8 at 3 m resolution, 0.4 at 20 m resolution	0.15–0.6 m for aggregated data	Highly variable	0.31 m for Snow depth

Note: GOES–Geostationary Operational Environmental Satellites; VIIRS–Visible Infrared Imaging Radiometer Suite; SSM/I–Special Sensor Microwave Imager; SMMR–Scanning Multichannel Microwave Radiometer; AMSR-E–Advanced Microwave Scanning Radiometer for EOS; AMSU-A/B–Advanced Microwave Sounding Unit-A/B; and ALOS–Advanced Land Observing Satellite-2.

Multispectral optical sensors provide information in visible and NIR wavelengths with spatial resolutions ranging from 50 cm to 1 km. These images are available only during daytime, and many times may have cloud cover. Sensors such as the advanced very high resolution radiometer (AVHRR), the moderate resolution imaging spectroradiometer (MODIS), or Landsats provide multispectral data to map the snow properties. The cloud-free optical images can provide information on snow cover area (SCA), albedo, snow micro-structure, and snow cover fraction (SCF), which is the percentage of snow coverage per pixel. Satellites (e.g., Pleiades-1A/1B and SPOT-5/6/7) providing stereoscopic images can cover a specific region with stereo-images using different view angles at very high resolution (50 cm to 2 m; Largeron et al., 2020). The acquisition of such stereo-images is generally upon request only. Tri-stereoscopic satellites, like Pleiades, can be used to retrieve snow depth of the order of 50–80 cm at 3 m spatial resolution.

Passive optical satellites with varying spatial resolution (1 m–1 km) and revisit times (16 days to daily) can also be very useful to map snow cover. Sensors with better radiometric resolutions, such as MODIS and AWiFS, have been used for generating snow products (Singh et al., 2014). In general, the higher spatial resolution data is merged (fused) with the lower revisit time data. Sensors, like MODIS or AVHRR with lower revisit time are combined with high-resolution satellites like SPOT-6/7, Sentinel-2, Landsat-8, and Pleiades-1A and 1B that make them ideal for snow cover mapping, since they provide long time series data. Monoscopic (i.e., one viewing angle) optical satellite data with the most frequent revisit times are currently MODIS, PROBA-V, Sentinel-3, and VIIRS, which also have the advantages to provide cloud-free images (Largeron et al., 2020). On the other hand, active optical sensors such as LiDAR (e.g., IceSat, IceSat-2) can be used for regional-scale snow depth mapping using the temporal images between snow-free and snow-covered dates. However, the revisit time for

IceSat-2 can be as high as ninety-one days. The higher revisit time along with the requirement of an accurate DEM limits the use of such sensors for mapping snow depth, especially in mid-latitude mountainous areas where the number of overpasses is very limited.

Thermal infrared sensors are useful to estimate the surface temperature of snow and ice covered surfaces. The advantage is that they are not restricted to daylight data acquisition, however their spatial resolution combined with the revisit time is not suitable for mountainous regions. During daylight, retrieval of snow surface temperature is complicated by the reflected sunlight. To overcome the problem, 1 km spatial resolution images (e.g., MODIS or Sentinel-3) are linked with sensors like Landsat-8 (30 m spatial resolution), which can provide higher resolution snow temperature maps with a lower revisit time.

Passive microwave sensors offer the advantage of providing data under all atmospheric conditions and during day and night for mapping wet snow and monitoring the snow cover at regional and global scales. The spectral energy measured by these sensors can be utilized to calculate brightness temperature values as well as estimate snow depth or SWE. The maximum snow depth that can be derived from passive microwave sensors depends on the wavelength. The 37 GHz channel has a wavelength of 0.8 cm that can be used to measure the snow depth up to ~100 cm (Dietz et al., 2012). The coarse spatial resolution of passive microwave sensors limits their applicability for mountainous areas, but these sensors are more suitable for global monitoring of snow properties. Although the resolution of these data may be the biggest disadvantage, their ability to map snow even in the presence of clouds and estimate the snow depth makes them a valuable tool for snow cover mapping (Largeron et al., 2020). The advancements in sensor technology have introduced systems like AVIRIS, Compact Airborne Spectrographic Imager (CASI), Digital Airborne Imaging Spectrometer (DIAS), and Airborne Prism Experiment (APEX), which are expected to provide improved results. The characteristics of passive microwave satellites used in snow cover studies are summarized by König et al. (2001).

The active microwave sensors are available with much higher spatial resolutions than the passive microwave sensors, reducing swath (50–500 km) and revisit frequency (24 days for RADARSAT SAR, 35 days for Envisat ASAR, 11 days for TerraSAR-X; Dietz et al., 2012). The active microwave RADAR can provide information on surface structure in terms of the snow cover roughness and microstructure. The radar signal is extremely sensitive to the presence of liquid water, leading to lower backscatter for wet snow conditions. Like passive microwaves, radars is not disturbed by cloud cover and can record during the day and night under any weather and illumination condition. The SAR can also penetrate clouds, allowing measurements of snow surfaces. These data are particularly valuable in snow-covered regions which are often covered by clouds, and high-latitudes regions which are affected by polar darkness during winter (Largeron et al., 2020). Radar frequencies relevant for snow cover monitoring typically vary from 1–40 Ghz, namely, L to Ka bands. Longer wavelengths of SAR can provide information about snowpack conditions, such as snow grain size and SWE, and can even penetrate the frozen layer on the top surface of snow.

With the advancements of spaceborne SAR sensors and image processing techniques based on interferometric SAR (InSAR) and polarimetric SAR (PolSAR), snow cover mapping under both dry and wet snow conditions can be monitored. InSAR has shown a great potential for retrieving snow depth with high accuracy in band C or band Ka. Figure 12.2 presents an overview of commonly used spaceborne SAR sensors, including their operation time span and band information. The C-band SAR sensors provide the longest available time series of continuous observations since 1992, featuring ESA's pioneering ERS-1/2, Envisat, CSA's Radarsat-1/2, and the ESA milestone mission of Sentinel-1. The availability of time series C-band imagery for more than twenty-nine years became the most commonly used dataset for snow detection and analyzing snow cover.

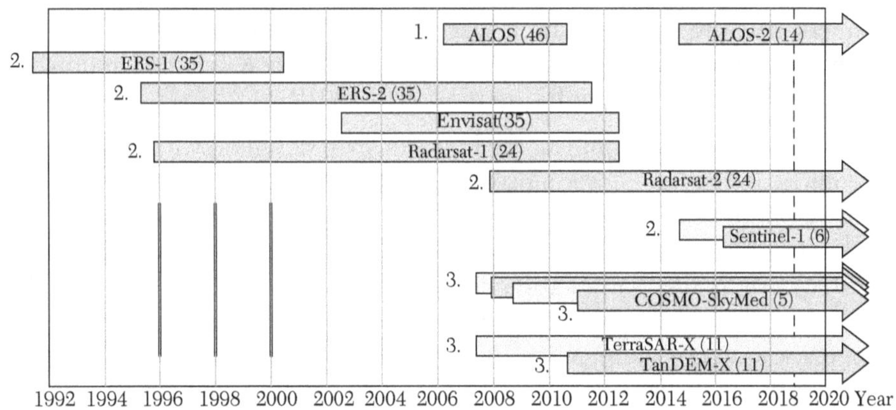

Figure 12.2. Available satellites equipped with SAR sensors and their band and revisit time (in brackets in days). Here, 1, refers to L-band, 2 denotes C-band, and 3 indicates X-band, having a frequency of 1–2 GHz, 4–8 GHz, and 8–12 GHz and a wavelength of 30–15 cm, 7.5–3.75 cm, and 3.75–2.5 cm, respectively.

The snow properties to which the backscattered signal are sensitive change with frequency and with the presence of liquid water in the snowpack. The backscatter signal of the SAR sensors allows the estimation of the snow depth, but some snow property information (such as liquid water content) must be known beforehand. Due to the longer wavelength of L-band SAR and subsequently deeper penetration of the L-band signal into the snowpack, the L-band data can not be used to detect snowpack (Largeron et al., 2020). Therefore, there are not many studies to analyze snow cover properties using L-band data, although it has been used longer than the X-band SAR. Longer wavelengths, in general, travel almost unaffected through dry snow. The X-bands with longer wavelengths (2.4–3.75 cm, 8.0–12.5 GHz) are not generally useful for detecting and mapping thin, dry snow because the size of snow particles is much smaller than the size of the wavelength. Thus, at these longer wavelengths, there is a little chance for a microwave signal to be attenuated and scattered by the relatively small ice crystals comprising a snowpack. Wavelengths longer than ~10–15 cm are not impeded as they move through most dry seasonal snowpacks (Meier, 1983). The X-band SAR has been facilitated more often than the L-band, which is due to the higher sensitivity of the X-band signal to the snowpack even when compared to the C-band. Due to the long-term preference of the C-band wavelength and its better capability to detect the snow as compared to L-band SAR, C-band SAR has been used longer for snow cover detection than any other sensor. Typically for bands L and C, bulk snow density or snow depth could be retrieved. However, many recent studies have proven that X-band is more suitable to detect dry snow (Largeron et al., 2020). Higher frequencies, such as X, Ku, and Ka, are more sensitive to snow microstructure and stratigraphy. Considering the new and planned X-band missions, the snow cover detection algorithms using X-band are required to be developed in the near future (Tsai et al., 2019).

Current optical/multispectral missions are designed with shorter, even daily revisiting time, for example, AVHRR, MODIS, Sentinel-3A/B, and so on. Even though the temporal resolution is yet insufficient to provide daily imagery, the spatial resolution of recent SAR missions, such as Sentinel-1 (5–20 m) is more than satisfactory. Present operational remote sensing satellite systems are very limited to determine the snow depth, snow quality, and snow-water equivalent, and also snow physical parameters are not directly measured by these systems. Most of the research in snow cover mapping is being done using multispectral remote sensing or hyperspectral images. The high spatial and temporal resolutions of SAR sensors make them potentially useful for snow hydrological applications. However, the oblique geometry viewing of SAR systems enhance the geometric distortions, which makes them particularly challenging to use in mountainous regions.

Different characteristics of SAR and optical sensors as well as advantages and drawbacks for snow cover monitoring are summarized in Table 12.2.

Table 12.2. Comparison of SAR and Optical/Multispectral Sensors Regarding Their Ability to Detect Snow Cover (Tsai et al., 2019)

Sensor	SAR	Optical/Multispectral
Sensing mode	Active	Passive
Wavelengths	0.01~0.3 m	0.3~1 μm
Spatial resolution	PALSAR-2: 3~10 m	Landsat-8: 15~30 m
	COSMO-SkyMed: 3~15 m	Sentinel-2: 10~20 m
	Sentinel-1: 5 × 20 m (Stripmap mode)	MODIS: 250~500 m (not including thermal band)
Temporal resolution	PALSAR-2: 14 days	Landsat-8: 16 days
	COSMO-SkyMed: ~5 days	Sentinel-2: 5 days
	Sentinel-1: 6 days	MODIS: 1 day
Recorded snow characteristics	Surface roughness, dielectric property	Surface reflection
Advantages	Day-and-night sensing under any weather condition; Possibility of interferometric and polarimetric information.	Visually natural to interpret; high temporal resolution; maturity of classification algorithms.
Drawbacks	Low temporal resolution; challenging to interpret due to its imaging geometry; significant geometric distortions and speckles.	Hindered by cloud, darkness; confusion between snow, ice, and cloud.

Hyperspectral remote sensing sensors are now being widely used for mapping land resources (Govender et al., 2007). These sensors acquire more than 100 contiguous spectral bands with narrow bandwidth (5–10 nm) with a wavelength region between 500–2500 nm. The conventional hyperspectral images, however, generally suffer from very low spatial resolution as compared to multispectral or panchromatic images. Various multiband reconstruction techniques, known as hyperspectral pan sharpening or multiband image fusion, have shown promising results to identify snow grain size as well as detection and identification of divergent surface targets, and topographical and geological features (Burke et al., 2016). Hyperspectral images own high spectral resolution and low spatial resolution, so a single pixel contains spectra of more than one material. The multispectral datasets have simple spectral information which makes them difficult to use for precise identification of similarly spectral features. Separating each spectrum from the mixed pixel becomes difficult. Specifically, spectral unmixing or spectral mixture analysis provides comprehensive descriptions of the hyperspectral measurements (Garg, 2020c). It consists of extracting the spectral signatures of the main materials present in the scene and quantifying their respective spatial distribution (abundance) over the image (Acito et al., 2010). To solve this problem, a method called end-member extraction algorithm (EEA) can be used to separate each single spectrum from the mixed pixels (Mozaffer et al., 2008). A good review of hyperspectral remote sensing is given by Govender et al. (2007) with a focus on use of hyperspectral imagery in water resources, flood detection and monitoring, detection of water quality, and vegetation applications.

12.4 CASE STUDY: SNOW CONTAMINATION FROM HYPERSPECTRAL IMAGES

A case study of a part of HP State is discussed as follows.

12.4.1 The Study Area and Method Used

The present work (Garg, 2020c) has been carried out to study the effect of contamination on snow cover using field spectra and hyperspectral images. The study is undertaken in two study areas in HP State: (*a*) Manali-Solang-Dhundi and nearby areas for snow studies and (*b*) Patsio Glacier for glacier studies, as shown in Figures 12.3 *a*) and *b*). The recent AVIRIS Hyperion scenes available for the study area include January 12, 2016, January 15, 2016, January 23, 2016 and January 11, 2017. Spectral libraries are created by using an SVC GER 1500 spectroradiometer (350–1050 nm)

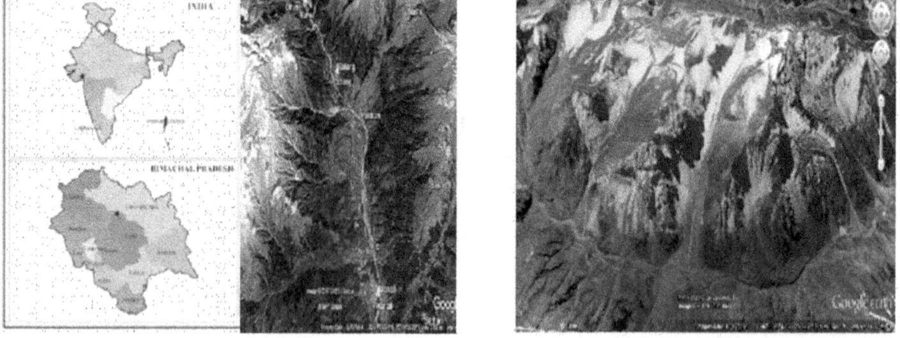

(a) (b)

Figure 12.3. The study area: (a) Manali, Solang, and Dhundi areas (HP) and (b) Patsio Glacier (please see color figures in companion files).

and collecting data for the samples for contaminant effects (Figure 12.4). The instrument collects the radiance, reflectance, and irradiance measurements. The flowchart of the methodology used for contamination study is shown in Figure 12.5.

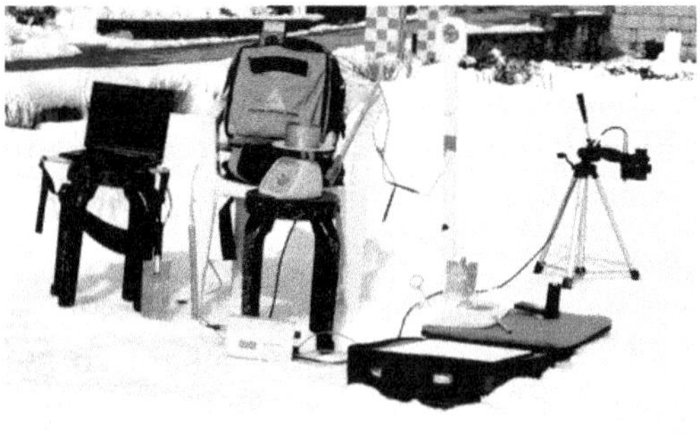

Figure 12.4. Field setup of spectroradiometer.

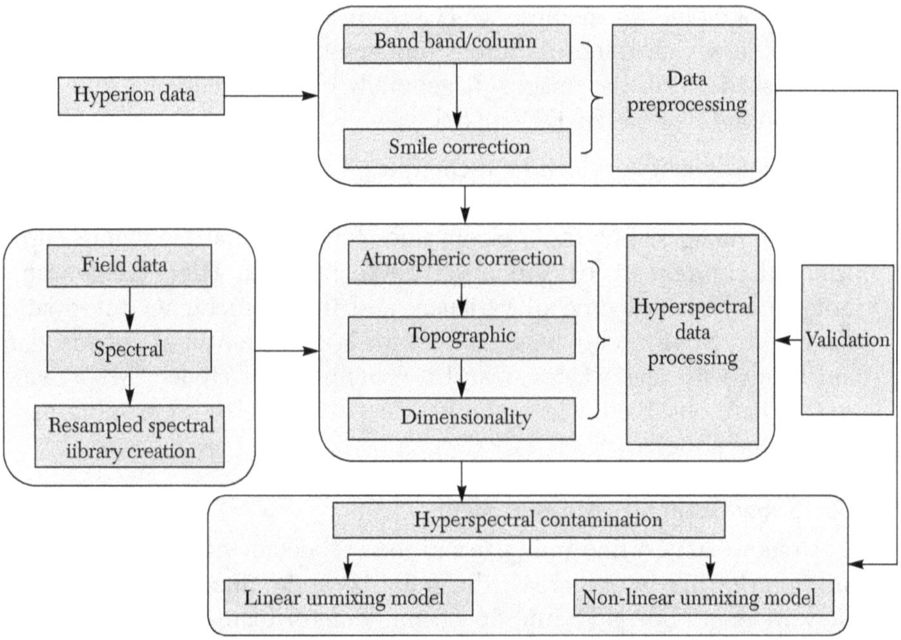

Figure 12.5. Flowchart of methodology used for Hyperion image classification.

Hyperion has 220 unique bands with a spectral range of 357–2576 nm at a 10 nm bandwidth. The Level 1 radiometric product has a total of 242 bands but only 198 bands are calibrated (bands 8–57 for the VNIR region and bands 77–224 in the SWIR region). The basic processing of hyperspectral images is done, which includes removal of bad bands, destripping, and atmospheric correction using ENVI's Fast Line-of-sight Atmospheric Analysis of Hypercubes (FLAASH). In each image, band numbers found suitable and bands having stripping errors are identified, and the error is removed with the average method. Atmospheric correction is applied to these hyperspectral data. In FLAASH, a mid-latitude winter atmospheric model is used, and in an aerosol model, no aerosol is found to be suitable. As per specifications given in the hyperion manual, a scaling factor of 400 is given to the bands up to 56, while 800 is assigned to other bands using FLAASH (Garg, 2020c).

Spectral profiles of snow and non-snow areas are compared between images before and after applying the atmospheric correction. It is observed that reflectance of every feature in corrected images increased significantly, as expected. An adjacency correction is performed on images while applying the atmospheric correction. For topographic error removal, Atmospheric and Topographic Corrections (ATCOR-3) software has been found to be suitable, which removes various topographic errors like shadow effect and hillshade from the images. It generally uses bidirectional reflectance distribution function (BRDF) model.

Dimensionality reduction techniques, that is, principal component analysis (PCA) and minimum noise function (MNF), have been applied on all the images. For PCA, eigenvalues for the initial few bands (up to principal component 10) are observed to be high. Higher eigenvalues indicate a higher degree of variance, and therefore higher information content. In MNF techniques, bands with large eigenvalues contain data, and bands with eigenvalues near 1.0 contain noise. Hence, MNF bands up to 10 are observed, which contain the data while the remaining bands contain the noise.

12.4.2 Snow Grain Size Measurement

Fresh snow has very fine grain size and good reflectance, but metamorphism and sintering in winter increase the grain size and reduce the reflectance in the wavelength beyond 8 μm. Spectroradiometer focuses on identification of surface materials by measuring the absorption wavelength and comparing it to the known wavelength of the absorption features. Increase in grain size as well as aging of the snow decreases the reflectance of snow in all wavelengths (Dozier et al., 1988). Snow grain size, that is, dry snow, small grain size snow, medium grain size snow, large grain size snow, and wet snow, is mainly estimated using the snow grain index (SGI), normalized difference snow index (NDSI), and spectral angle mapper (SAM) classification methods (Garg, 2020c). The SAM classification technique is used for snow grain size measurement by using spectral reflectance values of snow, and these spectral signatures are stored in a spectral library by using the maximum likelihood classification technique. A spectral library of different snow cover characteristics from both laboratory-based and in-situ experiments is essential as the reference spectra to select the end-members (Acito et al., 2010).

Negi et al. (2010) measured the snow reflectance in accordance with snow grain size and proposed the SGI as follows.

$$\text{SGI} = \frac{(\text{Reflectance @590 nm}) - (\text{Reflectance @1040} - 1050\text{ nm})}{(\text{Reflectance @590 nm}) + (\text{Reflectance @1040} - 1050\text{ nm})} \quad (12.1)$$

The NDSI uses reflectance data from both visible and SWIR, thereby taking the advantage of high contrast in reflectance between the two wavelengths. The NDSI is used to automatically distinguish between the clouds and snow; while the reflectance of clouds remains high for MODIS band 6 (7 for Aqua), the reflectance of snow drops to near zero in this spectral region. The NDSI is computed as:

$$\text{NDSI} = \frac{\text{Reflectance (Green)} - \text{Reflectance (SWIR)}}{\text{Reflectance (Green)} + \text{Reflectance (SWIR)}} \quad (12.2)$$

For Hyperion imagery, the suitable bands for measurement of SGI and NDSI are given as follows (where values within brackets indicate band number):

$$\text{NDSI} = \frac{\text{Ref}(15) - \text{Ref}(146)}{\text{Ref}(15) + \text{Ref}(146)}$$

$$\text{SGI} = \frac{\text{Ref}(24) - \text{Ref}(90)}{\text{Ref}(24) + \text{Ref}(90)}$$

To map snow extent automatically with NDSI, a threshold value of NDSI >0.4 is used to indicate snow coverage (Hall et al., 2004). For forested areas, the NDSI threshold must be decreased because forests tend to mask out snow-covered ground. Hall et al. (2004) found that NDSI values <0.4 also indicate snow if the NDVI is around 0.1. Small increases in the visible wavelengths would lead to NDSI values high enough to indicate snow. The NDSI is a useful technique in the mountainous region, as it can be applied to snow regions under shadow. One of the major challenges in snow mapping is the discrimination between clouds and snow. The characteristic decline of snow reflectance towards SWIR can be useful to distinguish between clouds and snow because most clouds reflect a higher proportion of the SWIR. The NDSI also overcomes the problem of clouds and snow under mountain shadow.

In the present study, snow grain size mapping is carried out using the grain index method based on the collected hyperspectral reflectance data. Hyperion band number 24 (central wavelength 589.62 nm) and band number 90 (central wavelength 1043.59 nm) are used for both the datasets. The snow map is generated for dry snow, small grain size snow, medium grain size snow, large grain size snow, and wet snow classes. Figures 12.6 and 12.7 show the spectral distribution of the snow grain sizes for two datasets of January 12, and January 23, 2016, calculated by GI and NDSI, respectively.

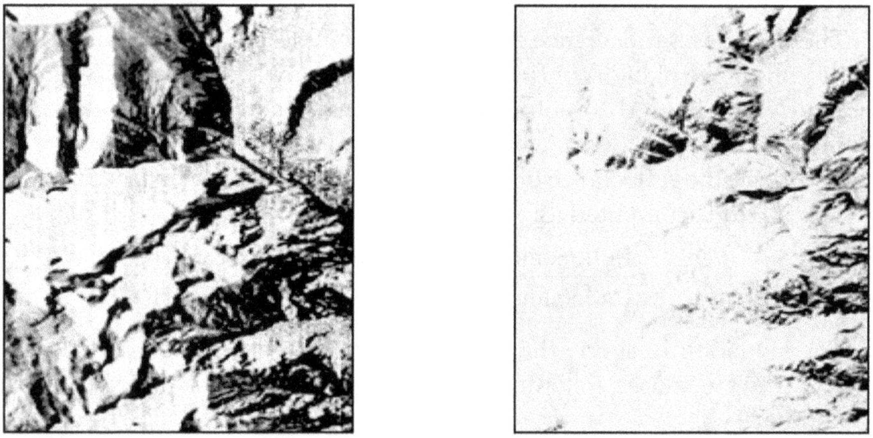

Figure 12.6. Snow grain size maps using GI of January 12 (left) and January 23 (right), 2016.

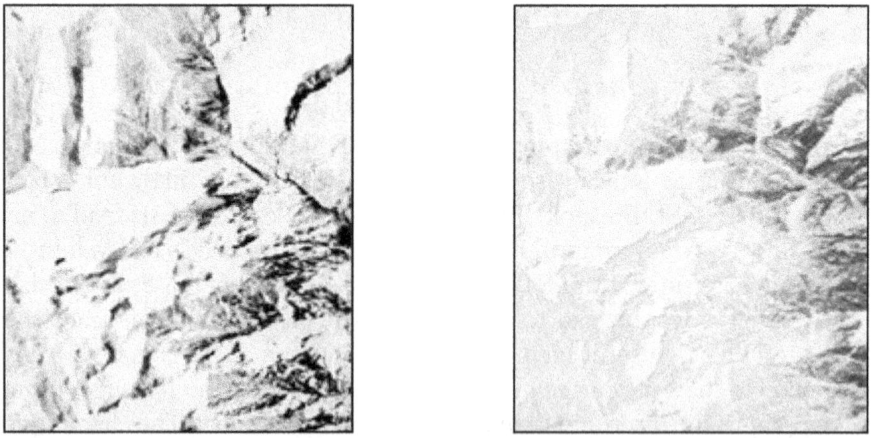

Figure 12.7. NDSI derived using Hyperion data of January 12 (left) and January 23 (right), 2016.

SAM classification is done using the spectral library approach. The different spectra are collected from the reflectance images of Hyperion and saved in a spectral library for SAM classification. An image spectra is generated for dry snow, small grain size snow, medium grain size snow, large grain size snow, and wet snow using simple z-profiling in ENVI software. The results of SAM classification are shown in Figure 12.8. Spectra are used to classify both images. In the SAM classification method, some areas, particularly at high altitude and shadowed hilly regions, are not classified because of low spectral signature, so such areas are named as unclassified in the output image. It was observed that the SAM is well suited for snow grain size mapping in the Himalayan region of varying altitude from 4,000 m and above *msl*. The matrix comparing different snow grain size classes using the results classification of the SAM method shows the overall classified area to be approximately 47% and the unclassified area as approximately 53% in the January 12, 2016 dataset. Another dataset defines 36.28% classified and 63.72% unclassified in January 23, 2016 (Table 12.3).

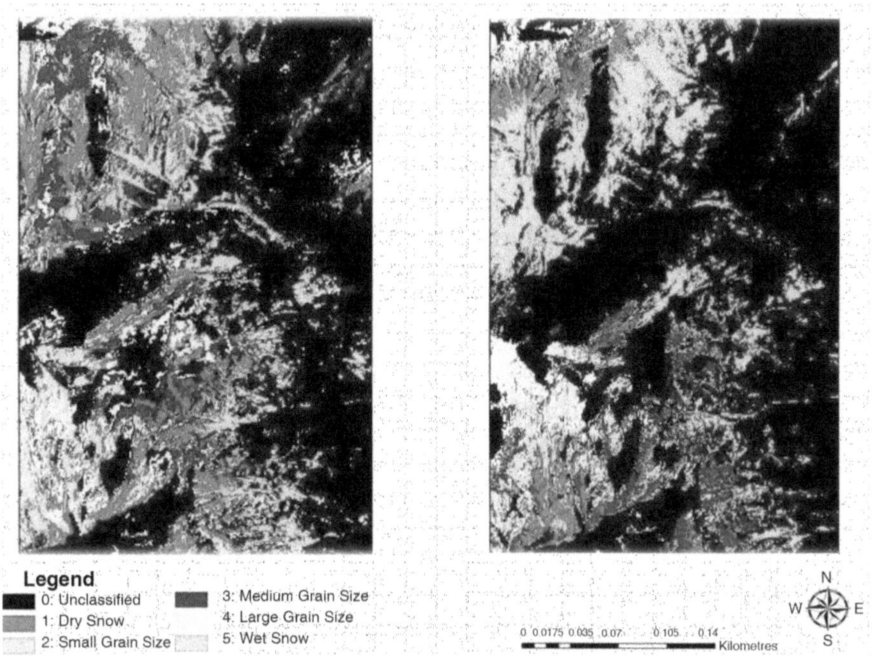

Figure 12.8. SAM classified maps (please see color figures in companion files).

Table 12.3. Grain Size Classification Results by SAM Using Images of January 12, 2016 and January 23, 2016

Snow grain class	GI method (% area)		SAM Classification (% area)	
	January 12, 2016	January 23, 2016	January 12, 2016	January 23, 2016
Fine (up to 0.25 mm)	7.89	15.44	6.38	13.64
Medium (0.2–0.50 mm)	22.64	17.68	10.43	7.21
Coarse (0.50–0.83 mm)	11.96	8.48	7.44	3.56
Dry snow	–	–	15.92	7.33
Wet snow	–	–	6.82	4.54
Unclassified	57.11	54.40	53.01	63.72

12.4.3 Spectra of Contamination in Snow

Contamination in snow is the presence of unwanted constituents or impurity in snow. Sometimes snow reflectance varies because snowpack contains liquid water content and impurities, such as continental dust, carbon soot, algae, vegetation, coal, soil, fire ash, and so forth. The contaminations are likely to have widespread effect on snow albedo (Warren, 1982). In all contaminants, soot is 50 times more effective than dust, and 250 times more effective than ash in reducing the snow albedo.

A small amount of contamination in snow cover distributed linearly through a larger grain size affects snow albedo significantly. Snowmelt is known to carry a higher concentration of contamination, which adversely affects the water quality. A small amount of impurities in snow can significantly reduce the snow albedo and affect the incoming energy balance. Contaminated snow shows a decrease in reflectance in the visible region, whereas grain size within the snowpack has more influence in the NIR region. For contaminated snow, reflectance reduces the maximum in the visible region (Warren, 1982), and less effect has been observed in longer wavelengths (Singh et al., 2011).

Spectral libraries of snow with contamination materials can be generated during the field campaign, as these spectral libraries are needed for applying spectral unmixing models. It is important to get the spectra of the snow with different grain sizes during the field observations. Using the spectroradiometer data, Negi et al. (2010) proposed the snow contamination index (SCI), given as follows:

$$\text{Snow contamination index (SCI)} = \frac{(\text{Reflectance @470 nm}) - (\text{Reflectance @590 nm})}{(\text{Reflectance @470 nm}) + (\text{Reflectance @590 nm})} \quad (12.3)$$

Negi et al. (2010) further observed that the SCI remains negative for contaminated snow, and in the case of clean snow a positive value of the index or close to zero exists. For soil and ash contamination, as the contamination increases, the SCI decreases; however, in case of coal contamination, a similar trend has not been observed, but overall, the value of SCI remains negative for all types of contaminations.

In the present study, the spectra of fresh and old snow are measured. To understand the effect of grain size on spectral reflectance, the field experiments are performed in two ways: a) collecting the snow reflectance and grain size observations in the top 10 cm of the snowpack everyday and then appropriately selecting reflectance plots for different ranges of grain size; and b) after the fresh snowfall (fine-grain snow), the snow reflectance and grain-size observations are taken at 30 minute intervals. The spectra of snow mixed with different contaminated classes with different grain sizes are also measured in the field using a spectroradiometer. These experiments are conducted with constant quantity of contaminant and varying quantity of contaminant on various grain sizes of snow to understand the effect of a varying amount of contamination on snow reflectance. These contaminants are added in the snowpack in: (*a*) small quantity, and (*b*) large quantity. To see the effect of a large amount of contaminants on snow reflectance, firstly, a small concentration of contamination is added to snow, and secondly, a large quantity of contamination is added.

For contamination like soil, coal, ash, carbon soot, and so forth to be added during the field experiment, these contaminants are oven dried, ground, and filtered, restricting the size of contaminated particles within 0.5 mm. For the first part, small sachets of 0.1 gm (low), 0.6 gm (medium), and 1.1 gm (high) for each contaminant are prepared to observe the subtle variations in snow reflectances while mixing them in snow. While spraying the contaminants over the snow cover area, it is ensured that it is evenly distributed in the instrument's view area. Proper care has been taken for other atmospheric parameters, like wind, ambient disturbance, and so on, to have their negligible effect. Snow was moist in nature with density of 0.24 gm/cm^3 and a grain size of 0.1–0.6 mm. The instrument sensor is kept at such a height that makes the view area of a 20 cm diameter on the snow surface. The sensor is targeted to collect the observations for clean snow, and after spraying the contaminants equally over the surface one by one. The different wavelength regions are chosen based on the influence of physical change in the snowpack characteristics. Position, strength, and shape of a spectral curve can provide the information on smoothly varying the spectral properties. The experiment has been carried out within 2 minutes to minimize the impact of other factors. To understand the effect of snow depth on spectral reflectance, the field setup used a black sheet of size 50 cm × 50 cm. The black sheet is inserted horizontally into the snow pack at different depths from the bottom in the upward direction. For the second part, the contaminants are weighed in 1 gm (low), 5 gm (medium), 20 gm (high), and 30 gm (very high) sachets which are equally distributed on the surface of the snow covered area.

For contaminated snow reflectance, measurements are carried out for soil, ash, carbon soot, and coal contaminants. Snow reflectance data are collected between 350 and 2500 nm binned at a 10 nm interval by an averaging method. A total of 120 samples are collected during the field visit during March 1–15, 2018 of Patsio Glacier. Measurements are collected at 1.0 m above the target with a field of view of 25° making a circle of a diameter of 20 cm. Contamination of 314 mg is spread on a 20 cm diameter circle, making the snow contamination of 1 mg/cm^2. By adding 314 mg of contamination successively, the concentration of contamination is increased. Each complete set of observations is taken in approximately ten to fifteen minutes, considering the negligible changes in the atmosphere. To understand the effect of moisture on reflectance, a dielectric moisture meter with a flat capacitive sensor is used. To study the influence of aging on reflectance, measurements are carried out daily on a plain ground. These experiments are performed at the same place and time to minimize the effects of other parameters. Each field experiment is repeated four to five times (Garg, 2020c).

12.4.3.1 Soil contamination

The spectrum of snow has been affected by soil contamination and shows a large drop in the visible region, but reflectance increases beyond the SWIR region, that is 1,500 nm (Negi et al., 2009). Hence, changes in reflectances in this region can be used for distinguishing between contaminated snow and patchy snow. This experiment is planned to study the spectral reflectance of snow and soil simultaneously with reference to the contribution of soil in the snow cover region. The reflectance observations are taken with a 100% snow covered region, 50% dry soil + 50% snow, 50% moisture soil + 50% snow, and a 100% soil covered region. A maximum drop in reflectance takes place in the visible region as the proportion of snow is reduced from 100% to 50%. After 50% increase of soil in snow, the area has marginal influence on reflectance. A shift of peak in reflectance is observed in the visible region, whereas an increase in reflectance is observed in the VNIR region to the SWIR region.

12.4.3.2 Coal contamination

The study of coal contamination and its effect on snow cover reflectance is a challenging task. Coal has the larger effect on the reflectance of snow much more than the soil, as coal reduces the albedo of snow significantly. Coal contamination absorbs the shortwave radiation, and it increases the snowmelt rate. Increase in contamination shifts the peak of reflectance in the visible range at higher wavelengths. The amount of coal is equally distributed one time to the snow area and the spectra recorded. This procedure is repeated to collect spectra of snow mixed with coal at different locations. The experiment is carried out to quantify the effect of coal contamination on snow reflectance. The amount of coal is added in fractions of 20%, 40%, and 50% to the specified target of an area 1 gm/m2. There is a drastic fall of reflectance in the visible region as the coal contaminant increases, and it remains constant to the short wavelength of the green band. As the amount of contamination increases in different fractions, the reflectance remains constant to the VNIR band.

12.4.3.3 Carbon soot contamination

Carbon soot is produced by incomplete combustion caused by humans as one of the major sources of soot, from burning coal and wood, oil for industries, and brush fires for agriculture. Natural forest fires can also contribute soot to atmospheric aerosol and deposits on snow cover. Soot is of great concern for its climatic effects as absorptive components of haze. The experiment is conducted to quantify the effect of carbon contamination on snow reflectance. The reflectance observations were taken with a 100% snow covered region, 50% carbon + 50% snow, 33% carbon + 33% coal + 33% snow, and a 100% soil covered region. A drop in spectral reflectance is more prominent (90% to 10%) in visible region and shows more reduction of reflectance in comparison to the same amount of carbon and coal contamination. The rate of drop of reflectance in the visible channel is less as it moves toward added coal contaminated snow.

12.4.3.4 Sparse mix vegetation contamination

This experiment is carried out in a snow and sparse vegetation mixed area and its effect on the reflectance values. The 80% snow and 20% sparse vegetation covered target area shows a reflectance pattern of snow, as there is a low quantity of vegetation. The reflectance pattern remains the same from the VNIR band to the SWIR band with drop in reflectance as compared to the reflectance of snow.

12.4.3.5 Ash contamination

Ash is visible in the sub-surface of the snow cover, which could reduce the snow albedo. Due to the episodic nature of fire, particularly in the Himalayan region, ash is mixed with new snowfall. Since the snow does not melt over immediately after the fall, the ash remains hidden from sunlight once it is covered by fresh snow. The effect of ash on snow albedo is thus unlikely to have long-term climatic significance.

12.4.3.6 Debris contamination

This experiment is conducted to quantify the effect of debris on snow reflectance. Gradual increase of reflectance in the visible to the SWIR region is observed, and then a little drop beyond the SWIR region is found. It shows high absorption in the blue band and low absorption from the green to SWIR region at different wavelengths. The reflectance of snow decreases in SWIR as compared to reflectance of debris in the same wavelength.

12.4.3.7 Mixed contamination on snow

The NDSI is found to be very helpful in snow cover monitoring. The NDSI is computed taking the green band as Hyperion's B15 (498 nm) and the SWIR band as B146 (1608 nm). The NDSI values for different contaminations are estimated, that is coal, carbon, soil, and sparse vegetation contamination, and given in Table 12.4. In this table, min and max reflectances in the VNIR and SWIR region are particularly shown. For the adjacency effect, min and max reflectances of debris at the VNIR and SWIR region are clearly shown in Table 12.14. In this experiment, the number of contaminations is added to a pure snow cover area of a 1 m radius. It has been observed that the NDSI values for coal contamination decreases in comparison to 20%, 40%, and 50% coal contamination. It indicates that as the area of vegetation mix increases, the NDSI value decreases. The NDSI values increase by 50% in moist soil contamination of snow in comparison with dry soil contamination of snow. These values for the snow, coal, and carbon mix (when mixed uniformly in the area) remain positive but significantly less than the NDSI value of snow alone. The NDSI values of 20% sparse vegetation contamination in snow are almost same as of snow because of a low amount of vegetation. These values for carbon contamination increase in comparison to the same amount of coal contamination.

Table 12.4. Minimum and Maximum Reflectance of Snow with Other Contaminants, Including NDSI and Contamination Index Values (Garg, 2020c)

Class	VNIR Region (350–850 nm)		SWIR Region (851–500 nm)		NDSI	Contamination Index
	Min reflectance	Max reflectance	Min reflectance	Max reflectance		
Snow	0.638	0.7231	0.0088	0.6375	0.893	0.0077
Snow (80%) + Coal (20%)	0.2752	0.2926	–0.0051	0.2805	0.864	0.0036
Snow (60%) + Coal (40%)	0.2151	0.2261	–0.0185	0.2186	0.809	–0.0018
Snow (50%) + Coal (50%)	0.0951	0.1054	–0.0144	0.1058	0.714	–0.0119
Snow (33%) + Carbon (33%) + coal (33%)	0.0408	0.044	–0.0072	0.0464	0.525	–0.0086
Snow (50%) + Carbon (50%)	0.1267	0.1358	–0.01	0.129	0.781	–0.0118
Snow (50%) + Dry soil (50%)	0.1587	0.2859	0.0398	0.2833	0.443	–0.1031
Snow (50%) + Moisture soil (50%)	0.2511	0.2787	0.003	0.2717	0.730	–0.0244
Snow (80%) + Sparse vegetation (20%)	0.6286	0.6567	–0.0047	0.628	0.867	–0.0074
Debris	0.1119	0.4425	0.2479	0.5257	–0.391	–0.3033

The snow contamination index (SCI) is very helpful in snow cover monitoring. Here, Hyperion bands used include B12 (467 nm) and B24 (589 nm) for computing the SCI and are given in Table 12.5. From the SCI, it has been observed that the SCI for contaminated snow remains negative, while it remains positive or close to zero for clean snow and sparse vegetation contaminated snow. For coal and carbon, as the contamination increases, the SCI decreases. The SCI values increase 50% in moist soil contamination of snow in comparison to dry soil contamination of the snow which remains negative. The value of SCI remains negative for all types of contaminations.

12.4.4 Spectral Unmixing Methods for Image Classification

The linear unmixing model (LUM) is a standard technique for spectral mixture analysis that infers a set of pure spectral signatures, called *end-members*, and the fractions of these end-members are, called *abundances*. The LUM has received considerable attention in the near-past, since it generally consists of an acceptable first-order approximation of the physical processes involved in most of the scenes. Consequently, it is used in several research works that aim at developing efficient end-member extraction algorithms (EEAs) and is able to recover pure component signatures in the image, and use inversion techniques to estimate the abundance coefficients for a given (estimated or a priori known) set of end-members.

One of the most popular EEAs is the pixel purity index (PPI), designed to search for a convex geometry in a given dataset that is supposed to represent pure signatures present in that data (Boardman et al., 1994). The PPI is the most common method to find extremely pure pixels in multispectral and hyperspectral images. Ideally, ground-based spectra are required to produce accurate end-members, since end-members taken from even very high spatial resolution imagery may contain multiple surface components. There exists a linear relationship between the fractional abundance of the substances comprising the area being imaged in the reflected image. The pixel value of this image indicates the fraction of the pixel that contains the end-member material corresponding to that image. A pixel from this abundance image with a value of 0.92 indicates that 92% of the pixel contains snow and 0.07 indicates that 7% of the pixel contains vegetation. If many pixels have values above 1.0 or below 0.0, it indicates that one or more end-member chosen for analysis are probably not well characterized.

The ENVI tool for linear mixing model (LMM) has been used on all images for pixel classification for these two classes. The hyperspectral images are classified into snow and non-snow areas with some errors, as spectra of shadowed areas of the snow area and non-snow area interfere with each other. The LMM is attempted using existing inbuilt spectral libraries of snow using ENVI software. End-members are extracted using the PPI method. Forward MNF is applied on AVIRIS subset data. A total of the first 10 bands are selected having high eigenvalues for further processing. Fast PPI is applied to MNF images with an iteration of 10,000 with a threshold factor of 5 (Garg, 2020c), selecting the minimum and maximum threshold values to fall in a specified range and export selected points to an n-D visualizer to collect end-members. As AVIRIS data has a full part covered with snow, a limited number of end-member is selected for unmixing. The first 10 MNF bands are used to collect the end-members from the 3D scatterplots through n-D visualization. Linear spectral unmixing is done from selected end-members, and fraction maps of end-members are obtained in MNF images. Figure 12.9 a) shows the PPI plot of AVIRIS data, while Figure 12.9 b) presents the pure pixels in AVIRIS data.

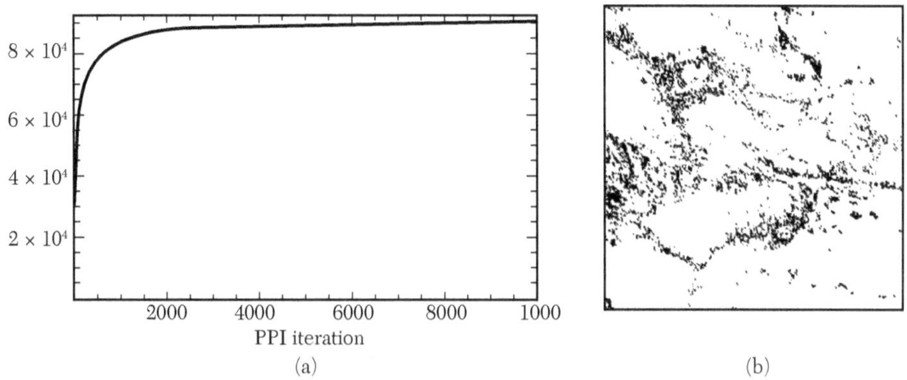

Figure 12.9. (a) PPI plot and (b) pure pixels in AVIRIS data.

As the grain size of snow increases, a decreasing trend can be seen in the spectral reflectance in NIR and higher wavelengths, except in the visible region. In the visible region, contamination decreases with reflectances. Due to an increase in the amount of contamination, a shift in peak is observed toward the higher wavelength. There is a significant change in reflectance which can be observed in the visible region (590–650 nm), and a minor decrease in reflectance can be observed beyond the NIR region. The SWIR region shows no change in reflectance due to the variation of snow depth.

Dust and coal contaminations are a major source of contamination influencing the snow and glaciated region through avalanche and atmospheric phenomenon. Dust concentration in the snow cover revealed the sporadic high concentrations frequently in spring and large year- to-year variations in the amount deposited from winter to spring. In case of coal mixing, as the proportion of coal increased the reflectance of the visible region reduced and reflectance decreased in the SWIR region. In the case of a snow-coal-carbon mix, the rate of drop of reflectance in the visible channel decreased less as it moves toward added coal contaminated snow. It suggests that it may be possible to detect the type and amount of mixing in snow pixels, if hyperspectral data is available. It will be helpful in the spectral unmixing problem in snow\bound areas to improve the accuracy of snow cover monitoring (Garg, 2020c).

More than two decades have passed after the first effort toward the application of spectral mixture analysis techniques to remotely sensed data for classification, but effective spectral unmixing still remains an elusive exploitation goal. Regardless of the available spatial resolution, the spectral signals collected in natural environments are invariably a mixture of the signatures of the various objects/materials found within the spatial extent of the ground instantaneous field view of the remote sensing imaging instruments.

12.5 CONCLUSION

The snow cover classification produced from remotely sensed data can be used for different applications, for example, the impact of climate change. Snow cover extent and snow mass are both sensible to temperature changes. Remote sensing methods are capable of analyzing the snow extent as well as SWE, which makes them an ideal tool to monitor the potential changes. Furthermore, results from snow cover analysis are useful for hydrological studies (Dietz et al., 2012), as they can be integrated into hydrologic models such as the snowmelt runoff model (Tsai et al., 2019). The results obtained from these models can be utilized by hydroelectric projects, especially during the melting season or for forecasting flooding events.

With improved spectral and spatial resolution, snow extent mapping would allow advances in the areas of snow wetness, snow albedo, and snow grain size estimation. Better techniques need to be developed to improve snow cover mapping under cloudy conditions. Microwave techniques for SWE monitoring are already in use, but active microwave techniques are greatly hindered by nonoptimal wavelengths and polarizations and complex geometries. Passive microwave techniques are already used for operational SWE maps in large flat areas where the existing lower spatial resolution is not a major problem. The launch of future microwave satellites will have better resolution which can extend the applications for areal SWE determination and contribute to solving the problem of snow extent mapping under cloudy conditions (Largeron et al., 2020). Areal snow extent and SWE thus obtained can be used as input to deterministic hydrology models. It is proposed to combine and integrate the data from different remote sensors and various types of low altitude and ground measurements.

Further research is required to analyze hyperspectral images along with the ground spectra for analysis of moisture content variation and its effect on reflectance, estimation of snow depth, snow density, and different spectral signatures for sensitive bands of snow cover. Estimation of snow properties such as fractional snow covered area, snow albedo, snow wetness, and absorbing impurities (dust and red algae) are also required to be studied from imaging spectrometry (Dietz et al., 2012). Analysis of spectral signatures of ambient objects like deep clear water, soil, coniferous trees, wet and dry grass, man-made features, spectral mixing of different objects, and radiances of different objects with respect to snow may also be carried out in the future.

The role of remotely sensed observations of snow will continue to play in important role in climate and hydrological forecasting. Future sensors will permit automated algorithms to be used to create snow maps that are consistent with existing old maps, and are amenable to compare with the long-term records of other geophysical parameters, such as global sea ice, and for input to general circulation models. It is expected that specially designed instruments with improved capabilities for snow and glacier studies will open up new applications of remotely sensed data.

CHAPTER 13

Feature/Object Extraction from Remote Sensing Algorithms

13.1 INTRODUCTION

Feature extraction is a very useful technique for extracting meaningful information. Feature extraction, in the context of remote sensing, can be defined as an image processing technique to identify specific image features. The approaches range from manual methods to semi-automated and automated feature extraction methods using single and multiple image frames. Feature or object detection in remote sensing images is carried out to determine the position, shape, size, and orientation of objects in the image. The term *object* refers to its generalized form, including man-made objects (e.g., vehicles, water ponds, buildings, bridges, etc.) that have sharp boundaries and are independent of the background environment, such as LULC parcels. Object detection in remote sensing plays an important role for a wide range of applications such as environmental monitoring, geological hazard detection, LULC mapping, GIS database update, precision agriculture, urban planning, forestry, and so forth. It is also used in medical imaging, traffic surveillance, geographical survey, image processing, object recognition, and so on.

The low spatial resolution of earlier satellite images (such as Landsat MSS) would not allow the detection of man-made or natural objects separately. Linear objects like roads, railroads, or river networks have long attracted researchers, but due to the limited resolution of satellite images they could not be successfully extracted for mapping at medium or large scales. Automatic object extraction has been a popular topic in the field of remote sensing for decades, but extracting buildings and other objects from remotely sensed images is still a challenge (Chen et al., 2018). Due to the rapid urbanization and growth of cities worldwide, urban mapping and change detection is increasingly important in modern city planning. For city planning, building extraction becomes an essential task. With high-resolution sensors between a 1–5 m spatial resolution, the possibilities to extract such objects have increased dramatically.

Generally, detection, recognition, and classification interpretations are based on features; hence, these features should be extracted in optimal ways to refine the accuracy of each method. The results of feature extraction methods are being used in various applications such as video tracking, gesture recognition, object matching, vegetation detection, navigation, 3D modeling, and so on. Feature extraction has an important application in spatial data management also for automatic updating of GIS databases from enormous satellite imagery as well as for production of maps (Garg, 2019). In many countries, up-to-date production of maps is done by automated methods which improve the speed and effectiveness of the map production process.

According to Guo et al. (2009), the targets/objects that can be recognized from remote sensing images have three levels of concepts (Figure 13.1): (*a*) terrain types, (*b*) individual objects, and (*c*) composite objects. The terrain types of the area covered by the images could be bare land, mountain, water, residential area, forest, and so forth. Individual objects that are recognizable in images include, individual buildings, road segments, road intersections, cars, trees, ponds, and so forth. The composite objects consist of several individual objects that form a new semantics concept, such as forest area.

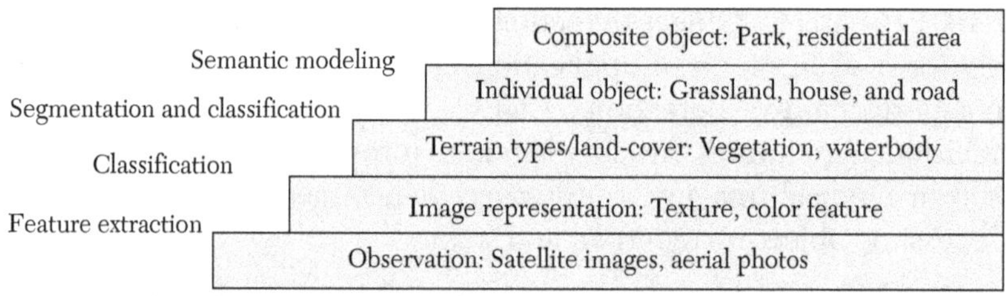

Figure 13.1. Information extracted from remote sensing images (Guo et al., 2009).

13.2 CHALLENGES IN OBJECT DETECTION ALGORITHMS

Automatic object detection is one of common methods in feature extraction; however, this method still has many issues. The development of an automated object detection algorithm still remains a challenge due to the complex structure of ground as well as remotely sensed imagery. In addition, it suffers from several increasing challenges including the large variations in the visual appearance of objects caused by viewpoint variation, occlusion, background clutter, illumination, shadow, and so on. To overcome this, a number of different statistical approaches have been widely used for classifying LULC information (Chavez et al., 1994). With the advancement of satellite sensors, more attention is being paid toward the novel robust classification approaches like ANN (Gislason et al., 2006).

The automated extraction of 3D objects like buildings, roads, bridges, street furniture, or trees is not yet widely used in practice (Chen et al., 2018). High-resolution images provide more geometric information but also have more variation within the pixel. Consequently, the increasing amount of textural information does not provide increased accuracy but rather it makes the object extraction from images complicated. Although many approaches have been developed, automatic object detection with high accuracy is a very challenging task. No single approach is available universally which provides a perfect solution to detect the objects from various images.

Building detection is the most popular application of feature extraction. Automated building detection still faces several challenges due to varying shapes and sizes of buildings, which make their extraction using a simple and uniform model. In addition, aerial and satellite images contain a number of other objects, for example trees, roads, ponds, rivers, and fields, which make the job difficult to pick only one feature. Some buildings may be occluded by vegetation cover or shadows or other buildings, which poses another challenge. With the rapid development of machine learning (ML), data-driven methods begin to attract increasing attention. The performance of ML largely depends on the quality of the feature extraction, which is quite challenging. Occlusion is another challenge in the particular case of building detection, and especially in residential areas, where some houses are often occluded by trees and only partial roofs can be seen on the images.

To distinguish buildings from nonbuilding objects in aerial or satellite images, a variety of clues including color, edges, corners, and shadows has been used in previous work. Due to the complex background of the images, a single cue is not sufficient to extract the buildings under different circumstances. Therefore, multiple clues are used in combination, and designing more useful feature descriptors becomes crucial to building detection (Mayer et al., 1999a). In recent years, extracting buildings from

LiDAR points has attracted substantial attention because LiDAR data can provide useful height information. Despite great advantages of LiDAR data, this approach is largely limited by meager data availability. The LiDAR data are not easily accessed and the cost is much more than that of high-resolution aerial or satellite imagery. Among the data sources for building extraction, high-resolution images are the most easily accessed and widely applicable.

As resolution of the data becomes higher, the complexity of feature extraction increases, and therefore it is a challenging job for any individual solution to be generic (Yadav et al., 2020). For feature extraction from high-resolution satellite data, pixel-based classifications are not successful due to the diversity in image pixels and the small size of pixels as compared to the size of features such as buildings, ponds, or trees. Therefore, different methods such as hard classifiers (maximum likelihood, thresholding, etc.); soft classifiers (fuzzy systems, neural network, etc.), and nearest neighbor algorithms have been used for feature extraction, but object based image analysis (OBIA) is found to be an effective method for complex areas. This method along with rule-based classification can be used to overcome the limitation of improved resolution data and difficulty in recognition of complex features. The OBIA utilizes a rules hierarchy integrating a huge range of object features like spectral, geometric, shape, and texture values. It is considered superior to pixel-based image analysis when classifying and measuring homogeneous features from high-resolution data.

13.3 VARIOUS OBJECT DETECTION ALGORITHMS

In the last few decades, a large number of methods have been developed for object detection from aerial and satellite images. According to Cheng and Han (2016), these methods can be divided into four main categories: template matching-based, knowledge-based, OBIA-based, and machine learning-based methods, as shown in Figure 13.2. These four categories are not necessarily independent and sometimes the same method exists with different categories. The knowledge-based object detection methods use prior knowledge. In machine learning-based methods, three crucial steps play important roles in the performance of object detection: feature extraction, optional feature fusion and dimension reduction, and classifier training. In feature extraction, five types of techniques include: the histogram of oriented gradients (HOG) feature, bag-of-words (BoW) feature, texture features, sparse representation (SR)-based features, and Haar-like approaches. In classifier training, six kinds of machine learning

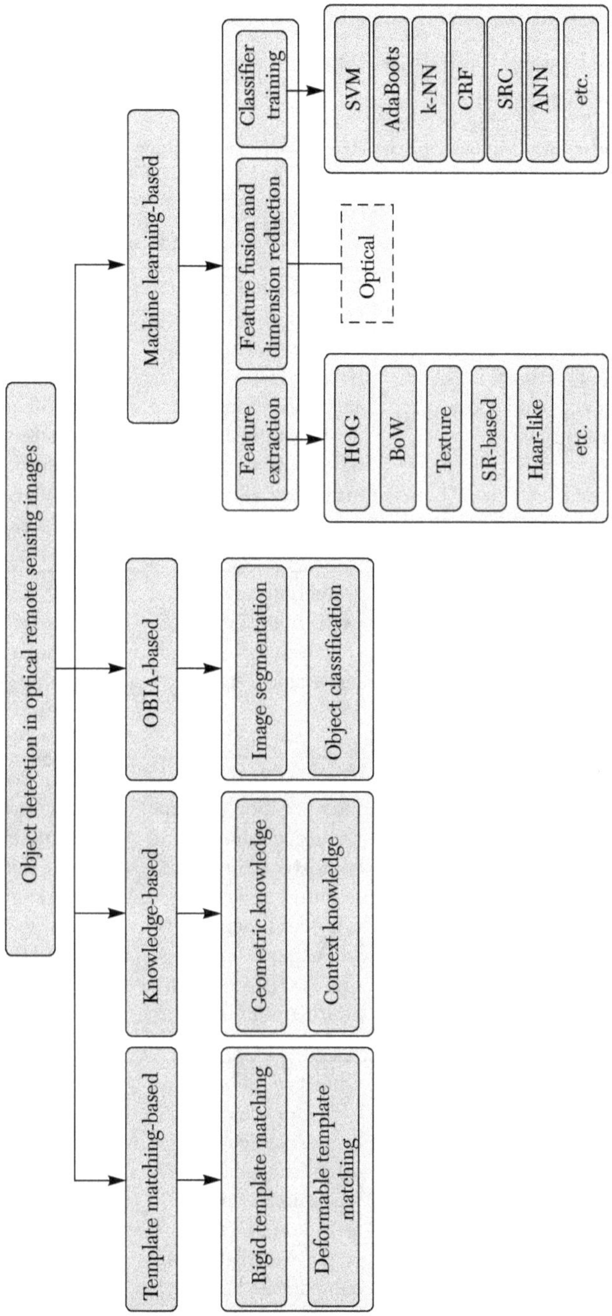

Figure13.2. Methods used for object detection (Cheng & Han, 2016).

algorithms include SVM, AdaBoost, kNN, conditional random field (CRF), sparse representation-based classification (SRC), and ANN.

Table 13.1 presents further details of these approaches along with their strengths and limitations.

Table 13.1. Various Object Detection Approaches Along with Their Strengths and Limitations (Cheng & Han, 2016)

Types of methods	Objects detected	Strengths	Limitations
Templates matching-based	Building, coastline, storage tank, road, airplane, ship, storage tank, airport, building, edges of oil slick, road, tree crown.	Simple and easy to implement. More powerful and flexible than rigid shape matching in dealing with shape deformations and intra-class variations.	Scale and rotation dependent; sensitive to shape and viewpoint change. Need more prior information and parameters of the geometrical shape for template designing; computationally expensive.
Knowledge-based	Bridge, building, water channel, crops, drainage forest, landslide, road, vehicle.	Detection can be performed through coarse-to-fine hierarchical structure.	How to define prior knowledge and detection rules is subjective; too loose rules cause false positives and vice versa.
OBIA-based	Change detection, LULC mapping including vegetation, tree, water, residential, etc., landslide mapping.	The flexible incorporation of shape, texture, geometry, and contextual semantic features, as well as GIS-like functionality and expert knowledge, makes OBIA context-aware and multi-source capable.	Generic solutions to the full automation of the segmentation process are still missing. The expert knowledge on how to define the classification rules are still subjective.
Machine learning-based	Change detection, forest, building, landslide, multi-class objects (e.g. airport, airplane, ship, storage tank, etc.), road, tree, LULC, vehicle.	Object model (detector) can be automatically established via machine learning technique; Detection system is scalable and compatible; have high detection accuracy.	Need a lot of training samples of objects and non-objects to learn classifiers. Detection accuracy depends on the training samples.

Feature extraction techniques are becoming more efficient and robust day-by-day. Several feature extraction methods have been developed, and their working patterns are quite different from each other. The performance of each method is optimum for a specific application or environment: for example, binary robust invariant scalable keypoints (BRISK) and features from accelerated segment test (FAST) are suitable for corner detection; speeded-up robust features (SURF) and scale-invariant feature transform (SIFT) are commendable for blob detection; and HOG and local binary patterns (LBP) are appropriate for local shape and texture analysis.

Easson and Momm (2010) have provided a detailed survey of the use of evolutionary algorithms to extract information from remotely sensed data and found that the majority of applications are based on genetic algorithms (GA) and genetic programming (GP). In this review, the different applications were classified into four categories according to the general research objective: image enhancement, image classification, modeling and feature extraction. Momm and Easson (2011) proposed the use of evolutionary algorithms, in the form of genetic programming, to aid the feature extraction process from IKONOS and QuickBird high-resolution satellite images. A novel framework involving genetic programming, standard image processing methods, and clustering algorithms is presented in Figure 13.3. The proposed framework develops the spectral indices, which maximize the performance of standard classification algorithms to separate the target feature (isolated building) from the remaining image background. The system works in a learn-from-examples approach where

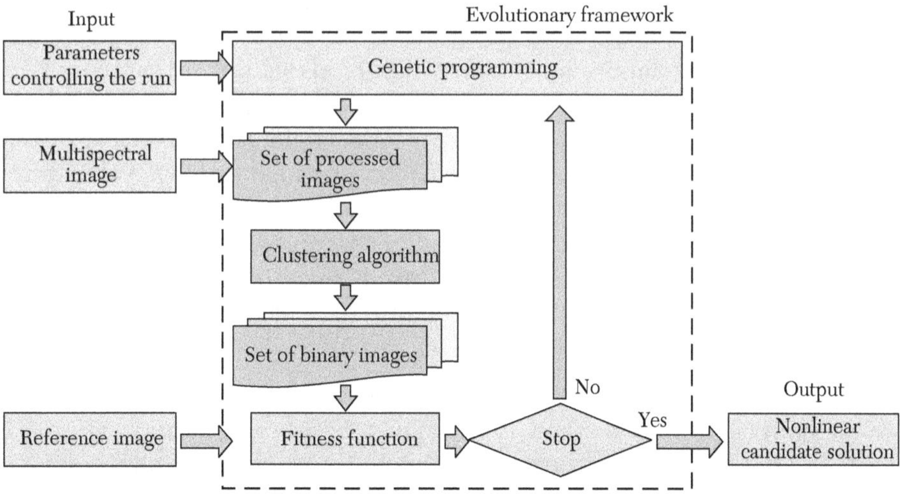

Figure 13.3. Flowchart showing components of the evolutionary framework (Momm & Easson, 2011) (please see color figures in companion files).

positive and negative samples are used by the GA to evolve candidate solutions through an optimized iterative search. In the development mode, the system requires three inputs: original image, parameters controlling the run, and reference image.

Crommelinck et al. (2016) proposed several methods for object detection from optical remote sensing images. They have grouped the methods into two categories: pixel-based or object-based. The pixel-based approaches analyze single pixels, optionally taking into account the context of pixels, which can be considered through moving windows or implicitly through modeling. These data-driven approaches, such as CNNs, are often employed when the object to be extracted is either smaller or similar in size as the spatial resolution of the image. The pixel-based approaches are often used to extract low-level features, which do not consider information about spatial relationships. Low-level features are extracted directly from the raw, possibly noisy pixels with edge detection being the most prominent algorithms. The lack of an explicit object topology is the main drawback that might lead to inferior results from these algorithms, particularly for topographic mapping applications.

The object-based approaches are employed to explicitly integrate knowledge of object appearance and topology into the object extraction process. If the spatial resolution of an image is finer than the object of interest, these approaches provide good results. Pixels with similar characteristics such as color, tone, texture, shape, context, shadow, or semantics are grouped to objects for classification. Such approaches are referred to as OBIA, which are used to extract high-level features that represent shapes in images that are detected invariant of illumination, translation, orientation, and scale. High-level feature extraction, aimed at automated object detection and extraction, is currently achieved in a stepwise manner, and is still an active research field. High-level features are mostly extracted based on the information provided by low-level features. Algorithms for high-level feature extraction often need to be interlinked to a processing workflow and do not lead to appropriate results when applied solely. They are considered model-driven, since knowledge about scene understanding is incorporated to structure the image content spatially and semantically.

Lu et al. (2014) have proposed two main spectral-spatial analysis methods. The first method is the object-based classification approach, where the image is first segmented into a set of objects which consist of adjacent pixels with similar spectral-spatial properties, using a segmentation algorithm such as mean-shift, fractal net evolution approach (FNEA), watershed, and so on. The segments/objects are viewed as the minimum image processing units for the subsequent classification. This method has been proved to be an effective approach in using high-resolution remote sensing images. On the other hand, the classification with spectral-spatial features, which incorporates the spatial features into image analysis, has received increased attention since it is an effective way to complement the spectral information for image classification. The gray-level co-occurrence matrix (GLCM) is one of the commonly used features for spectral-spatial classification. With the GLCM, the textural information of each pixel is computed by the spatial correlation between the neighboring pixels in a defined window. Differential morphological profiles (DMPs) are constructed by mathematical morphological transformation, and are another well-known feature for high-resolution image classification. The DMP is a multiscale approach that adopts a series of morphological filters and generates a series of features with different structural elements. The use of morphological reconstruction after opening and closing can preserve the shape of the objects and suppress the undesired noisy signals. More recently, a 3D discrete wavelet transformation (WT) has been used for urban mapping, which is suitable for describing complex urban scenes and can distinguish different information classes. In addition, the shape characteristics, pixel shape index, and height information extracted from LiDAR data have also been considered as complementary information for the spectral signals in object classification.

Straight line extraction is mostly done with the Hough transform. This is a connected component analysis for line, circle, and ellipse detection in a parameter space, referred to as Hough space. Each candidate object point is transformed into Hough space in order to detect clusters within that space that represent the object to be detected. The standard Hough transform detects analytic curves, while a generalized Hough transform can be used to detect arbitrary shaped templates. As an alternative, the line segment detector (LSD) algorithm could be applied where the gradient orientation that represents the local direction of the intensity value, and the global context of the intensity variations, are utilized to group pixels into line-support regions and to determine the location and properties of edges.

Yadav et al. (2020) proposed an OBIA rule-based classification approach for automatic road extraction using eCognition software, as shown in Figure 13.4. The WorldView-2 satellite images, having a resolution of 0.5 m for PAN and 2 m for multispectral, is pan-sharpened using PCA. Furthermore,

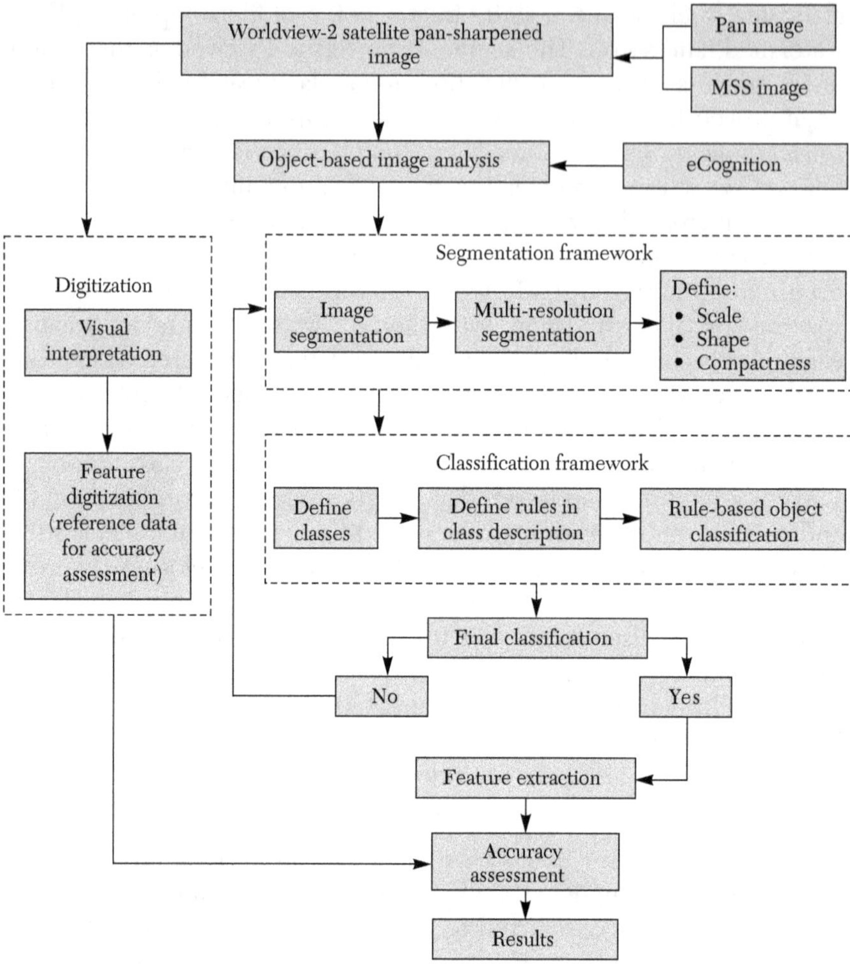

Figure 13.4. Methodology for automatic extraction of road (Yadav et al., 2020).

image segmentation is done using multi-resolution segmentation and, after defining the values for scale ratio, shape, and compaction parameters, the fuzzy rule-based classification is performed by defining classes and applying tested values to extract the road feature. Object-oriented classification in eCognition is a fuzzy logic-based process where various object features such as shape, texture, or spectral values are used for classification. A hierarchy of the number of segmentation levels helps to extract image objects which represent image information in different scales simultaneously. This optimized segmentation minimizes the heterogeneity of resulting image objects. The pan-sharpened data is visually interpreted to digitize the road feature to compute the accuracy of the extracted road feature. Its accuracy assessment attains the completeness of 71.65%, correctness of 70.33%, and quality of 59.98%.

Mayer (1999a, 1999b) carried out a detailed survey on automated methods for building extraction. He identified five trends for automatic GIS data collection: scale, context for knowledge structuring, 3D structure a key issue for their recognition, fusion of data from different sources, and usage of GIS data for improved extraction of roads. Mayer (1999b) presented an analysis of strengths and weaknesses of some existing approaches on automated road extraction. Automatic methods for building extraction have led to promising results, but the methods show a definite lack in performance needed for practical application, whereas semiautomatic methods appear to be mature for practical implementations. Ziaei (2014) compared three algorithms, that is, SVM, nearest neighbor, and the rule-based system, for extraction of buildings and roads from WorldView-2 satellite images with an object-based classification approach, and concluded that the rule-based approach is satisfactory with the highest accuracy of 92.92%. This approach works on user-defined rules that are developed with relevance to features to be extracted, which depends upon knowledge and experience for the development of a ruleset.

Chen et al. (2018) proposed a methodology for extraction of buildings, as shown in Figure 13.5. The left part of the flowchart displays the training

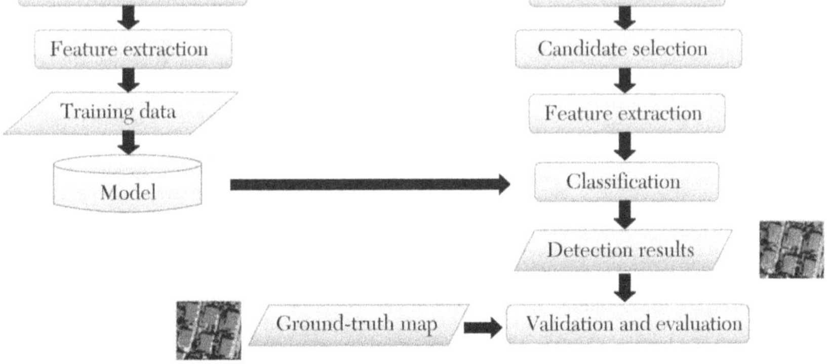

Figure 13.5. Flowchart of the proposed approach for building extraction (Chen et al., 2018).

procedure. The sample database contains all training orthophotos along with corresponding building mask images that provide the exact location and shape of all buildings on an orthophoto. Thereafter, the input image is segmented into small homogeneous regions. In the candidate selection procedure, some definite nonbuilding objects are eliminated and only candidate objects are left, while in the feature extraction step, multiple features are extracted to describe each candidate object. Finally, the classifier model is trained and saved for future use. The right part describes the detection (predicting) procedure. Segmentation, candidate selection, and feature extraction are first applied to test the orthophotos. Once the features are generated, the previously trained classifier model is employed to classify the candidate objects into two categories: building and nonbuilding. At the end, the detection results are validated and evaluated.

According to You et al. (2020), the object detection and change detection can be divided into three broad levels: scene, region, and target based, as shown in Figure 13.6. In these levels, two factors are the most influential. One is the basic processing unit for final result analysis, namely patches, pixels, and objects; and the other is the targeted optimization technology for different target change applications. For example, the patch-based methods are capable of discriminating rough scene changes in large areas. The pixel-based method emphasizes the difference of each pixel and considers internal relevance; however, the high spectral differences of the unchanged targets lead to the inevitable omission and error detection. In addition, for cases with complex details and stability of local features, object-based methods are being developed, which achieve a compromise between the pixel-based method and the patch-based methods. In reality, it is most desirable to determine the basic unit and change extraction method according to the actual change detection situation.

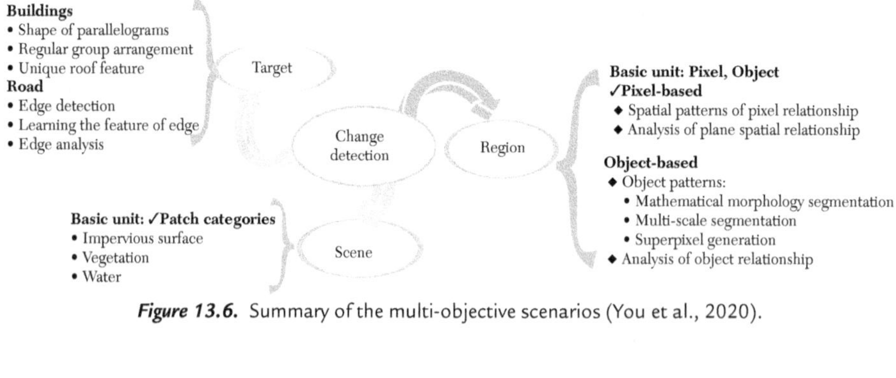

Figure 13.6. Summary of the multi-objective scenarios (You et al., 2020).

13.4 CASE STUDIES

Three case studies are presented here to demonstrate the utility of automated feature extraction techniques.

13.4.1 Extraction of Riverine Features

In this research work (Kuldeep et al., 2021), riverine features (mainly islands) are extracted based on the optimal feature selection approach which enhances the classification performance.

The part of the Yamuna river (Haryana district) which contains many small size islands and other riverine features was selected for the study with upper-left corner coordinates as 30° 21' 40.65" N, 77° 34' 12.60" E and lower-right corner coordinates as 30° 05' 0.60" N, 77° 27' 20.08" E. The IRS PAN (Cartosat-1) image of October 2009 at a 2.5 m spatial resolution and the IRS (Resorcesat-1) multispectral (MS) image of October 2009 at a 5.8 m spatial resolution are used in this study.

A flow diagram of the proposed methodology is shown in Figure 13.7. A number of textural features from three different texture models, gray level co-occurrence matrix (GLCM), local binary pattern (LBP), and wavelet transform (WT) are computed. From the GLCM texture model, 8 different textural features, namely. mean, entropy, angular second moment, dissimilarity, homogeneity, variance, contrast, and correlation have been derived. The island areas appeared as mixed shades of gray in all the GLCM feature images because of more variation in land cover features in these areas. From the WT model, three textural features namely, mean, variance, and entropy, have been derived for all three levels of WT decomposition with 3 wavelet families. Three types of wavelets, namely, Coiflet, Daubechies, and symlets with 5 variants in each wavelet family, are used for island extraction. The wavelet features at each level of decomposition produced 45 textural features, extracting a total of 135 features from the input PAN imagery. The main advantages with the WT in extraction of riverine features are the availability of a wide range of wavelet functions and the choice of wavelet which best suits texture analysis. From the LBP model, 3 textural features namely, mean, variance and entropy with 3 variants have been derived as uniform LBP, rotation invariant LBP, and uniform rotation invariant LBP. Each variant of LBP has been used to extract three textural features from PAN imagery, extracting a total of 9 features.

The textural features from all three texture models have been combined to obtain a hybrid textural model with a total of 152 textural features. Most of the texture-based image classification approaches use a feature extraction

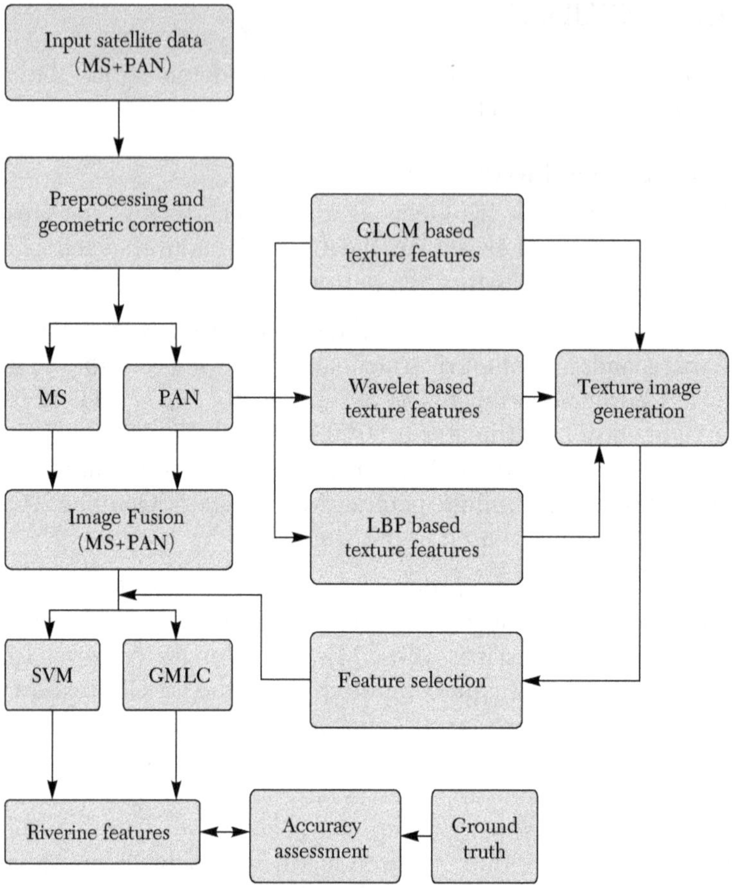

Figure 13.7. Flowchart of the method for riverine feature extraction.

method that belongs to same class. This leads to the high computational cost of the process. That is why textural features from different textural models belonging to different classes are combined to use for further classification.

The sequential floating forward selection (SFFS) method is adopted to obtain the optimal set of features along with the classification score of each feature. Optimal feature selection and feature evaluation is based on the k-NN classifier. Out of 152 textural features, the SFFS method selected a total of 6, namely, correlation, homogeneity, uniform RI variance, variance features derived from Coiflet1 at the second level of wavelet decomposition, a variance feature derived from Daubechies 2 at the first level of wavelet decomposition, and a variance feature derived from Coiflet5 at the second level of wavelet decomposition as the optimal textural features set, based on the maximum score.

The optimal features obtained from each texture model have been used further for classification using two different classifiers , namely, the Gaussian maximum likelihood classifier (GMLC) and the support vector machine (SVM). The WT-derived texture-based classification outperformed both the GLCM and LBP texture-based classification methods, and produced output with overall accuracy of 80.14% and 84.14% using GMLC and SVM methods, respectively. The hybrid texture-based model using SVM produces the classified map with an overall accuracy of 93.03% with a kappa coefficient of 0.89, whereas the same model using GMLC method produces the classified map with an overall accuracy of 89.85% with Kappa coefficient of 0.84. The hybrid texture features using the SVM method for river island extraction appeared to be correctly classified, as shown in Figure 13.8. The experimental results suggest that spatial features play a more important role than the spectral features for island extraction, especially when there are high spectral similarities among different land features present. To utilize the islands for tourism development or greenery development, it is necessary to assess the vulnerability of these islands.

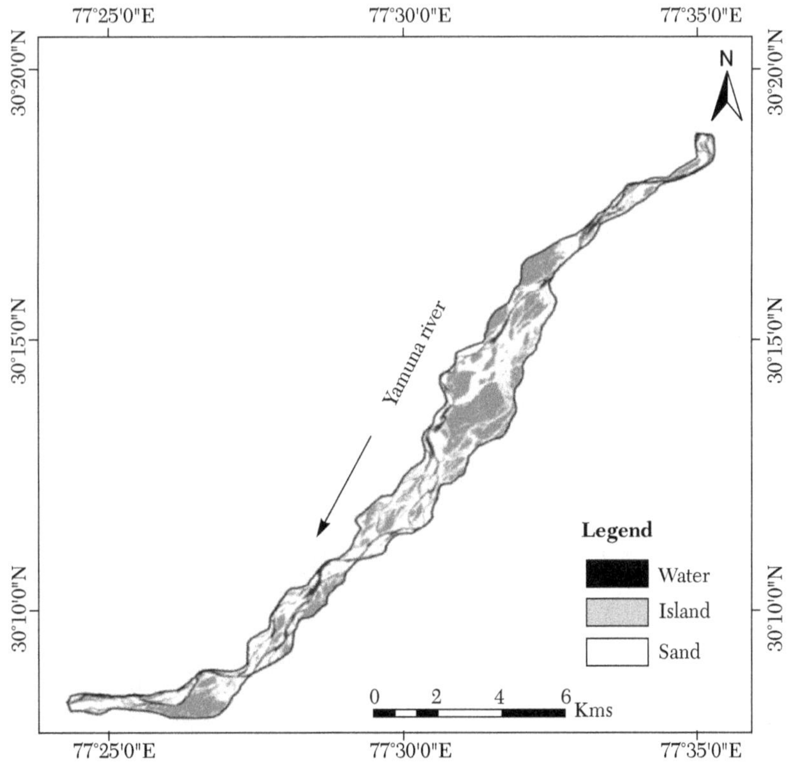

Figure 13.8. Hybrid texture-based classification for river islands using SVM.

13.4.2 Automated Building Extraction

Identification of buildings from high-resolution satellite images is an important aspect in remote sensing-based applications. While detecting buildings from satellite images, spatial resolution (i.e., smaller pixel size) is considered to be more important than spectral resolution. In India, building extraction from satellite images is a difficult task because the buildings don't follow a specific pattern and the individual buildings are smaller in size. In addition, the reflectance of buildings and roads are similar on many satellite images, which results in error in digital classification. In such cases, differentiation between buildings and roads by digital classification of satellite images becomes a difficult task. Because of this reason, some additional characteristics of buildings (like area, shape, etc.) are also required to be considered for enhancing the accuracy of extracted buildings from satellite images.

In a study by Dahiya et al. (2013, 2014), an object-oriented approach in ERDAS Imagine software for automatic building extraction from three high-resolution images is used. The flow chart of methodology proposed is presented in Figure 13.9. One image is taken from an example image given in ERDAS Imagine Objective, and the other two are IKONOS satellite images covering a part of Dehradun city, which are shown in Figures 13.10 *a*), *c*), and *e*), respectively. The single feature classification (SFC) is used to perform pixel-based classification to compute a probability metric in which the values lie between 0 and 1 to each pixel of the input image based on its pixel value and training samples. The buildings to be extracted are selected by the user as training samples, and based on the pixel values of the training samples, the probability values are assigned. The pixels whose values are similar to pixels in the training samples are assigned higher probability values. Lower probability values are assigned to pixels whose values are significantly different from the values of pixels in the training samples.

The images are segmented by using split-and-merge segmentation so that the pixels that are grouped as raster objects have probability attributes associated with them. The split-and-merge segmentation along with the edge detection is used to identify segments with similar characteristics. The parameters used for image segmentation are Euclidean distance, minimum value difference, variation factor, threshold, and minimum number of pixels representing an edge length. Different filters are applied on the image to remove the objects which are not required, and the output raster image is converted into a vector image. The value taken as input for minimum object size is 100 pixels and for maximum object size is 2,000 pixels. The

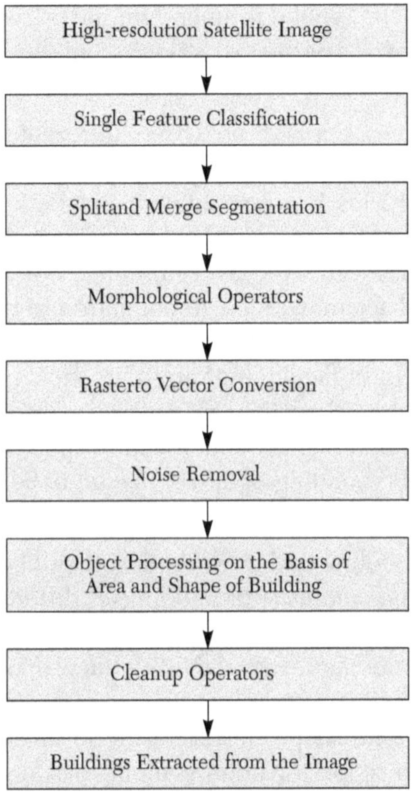

Figure 13.9. Flowchart of the approach used.

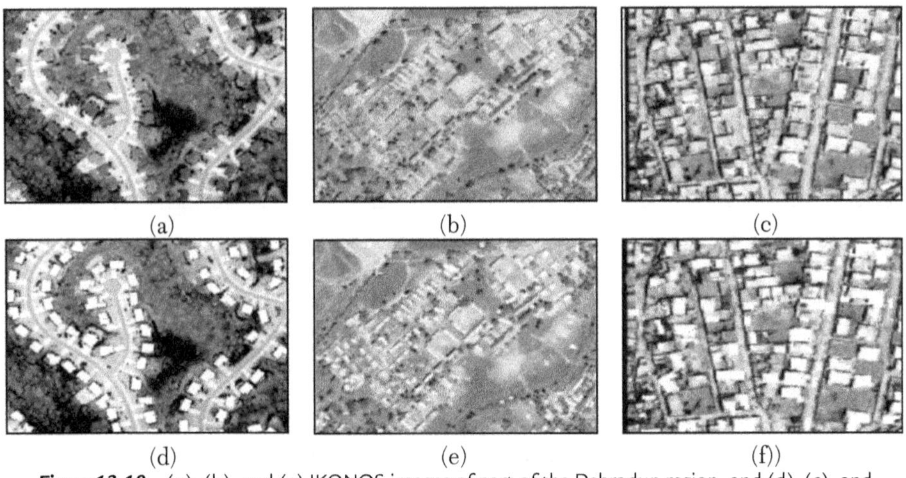

Figure 13.10. (a), (b), and (c) IKONOS images of part of the Dehradun region, and (d), (c), and (f) images superimposed after building extraction (please see color figures in companion files).

clean-up methods are applied to smoothen the extracted buildings and also to increase the accuracy of the extraction of buildings. Figures 13.10 *b*), *d*), and *f*), show automated building extraction using the proposed approach.

The approach is applied on three different satellite images. The extracted buildings are compared with the manually digitized buildings as shown in Figure 13.11. For one satellite image, it has picked up all the buildings with a slight change in the area of footprints of buildings. Only one patch of road is extracted as a building. For the other two satellite images, the overall accuracy is low as compared to the first satellite image. Some patches of road and ground are also extracted as buildings. The branching factor, miss factor, building detection percentage and quality percentage, were also calculated for accuracy assessment. Nonetheless, the overall accuracy of building extraction with respect to area was found to be 85.38% in a set of 66 buildings, 73.81% in a set of 94 buildings, and 70.64% in a set of 102 buildings.

The proposed technique has worked well on high-resolution images in which the reflectance intensity of buildings is different from other objects. It clearly indicated that some refinement is still required to get the correct shape and size of buildings extracted using the proposed methodology. Also there is a need to eliminate the shadow from the image and to add some more parameters, like shape, texture, and so on to the given approach. There is still much scope for improvement with respect to increasing the extraction rate, refining the delineation of buildings, and decreasing the false extraction rate.

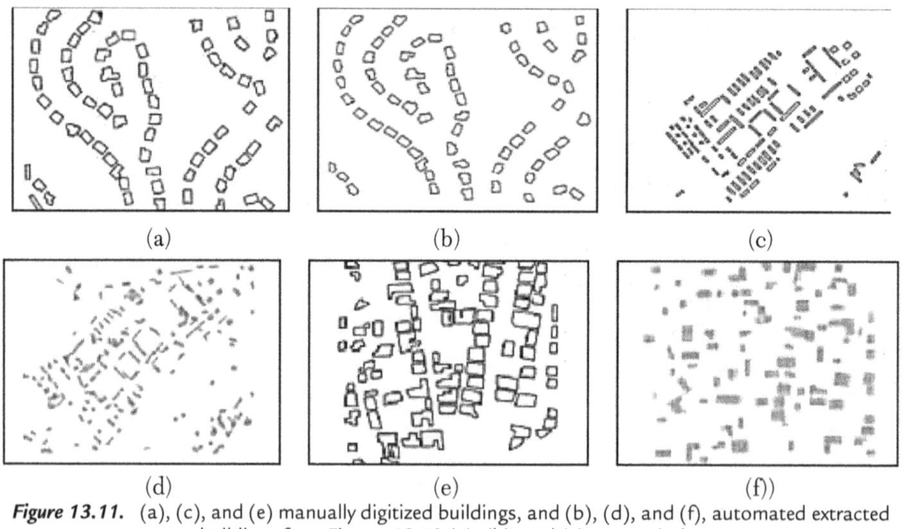

Figure 13.11. (a), (c), and (e) manually digitized buildings, and (b), (d), and (f), automated extracted buildings from Figures 13.10 (a), (b), and (c), respectively.

13.4.3 Detection of Pavement Cracks

The work carried out by Kumar et al. (2019) involves the development of tools for the evaluation of pavement surface conditions in terms of length and area of cracked pavement surface, and area of potholes, which could be used along with the depth of potholes, and international roughness index for prioritization of road networks based on pavement surface condition data. Road sections between Noida Greater Noida Expressway, Delhi-Agra-Etawah section of National Highway 2 and Inner Ring Road, Delhi, have been selected, and data collected during 2012–2015 using an Automated Road Survey System (ARSS). The ARRS consists of two-line scan cameras for collection of pavement surface images. The digital camera, which is mounted at a distance of 2.2 m with its optical axis vertical, consists of images of size 2048 × 1536 pixels. One image/camera covers 2.0 m × 1.75 m half-lane width, that is, a 3.5 m^2 area of the pavement surface. To cover a full lane width of 7.0 m, a combination of two cameras has been used, which captured 500 frames each in one km length, that is a total of 1,000 images are captured to cover a 3.5 m wide lane in a 1 km road length. In addition, the ARSS carries a GPS or a DGPS receiver which collects survey position data using international GPS, enabling referencing of road data to geographical coordinates.

Full automation of crack monitoring is a challenging image processing problem, because in most cases the cracks appear as thin, irregular lines, buried into strong and textured noise. A methodology has been proposed by Kumar et al. (2019), as shown in Figure 13.12, to measure the area of pavement surface images having a cracked surface, and potholes based on image processing tools. The proposed methodology consists of five stages:contrast stretching, edge detection, Otsu thresholding, morphological dilation, and measurement of pixels representing the length and surface area of the cracking and potholes. The results of the pavement surface images are processed to compute the area, and are compared with the manually processed images using the Hawkeye processing tool kit software of ARSS.

For automatic analysis of cracks and potholes and determination of area, a software has been developed using OpenCV-Python software. Figure 13.13 shows the manual (using the Hawkeye processing tool kit) and automatic processing (using the developed methodology) of the alligator type of pavement surface distress (cracks and potholes).

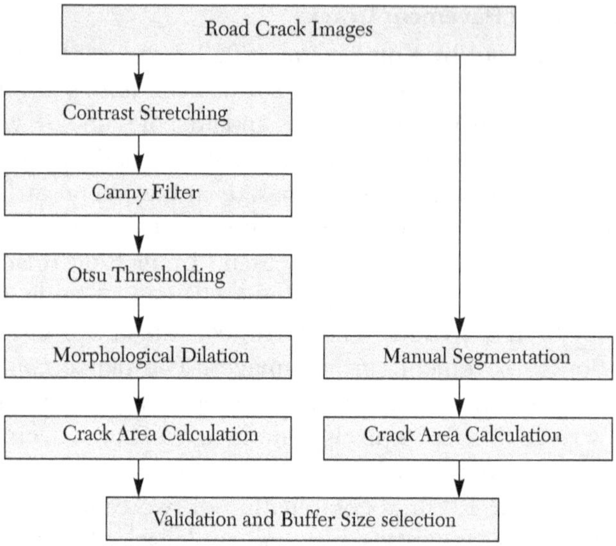

Figure 13.12. Methodology for detection of cracks from pavement surface images.

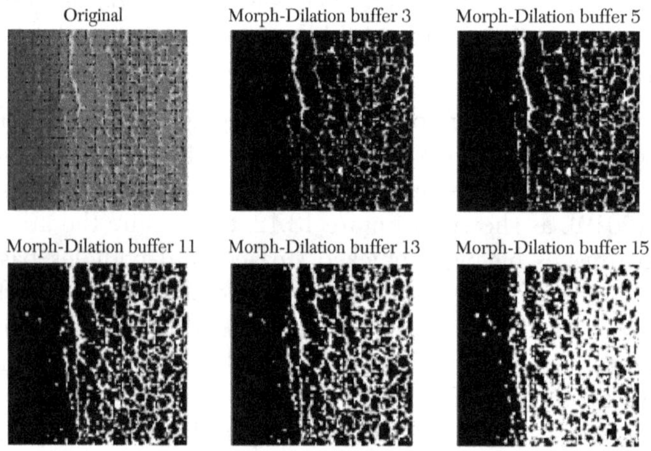

Measurement area (m^2)	
Buffer 3	0.34
Buffer 5	0.51
Buffer 11	0.95
Buffer 13	1.10
Buffer 15	2.22
Manual processing	2.43

Figure 13.13. Measurement of alligator cracked pavement surface area using manual and automatic analysis of pavement surface images (Image ID-Noida 90) with buffer size of 3, 5, 11, 13, and 15.

The results obtained from manual and automatic processing of various pavement surface images having different types of distresses with respect to cracks and potholes have been summarized and compared. These results are summarized in Tables 13.2 and 13.3.

Table 13.2. Area of Alligator Cracks

(Recommended buffer/structure element size for morphological dilation 15)							
Section ID	**Area with different buffer/structure element (m²)**					**Manually measured area (m²)**	**Error %**
	3	**5**	**11**	**13**	**15**		
Noida90 (Noida Section Frame No. 90)	0.34	0.51	0.95	1.1	2.22	2.43	8.6
NH0272968 (NH02 Frame No. 72968)	0.26	0.42	0.87	1.04	2.43	2.31	-5.2
NH0272920 (NH02 Frame No. 72920)	0.27	0.41	0.77	0.91	1.97	2.09	5.7

Table 13.3. Measurement of Area of Potholes

(Recommended buffer/structure element size for morphological dilation 11)							
Section ID	**Area with different buffer/structure element (m²)**					**Manually measured area (m²)**	**Error**
	3	**5**	**11**	**13**	**15**		
RR4413 (Ring Road Frame No. 4413)	0.19	0.29	0.59	0.71	3.37	0.56	–5.4
RR4257 (Ring Road Frame No. 4257)	0.21	0.32	0.60	0.71	2.92	0.57	–5.3
RR4426 (Ring Road Frame No. 4426)	0.20	0.28	0.50	0.58	2.61	0.48	–4.2

The results show that the morphological dilation operations with different structure element size are suitable to access the area of cracks and potholes. The area of alligator cracks can be accurately accessed by performing a morphological dilation operation using structural element size 15. However, for other cracks (longitudinal or transverse cracks) a morphological dilation operation with structural element size 5 gives accurate results. The area of potholes can be accurately accessed by performing a morphological dilation operation using structural element size 11.

13.5 CONCLUSION

Automatic object extraction has been a popular topic in the field of remote sensing for decades, but extracting objects and features from remotely sensed images is still a challenge. Object detection in optical remote sensing has always been a fundamental but challenging issue in the field of aerial and satellite image analysis. During the last decades, considerable efforts have been made to develop various methods for the extraction of different types of objects. An automated building extraction approach uses structural, contextual, and spectral information from high-resolution satellite imagery. In the future, the work using other information sources and more complex decision-level fusion strategies could improve the overall extraction performance.

Different kinds of objects may be investigated concerning their scale-space behavior from various scale-space images. The investigation may be accompanied by a further exploration of sophisticated algorithms, especially evaluating their performance for different applications. To improve the versatility of object extraction, machine learning techniques such as evolutionary algorithms can be useful. Data from imaging laser scanners have opened up new ways for data fusion and extraction of features. Since object extraction in aerial imagery has received much attention recently, automatic object extraction will become a reality for practical applications.

CHAPTER 14

Applying Remote Sensing for Smart Cities

14.1 Introduction

Today, more than half of the world's population, that is about 3.5 billion people, lives in urban areas, and by 2030 it is expected to rise to 60%. By 2050, the population of the world's cities is expected to double, while the global population will grow from 7 billion to about 10 billion, bringing significant changes in the size and distribution of the world's population (Sarkar, 2019). According to the United Nations (UN), about half of the urban population lives in cities with less than 0.5 million inhabitants, but one in eight lives in twenty-eight megacities, such as Tokyo, Delhi, Shanghai, São Paulo, and London, having more than 10 million inhabitants. The fastest growing cities having 0.5–1 million inhabitants are located in Asia and Africa (United Nations, 2015). In these cities, there is also a shortage of resources, creating huge pressure on existing systems.

In 1992, the concept of *smart growth* emerged which proposed an alternative paradigm to urban sprawl, individual housing, and transport systems. This concept suggested that the concentration of growth in a city mostly happens in mixed land use and walkable urban centers, where the community participates and makes fair and cost-effective development decisions. During those days, creative ways of urban planning and design also emerged which gained immense popularity in the 1990s. Later, a new concept called *intelligent cities* emerged, which included the use of data and information and communication technology (ICT) for the functioning of the cities. From the feedback and debates on *smart growth* and *intelligent cities*, the idea of a *smart city* has emerged. The "intelligent" and "smart" enterprises, such as IBM, CISCO, and Siemens, also started working toward smart cities, which were later joined by the technology giants like Hitachi and Microsoft and academic institutes, like the Massachusetts Institute of Technology (MIT) laboratories (Bull, 2016).

With the concept of a smart city, urbanization gained momentum and people started migrating to cities for a better quality of life and employment opportunities. The migration created a huge pressure on the cities in order to fulfill the aspirations of people (Bhattacharya et al., 2015). It literally forced cities to adopt the technologies that could enhance efficiency, generate more jobs, improve quality of life, and create sustainable living conditions and economic growth in the cities. During this journey of smart cities, several cities have adopted the advanced technologies to make them livable and sustainable (DuPuis & Stahl, 2016).

Globally, there are multiple ideas, definitions, and approaches to smart cities. Smart city is an umbrella term to describe a city that makes use of technology to enhance the quality of life of citizens by improving services and communication. The World Bank defines a *smart city* as a technology-intensive city that has sensors installed everywhere and offers highly efficient public services using information gathered in real time by thousands of interconnected devices. In essence, the smart city is all about the use of advanced technology which connects every all citizens and infrastructure, and provides intelligent solutions to its citizens. It puts in place mechanisms to gather this information, and would greatly rely on feedback from citizens to help improve service delivery.

A smart city provides better living conditions to the citizens while making it more sustainable, resilient, and livable. A study by DeJong et al. (2015) carried out the available literature survey from 1996–2013, and highlighted twelve different categories of cities: "sustainable city," "eco city," "low carbon city," "liveable city," "green city," "smart city," "digital city," "ubiquitous city," "intelligent city," "information city," "knowledge city," and "resilient city." They found "sustainable city" had the highest number of occurrences followed by "smart city," and that the "smart city" is linked more to "digital city," "intelligent city," "eco city," and "low carbon city." "Low-carbon city" is viewed as a subset of "sustainable city," whereas "smart city" is viewed as a new concept with particular connotations around integrated building and technological fixes. The majority of definitions of smart cities suggest one common thing, that is, the use of ICT as the prime important factor. The ICT would integrate the systems and frameworks on which a city functionally operates. Figure 14.1 provides a representation of the different parameters mentioned in definitions pertaining to smart cities and the weightages assigned to them (Bhattacharya et al., 2015).

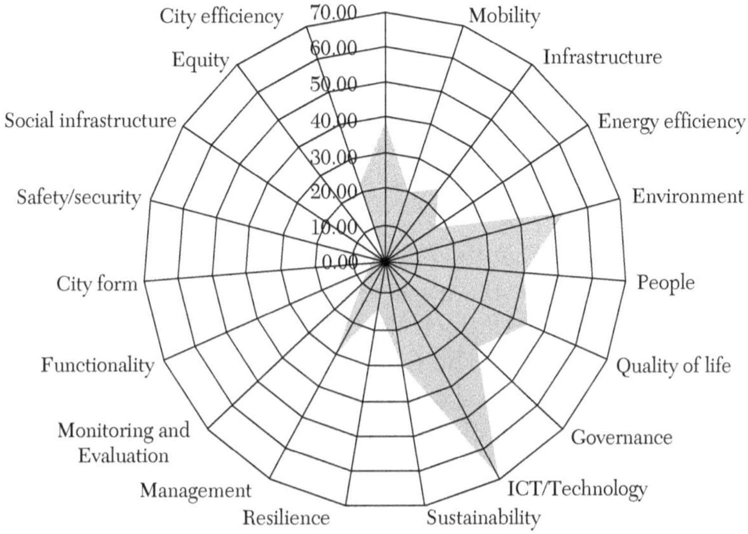

Figure 14.1. Combination of various aspects in smart cities and their relative weightage (Bhattacharya et al., 2015).

14.2 DATA: THE FOUNDATION OF SMART CITIES

Smart city application will inevitably be shaped by a unique combination of social, geographical, environmental, technological, and financial factors of the area. To provide smart services in cities, data from many sources are collected and collated. Smart city operations require a decentralized approach which can serve the areas of common needs. It would allow citizens to get all information of city services at one digital touch, so that they can interact with local government and give their feedback (Choudhary, 2018). This requires a platform that provides all information at a one stop shop and which acts as a two-way communication channel between citizens and governments as well as between government and government so that smart decisions could be taken jointly. The GIS-based platform can provide all information to citizens about the city, like land availability, parking availability, local transport system information, and so on. (Bull, 2016). The GIS platform can also give an insight to the government of a smart city in a holistic way so that they can work in a better way.

The smart city initiative relies on the quality and quantity of data. The most important thing in making any decision for planning a smart city is the availability of accurate and updated data, as the decisions are only going to be as good as the data (Garg, 2021b). The smart city requires technological infrastructure to collect, aggregate, and analyze the real-time data to improve the lives of its citizens. A smart city must build a "data culture" across various stakeholders, keeping data privacy a core priority. It may have an open data platform with adequate storage and a supporting data governance framework. If the data is shared openly, smart cities can fulfill their objectives with interconnections between the systems and devices to share the data effectively.

Data collection is not the only objective of a smart city, but it has to be relevant and useful for the applications. Data processing is as important as the data collection. All the collected data from the entire network may not be useful for a task. The data not being used can be stored on the cloud, while the other useful data can be processed on local data processing resources (DuPuis & Stahl, 2016). Distributed processing systems provide the ability to process the large amount of data much faster.

The advances in processing and analytics can help in understanding the data requirements, data collection approaches, data storage, and data transfer methods. Historical data which can be stored in the cloud are particularly important for urban planning and development to study the trend and plan sustainability. On the other hand, there may be redundant backups in the cloud, which may not be required anymore. It is therefore important to know what data is needed and for how long it is needed. For example, a modern smart car can generate up to four terabytes of data per day, but perhaps it is not required to keep all that data in perpetuity. So, setting up local storage for a short period, and periodically sorting out the data for longer term use on the cloud network are essential activities in order to make the optimum use of cloud resources (Alavi et al., 2018).

The information derived from the analysis of this data can also be used for real-time problem-solving. The local governments, depending upon the public problems identified as priorities, may address how the data would be collected to resolve those problems. The governments also need to consider how new data will influence the policies and what kind of administrative and technical expertise is required for effective analysis and use of data. According to the International Data Corporation (IDC, 2020), the amount of digital information created and replicated in a year increased by 62% in 2009, reaching 1.2 million petabytes in 2010. By 2020, the IDC estimated that the digital universe would be 44 times larger than in 2009.

14.3 KEY ENABLING TECHNOLOGIES FOR SMART CITIES

Technology has always been a critical force deeply integrated with the evolution of cities and development of infrastructures. Technology is the backbone of a smart city. From the earlier human settlement to the industrial revolution to the present age (Industry 4.0), technology has greatly impacted everyone; the way we get around, live, and work in the urban space (IBM, 2013). In all smart city definitions the use of technology is one of the common elements required to engage the citizens, deliver services, and enhance infrastructure systems. Technologies, the rapidly accelerating drivers of smart cities and smart citizens, facilitate the modernization of approaches. A smart city is expected to deploy ICT and several other essential technologies to optimize the resources. Effective information management is at the heart of this capability, and integration and analytics are the key enablers (IBM, 2013).

There are six major technologies that are needed for a smart city application. The details are summarized in Figure 14.2, as well as discussed as follows:

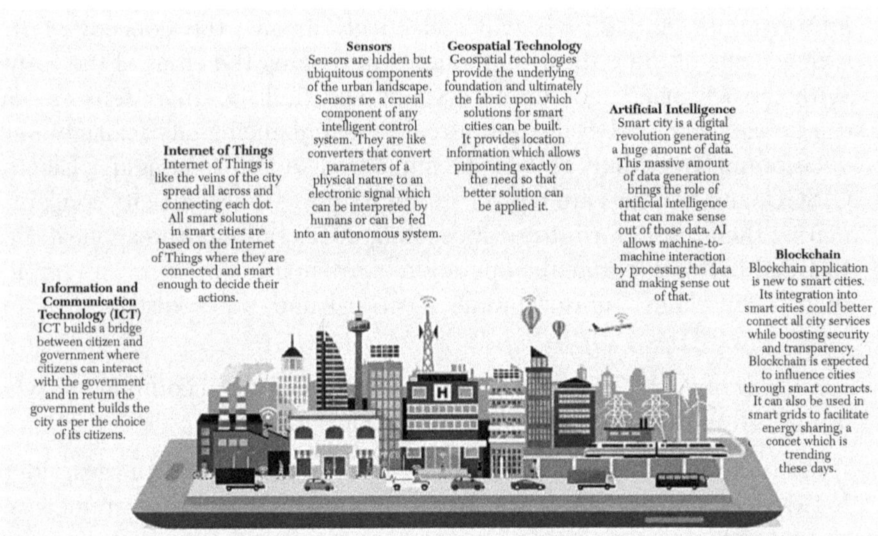

Figure 14.2. Six major technologies used in smart cities (Choudhary, 2018).

14.3.1 ICT and IoT Technology

The ICT infrastructure forms the foundation of key layers of planning and development of a smart city. It is the fundamental layer on which all other components rely, and comprises high-speed wired and wireless network connectivity, high-end data centers, and physical space enrichment with smart devices, sensors, actuators, and so on. (Sarkar, 2019).

If data is the foundation on which smart cities are built, then the network is definitely the structure of a smart city. The data will only be useful if it can be analyzed in real time, and information derived quickly from the network (Beck, 2018). Networks are much faster and more cost-effective to deploy. Not all networks are created equal. The organizations will calibrate variables, such as range, bandwidth, power usage, intermittent connectivity, interoperability, and security in their own way. In addition, each organization prioritizes the factors which ultimately decide the network to be adopted for efficient delivery of connectivity (Simpson, 2018).

A smart city is an economically vibrant city that deploys diverse systems of technologies. These technologies are deployed to automate all the processes required for a smart city, from bringing WiFi and wireless networks to collecting data via IoT-enabled devices that can be used to make real-time decisions. For example, the emergency response system can be focused and fast during an emergent situation with the use of the IoT and sensors. Various tools like sensors, gateways, communication infrastructure, and servers will collectively involve the concept of IoT, which is one of the critical components in shaping the cities of the future. With the advanced monitoring systems and built-in smart sensors, data can be collected and evaluated in real time, enhancing city management's decision-making. Many IT-based companies like IBM, Schneider Electric, CISCO, and Siemens are more focused on using the smart city concept to market their vision for future cities (Bull, 2016) through the application of ICT and IoT to integrate the operation of urban infrastructure and services such as buildings, transportation, electrical and water distribution, and public safety (IDC, 2020).

The framework for smart cities consists of four main components when using ICT and IoT technology, as shown in Figure 14.3.

Sensing layers–consist of a network of machines, sensors, images, videos, IoT devices, and human sensors to continuously monitor the parameters of city subsystems.

Data layers–consist of databases related to the functioning of urban bodies and municipalities along with the departments supporting the city infrastructure, like transportation, utilities, and so on.

Business layers–consist of models related to analytics, visualization, business logic, semantics, data catalogs, metadata, and so forth.

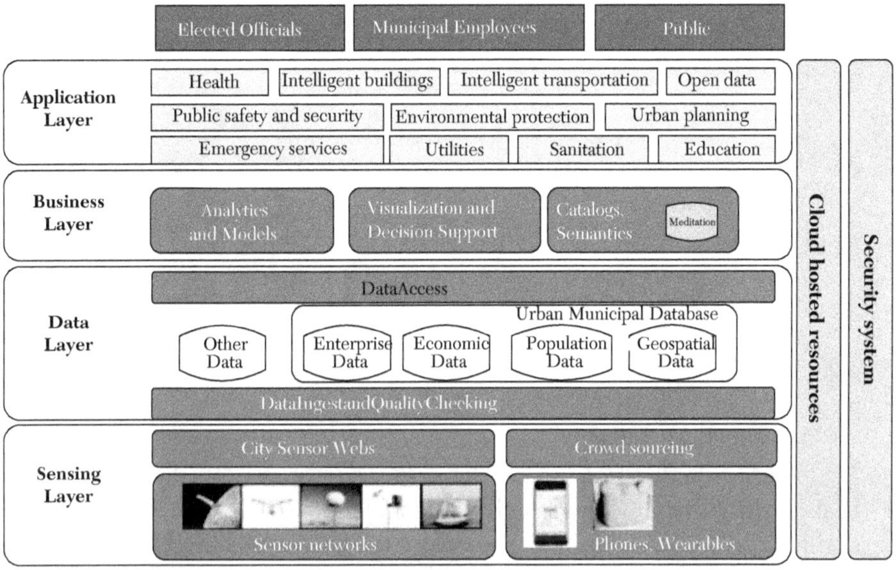

Figure 14.3. Four main components of the framework for smart cities.

Application layers–consist of various applications used by citizens, municipal staff, and administrators related to various departments of the smart city.

Any event with a large gathering of people in an area may not have adequate infrastructure to support cellular connectivity to send images, text, or even make phone calls at the same time. People, businesses, and government are required to connect quickly through the deployment of wireless connectivity infrastructure in a secured manner in order to share and transfer the data and improve activities. The centralized cloud infrastructure can be used to create interoperability through intense networking technology between widespread devices throughout the city (Simpson, 2018). It will facilitate the development of compatible software and hardware and devices to support present needs and planning for the future needs.

The IoT is like a lifeline for urban living, and can expedite the development of smart cities (Beck, 2018). In general, IoT is technology-oriented while smart cities are user-oriented. Each are moving toward each other with a common goal of providing better services for modern cities. A smart city consists of six major components, including smart governance, smart economy, smart citizens, smart mobility, smart environment, and smart living. Figure 14.4 shows the relationship between IoT and key smart city components. Indeed, a smart city is a knowledge-based system

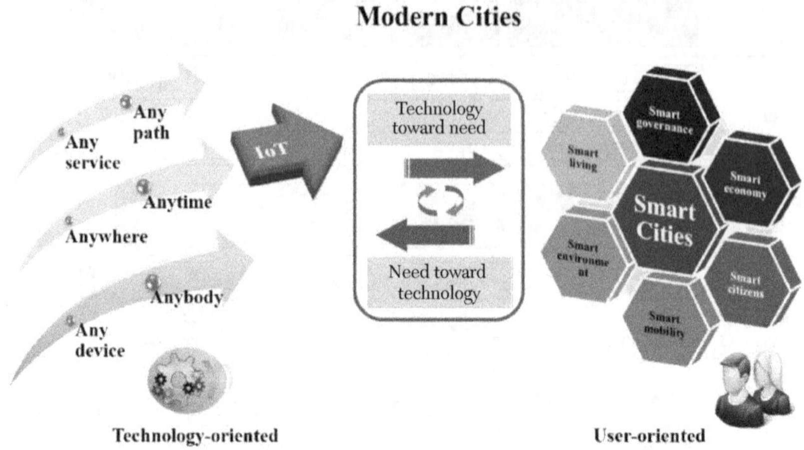

Figure 14.4. Relationship between IoT and smart city components (Alavi et al., 2018).

that provides real-time insights to stakeholders to manage the city's components. Smart cities are employing the IoT technology to connect their utility, infrastructure, and public service grids, as well as generating real-time data. The IoT transmits data to the command, and control center analysis is carried out for planning as well as responding to the needs of the city and its citizens. It is revolutionizing not only the collection of data for planning and decision-making but also delivering solutions for greater safety, reducing pollution, and energy-efficient usage for better quality of life in urban areas. To make this technology work for all, government has to ensure that new policies and systems are in place that can help people transition to this new connected world.

The IoT and the Internet of everything (IoE) are the key catalysts in the evolution of smart cities (Vince, 2020). The IoT is already making a difference to urban lives in cities across the globe, and one big role the IoT plays is to digitally empower the citizens. A smart city will no longer be a static accumulator of data but will use dynamic data through mobile technology with the capability to interact with each other and instantaneously make decisions based on this interaction without human intervention (Williams, 2018).

Use of the IoT and AI is further enhancing real-time application domains. The latest IoT and analytics technologies will not only simply connect the citizens but also allow predictive and prescriptive analytics decision-making for a variety of urban services. The future city is an interconnected one, where devices communicate with each other to provide real-time data continuously to the public and the municipality. According to the research firm Gartner, about 20.4 billion devices worldwide are connected to the Internet by 2020. The IoT will be useful to pave the way for future smart cities, such as lighting, parking, traffic and waste management, citizen engagement, and safety and security, as well as to empower policy and infrastructure planners and city administrators.

Ubiquitous sensors in mobile robots, drones, and autonomous vehicles as well as connection to city infrastructure through IoT offer more efficient delivery of utilities and services. As the wireless, bluetooth, and sensor technology has become cost-effective and increasingly used, modern devices are being redesigned accordingly (DuPuis & Stahl, 2016). These along with the IoT provide the capabilities to products or infrastructure to sense the surrounding environment and communicate with each other. Huge data is captured by sensors without being specific about its application and utility. The need for physical storage and computing power of the devices will be minimized using the cloud-based technology. The more smart devices there are the more data will be generated about the product or infrastructure and the better the communication link. For example, smart parking meters can communicate with the parking manager as soon as the time is out.

14.3.2 Geospatial Technology

Smart cities are knowledge-based cities with the capability to uplift growth by planning, management, governance, and development systems where geospatial technology is playing a key role enabling smart cities and helping decision-makers (Choudhary, 2018). Geospatial technology includes the tools and methods used in land surveying, remote sensing (LiDAR, UAV), cartography, GIS, global navigation satellite systems (GPS, GLONASS, Galileo, Compass), photogrammetry, and geography and related forms of Earth mapping. These technologies are used to channelize all data of the city, such as traffic, utilities, finance, health, or waste management to monitor and deliver improved services, improve e-governance, and offer smart and intelligent solutions.

Geospatial data, particularly from satellite images, UAV images, and GPS data, available at various scales and accuracies, are needed for the management and maintenance of smart cities. For example, the data is needed for city planning; creation of smart city master plans; and physical infrastructure such as roads and bridges, city furniture, public spaces, and data of individual buildings for energy efficient and green building design. Geospatial data and technologies have a key role to play for smart city applications such as energy, water, transportation, public safety, citizen services, city governance, healthcare, education, and so on. These technologies can provide the underlying foundation for delivery of good governance in a smart city, ranging from smart safety to smart infrastructure to smart planning. The ultimate aim is to not only improve the efficiency in managing the city infrastructure, but also enable collaboration during the planning, construction, monitoring and management of the city infrastructure.

Smart solutions require the use of IoT, ICT, sensors, geospatial data and geographical location. The GIS services can help developing and delivering state-of-art technology and services to the enterprise (citywide). In addition, the GIS technology provides city officials various services to develop, deploy, and maintain spatial data as well as 3D geospatial Web services. Figure 14.5 illustrates a paradigm for the smart city enabled by real time GIS that is designed for the acquisition, storage, analysis, and visualization of geospatial data in real-time (Li et al., 2020). The real time GIS emphasizes the provision of high-resolution and high-speed processing of GIS data streams, which can be generated continuously from sensing devices connected with the IoT. These frameworks can be used to do things

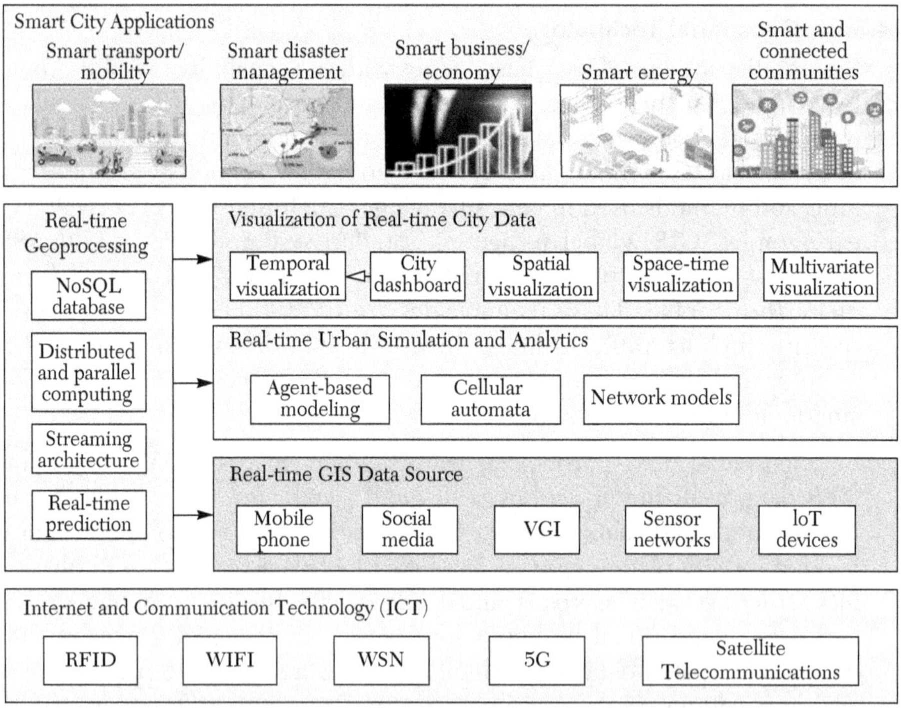

Figure 14.5. A smart city-enabled real-time GIS (Li et al., 2020).

that are incredibly valuable for a city, like identifying where investments should be made, where people are in need of help, and subsequently being able to monitor those areas to see the progress being made. GIS provides precise and accurate information, so mistakes won't be repeated and successes can be reproduced.

Location intelligence is at the core of smart city planning and design. Location data is very critical and important for smart and sustainable planning. Accurate location data allows the pinpointing of an object or a facility, for example, tracking a vehicle's location in real time. There is a set of common patterns unique to GIS, which are as follows (Chaturvedi, 2018b):

1. Mapping and visualization that enhance the understanding of locations and relationships with maps and visual representation.
2. Data management tools for collecting, organizing, and maintaining accurate locations and details about assets and resources.
3. Field mobility for managing and enabling a mobile workforce to collect and access information in the field.
4. Monitoring of live data feeds to track and manage assets and resources in real time.
5. Spatial analytics for discovering, quantifying, and predicting trends and patterns to improve outcomes.
6. Design and planning tools for evaluating alternative solutions and creating optimal designs.
7. Decision support that provides situational awareness and enables information-driven decisions.
8. Constituent engagement channels for communicating and collaborating with the citizens and external communities of interest.

Today, a growing number of cities rely on geospatial technology to derive useful insights from collected data. The technology has the capability to involve large number of variables to be used for stimulation, planning, management, and development of a city. Geospatial technology is being applied to various elements of the smart city, such as energy, water, transportation, public safety, citizen services, city governance, healthcare, education, and so forth (Choudhary, 2018). It is an accurate, timely, and cost-effective tool used for measurement, analysis, and visualization of phenomena on the Earth's surface by combining geospatial analysis, geospatial models, geospatial databases, and human-computer interaction.

14.3.3 Sensor Technology

Ubiquitous sensors and their subsequent applications are the driving force for the growth of smart cities (Williams, 2018). Some sensors may be required to collect the data once a day (e.g., water level in a well), while others may be needed to collect the data all the time (e.g., a camera at a busy road intersection or rainfall data); each one may have its different requirements. The biggest challenge in such a digital environment is to create a low-latency environment that keeps up the data and decision-making processes. For those devices/sensors that run on batteries, a low-power wide-area network (LPWAN) protocol is required. This would allow them to continue work without requiring constant replacement. A camera, on the other hand, needs the bandwidth to support moving large files around as quickly and seamlessly as possible. There are several options available for effective networking, such as 5G/LTE/WiMax, LPWAN, the low-power, low-cost IoT network LoRaWan, and the Wi-SUN Field Area Network (Vince, 2020).

Ubiquitous sensors in mobile robots, drones, and autonomous vehicles as well as connection to city infrastructure through IoT offer more efficient delivery of utilities and services. As the wireless, bluetooth, and sensor technology has become cost-effective and increasingly used, the modern sensors and devices are being redesigned accordingly (DuPuis & Stahl, 2016). These along with the IoT provide the capabilities to products or infrastructure to sense the surrounding environment and communicate with each other. As the huge data is collected, there is a need to minimize the physical storage and increase the computing power of the devices using cloud-based technology. The more smart devices there are the more data is generated about the product or infrastructure, and the better the communication link. For example, smart parking meters can communicate with the parking manager as soon as the time is out.

While a variety of sensors and applications for smart cities have been created in recent years, a lot of work still remains to be done, especially in the areas of AI and machine learning (ML) to quickly analyze and interpret the huge data from these sensors. The main challenge in using ML is that it does not have the capacity to handle the large quantity of data, and is not an automatic technique to analyze the data without human intervention.

Another challenge lies in integrating different physical and virtual technologies necessary to make a city smart. With increasing demands of using ML by smart city applications, these challenges can be overcome by developing a holistic and integrative approach (Sarkar, 2019). Another step toward overcoming existing challenges is being more aware that the physical technology, like robots, is an integral aspect of ML and vice versa.

14.3.4 Artificial Intelligence Technology

AI is the science and engineering of making intelligent machines, especially intelligent computer programs. It is related to the similar task of using computers to understand human intelligence. In smart cities, sensors and actuators, video cameras, environment sensing, traffic sensing, smart meters, smart vehicles, as well as smartphones are all collecting huge data, called big data (Allam & Dhunny, 2019). According to the forecasts by CISCO (2018), nearly two-thirds of the global population will have Internet access by 2023, and there will be 5.3 billion total Internet users (66% of global population) by 2023. In addition, the number of devices connected to IP networks will be more than three times the global population by 2023. There will be 29.3 billion networked devices by 2023, and 3.6 networked devices per capita by 2023.

A smart city has lots of applications for AI-powered and IoT-enabled technology, from maintaining a healthier environment to enhancing public transport and safety of people. One of the key applications of AI is to make quick and optimal decisions according to real-time situations, and it has been demonstrated recently that AI can outperform human beings in many areas. In general, the AI application can be divided into two modules, the learning module and the predicting module (DuPuis & Stahl, 2016). The learning module is mainly responsible for effective data collection, data training, and modeling. The predicting module is responsible for making actions to respond to the current situation. AI uses various learning techniques to facilitate automatic resource provision and judicious decision-making.

The AI and IoT can implement smart traffic solutions to ensure that residents of a smart city travel within the city as safely and efficiently as possible (Bull, 2016). Many of the problems related with the administration and management to maintain hygiene, minimize traffic congestion, reduce crime, and so on can be resolved by the use of AI and IoT. Using the power of AI, smart city systems can create municipal systems and services that operate more efficiently and provide significant benefits to residents. These benefits may be derived in many forms, including cleaner air, orderly traffic flow, and efficient government services. Adopting AI for smart recycling and waste management can provide a sustainable waste management system.

The technology is to be a fundamental core of smart cities where big data can emerge through IoT, and AI is used to process, analyze, and interpret the generated data. Allam and Dhunny (2018) have proposed the framework to integrate the use of AI technology, taking the input of big data and AI at the core of smart cities (shown in Figure 14.6). By calibration of three key dimensions of Culture, Metabolism, and Governance, the proposed model is aimed to ensure that societal livability is preserved by using the technological input. The advent of 4G and 5G IoT-based connectivity is spurring the online migration of smart city applications, helping to generate a more than sevenfold increase in smart-city AI software revenue by 2025. The global smart-city AI software market is set to soar to $4.9 billion in 2025, up from $673.8 million in 2019.

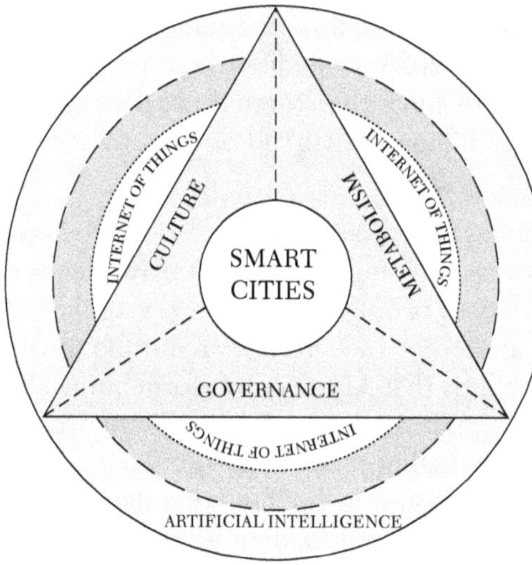

Figure 14.6. Framework for the integration of AI and big data in smart cities to ensure livability (Allam & Dhunny, 2018).

14.3.5 Blockchain Technology

Blockchain is a collaborative ecosystem that has the ability to resolve the trust issues between all the stakeholders involved, and enhances transparency in the processes (ESA, 2019b). Blockchain technology offers a decentralized registry and distributed database on a computer network with encrypted documentation so that the information can be shared to authorized users in a highly secured manner. It can easily get connected with other modern technologies in order to automate the basic services in a smart city. Blockchain may provide a new way for cities to share transactions with little or no costs, which are safe, transparent, and instantaneous, for example, parking revenue in cities (Moore et al., 2018).

Blockchain consists of a distributed peer-to-peer computer network in which interconnected nodes (peers) share the resources among each other without using a centralized administrative system (Mire, 2018). For every new register, a new block is created, adding the new register to the previous information which is then updated on the whole network. In a smart city, the platform providing the public services structured upon blockchain has some unique components, such as the digital wallet, DApps, tokens, and so on which are not present in governance models without blockchain (Pincheira et al., 2020). The use of blockchain's distributed ledger technology increases the transparency of government actions, automates the bureaucratic processes, and brings citizens closer to their government; it also helps the people addressing the challenges of urbanization through better implementation of the smart city framework. The most important benefits of blockchain for smart cities include: increased transparency, enhanced connectivity, direct communication, integrity over information, and efficient management.

In remote sensing, very little work has explored the benefits of blockchain (ESA, 2019b). Blockchain enables the transfer of value through a public network to share and retrieve data without a central authority. The short-range sensors acquire a large amount of data everyday that are useful for Earth observations. Blockchain is the solution to build a system to generate trust among users to share the data without an intermediary. It keeps track of the flow of information. Blockchain technology is verifiable and immutable by default—all the actors can access the information and its changes over time. For this reason, blockchain is suggested as a distributed database to share the knowledge on land ownership, and to share geodata in an open way, for example, public map, without relying on a central authority, for example, Google Maps or OpenStreetMaps (Mire, 2018).

Pincheria et al. (2020) proposed a blockchain-based system to share and retrieve data without the need for a central authority. The proposed architecture (*a*) allows sharing data, (*b*) maintains the data history (origin and updates), and (*c*) allows retrieving and evaluating the data with added trustworthiness. It is shown in Figure 14.7. The data are received from different sources that are trusted, for example, space agencies and universities, or untrusted, for example, private companies and volunteers. They consider as untrusted actors those that voluntarily collect and share

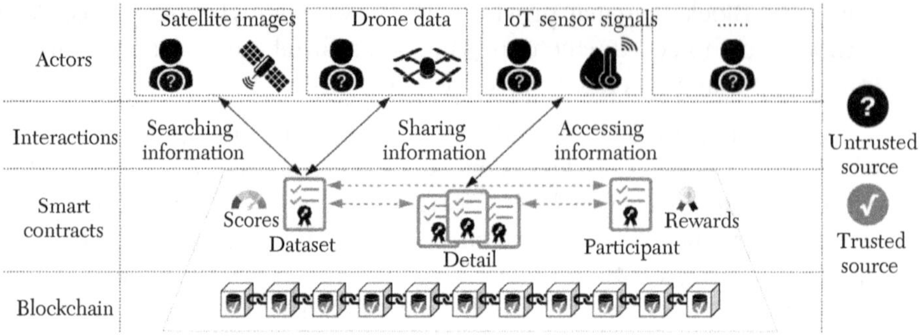

Figure 14.7. Structure of the proposed blockchain architecture (Pincheria et al., 2020).

the data. Some examples of common actors are: the company sharing data acquired by IoT sensors on the temperature and the humidity of a crop field; the research group sharing a database on multi-temporal images acquired by optical and SAR sensors; and finally, the people that share pictures taken by a drone or a cellphone.

Blockchain together with IoT can bring big changes in the smart city. It can increase the speed of public services as well as widen their capabilities, and in some cases even reduce the costs. To enhance the interest in the use of blockchain for public services, many cities of the world have started to develop blockchain applications (Moore et al., 2018). However, almost all the blockchain applications currently developed are isolated and no city has a clear strategy as to how to integrate them within a complete blockchain-based governance model for providing efficient public services. There are also efforts to make policing, security, governance, healthcare, and public services more autonomous through blockchain technology and AI. Figure 14.8 presents twelve important applications where blockchain may serve as the interoperable platform that gives citizens of smart cities a greater flexibility (Mire, 2018).

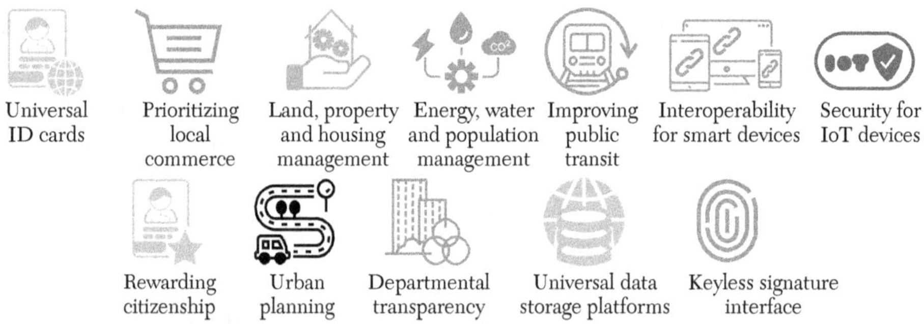

Figure 14.8. Twelve important applications of blockchain in smart cities.

Open data policies like blockchain are also expected to stimulate local software developers and innovators to find patterns in the data which may lead to development of better solutions that offer conveniences and better public services for the citizens who depend on them. The blockchain-based governance model could be used for public services in smart cities ranging from clean environment, transportation, the energy to healthcare, and their performance evaluated from technical, legal, social, and economic perspectives. Dubai, which aimed at creating the world's first blockchain-powered government, has an autonomous transportation strategy that seeks to make 25% of all transportation in the city autonomous by 2030 (Vince, 2020).

14.4 CASE STUDIES

Two case studies are presented here:

14.4.1 Extraction of Urban Area Using Deep Learning

In the past few decades, urbanization has occurred at a rapid pace worldwide, mainly to steep population growth. Infrastructural growth has resulted in the increase of built-up areas (BUA). Information about BUA serves as an input to smart cities development as well as modeling the impacts of urbanization on agriculture, water, temperature, and carbon cycles for planning sustainable development. The composition of any BUA consists of multiple classes, such as open area, vegetation, buildings, and roads. Therefore, it is still a challenging task to achieve precision in extracting the BUA from satellite images by employing existing built-up detection approaches. Frequently available remote sensing datasets and advanced ML techniques provide an effective way for extraction of BUA, which are not only cost-effective but require the least human effort. These methods require labeled data for training the model where model parameters are iteratively learned.

Brahme et al. (2018) used several well-known ML techniques like k-nearest neighbour (k-NN), ANN, and SVM with spectral and spectral-textural features using multispectral IRS LISS-IV, Sentinel-2A images for extraction of BUA in Haridwar *tehsil* situated in Uttarakhand State, India, covering an area of 1166.5 km^2. The Deep Learning (DL) classifiers are well known to the computer vision community, as these methods give impressive results in the field of pattern recognition. This study also examines the potential of CNN for the extraction of BUA using high spatial resolution satellite data. The complete methodology used is presented in Figure 14.9.

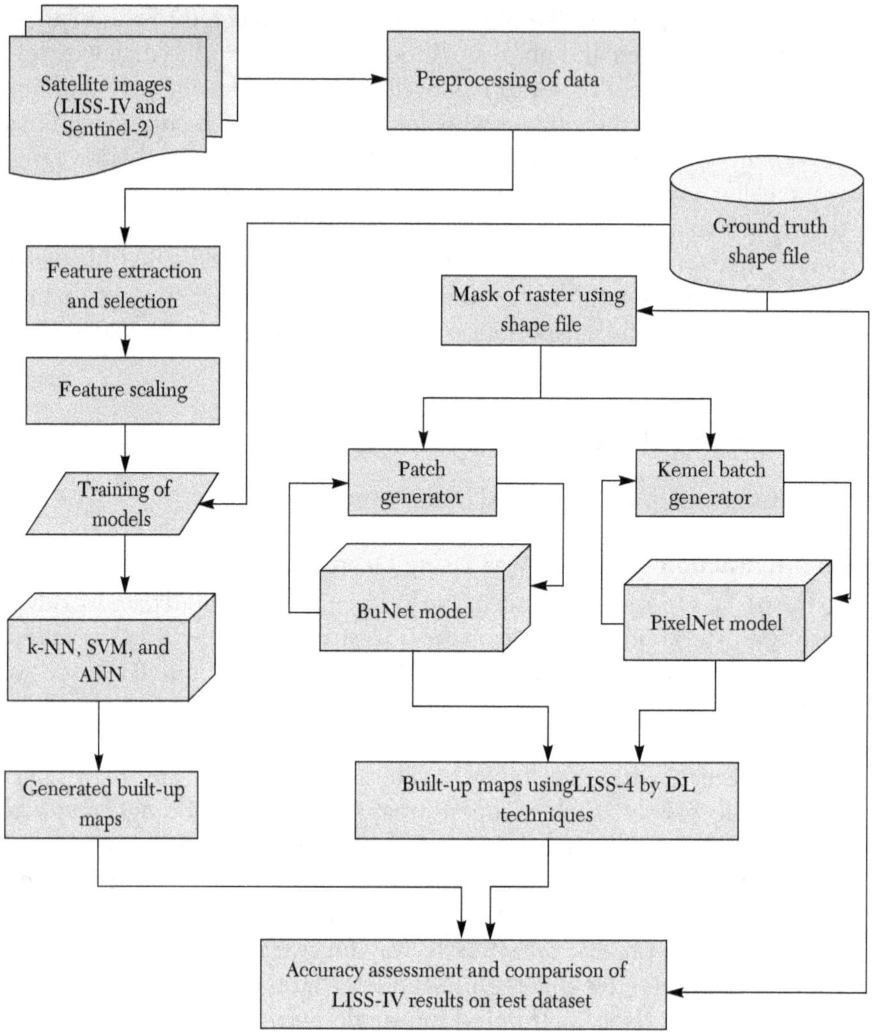

Figure 14.9. Flow diagram of proposed methodology for built-up area extraction.

The ML techniques used spectral features as input to extract the BUA from both Sentinel-2A and LISS-IV datasets. Since these algorithms can perform better with additional spatial features, which are able to delineate classes accurately in the feature space, the gray-level covariance matrix (GLCM) textural features are generated. These spectral-textural features are used to train k-NN, SVM, and ANN models. Finally, BUA maps of test images are generated using trained models, and accuracy metrics are calculated to access the performance of the classification.

The proposed DL models (i.e., BUNet and pixelNet) used LISS-IV data only, mainly due to its higher resolution (Brahme et al., 2018). Training of both DL models has been done by generating training data from feature masks obtained from ground truth shapefiles. Once the models have been trained, test images are classified using these models. The accuracy of predicted data is compared with the reference data in terms of statistical measure, such as overall accuracy, kappa coefficient, and F-1 measures. The results of BUA from deep pixelNet and BUNet models are shown in Figure 14.10.

Incorporation of textural features with spectral features provided approximately 3–5% better overall accuracy than the spectral features alone, in the case of Sentinel-2A data. However, an improvement in overall accuracy of approximately 4–5% is obtained for the LISS-IV dataset using spectral-textural data. The maximum rise of about 5% average accuracy has been achieved for the k-NN classifier using Sentinel-2A. However, for the LISS-IV data, ANN achieved a maximum increase of 5.08% in average accuracy. It has been observed that the use of spectral-textural features also resulted in an increasing F1-score for all the classifiers. The highest increase in overall score for Sentinel-2A is 0.113 using the k-NN classifier, whereas a maximum rise of has been observed for LISS-IV data using linear-SVM. The improvement in overall accuracy by radial basis function (RBF)-SVM, in comparison to k-NN, using Sentinel-2A for spectral and spectral-textural features is 9.36% and 7.33%, respectively. Similarly, RBF-SVM resulted in improving an overall accuracy of 10.72% for spectral features and 10.92% for spectral-textural features, in comparison to k-NN. It has been observed that the impact of texture resulted in increasing classification accuracy on the places where neighborhoods around the center pixel have similar characteristics.

(a) (b) (c)

Figure 14.10. (a) LISS IV image, (b) results of BUA from deep pixelNet, and (c) BUNet model (Brahme et al., 2019).

In the case of the deep pixelNet model, a kernel size of 5 × 5 provided minimum losses of 0.012 and 0.013 for cross-entropy and Jaccard dissimilarity, respectively. However, a Patch size of 28 × 28 gives a minimum cross-entropy and Jaccard dissimilarity value of 0.007 and 0.010, respectively for the BUNet model. The mean overall accuracy obtained using deep pixelNet and BUNet are 91.41% and 93.69%, respectively, which is higher than all ML classifiers. It suggests that the BUNet model outperforms all other classifiers in extraction of BUA from LISS-IV images. The second best results are provided by the deep pixelNet model. An increase of greater than 5% overall accuracy has been observed by proposed CNN models in comparison to other ML models (Brahme et al., 2019).

Althugh results are satisfactory, the scalability of proposed deep pixelNet and BUNet models for other classes and other datasets are also required to be explored further. Development of data visualization technique for DL models is required, as DL models consist of multiple layers and the tracking of change in weights or forming of classes while training is very difficult. Also, learning DL models from scratch requires a lot of training data and computational efforts. Such maps could be used to assess the present area under urbanization, and other remaining areas for better planning of smart cities.

14.4.2 Mapping Urban Dynamics

With industrial and economic development, urban sprawl resulted in the conversion of natural forests and vegetation to urbanized regions with highly built-up areas and infrastructure. Urbanization refers to lateral and vertical growth of urban pockets as a result of population growth, industrialization, political, cultural, and other socioeconomic factors. The relation of urban expansion and temperature patterns of cities varies spatially and temporally. The UHI (urban heat island) effect on cities can be determined in the horizontal extremities of cities, while the UGS (urban green space) concept helps in the study of LST (land surface temperature) in relation to the expansion of cities with the effect of building heights and other related important parameters, like density of buildings or greenery and nearness to patches of buildings and greenery (Verma & Garg, 2021). The distribution pattern of cities and their growth have been studied through open-source coding (traditional SLEUTH or its modified versions) and different neural networks coupled with cellular automat-Markov chain (CA-MC) using satellite images. Individual building-level extraction of parameters, such as access to greenery, LST, or pollution, help in planning cities for reducing the effect of UHI and also site selection for smart city planning and maintaining better health conditions of city dwellers. Figure 14.11 presents a flowchart for studying urban sprawl with LST for UHI (Verma & Garg, 2021).

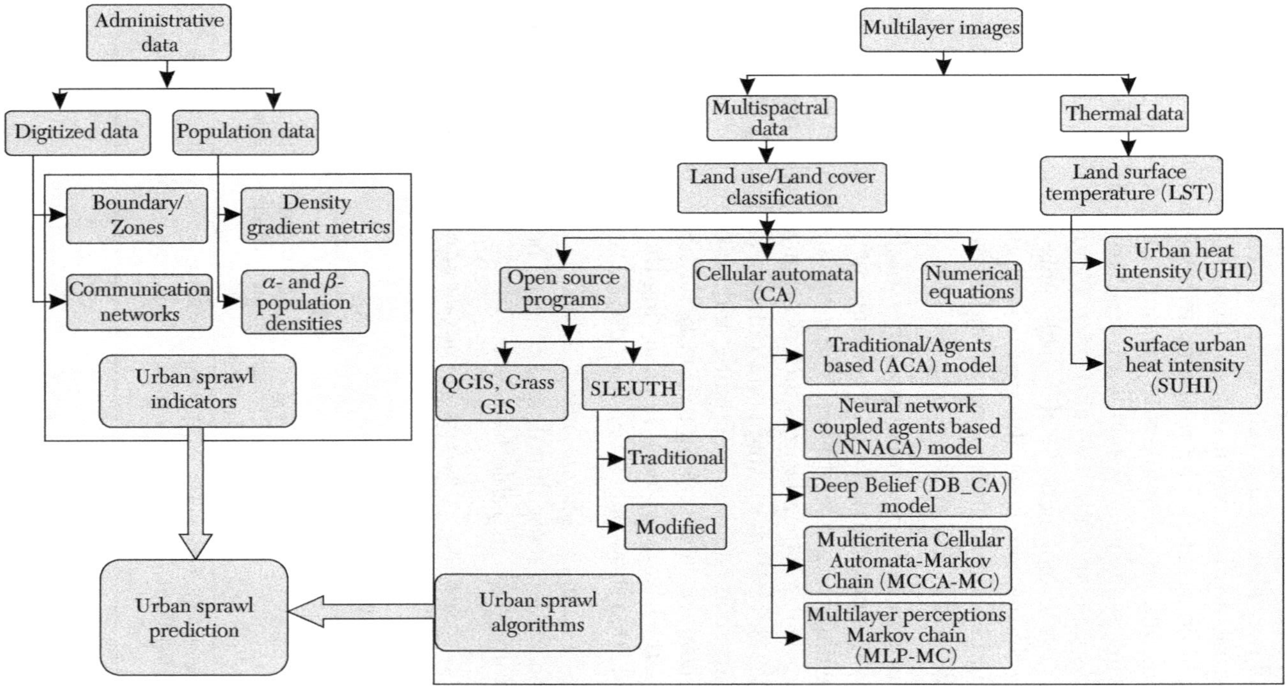

Figure 14.11. Flowchart for studying urban sprawl with LST for UHI (Verma & Garg, 2021).

Urban green spaces may include places with natural surfaces or natural settings, but may also include specific types of urban greenery, such as street trees, and may also include blue space, which represents water elements. The concept of urban green space, urban blue space, urban heat islands, urban pollution islands, and urban growth measurement techniques help in defining various parameters to measure the pressure over the area by the growing population (Chaudhuri & Clarke, 2013). Improving access to green spaces in cities has also been included in the United Nations Sustainable Development Goal, which aims to provide universal access to safe, inclusive, and accessible green and public spaces, in particular for women and children, older persons, and persons with disabilities by 2030 (WHO, 2016).

Urban sprawl in metropolitan cities of India is a great concern for planners. A study carried out by Verma et al. (2019) attempts to help understanding the gap by studying the urban sprawl of Lucknow city, India, through urban growth indicators such as urban growth types (infilling, edge-expansion, and outlying), shannon's entropy value, and landscape metrics (complexity, compactness, centrality, and porosity) from year 1985–2017. Decadal land use land cover for years 1985, 1995, 2005 and Landsat-8 data for year 2017 have been used in achieving this. Classified images of 1985, 1995, and 2005 show only seven land use classes of Lucknow city.

These seven classes are merged into five classes, namely: built-up, vegetation, water, agricultural, and other. The landscape metrics used in the study are given in Table 14.1.

Table 14.1. The Landscape Metrics and Relationships Used

Landscape metric	**Formula**
Landscape Shape Index (LSI; LSI is one if patch is square but it increases without limits edges increase in patches)	$LSI = \frac{0.25E}{\sqrt{A}}$; Range ≥ 1 without limit where, E is A sum of edge of patch (true edge or segment of edges doesn't matter) and A is total area of landscape.
Number of Patches (NP; total number of adjacent patches if patches of class alike)	$NP = n_i$; Range ≥ 1 without limit where, n_i is number of patches for class i in landscape.
Mean Euclidean Nearest-Neighbor distance (ENN_MN; it increases when distance between respective patches keep increasing)	$ENN_{MN} = \frac{\sum_{j=1}^{n} h_{ij}}{n_i}$; Range ≥ 1 without limit where h_{ij} is Euclidean distance between patches i and j and n_i is total no. of patches present in landscape.

Landscape metric	Formula
Aggregation Index (AI; it is 0 when patch have no like adjacency and reaches to 100 when landscape consists of single patch)	$AI = \frac{a_{ij}}{a\_\max_{ij}}$; Range from 0 to 100 where a_{ij} is number of like adjacencies existing between pixels of patch j to that of i and $a\_\max_{ij}$ is maximum no. of like adjacency existing in landscape.

Results of the study (Verma et al., 2109) show that the city has gone under a large amount of urban growth between 1985–2017. The amount of edge expansion during the 2005–2017 period is decreasing from the previous duration of changes in the years 1985–1995 and 1995-2005. The outlying growth of the built-up area, showing formation of new patches, increases 120.78 km^2 from 1985–2017, which was at first cycle (1985–1995) caused by a decrease in vegetation and other land use classes, but after that in 1995–2005 and 2005–2017, it is mainly due to a decrease in agricultural land. Among landscape metrics, NP (number of patches) and LSI (landscape shape index) show an upward trend from 1985–2017, indicating an increase in compactness and complexity of patches of built-up areas, but a decrease in ENN_MN (mean Euclidean nearest neighbour distance) and AI (aggregation index) shows a negative trend. Shannon's entropy value (H_n) is used for analysis for urban sprawl in the study area. In all the years (1985–2017), Lucknow has an entropy value log(n) pf 1.609. An overall accuracy of classification for years 1985, 1995, and 2005 was found to be about 94% with a kappa coefficient of 0.94, and for year 2017 as 84.85% with a kappa coefficient of 0.8138. The entire study area is found to be 78.6% urbanized in year 2017, which is a problematic situation for planners to counteract the ever-increasing population in the city by providing settlements and other amenities. The sudden presence of outlying built-up patches in duration 2005–2017 indicates the start of more dispersed growth in an already urbanized area. Dispersed urban growth will be helpful in mitigation of several factors of urban dynamics, like the UHI effect and better access to greenery in the form of UGS, but will also make it difficult to lay down the communication networks all over the city for municipal corporations or planning departments of cities.

14.5 CONCLUSION

Cities are referred to as the engines of economic growth. Accordingly, there is a growing need for the cities to get smarter in order to handle the large scale urbanization as well as finding new ways to manage complexity, increase efficiency, reduce expenses, and improve quality of life (Vince, 2020). The smart cities should be able to provide good infrastructure facilities such as water, sanitation, utility services, education, healthcare, good governance, and so on, while the increasing role of geospatial databases and technologies enhance the citizens' experience (shopping or healthcare or mobility). The geospatial technologies help in providing possible solutions for fundamental issues like water, environment, energy, sanitation, housing, waste management, and mobility for smart cities. With the unprecedented rate of urbanization, new geospatial tools and technologies can help managing the complex challenges of smart city systems, while maximizing the services to its citizens (Simpson, 2018).

The smart cities require modern digital technology to make urban systems more efficient, cost effective, and environmentally sustainable. These cities are considered to be the centers of innovation and creativity, but they face great challenges, such as migration of people to urban centers, rapid urbanization, increased pollution, climate change, impact on flora and fauna, and increased pressure on city resources, such as water, food, housing, transport, and healthcare (Sarkar, 2019). The ever-increasing consumption rate of physical and social resources may affect the sustainable growth of cities. The ICT, IoT, GPS, remote sensing, various sensors, and GIS are the key catalysts in the evolution of smart city applications. With the advancements of these technologies, cities are deriving benefits from connecting people, processes, and data.

Ubiquitous connectivity and enhanced use of sensors and digital devices are contributing to the development of smart urban infrastructure. In addition, tools are being developed to better monitor performance, detect patterns, predict trends, and visualize information with improvement in analytics and cognitive intelligence. In the future, IoT-based sensors will collect huge data uninterruptedly, and computers will augment visual insights, automating the process of collecting observations, counting traffic, identifying cracks on bridges, recognizing inventory on construction sites, and applications where human intervention is not required (Williams, 2018). Sensors will get smaller, better, and cheaper, and satellite images will be available at much higher resolution, which will allow more applications with higher productivity. It is expected that more machine learning-enhanced services will emerge, ranging from governance to the service industry.

CHAPTER 15

THE FUTURE OF REMOTE SENSING

15.1 INTRODUCTION

Having up-to-date Earth information is a foundation for prosperity. Urban planning, natural hazards management, precision farming, climate study, weather forecast, energy use, and optimum utilization of natural resources all require reliable and detailed information about the Earth's dynamic environment. Due to technological advances in remote sensing sensors, and the computing capabilities to combine and analyze sensors' information and the communications tools for sharing this knowledge, it has become possible to collect the latest and near-real information (Gail, 2007). Besides data collection and mapping, applications range from autonomous vehicles to smart city implementations, human/machine collaborative manufacturing to disaster and environmental management, medical and geriatric assistive-functions to defense operations and surveillance, and so on. For example, city municipalities are the emerging users of remote sensing information, particularly for smart city solutions as well as monitoring and simulating scenarios related to flooding events while mitigating the risks and increasing the infrastructure resilience. For the past three decades, digital geospatial data is increasingly being used by governments, organizations and businesses to make important decisions (Walter, 2020).

The commercial remote sensing industry addresses a variety of global applications including food security, global conflict, environmental issues, land sustainability, and more. This evolution can primarily be attributed to technological developments in the remote sensing industry as well as development of critical technologies such as increased computing power, mobile technology adoption, efficient distributed computing and dissemination, advanced pattern recognition technologies for medical imaging, robotics and machine learning, and others. As a result, mass-market consumer applications are emerging that are expected to radically transform the use of the Earth's information over the next decade. The transformation of the industry by remote sensing approaches can be captured by four trends, resolution, accuracy, speed and analytics, as shown in Figure 15.1 (DigitalGlobe, 2020).

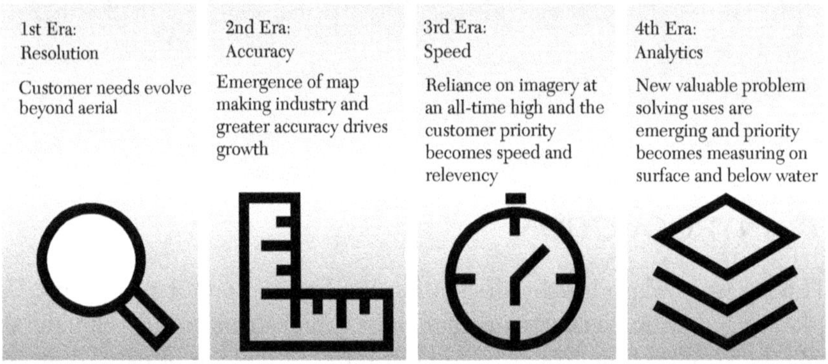

Figure 15.1. Four trends in the remote sensing-based industry.

Remote sensing provides tools for gathering a large quantity of repetitive data and solving real-world problems. Data collected by remote sensing systems can provide more accurate and up-to-date inputs for a variety of models. Advances in sensors and data analysis techniques are making remote sensed data extremely useful in monitoring and modeling the changes in the Earth's ecosystem. The speed at which innovation occurs represents great opportunities and challenges to use remote sensing technology. Automation, AI, and connectivity through 5G is expected to have the greatest impact over the use of this technology (Jung et al., 2021). The variety and amount of remotely sensed data available has increased dramatically in recent years, and it is expected to continue to increase in the future. In last ten years, freely available images for analysis from multiple public and free remote sensing platforms have completely enhanced and changed the way the data were utilized for various applications.

Climate change and naturally occurring hazards have more of an impact on developing nations. These nations often are not in a position to afford costly high-resolution data for analysis. Today, freely available data sources can effectively be used by the developing nations to monitor, study, and model the characteristics and impacts of events such as flooding, landslides, and other phenomena. The ability to collect data at higher resolutions and enhance the resolution of data through data fusion techniques as well as use of AI processes will benefit the developing nations to use the data for various applications. Combined with AI and computational capabilities, developed and developing nations will witness a productivity increase in the processes of data creation, maintenance, and management (Jung et al., 2021).

15.2 FUTURE APPLICATIONS

There are a large number of highly important applications for atmosphere, ocean, and land that currently rely upon data from various satellites (Dubovik et al., 2021). Remote sensing-based observations of the atmosphere are used for weather prediction, monitoring of environmental pollution, climatic change, and so on. For ocean surfaces, remote sensing is used for monitoring coastline dynamics, sea surface temperature (SST) and salinity, ocean ecosystem and carbon biomass, sea level change, marine traffic and fisheries, mapping of water current and underlying topography in shallow waters, and so forth. Remote sensing of the land greatly contributes to the exploration of resources, monitoring of floods and droughts, soil moisture, vegetation, deforestation, forest fires, agricultural monitoring, urban planning, and so on. The social science survey for global crises (such as the COVID-19 pandemic) are also benefiting from satellite remote sensing datasets that create visualizations to classify human environments and then relate these observations to various socioeconomic datasets, and so forth (Diffenbaugh et al., 2020).

The rapid growth of remote sensing satellites is generating a massive amount of data for Earth-related applications. The science and technology of remote sensing has reached the era of big data. This brings a paradigm change in the way remote sensing data is processed to extract the information for a variety of environmental and societal applications (Xu et al., 2020). The extraction of information from large data derived from various satellites, regardless of the application, can be classified as big remote sensing data, while simultaneously meeting the volume, variety, and data growth rates. The challenges involved in handling the big data have led to the emergence of cloud-based platforms. The classification approaches that are implemented for large geographic regions use cloud-based platforms, and generally reduce the dimensionality of the input satellite data.

Table 15.1. Directions of Future Research in Urban Sustainability (Kadhim et al., 2016)

	Integration of heterogeneous data	Algorithms to identify urban features	Improve accuracy for spectral classification algorithms
Objective	Improve the spatial and spectral resolution. Improve the classification accuracy. Enhance the ability of features detection and display. Promote the geometric precision. Increase the capability of the change detection. Refine, replace, or repair the defects of image data. Handle multisource remote sensing data at the pixel, feature, and decision of level fusion.	Accelerate future processing and improve classification accuracy. Automated processes for detecting, extracting, simulating, classifying and modeling urban features. Capability of handling and fusing the large number of datasets. Improve image object segmentation. Increase the reliability and precision of the feature extraction. Mitigate the ambiguous and uncertainty.	Capability of separating urban LULC classes in an adequate manner. For a better visualization and interpretation of urban landscape. Performing a change detection analysis and pattern recognition. Accurately characterize the model parameters for different urban remote sensing applications.
Problems requiring solutions	A rescaling of multi-source data of divers EO instruments. Assessing the distortion of spatial and spectral resolution. Time consuming and subjective. Complexity and/or availability of large training datasets for deep learning features. Manual or semi-automated post-processing process. Computation efficiency and effectiveness. Quality of distinguishing features. With fusion schemes, the optimal combining strategy of current fusion algorithms is a challenging task and promising research that requires further investigations in the near future.	Further developments for fusing LiDAR data with thermal, multispectral, and hyperspectral imagery. Reliable determination of boundaries of urban objects in an automated manner. Further improvements for addressing different characteristics of EO data. Ability to cope with unpredictable environmental and illumination factors of diverse datasets. Similarity in the characteristics of spectral, textural, and geometrical-based between the urban objects and their background. Problems in occlusions and relief distortions. Computational time to perform a task.	Mixed pixels. Uncertainty in urban LULC classes. Mixed objects. Promote pixel-based and object-based classification using contextual information. Over-fitting can cause speckled results. An automatic labeling strategy is required for actual label sets in several applications. Further refinements for the fusion of diverse data sources. Integrate the derived urban features (e.g., building shadows) and/or spatial matrices with various classifiers. Investigate the recent computer vision techniques for improving the accuracy.

The geospatial industry has undergone significant changes in terms of map generation technologies, business models, and user requirements. The scope of geospatial data using remote sensing images is vast, such as geography, ecology, tourism, marine sciences, agriculture, forestry, marketing and advertising, military forces, aircraft, law enforcement, logistics and transportation, astronomy, demography, healthcare, meteorology, and many others. Table 15.1 presents the challenges and/or opportunities in three key research trends in urban sustainability (Kadhim et al., 2016): (*a*) the integration of heterogeneous remote sensing data; (*b*) algorithms for extracting urban features, and (*c*) accuracy of urban land-cover and land-use classification. It highlights the main benefits and limitations in each research trend for further investigations in the future.

15.3 CHALLENGES AND PROBLEMS

A big challenge in the advancement of satellite remote sensing is discovering innovative and affordable technologies and measurement concepts to take up new applications (Dubovik et al., 2021). While single image data can provide significant information, combining several image data sources will often increase and improve the information content. This is especially important as the Earth's systems are quite complex and interconnected. The National Research Council (2010) had identified ten research challenges using remote sensing data. These areas include, visual analytics, integrating sensors, human terrain/behavior, participatory sensing, improved models of space-time, development of new paradigms for conveying certainty, improved geodetic, photogrammetric, and remote sensing positioning, geospatial information retrieval and extraction from text, database technology, and spatial data infrastructure and geospatial narratives.

Various remote sensing data products are available for site-specific applications, but their resolutions vary from each other. For example, the assessment of a forest fire would require high spatial and temporal resolution data, but this is not possible from one sensor. If MODIS data is utilized for assessing the forest fire, it provides high temporal resolution of one day but will have poor spatial resolution. Similarly, real time flood assessment needs high temporal and spectral resolution with cloud free images for assessing the water spread and damages. Hyperspectral data is required to be specifically enhanced for better spatial resolution to get enhanced results. This will also reduce the processing time for various spectral bands. Hyperspectral remote sensing data with high spectral resolution can provide promising solutions to the future challenges of remote sensing technology. However, these images need large storage and heavy processing support systems which might limit their use for digital image processing.

While there are many opportunities for using remote sensing in earthquake-related studies and other disaster studies, challenges remain that inhibit its widespread use (Rathje & Adams, 2008). These challenges are summarized in Table 15.2, with specific emphasis on the use of satellite imagery. Various challenges are related to the simple process of acquiring the appropriate imagery and gaining access to it. A satellite may not be available to capture an area shortly after an earthquake. For optical imagery, cloud cover is also another issue. Nonetheless, with more satellites and improved sensors being deployed, it may be possible to overcome most of these problems. Currently, image access and cost are some issues when using imagery for rapid response to an earthquake, and also it may take several days for processing the satellite images before the data are given to administrators. Although the prices of satellite images continue to be cheaper, it will always be expensive to procure high-resolution images over a large area. Additionally, the large quantity of data can be difficult to analyze when dealing with such images. In this case, low-resolution and medium-resolution images are used to identify the heavily damaged areas across the region, and high-resolution data are used for more detailed analysis of such areas (DigitalGlobe, 2020). The International Charter for Space and Natural Disasters (*http://www.disasterscharter.org/*), which provides access to remote sensing data for natural and man-made disasters, is helpful in this regard. However, high-resolution images and current images are not available here, and current images have only been available to governmental agencies.

Table 15.2. Challenges for the Use of Remote Sensing in Earthquake Engineering (Rathje and Adams, 2008)

Challenge	Issue	Remedy
Image acquisition	Satellite location. Acquisition angle. Cloud cover.	Improved sensors. More satellites.
Image access	Time for users to obtain data. Cost of data.	Work with satellite companies.
Data quantity	High-resolution data too massive for large regions.	Tiered approach using low-, medium-, and high-resolution data. Improved computational resources.
Experience of users	Many earthquake professionals are unfamiliar with remote sensing data.	Education. Use of visual products.

Challenge	Issue	Remedy
Damage level resolution	Can only identify collapse or partial collapse, larger failures.	Improved sensors. Integration of field data.
Damage scales	Many different scales and approaches are used.	Consensus within user community.

There is always a trade-off between spatial and temporal resolutions as well as between spatial and spectral resolutions. High temporal resolution data usually have low spatial resolution (such as MODIS), while low spatial resolution can hardly discriminate smaller objects on the ground, resulting in lower classification accuracy (Morisette et al., 2002). In general, finer spatial resolution increases the classification accuracy, but at the same time, smaller pixels increase the spectral variance, resulting in decreased spectral separability of classes. Data with both extensive spatial-temporal coverage and high spatial resolution is desirable for many applications, but it is highly challenging (Dubovik et al., 2021). Hence, the design of satellite observations may need new innovations, ancillary data, and synergies of complimentary observations to address the specific objects and relevant problems.

Remote sensing data are normally limited by their short time span availability, so, their contributions to modeling future projections (such as climate change) are limited. However, current archives of free remote sensing data may provide important baseline information for such studies. New sensors with high temporal resolutions will become an integral part of monitoring the Earth's surface. The use of free remote sensing data with global coverage can also be challenging, particularly for developing countries, where data processing, storage, and sharing are still hampered by information technology and archiving capability (Global Open Data, 2016). To fully utilize the potential of remote sensing data, expertise in data processing and new algorithm development is also required. However, in recent years, this has been facilitated to some extent by available open-source algorithms and software.

The availability of low-cost, high quality, high frequency satellite data has contributed to the ever-increasing volumes of data (Parente et al., 2019). Further, the use of low-cost autonomous drones/UAVs and remotely controlled UAVs can be used in several real time applications. But there are substantial regulatory and economic challenges that limit their adoption. The operating rules for UAVs need to be further relaxed, keeping in view the safety and security of the people, infrastructure, and country. Once these challenges are resolved, UAVs will find their use in many applications, and will be considered as smarter systems to collect information from space in inaccessible and not economically viable areas.

The disciplines associated with the use of geospatial tools and technology have diversified significantly over the last decade. While geographers, cartographers, GIS analysts, remote sensing analysts, surveyors, photogrammetrists, and Earth observation scientists will constitute as the dominated workforce, the range of geospatial career paths has diversified, incorporating experts from other domains also. The increased use of computers in imaging and geospatial tools and technology has tremendously contributed toward growth in disciplines such as big data, data science and analytics, computer vision, and data visualization. Data science is an important component of disruptive technologies, namely AI, machine learning and deep learning, cloud computing, sensor networks, or blockchain. It is estimated that around 50 percent of the working population in many parts of the world will need reskilling in the next five years due to the impact of this digital transformation (Walter, 2020). Due to rapid changes in digital methods, this is a big challenge for companies/industry and governments to meet the growing demand in the geospatial area. The lack of attention might contribute to hampering economic growth.

At the moment, many of the benefits of blockchain are theoretical. The Earth observations can focus efforts first on clearly identifying the problems

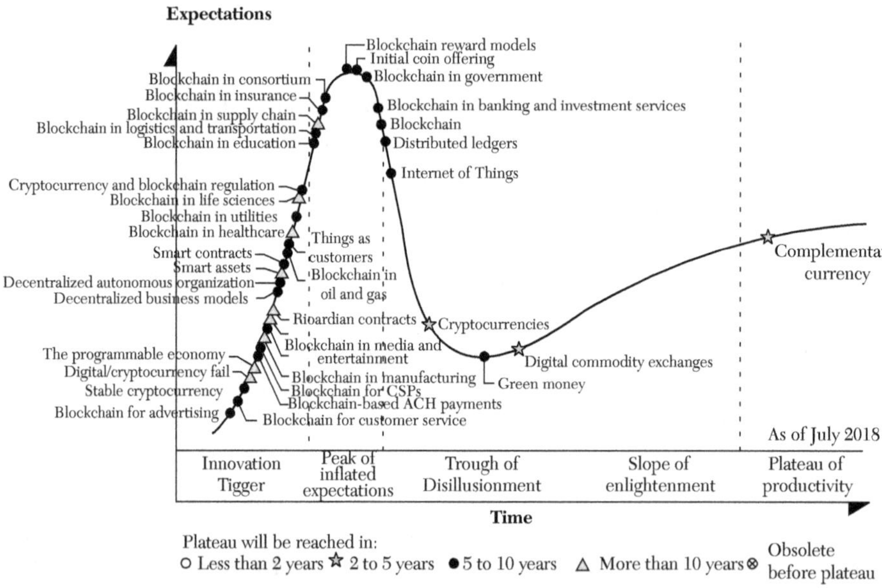

Figure 15.2. Hype cycle for blockchain business, 2018 (Kandaswamy & Furlonger, 2018).

to solve, and then devising appropriate solutions (ESA, 2019b). It is therefore important to navigate the expectations associated with distributed ledger technology carefully, and to evaluate the maturity and state of adoption of distributed ledgers, including their limitations, as they move through the "hype cycle" (as shown in Figure 15.2). Overall, it has been well established that various blockchains can be difficult to work with and scale, expensive to maintain, controversial from the environmental sustainability perspective, and hard to upgrade. The four most widely quoted limitations of blockchain are cost of creating and using such systems, scalability, governance, and security of private key management.

15.4 OPPORTUNITIES

The increasing availability of high-resolution satellite imagery has transformed the use of remote sensing by improving accessibility and frequency, thus enabling better evidence-based decision-making and service delivery (DigitalGlobe, 2020). Today, the volume, size, speed, and diversity in which remote sensing data is generated requires changes in (*a*) approaches to process these data; and (*b*) workforces that are capable of searching, merging, and analyzing these large amounts of data in almost real time (Walter, 2020). Expertise in consolidating large numbers of data sources, understanding mapping requirements, and new toolsets developed for automated map creation will be critical for the future.

Remote sensing can be used to increase the prediction capabilities of several models. Results from remote sensing-based models require calibration and validation, which may be a complex task. All mathematical models don't have compatibility with all the sensors, and these also don't work properly in all geographical locations. Therefore, validation of the model is essential with in situ data. The ground data can also be utilized for model calibration/improvement through its integration with remote sensing data. This will improve the accuracy of results obtained by remote sensing based models.

The IoT, AI, neural networks, cloud computing, blockchain, and so on play an important role in acquisition and processing of large data (big data). Due to exponential growth in sensing platforms, the geospatial sensors started to produce an ever-increasing amount of big data. For processing these data, cloud computing has gained dominance. Geospatial data is one of the big beneficiaries of the ongoing developments in big data (Parente et al., 2019). Data streams at petabyte (10^{15} bytes) and exabyte (10^{18} bytes) scales are becoming common and widely available via warehouse-scale computers and the Internet. Figure 15.3 shows the trend of information creation worldwide. Besides the highly automated information processing, the real potential of harnessing the big data is the capability to extract additional information that has not been feasible in the past. For example, using temporal big remote sensing data, a correlation between the spread of disease and vegetation condition can be established in a region which is affected by some health epidemic. Besides mapping, big geodata are expected to contribute to society by offering massive datasets on human behavior containing information relevant to understanding the big problems being faced by society and individuals today.

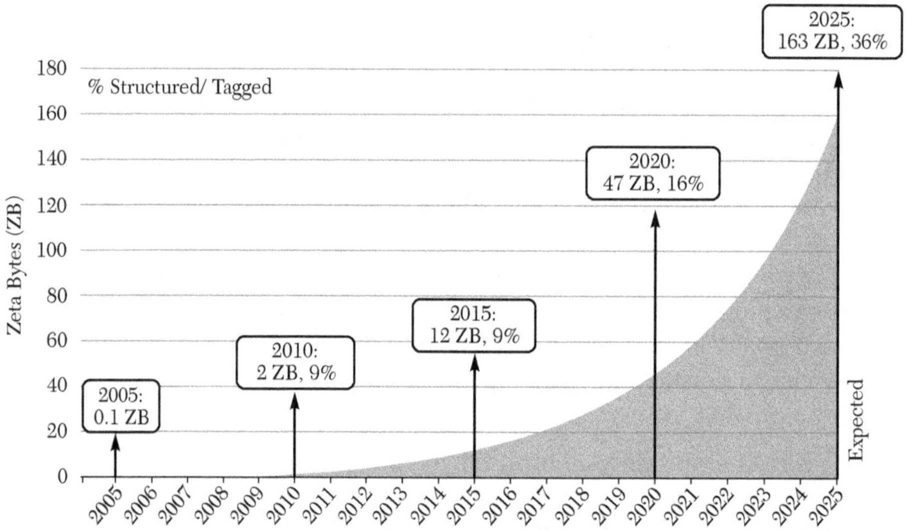

Figure 15.3. Increasing data creation worldwide.

For collection and making use of geospatial data, there are different national legal systems and levels of maturity of national geospatial information management policy. Governments are increasingly developing national geospatial strategies or master plans to access geospatial data for social and economic benefits. These policies aim to foster economic growth, generate employment opportunities, and combat societal challenges related to climate change, floods, calamities, urbanization, and resources and infrastructure management (Walter, 2020). The digitization process around the globe is expected to provide new opportunities for knowledge creation and sharing, data sharing, collaboration, and interaction between various stakeholders. Many nations have already estimated the potential economic value to be derived using the relaxed geospatial policy, and therefore they have implemented it.

15.5 TECHNOLOGICAL DEVELOPMENTS

Digital transformation has been instrumental in mapping information, and geospatial technology is available to all by using location-enabled devices/sensors. In an interconnected world, integrated geospatial information can be used for many works such as location-based services through smartphones, the development of self-driving cars through the IoT, and satellite sensors (Alavi et al., 2018). GPS-based devices are now widely employed for car navigation and location-based services. Mobile phones now provide a ubiquitous distribution platform for location-based information. Mobile data collection, crowdsourcing, and social media enable accurate and near real time applications that are increasingly adopted by various users of geospatial data. Internet maps and virtual worlds (such as Google Earth and Microsoft Virtual Earth) can be used for a large number of applications, such as tourism planning, reconnaissance survey, real estate searching, forest planning, and so forth. It is anticipated that these forms of data collection are likely to have the greatest impact over the coming decade.

Driven by the technological developments, sensors have become ubiquitous and can be found in almost every facet of our lives. In particular, imaging sensors have shown dramatic improvements in performance, providing a broad range of spatial and temporal resolutions such as panchromatic, multispectral, thermal, microwave, and hyperspectral bands. These developments have also given rise to a number of remote sensing platforms, which are now available from space to air to ground to sub-surface. Several trends are expected to continue in the future, such as decreasing the size of sensors, increase in computing power, increase in data size, and so on. As technology advances, the spatial, spectral, radiometric, and temporal characteristics of satellite sensors will further improve. The developments in remote sensing sensors (e.g., consumer grade sensors) and platforms (e.g., small mobile platforms) are continuing, which make them less expensive. For example, autonomous and assisted vehicle technology would require simple and reliable sensing capabilities. In comparison, optical sensing has become extremely advanced, mainly due to widespread smartphone applications. With the growing demand, novel sensor approaches are also likely to appear. One possibility is "interactive remote sensing," such as farmers genetically "tagging" their crops to enhance the remotely detectable spectral signature for crop distress or optimal harvesting.

Remote sensing tools support daily activities vital for humanity (Dubovik et al., 2021). The satellite sensors and the quality and quantity of information collected by satellites is constantly improving. The enhanced spatial and temporal resolutions as well as fast data processing approaches will further increase the value of data. In the future, the observations may be increased by deploying the satellite sensors with enhanced capabilities as well as through combined observations of passive sensor images with active sensors' vertical profiling of the atmosphere and hyperspectral spectrometry, combining observations of different sensitivities obtained in different spectral ranges, or at different time or spatial scales. The next generation state-of-art data processing approaches, such as deep learning and neural networks that rely on rigorous forward modeling and numerical inversion methodologies, and consider extensive sets of state parameters, are required to be developed in future (Parente et al., 2019). Achieving continuity in consistent satellite observations and long-term datasets by assuring sufficient compatibility and agreement of past, present, and future datasets are also required.

In general, achieving better mobility and autonomy drives the developments in platforms for data collection. Piloted aircraft for data collection are generally more expensive than other platforms. Small unmanned aerial vehicle (UAVs) are lightweight under 55 pounds, and are designed for commercial and civilian uses. Micro-UAVs are even further smaller systems that weigh less than 4.4 pounds (Zolnai, 2016). The UAV systems have shown remarkable progress in terms of performance, reliability, increased capabilities, and ease of use. The enabling technology in UAV is remote sensing combined with advanced data processing methods. The high spatial revolution and high temporal resolution make the UAV an ideal platform to use for various applications as well as conduct a variety of research in remote sensing. In coming years, the UAVs will establish themselves as a viable alternative to conventional mapping for small area surveys.

Another trend in development is the smaller satellites which are launched into lower orbits. In the 1990s, the commercial small remote sensing satellite industry emerged. These satellites are capable of producing very high spatial resolution imagery. Small remote sensing satellites are significantly smaller than the traditional satellites, and are therefore economical to build and launch due to their small size and weight. For example, the Dove satellites developed by Planet Labs are nano-satellites that weigh about 5 kg. Nearly 100 Dove satellites have been launched over the past couple of years, providing a constellation of satellites to collect data. These satellites are capable of producing daily images at a 3–5 m spatial resolution (Behrens & Lal, 2019).

Small satellite and nano-satellite technology have made significant advances in recent years, and have found strong uptake by developing countries. Emerging low-cost satellite constellations have changed data collection techniques by providing several images together. As a low-cost application, the nations with limited budgets will be able to develop geospatial capabilities that can support a variety of sectors (Walter, 2020). For example, very high-resolution imagery from these sources is currently being used for precision agriculture, and local and regional agriculture applications. Table 15.3 presents the basic differences of details of UAV and small satellites for the collection of remote sensing data (Zolnai, 2016).

Table 15.3. Details of UAVs and Small Satellite Remote Sensing Data (Zolnai, 2016)

Details	UAV	Satellite
Resolution limitation	No	Yes
Need permit to fly	Yes (mostly)	No
Temporal resolution	High, but not practical	High
Build high-resolution DEM	Yes	Via stereo-pairs
Data ownership	With UAV	Pay per image
Regulated	Not as yet	Highly
Flight passes	Daily @best	When conditions are best
Aerial coverage	Small	Large
Security clearance	Depends on communications	Established

Recently, significant improvements in computer vision, image classification, and object recognition have taken place. Providing timely, dynamic, and accurate mapping-based solutions are crucial in many critical applications, such as disasters. For example, initiatives like the Humanitarian OpenStreetMap Team increasingly apply computer vision techniques to speed up and improve the accuracy of map data generation, and subsequently its timely dissemination for large geographic regions (Walter, 2020). Continued developments in image recognition and feature extraction, coupled with reduced storage costs, will provide further opportunities for faster data capture and creation of quality geospatial information.

Normally, a single sensor can't provide comprehensive information about a targeted object in a complex environment, so the support from complementary observations is required to be explored. The importance of observations from coordinated passive and active sensors has been accounted for in planning satellite missions, as illustrated by Figure 15.4. For example, it has been clearly recognized that multi-angular polarimeters (MAPs) will provide the most appropriate data for characterizing the detailed columnar properties of atmospheric aerosols and clouds, so their use is expected to significantly increase in the next decade (Dubovik et al., 2021). Several advanced polarimetric missions are planned in the coming few years by European and U.S. space agencies, which include 3MI (Multi-View Multi-Channel Multi-Polarization Imaging mission) on the MetOp-SG satellite, the Multi-Angle Imager for Aerosols (MAIA) instrument, Spex (Spectropolarimeter for Planetary Exploration) and Hyper-Angular Rainbow Polarimeter (HARP) as part of the NASA PACE mission, Multispectral Imaging Polarimeter (MSIP)/Aerosol-UA, and MAP instruments as part of the Copernicus CO_2M Mission (Dubovik et al., 2021). The number of satellite-based LiDARs and RADARs is also expected to increase in the future, since active remote sensing instruments provide detailed information about the vertical variability of the atmosphere. The CNSA (China National Space Administration) will launch the DPC-LiDAR

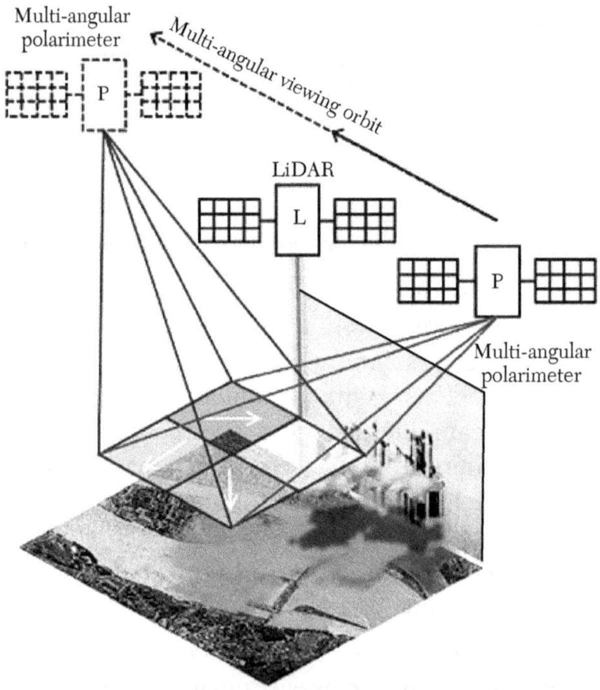

Figure 15.4. Combined observations from the LiDAR and MAP instrument to provide 3D characterization of the atmosphere (Dubovik et al., 2021).

onboard the CM-1 satellite in 2021, and the joint European/Japanese EarthCARE satellite, projected to be launched in 2023, will include high-performance LiDAR and RADAR technology for the first time (Illingsworth et al., 2015). The success of these missions will make LiDAR an essential component of future observing systems.

Satellite sensors are hardly used alone on any current platform, and multi-sensor data acquisition is very common in remote sensing. For example, an airborne mapping system usually includes optical sensors such as digital cameras and/or LiDAR sensors, and a GPS/IMU sensor for accurate sensor georeferencing. Sensor integration represents the methods to combine data streams produced by various sensors. Before performing any data integration from various sensors, georeferencing of the image data is essential, which can be done by using the GPS/IMU navigation solution. Smaller sensors can also be mounted on a variety of platforms, allowing for a greater flexibility in accurate and precise data collection. New technologies for linking sensors through wireless and traditional means into sensor networks are required to be developed (Gail, 2007). This will allow the information to be combined so as to support rapid decisions in complex situations. In addition, the output of one or several sensors is expected to be used to trigger observations from others, or even to rapidly reconfigure the other sensors so as to optimize the observations of an event.

Remote sensing data can also be used to create visual models and representations of Earth's systems having complex data relationships. The two most widely used immersive technologies for this purpose are (Walter, 2020): (*a*) virtual reality (VR), a fully computer-generated simulation of a 3D environment, and (*b*) augmented reality (AR), a reality-based environment that is over-layered with computer-generated effects, display, or text that enhances the users' real-world experience. These technologies enable the users to interact with simulations and visually relate to the information that remote sensing sensors provide. The application of AR has progressed in recent years, such as Google and Microsoft extensively investing in the AR headset technology.

Both AR and VR systems are based on spatial data and its processing, and can be used for 3D visualizations of time-series geospatial datasets. These systems used for visualization applications require accurate georeferencing of time-series data. Visualizations and immersive technology enhance the way in which people interact with the environment and increasingly use it for improved analysis and decision-making. The visualizations can be used for scientific purposes and to better inform the general public. For example, the website (*http://earth.nullschool.net/*) provides a visualization of global weather conditions which are updated every three hours. The data is collected from a variety of remote sensing satellites. Free data from programs like Copernicus and USGS ensure a long-term, consistent data backbone upon which the ecology of platform and solution providers can evolve.

Technology, ranging from automation to the IoT, big data, AI, cloud computing, and so on plays a prime role in the geospatial-based applications. Automation and improvements in machine learning may lead to higher production efficiency. For big data processing, machine learning and deep learning techniques play a key role in production of geospatial information (Parente et al., 2019). As a big volume of satellite imagery may be made available in the cloud, it can be used for global spatial analysis with increased spatial and temporal resolution. Both automatic feature extraction and change detection methods can be employed through neural network algorithms to analyze historic time-series images. As cloud processing services become increasingly cost-effective and accessible, it is expected that they will provide improved efficiency and productivity in the future. In addition, quantum computing will enable more intensive processing of an increased volume of geospatial data.

During the past five decades, remote sensing technologies have developed into a complete system, which will provide huge data for Digital Earth (Guo, 2018). Leading countries and regions such as the United States and Europe with Earth observation technologies have formulated Earth observation plans for long-term development. Figure 15.5 presents the

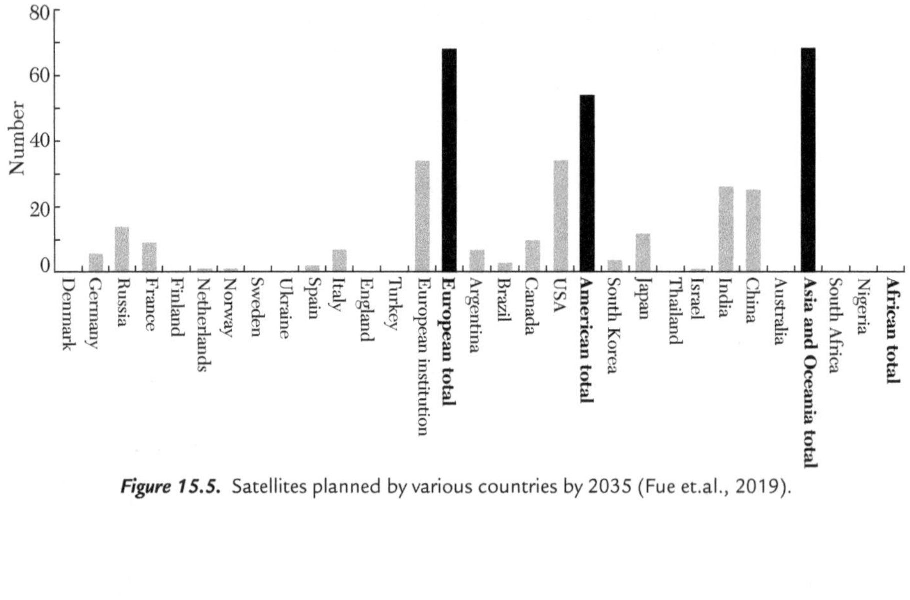

Figure 15.5. Satellites planned by various countries by 2035 (Fue et.al., 2019).

plan of satellite launchs by various countries by 2035 (Fu et al., 2019). It is observed that the European organizations will launch significantly more satellites with a greater emphasis on cooperation and coordination between them. The United States will remain a leading force, and Canada will occupy a secondary role. In Asia, the existing trend will continue, with China, India, Japan, and South Korea continuing to be the major contributors. These future Earth observation satellites will mainly focus on program continuity, development potential, and the capacity for comprehensive and coordinated applications. Therefore, long-term observation programs will be developed to emphasize the coordinated use of Earth observation platforms and data to better meet the requirements of various fields that may benefit the goals of countries and regions (Guo, 2018).

15.6 GLOBAL MARKET POTENTIAL

Spatial thinking has greatly been enhanced by the opportunities provided by digital technologies, such as easily accessible Google Maps. It is estimated that worldwide over two billion people are using remotely sensed data and geospatial datasets, which include data from a variety of geospatial technologies such as GPS, GIS, remote sensing, UAV, and CAD/BIM. Global adoption of mapping and navigational applications by these users has transformed the growth of remote sensing in the past decade. The key catalysts include: widespread acceptance of GPS technology, smartphones, and mapping services offered by a variety of mapping portals, such as Bing Maps, Google Earth/Maps, Nokia Maps, Baidu, and others (DigitalGlobe, 2020). Research and Markets (2019) mentioned that the geospatial technologies are an integral part of the global economy, affecting 10% of the global GDP.

The remote sensing technology market is proliferating across the globe, North America being the dominating leader in the market (Market Research Future, 2021). In addition to substantial advancements in remote sensing technology and expansion, inland assessment techniques drive the growth of the regional market. Europe stands second in the global remote sensing technology market. The vast advancements drive the market in sensing technology, improvements in healthcare standards, and improved infrastructure. The Asia Pacific remote sensing technology market has emerged as a profitable market, and factors such as the increasing uptake of smart technology and government initiatives to improve infrastructure boost the regional market growth. China, Japan, and India are major markets for remote sensing technology in this region. Considerable advancements in

AI and other similar technologies are pushing the growth of the remote sensing technology market.

According to a Markets & Market (2017) report, the remote sensing services market is projected to grow from USD 10.68 billion in 2017 to USD 21.62 billion by 2022, at a CAGR (cumulative annual growth rate) of 15.14% during 2017–2022. The increased use of remote sensing services in defense and commercial applications is one of the most significant factors driving the growth of this services market. Furthermore, the introduction of big data analytics in remote sensing and the growth of the cloud computing market are anticipated to increase the demand for remote sensing services. However, factors such as interoperability issues and huge initial investments are acting as restraints to the growth of the remote sensing services market. Based on platform, the satellites segment is projected to lead this market during 2017–2022. The growth of the satellites segment can be attributed to the increased demand for Earth observation satellites to obtain valuable information for mapping, mineral exploration, land use planning, resource management, and other activities. Based on platform, the UAVs segment of the remote sensing services market is projected to grow at the highest CAGR from 2017–2022. The UAVs are considered to be the most preferred remote sensing platform, and their services will be mainly used for large-scale mapping, real time assessment, and monitoring activities of various applications, such as precision farming, 3D terrain model construction, damage assessment, geohazard mapping, and mineral exploration.

The remote sensing service market globally was valued at USD 13.02 billion in 2019 and is projected to reach USD 37.65 billion by 2027 with a CAGR of 14.8% from 2020–2027 (Research and Markets, 2019). The penetration of remote sensing in Asia will increase to monitor the change in land usage for agriculture, 3D terrain model construction, damage assessment, pipeline monitoring, geohazard mapping, mineral exploration, and defense and intelligence. Moreover, the European agriculture sector holds significant potential for utilization of remote sensing technologies, especially for smart farming. These sensing services would help improve crop health and soil quality, and reduce the costs of farming.

In business, service-oriented architecture is adopted in cloud computing that will reduce the investment in hardware through development of a range of facilitating services, including Infrastructure as a Service (IaaS), Platform as a Service (PaaS), and Software as a Service (SaaS). In addition to the Platform-as-a-Service (PaaS) offer, the current shift in technology is toward Data-as-a-Service (DaaS) and shared Infrastructure-as-a-Service

(IaaS) that will continue to grow, facilitating easier, more dynamic sharing of data. The success of DaaS is based on the delivery of geospatial data for search and download through simple APIs.

15.7 CONCLUSION

In recent years, the potential of remote sensing to support several researches and applications has been boosted by the prospects of new technological developments and new space missions, including a number of very high spatial and spectral resolution passive optical satellites as well as active optical (LiDAR) and RADAR imaging systems equipped with state-of-the-art technology. Technological advancements in remote sensing coupled with advances in IT, cloud computing, mobile technology, GPS, and digital technologies have created a unique opportunity for various real time applications, globally. The UAV, GPS, and camera systems have made the collection of remotely sensed data more affordable and accessible than ever before. These data, along with digital developments and industry innovations, will enable faster, more accurate and trustworthy information available to decision-makers to take informed decisions, monitor progress, and assess the impact of interventions. With the growing demand, novel sensor approaches are also likely to appear in the future.

The main advantage of using remote sensing technology is its capability to integrate with multiple sensors of varying spatial, spectral, or temporal resolutions to collate the information for a common application. Robust algorithms are required to be developed to fully automate the geo-referencing of data captured with several sensors, such as optical and radar, hyperspectral imagery, and LiDAR data, which provide data at different resolutions by using diverse data acquisition approaches. Although the derived information from such sensors are potentially very useful for diverse applications, fusing the data from two different sensors for improved results is a real challenge. Additionally, techniques for object recognition, classification, segmentation, and change detection from the images of data fusion are still not fully developed. Therefore, to take full advantage of the diversity of remote sensing data, there is a need to develop new strategies and further refine the existing techniques and approaches (Kadhim et al., 2016).

In the digital age, geospatial information and technologies play a crucial role in leveraging the potential of the "Fourth Industrial Revolution" to help solve global challenges. The next five to ten years will see significant developments in the maturity and application of already well-established technologies across the geospatial industry. AI, sensor technology, and the IoT will drastically change the way in which data is collected, managed, used, and maintained. The growing amount of geospatial data, processed by new big data methods, are already driving developments and will create new applications that will have major societal impact over the next decade and beyond. Increased demand, advanced technologies, and limited resources will bring better, faster, and smarter technology to the remote sensing community in the future.

INDEX

J

K

L

M

S

W

Y

www.ingramcontent.com/pod-product-compliance
Lightning Source LLC
LaVergne TN
LVHW082010200926
848481LV00010B/2829